Lecture Notes in Computer Science   12590

More information about this subseries at http://www.springer.com/series/7409

David Auber · Pavel Valtr (Eds.)

Graph Drawing and Network Visualization

28th International Symposium, GD 2020
Vancouver, BC, Canada, September 16–18, 2020
Revised Selected Papers

 Springer

Editors
David Auber (ID)
LaBRI, University of Bordeaux
Talence, France

Pavel Valtr
Charles University
Prague, Czech Republic

ISSN 0302-9743   ISSN 1611-3349   (electronic)
Lecture Notes in Computer Science
ISBN 978-3-030-68765-6   ISBN 978-3-030-68766-3   (eBook)
https://doi.org/10.1007/978-3-030-68766-3

LNCS Sublibrary: SL3 – Information Systems and Applications, incl. Internet/Web, and HCI

This Springer imprint is published by the registered company Springer Nature Switzerland AG
The registered company address is: Gewerbestrasse 11, 6330 Cham, Switzerland

Preface

This volume contains the papers presented at GD 2020, the 28th International Symposium on Graph Drawing and Network Visualization, held on September 16–18, 2020 online. Graph drawing is concerned with the geometric representation of graphs and constitutes the algorithmic core of network visualization. Graph drawing and network visualization are motivated by applications where it is crucial to visually analyse and interact with relational datasets. Information about the conference series and past symposia is maintained at http://www.graphdrawing.org. The 2020 edition of the conference was hosted by The University of British Columbia, with Will Evans as chair of the Organizing Committee. A total of 251 participants attended the conference.

Regular papers could be submitted to one of two distinct tracks: Track 1 for papers on combinatorial and algorithmic aspects of graph drawing and Track 2 for papers on experimental, applied, and network visualization aspects. Short papers were given a separate category, which welcomed both theoretical and applied contributions. An additional track was devoted to poster submissions. All the tracks were handled by a single Program Committee. In response to the call for papers, the Program Committee received a total of 82 submissions, consisting of 75 papers (40 in Track 1, 14 in Track 2, and 21 in the short-paper category) and 7 posters. More than 250 single-blind reviews were provided, more than a third of which were contributed by external sub-reviewers. After extensive electronic discussions via EasyChair, the Program Committee selected 38 papers and 7 posters for inclusion in the scientific program of GD 2020. This resulted in an overall paper acceptance rate of just over 50% (55% in Track 1, 50% in Track 2, and 43% in the short-paper category). Authors were invited to publish an electronic version of their accepted papers on the arXiv e-print repository and also to provide a recorded presentation of their work. These contributions were made available before the conference to all the online participants. There were two invited lectures at GD 2020, one on each track of the scientific program. Sheelagh Carpendale, Simon Fraser University, CA, presented "An Alternate Look at Aesthetics" and Jeff Erickson, from University of Illinois at Urbana-Champagne, talked about "Fun with Toroidal Spring Embeddings". Abstracts of all invited lectures are included in these proceedings.

The conference gave out best paper awards in Track 1 and Track 2, as well as a best presentation award and a best poster award. As decided by a majority vote of the Program Committee, the award for the best paper in Track 1 was assigned to "Crossings between non-homotopic edges" by János Pach, Gábor Tardos, and Géza Tóth, and the award for the best paper in Track 2 was assigned to "Graph drawing via gradient descent, $(GD)^2$" by Reyan Ahmed, Felice De Luca, Sabin Devkota, Stephen Kobourov, and Mingwei Li. Based on a majority vote of conference participants, the best presentation award was given to Johannes Obenaus, Rosna Paul, and Alexandra Weinberger for their presentation of the paper "Plane Spanning Trees in Edge-Colored Simple Drawings of K_n and the best poster award was given to "MetroSets: Visualizing

Hypergraphs as MetroMaps" by Ben Jacobsen, Markus Wallinger, Stephen Kobourov, and Martin Nöllenburg.

Congratulations to all the award winners for their excellent contributions, and many thanks to Springer whose sponsorship funded the prize money for these awards.

Following tradition, the 27th Annual Graph Drawing Contest was held during the conference. The contest was divided into two parts, creative topics and the live challenge.

The creative topics featured two graphs, the Hrafnkels Saga graph and the K-pop graph.

The live challenge focused on minimizing the number of crossings in an upward drawing on a fixed grid, and had two categories: manual and automatic. Awards were given in each of the four categories. We thank the Contest Committee, chaired by Philipp Kindermann, for preparing interesting and challenging contest problems. A report about the contest is included in these proceedings.

Many people and organizations contributed to the success of GD 2020. We would like to thank all members of the Program Committee and the external reviewers for carefully reviewing and discussing the submitted papers and posters; this was crucial for putting together a strong and interesting program.

Thanks to all authors who chose GD 2020 as the publication venue for their research.

We are grateful for the support of the sponsors Springer, yWorks, the Pacific Institute for the Mathematical Sciences, and The University of British Columbia. Their generosity helped make this symposium a memorable event for all participants. Last but not least, we would like to express our appreciation of the organizing team, William Evans, Holly Kwan, and Ruth Situma, as well as all the student volunteers: Ben Chugg, Kyle Clarkson, Rebecca Lin, Noushin Saeedi, Lucca Siaudzionis, Matthew Tang, Kelvin Wong, and David Zheng.

The 29th International Symposium on Graph Drawing and Network Visualization (GD 2021) will take place from September 13–17, 2021. We hope to hold the conference at the University of Tübingen, Germany, but in the event that we are unable to have a physical conference, GD 2021 will, like GD 2020, be held virtually. Helen Purchase and Ignaz Rutter will co-chair the Program Committee, and Michael Bekos and Michael Kaufmann will co-chair the Organizing Committee.

October 2020   David Auber
                                                               Pavel Valtr

Organization

Program Committee

Daniel Archambault	Swansea University, UK
David Auber (co-chair)	University Bordeaux, France
Benjamin Bach	The University of Edinburgh, UK
Fabian Beck	University of Duisburg-Essen, Germany
Romain Bourqui	University Bordeaux, France
Steve Chaplick	University of Maastricht, Netherlands
Markus Chimani	Osnabrück University, Germany
Sabine Cornelsen	University of Konstanz, Germany
Walter Didimo	University of Perugia, Italy
Stefan Felsner	TU Berlin, Germany
Radoslav Fulek	IST Austria, Austria
Fabrizio Frati	Roma Tre University, Italy
Seokhee Hong	University of Sydney, Australia
Yifan Hu	Yahoo!, USA
Katherine Isaacs	University of Arizona, USA
Philipp Kindermann	Universität Passau, Germany
Kwan-Liu Ma	University of California, Davis, USA
Tamara Mchedlidze	Karlsruhe Institute of Technology, Germany
Fabrizio Montecchiani	University of Perugia, Italy
Tamara Munzner	The University of British Columbia, Canada
Martin Nöllenburg	TU Vienna, Austria
Arnaud Sallaberry	University of Montpellier, France
Alexandru Telea	Utrecht University, Netherlands
Ioannis Tollis	University of Crete, Greece
Csaba Tóth	California State University, Northridge, USA
Pavel Valtr (co-chair)	Charles University, Czech Republic
Alexander Wolff	Universität Würzburg, Germany

Additional Reviewers

Ahmed, Abu Reyan	Bekos, Michael
Alegría, Carlos	Biniaz, Ahmad
Angelini, Patrizio	Binucci, Carla
Arroyo, Alan	Blum, Johannes
Arseneva, Elena	Borrazzo, Manuel
Beck, Moritz	Brückner, Guido
Behr, Timon	Chandrasegaran, Senthil

Da Lozzo, Giordano
Damásdi, Gábor
Di Giacomo, Emilio
Evans, William
Firman, Oksana
Fleszar, Krzysztof
Frank, Fabian
Fujiwara, Takanori
Förster, Henry
Ganian, Robert
Grilli, Luca
Gronemann, Martin
Huroyan, Vahan
Kesavan, Suraj P.
Klawitter, Jonathan
Kleist, Linda
Klemz, Boris
Kritikakis, Giorgos
Kryven, Myroslav
Kwon, Oh-Hyun
Kynčl, Jan
Le, Van Bang
Lionakis, Panagiotis
Löffler, Andre
Malic, Goran

Morin, Pat
Ortali, Giacomo
Parada, Irene
Patrignani, Maurizio
Patáková, Zuzana
Pupyrev, Sergey
Radermacher, Marcel
Raftopoulou, Chrysanthi
Saeedi, Noushin
Scheucher, Manfred
Schindl, David
Schlipf, Lena
Schröder, Felix
Servatius, Brigitte
Spillner, Andreas
Steiner, Raphael
Storandt, Sabine
Tappini, Alessandra
Ueckerdt, Torsten
van der Hoog, Ivor
Van Dijk, Thomas C.
van Goethem, Arthur
Villedieu, Anaïs
Zink, Johannes

Abstracts of the Invited Talks

Looking at Alternative Aesthetics

Sheelagh Carpendale

Simon Fraser University, Canada

Abstract. There have been many discussions in graph drawing, information design and human computer interaction communities about the importance of aesthetics in design, mentioning advantages such as people paying closer attention, people being more willing to engage longer, and generally enhancing usability. The Graph Drawing community is particularly aware of this, having developed their own perspective on aesthetics. In this talk, I laud the current Graph Drawing aesthetics, and note that alternatives are possible. I describe four such possible directions: 1) looking beyond our culture – the world is full of wonderful diversity; 2) data – comes with many variations in structure that may contain sources of design ideas; 3) interaction – these possibilities continue to expand and now include notions of agency, and serendipity; and 4) biomimicry – nature contains fabulous inspirations for design.

Fun with Toroidal Spring Embeddings

Jeff Erickson

Department of Computer Science,
University of Illinois at Urbana-Champaign, USA

Abstract. Tutte's classical spring embedding theorem is the foundation of hundreds of algorithms for drawing and manipulating planar graphs. A somewhat less well-known generalization of Tutte's theorem, first proved by Yves Colin de Verdière in 1990, applies to graphs on more complex surfaces. I will describe two recent applications of this more general theorem to graphs on the Euclidean flat torus. The first is a natural toroidal analogue of the Maxwell-Cremona correspondence, which relates equilibrium stresses, orthogonal dual embeddingss, and weighted Delaunay complexes. The second is an efficient algorithm to morph between geodesic torus graphs using a small number of parallel linear morphing steps, matching (and slightly simplifying) recent planar moprhing algorithms.

This talk includes joint work with Erin Chambers, Patrick Lin, and Salman Parsa, available at:

- https://arxiv.org/abs/2003.10057
- https://arxiv.org/abs/2007.07927

Contents

Crossings, k-Planar Graphs

Planarity

Gradient Descent and Queue Layouts

Graph Drawing via Gradient Descent, $(GD)^2$

Reyan Ahmed$^{(\boxtimes)}$ (iD), Felice De Luca (iD), Sabin Devkota (iD), Stephen Kobourov (iD),
and Mingwei Li (iD)

Department of Computer Science, University of Arizona, Tucson, USA
abureyanahmed@email.arizona.edu

Abstract. Readability criteria, such as distance or neighborhood preservation, are often used to optimize node-link representations of graphs to enable the comprehension of the underlying data. With few exceptions, graph drawing algorithms typically optimize one such criterion, usually at the expense of others. We propose a layout approach, Graph Drawing via Gradient Descent, $(GD)^2$, that can handle multiple readability criteria. $(GD)^2$ can optimize any criterion that can be described by a smooth function. If the criterion cannot be captured by a smooth function, a non-smooth function for the criterion is combined with another smooth function, or auto-differentiation tools are used for the optimization. Our approach is flexible and can be used to optimize several criteria that have already been considered earlier (e.g., obtaining ideal edge lengths, stress, neighborhood preservation) as well as other criteria which have not yet been explicitly optimized in such fashion (e.g., vertex resolution, angular resolution, aspect ratio). We provide quantitative and qualitative evidence of the effectiveness of $(GD)^2$ with experimental data and a functional prototype: http://hdc.cs.arizona.edu/~mwli/graph-drawing/.

1 Introduction

Graphs represent relationships between entities and visualization of this information is relevant in many domains. Several criteria have been proposed to evaluate the readability of graph drawings, including the number of edge crossings, distance preservation, and neighborhood preservation. Such criteria evaluate different aspects of the drawing and different layout algorithms optimize different criteria. It is challenging to optimize multiple readability criteria at once and there are few approaches that can support this. Examples of approaches that can handle a small number of related criteria include the stress majorization framework of Wang et al. [36], which optimizes distance preservation via stress as well as ideal edge length preservation. The Stress Plus X (SPX) framework of Devkota et al. [14] can minimize the number of crossings, or maximize the minimum angle of edge crossings. While these frameworks can handle a limited set of related criteria, it is not clear how to extend them to arbitrary optimization goals. The reason for this limitation is that these frameworks are dependent

© Springer Nature Switzerland AG 2020
D. Auber and P. Valtr (Eds.): GD 2020, LNCS 12590, pp. 3–17, 2020.
https://doi.org/10.1007/978-3-030-68766-3_1

on a particular mathematical formulation. For example, the SPX framework was designed for crossing minimization, which can be easily modified to handle crossing angle maximization (by adding a cosine factor to the optimization function). This "trick" can be applied only to a limited set of criteria but not the majority of other criteria that are incompatible with the basic formulation.

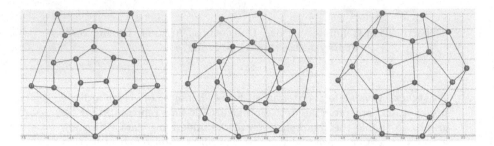

Fig. 1. Three $(GD)^2$ layouts of the dodecahedron: (a) optimizing the number of crossings, (b) optimizing uniform edge lengths, and (c) optimizing stress.

In this paper, we propose a general approach, Graph Drawing via Gradient Descent, $(GD)^2$, that can optimize a large set of drawing criteria, provided that the corresponding metrics that evaluate the criteria are smooth functions. If the function is not smooth, $(GD)^2$ either combines it with another smooth function and partially optimizes based on the desired criterion, or uses modern auto-differentiation tools to optimize. As a result, the proposed $(GD)^2$ framework is simple: it only requires a function that captures a desired drawing criterion. To demonstrate the flexibility of the approach, we consider an initial set of nine criteria: minimizing stress, maximizing vertex resolution, obtaining ideal edge lengths, maximizing neighborhood preservation, maximizing crossing angle, optimizing total angular resolution, minimizing aspect ratio, optimizing the Gabriel graph property, and minimizing edge crossings. A functional prototype is available on http://hdc.cs.arizona.edu/~mwli/graph-drawing/. This is an interactive system that allows vertices to be moved manually. Combinations of criteria can be optimized by selecting a weight for each; see Fig. 1.

2   Related Work

Many criteria associated with the readability of graph drawings have been proposed [37]. Most of graph layout algorithms are designed to (explicitly or implicitly) optimize a single criterion. For instance, a classic layout criterion is stress minimization [26], where stress is defined by $\sum_{i<j} w_{ij}(|X_i - X_j| - d_{ij})^2$. Here, X is a $n \times 2$ matrix containing coordinates for the n nodes, d_{ij} is typically the graph-theoretical distance between two nodes i and j and $w_{ij} = d_{ij}^{-\alpha}$ is a normalization factor with α equal to $0, 1$ or 2. Thus reducing the stress in a layout corresponds

to computing node positions so that the actual distance between pairs of nodes is proportional to the graph theoretic distance between them. Optimizing stress can be accomplished by stress minimization, or stress majorization, which can speed up the computation [22]. In this paper we only consider drawing in the Euclidean plane, however, stress can be also optimized in other spaces such as the torus [9].

Stress minimization corresponds to optimizing the global structure of the layout, as the stress metric takes into account all pairwise distances in the graph. The t-SNET algorithm of Kruiger et al. [27] directly optimizes neighborhood preservation, which captures the local structure of a graph, as the neighborhood preservation metric only considers distances between pairs of nodes that are close to each other. Optimizing local or global distance preservation can be seen as special cases of the more general dimensionality reduction approaches such as multi-dimensional scaling [28, 34].

Purchase et al. [30] showed that the readability of graphs increases if a layout has fewer edge crossings. The underlying optimization problem is NP-hard and several graph drawing contests have been organized with the objective of minimizing the number of crossings in the graph drawings [2, 8]. Recently several algorithms that directly minimize crossings have been proposed [31, 33].

The negative impact on graph readability due to edge crossings can be mitigated if crossing pairs of edges have a large crossings angle [4, 15, 24, 25]. Formally, the crossing angle of a straight-line drawing of a graph is the minimum angle between two crossing edges in the layout, and optimizing this property is also NP-hard. Recent graph drawing contests have been organized with the objective of maximizing the crossings angle in graph drawings and this has led to several heuristics for this problem [5, 12].

The algorithms above are very effective at optimizing the specific readability criterion they are designed for, but they cannot be directly used to optimize additional criteria. This is a desirable goal, since optimizing one criterion often leads to poor layouts with respect to one or more other criteria: for example, algorithms that optimize the crossing angle tend to create drawings with high stress and no neighborhood preservation [14].

Davidson and Harel [11] used simulated annealing to optimize different graph readability criteria (keeping nodes away from other nodes and edges, uniform edge lengths, minimizing edge crossings). Recently, several approaches have been proposed to simultaneously improve multiple layout criteria. Wang et al. [36] propose a revised formulation of stress that can be used to specify ideal edge direction in addition to ideal edge lengths in a graph drawing. Devkota et al. [14] also use a stress-based approach to minimize edge crossings and maximize crossing angles. Eades et al. [19] provided a technique to draw large graphs while optimizing different geometric criteria, including the Gabriel graph property. Although the approaches above are designed to optimize multiple criteria, they cannot be naturally extended to handle other optimization goals.

Constraint-based layout algorithms such as COLA [17, 18], can be used to enforce separation constraints on pairs of nodes to support properties such as

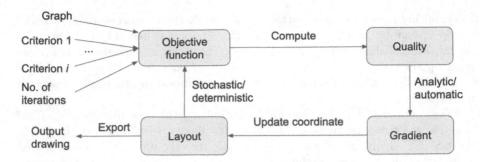

Fig. 2. The $(GD)^2$ framework: given a graph and a set of criteria (with weights), formulate an objective function based on the selected set of criteria and weights. Then compute the quality (value) of the objective function of the current layout of the graph. Next, generate the gradient (analytically or automatically). Using the gradient information, update the coordinates of the layout. Finally, update the objective function based on the layout via regular or stochastic gradient descent. This process is repeated for a fixed number of iterations.

customized node ordering or downward pointing edges. The coordinates of two nodes are related by inequalities in the form of $x_i \geq x_j + gap$ for a node pair (i, j). These kinds of constraints are known as hard constraints and are different from the soft constrains in our $(GD)^2$ framework.

3   The $(GD)^2$ Framework

The $(GD)^2$ framework is a general optimization approach to generate a layout with any desired set of aesthetic metrics, provided that they can be expressed by a smooth function. The basic principles underlying this framework are simple. The first step is to select a set of layout readability criteria and a loss functions that measures them. Then we define the function to optimize as a linear combination of the loss functions for each individual criterion. Finally, we iterate the gradient descent steps, from which we obtain a slightly better drawing at each iteration. Figure 2 depicts the framework of $(GD)^2$: Given any graph G and readability criterion Q, we find a loss function $L_{Q,G}$ which maps from the current layout X (i.e. a $n \times 2$ matrix containing the positions of nodes in the drawing) to a real value that quantifies the current drawing. Note that some of the readability criteria naturally correspond to functions that should be minimized (e.g., stress, crossings), while others to functions that should be maximized (e.g., neighborhood preservation, angular resolution). Given a loss function $L_{Q,G}$ of X where a lower value is always desirable, at each iteration, a slightly better layout can be found by taking a small (ϵ) step along the (negative) gradient direction: $X^{(new)} = X - \epsilon \cdot \nabla_X L_{Q,G}$.

To optimize multiple quality measures simultaneously, we take a weighted sum of their loss functions and update the layout by the gradient of the sum.

3.1 Gradient Descent Optimization

There are different kinds of gradient descent algorithms. The standard method considers all vertices, computes the gradient of the objective function, and updates vertex coordinates based on the gradient. For some objectives, we need to consider all the vertices in every step. For example, the basic stress formulation [26] falls in this category. On the other hand, there are some problems where the objective can be optimized only using a subset of vertices. For example, consider stress minimization again. If we select a set of vertices randomly and minimize the stress of the induced graph, the stress of the whole graph is also minimized [38]. This type of gradient descent is called stochastic gradient descent. However, not all objective functions are smooth and we cannot compute the gradient of a non-smooth function. In that scenario, we can compute the subgradient, and update the objective based on the subgradient. Hence, as long as the function is continuously defined on a connected component in the domain, we can apply the subgradient descent algorithm. In [3], we give a list of loss functions we used to optimize 9 graph drawing properties with gradient descent variants. In Sect. 4, we specify the loss functions we used in detail.

When a function is not defined in a connected domain, we can introduce a surrogate loss function to 'connect the pieces'. For example, when optimizing neighborhood preservation we maximize the Jaccard similarity between graph neighbors and nearest neighbors in graph layout. However, Jaccard similarity is only defined between two binary vectors. To solve this problem we extend Jaccard similarity to all real vectors by its Lovász extension [6] and apply that to optimize neighborhood preservation. An essential part of gradient descent based algorithms is to compute the gradient/subgradient of the objective function. In practice, it is always not necessary to write down the gradient analytically as it can be computed automatically via automatic differentiation [23]. Deep learning packages such as Tensorflow [1] and PyTorch [29] apply automatic differentiation to compute the gradient of complicated functions.

When optimizing multiple criteria simultaneously, we combine them via a weighted sum. However, choosing a proper weight for each criterion can be tricky. Consider, for example, maximizing crossing angles and minimize stress simultaneously with a fixed pair of weights. At the very early stage, the initial drawing may have many crossings and stress minimization often removes most of the early crossings. As a result, maximizing crossing angles in the early stages can be harmful as it move nodes in directions that contradict those that come from stress minimization. Therefore, a well-tailored *weight scheduling* is needed for a successful outcome. Continuing with the same example, a better outcome can be achieved by first optimizing stress until it converges, and later adding weights for the crossing angle maximization. To explore different ways of scheduling, we provide an interface that allows manual tuning of the weights.

3.2 Implementation

We implemented the $(GD)^2$ framework in JavaScript. In particular we used the automatic differentiation tools in tensorflow.js [35] and the drawing library d3.js [7]. The prototype is available at http://hdc.cs.arizona.edu/~mwli/graph-drawing/.

4   Properties and Measures

In this section we specify the aesthetic goals, definitions, quality measures and loss functions for each of the 9 graph drawing properties we optimized: stress, vertex resolution, edge uniformity, neighborhood preservation, crossing angle, aspect ratio, total angular resolution, Gabriel graph property, and crossing number. In the following discussion, since only one (arbitrary) graph is considered, we omit the subscript G in our definitions of loss function $L_{Q,G}$ and write L_Q for short. Other standard graph notation is summarized in Table 1.

Table 1. Graph notation used in this paper.

Notation	Description
G	Graph
V	The set of nodes in G, indexed by i, j or k
E	The set of edges in G, indexed by a pair of nodes (i,j) in V
$n = \|V\|$	Number of nodes in G
$\|E\|$	Number of edges in G
$Adj_{n \times n}$ and $A_{i,j}$	Adjacency matrix of G and its (i,j)-th entry
$D_{n \times n}$ and d_{ij}	Graph-theoretic distances between pairs of nodes and the (i,j)-th entry
$X_{n \times 2}$	2D-coordinates of nodes in the drawing
$\|X_i - X_j\|$	The Euclidean distance between nodes i and j in the drawing
θ_i	i^{th} crossing angle
φ_{ijk}	Angle between incident edges (i,j) and (j,k)

4.1   Stress

We use stress minimization to draw a graph such that the Euclidean distance between pairs of nodes is proportional to their graph theoretic distance. Following the ordinary definition of stress [26], we minimize

$$L_{ST} = \sum_{i<j} w_{ij}(|X_i - X_j|_2 - d_{ij})^2 \tag{1}$$

Where d_{ij} is the graph-theoretical distance between nodes i and j, X_i and X_j are the 2D coordinates of nodes i and j in the layout. The normalization factor, $w_{ij} = d_{ij}^{-2}$, balances the influence of short and long distances: the longer the graph theoretic distance, the more tolerance we give to the discrepancy between two distances. When comparing two drawings of the same graph with respect to stress, a smaller value (lower bounded by 0) corresponds to a better drawing.

4.2   Ideal Edge Length

When given a set of ideal edge lengths $\{l_{ij} : (i,j) \in E\}$ we minimize the average deviation from the ideal lengths:

$$L_{IL} = \sqrt{\frac{1}{|E|} \sum_{(i,j)\in E} (\frac{||X_i - X_j|| - l_{ij}}{l_{ij}})^2} \tag{2}$$

For unweighted graphs, by default we take the average edge length in the current drawing as the ideal edge length for all edges. $l_{ij} = l_{avg} = \frac{1}{|E|} \sum_{(i,j)\in E} ||X_i - X_j||$ for all $(i,j) \in E$. The quality measure $Q_{IL} = L_{IL}$ is lower bounded by 0 and a lower score yields a better layout.

4.3   Neighborhood Preservation

Neighborhood preservation aims to keep adjacent nodes close to each other in the layout. Similar to Kruiger et al. [27], the idea is to have the k-nearest (Euclidean) neighbors (k-NN) of node i in the drawing to align with the k nearest nodes (in terms of graph distance from i). A natural quality measure for the alignment is the Jaccard index between the two pieces of information. Let, $Q_{NP} = JaccardIndex(K, Adj) = \frac{|\{(i,j):K_{ij}=1 \text{ and } A_{ij}=1\}|}{|\{(i,j):K_{ij}=1 \text{ or } A_{ij}=1\}|}$, where Adj denotes the adjacency matrix and the i-th row in K denotes the k-nearest neighborhood information of i: $K_{ij} = 1$ if j is one of the k-nearest neighbors of i and $K_{ij} = 0$ otherwise.

To express the Jaccard index as a differentiable minimization problem, first, we express the neighborhood information in the drawing as a smooth function of node positions X_i and store it in a matrix $\hat{K}$. In $\hat{K}$, a positive entry $\hat{K}_{i,j}$ means node j is one of the k-nearest neighbors of i, otherwise the entry is negative. Next, we take a differentiable surrogate function of the Jaccard index, the Lovász hinge loss (LHL) [6], to make the Jaccard loss optimizable via gradient descent. We minimize

$$L_{NP} = LHL(\hat{K}, Adj) \tag{3}$$

where LHL is given by Berman et al. [6], $\hat{K}$ denotes the k-nearest neighbor prediction:

$$\hat{K}_{i,j} = \begin{cases} -(||X_i - X_j|| - \frac{d_{i,\pi_k}+d_{i,\pi_{k+1}}}{2}) & \text{if } i \neq j \\ 0 & \text{if } i = j \end{cases} \tag{4}$$

where d_{i,π_k} is the Euclidean distance between node i and its kth nearest neighbor and Adj denotes the adjacency matrix. Note that $\hat{K}_{i,j}$ is positive if j is a k-NN of i, otherwise it is negative, as is required by LHL [6].

4.4   Crossing Number

Reducing the number of edge crossings is one of the classic optimization goals in graph drawing, known to affect readability [30]. Following Shabbeer et al. [33], we employ an expectation-maximization (EM)-like algorithm to minimize the number of crossings. Two edges do not cross if and only if there exists a line that separates their extreme points. With this in mind, we want to separate every pair of edges (the M step) and use the decision boundaries to guide the movement of nodes in the drawing (the E step). Formally, given any two edges $e_1 = (i,j), e_2 = (k,l)$ that do not share any nodes (i.e., i, j, k and l are all distinct), they do not intersect in a drawing (where nodes are drawn at $X_i = (x_i, y_i)$, a row vector) if and only if there exists a decision boundary $w = w_{(e_1,e_2)}$ (a 2-by-1 column vector) together with a bias $b = b_{(e_1,e_2)}$ (a scalar) such that: $L_{CN,(e_1,e_2)} = \sum_{\alpha=i,j,k \text{ or } l} ReLU(1 - t_\alpha \cdot (X_\alpha w + b)) = 0.$

Here we use (e_1, e_2) to denote the subgraph of G which only has two edges e_1 and e_2, $t_i = t_j = 1$ and $t_k = t_l = -1$. The loss reaches its minimum at 0 when the SVM classifier $f_{w,b} : x \mapsto xw + b$ predicts node i and j to be greater than 1 and node k and l to be less than -1. The total loss for the crossing number is therefore the sum over all possible pairs of edges. Similar to (soft) margin SVM, we add a term $|w_{(e_1,e_2)}|^2$ to maximize the margin of the decision boundary: $L_{CN} = \sum_{\substack{e_1=(i,j),\ e_2=(k,l)\in E \\ i,\ j,\ k \text{ and } l \text{ all distinct}}} L_{CN,(e_1,e_2)} + |w_{(e_1,e_2)}|^2.$ For the E and M steps, we used the same loss function L_{CN} to update the boundaries $w_{(e_1,e_2)}, b_{(e_1,e_2)}$ and node positions X:

$$w^{(new)} = w - \epsilon \nabla_w L_{CN} \tag{M step 1}$$

$$b^{(new)} = b - \epsilon \nabla_b L_{CN} \tag{M step 2}$$

$$X^{(new)} = X - \epsilon \nabla_X L_{CN}(X;\ w^{(new)}, b^{(new)}) \tag{E step}$$

To evaluate the quality we simply count the number of crossings.

4.5   Crossing Angle Maximization

When edge crossings are unavoidable, the graph drawing can still be easier to read when edges cross at angles close to $90°$ [37]. Heuristics such as those by Demel et al. [12] and Bekos et al. [5] have been proposed and have been successful in graph drawing challenges [13]. We use an approach similar to the force-directed algorithm given by Eades et al. [20] and minimize the squared cosine of crossing angles: $L_{CAM} = \sum_{\substack{\text{all crossed edge pairs} \\ (i,j),(k,l)\in E}} (\frac{\langle X_i - X_j, X_k - X_l \rangle}{|X_i - X_j| \cdot |X_k - X_l|})^2$. We evaluate quality by measuring the worst (normalized) absolute discrepancy between each crossing angle θ and the target crossing angle (i.e. $90°$): $Q_{CAM} = \max_\theta |\theta - \frac{\pi}{2}|/\frac{\pi}{2}$.

4.6   Aspect Ratio

Good use of drawing area is often measured by the aspect ratio [16] of the bounding box of the drawing, with $1:1$ as the optimum. We consider multiple rotations of the current drawing and optimize their bounding boxes simultaneously. Let $AR = \min_\theta \frac{\min(w_\theta, h_\theta)}{\max(w_\theta, h_\theta)}$, where w_θ and h_θ denote the width and height of the bounding box when the drawing is rotated by θ degrees. A naive approach to optimize aspect ratio, which scales the x and y coordinates of the drawing by certain factors, may worsen other criteria we wish to optimize and is therefore not suitable for our purposes. To make aspect ratio differentiable and compatible with other objectives, we approximate aspect ratio based on 4 (soft) boundaries (top, bottom, left and right) of the drawing. Next, we turn this approximation and the target $(1:1)$ into a loss function using cross entropy loss. We minimize

$$L_{AR} = \sum_{\theta \in \{\frac{2\pi k}{N}, \text{ for } k=0, \cdots (N-1)\}} crossEntropy([\frac{w_\theta}{w_\theta + h_\theta}, \frac{h_\theta}{w_\theta + h_\theta}], [0.5, 0.5])$$

(5)

where N is the number of rotations sampled (e.g., $N = 7$), and w_θ, h_θ are the (approximate) width and height of the bounding box when rotating the drawing around its center by an angle θ. For any given θ-rotated drawing, w_θ is defined to be the difference between the current (soft) right and left boundaries, $w_\theta = right - left = \langle softmax(x_\theta), x_\theta \rangle - \langle softmax(-x_\theta), x_\theta \rangle$, where x_θ is a collection of the x coordinates of all nodes in the θ-rotated drawing, and softmax returns a vector of weights $(\ldots w_k, \ldots)$ given by $softmax(x) = (\ldots w_k, \ldots) = \frac{e^{x_k}}{\sum_i e^{x_i}}$. Note that the approximate right boundary is a weighted sum of the x coordinates of all nodes and it is designed to be close to the x coordinate of the rightmost node, while keeping other nodes involved. Optimizing aspect ratio with the softened boundaries will stretch all nodes instead of moving the extreme points. Similarly, $h_\theta = top - bottom = \langle softmax(y_\theta), y_\theta \rangle - \langle softmax(-y_\theta), y_\theta \rangle$ Finally, we evaluate the drawing quality by measuring the worst aspect ratio on a finite set of rotations. The quality score ranges from 0 to 1 (where 1 is optimal): $Q_{AR} = \min_{\theta \in \{\frac{2\pi k}{N}, \text{ for } k=0, \cdots (N-1)\}} \frac{\min(w_\theta, h_\theta)}{\max(w_\theta, h_\theta)}$

4.7   Angular Resolution

Distributing edges adjacent to a node makes it easier to perceive the information presented in a node-link diagram [25]. Angular resolution [4], defined as the minimum angle between incident edges, is one way to quantify this goal. Formally, $ANR = \min_{j \in V} \min_{(i,j),(j,k) \in E} \varphi_{ijk}$, where φ_{ijk} is the angle formed by between edges (i, j) and (j, k). Note that for any given graph, an upper bound of this quantity is $\frac{2\pi}{d_{max}}$ where d_{max} is the maximum degree of nodes in the graph. Therefore in the evaluation, we will use this upper bound to normalize our quality measure to $[0, 1]$, i.e. $Q_{ANR} = \frac{ANR}{2\pi/d_{max}}$. To achieve a better drawing quality via gradient descent, we define the angular energy of an angle φ to be

$\cdot e^{-s\cdot\varphi}$, where s is a constant controlling the sensitivity of angular energy with respect to the angle (by default $s = 1$), and minimize the total angular energy over all incident edges:

$$L_{ANR} = \sum_{(i,j),(j,k)\in E} e^{-s\cdot\varphi_{ijk}} \tag{6}$$

4.8 Vertex Resolution

Good vertex resolution is associated with the ability to distinguish different vertices by preventing nodes from occluding each other. Vertex resolution is typically defined as the minimum Euclidean distance between two vertices in the drawing [10, 32]. However, in order to align with the units in other objectives such as stress, we normalize the minimum Euclidean distance with respect to a reference value. Hence we define the vertex resolution to be the ratio between the shortest and longest distances between pairs of nodes in the drawing, $VR = \frac{\min_{i\neq j}||X_i-X_j||}{d_{max}}$, where $d_{max} = \max_{k,l}||X_k - X_l||$. To achieve a certain target resolution $r \in [0, 1]$ by minimizing a loss function, we minimize

$$L_{VR} = \sum_{i,j\in V, i\neq j} ReLU(1 - \frac{||X_i - X_j||}{r \cdot d_{max}})^2 \tag{7}$$

In practice, we set the target resolution to be $r = \frac{1}{\sqrt{|V|}}$, where $|V|$ is the number of vertices in the graph. In this way, an optimal drawing will distribute nodes uniformly in the drawing area. The purpose of the ReLU is to output zero when the argument is negative, as when the argument is negative the constraint is already satisfied. In the evaluation, we report, as a quality measure, the ratio between the actual and target resolution and cap its value between 0 (worst) and 1 (best).

$$Q_{VR} = \min(1.0, \frac{\min_{i,j}||X_i - X_j||}{r \cdot d_{max}}) \tag{8}$$

4.9 Gabriel Graph Property

A graph is a Gabriel graph if it can be drawn in such a way that any disk formed by using an edge in the graph as its diameter contains no other nodes. Not all graphs are Gabriel graphs, but drawing a graph so that as many of these edge-based disks are empty of other nodes has been associated with good readability [19]. This property can be enforced by a repulsive force around the midpoints of edges. Formally, we establish a repulsive field with radius r_{ij} equal to half of the edge length, around the midpoint c_{ij} of each edge $(i, j) \in E$, and we minimize the total potential energy:

$$L_{GA} = \sum_{\substack{(i,j)\in E, \\ k\in V\setminus\{i,j\}}} ReLU(r_{ij} - |X_k - c_{ij}|)^2 \tag{9}$$

where $c_{ij} = \frac{X_i + X_j}{2}$ and $r_{ij} = \frac{|X_i - X_j|}{2}$. We use the (normalized) minimum distance from nodes to centers to characterize the quality of a drawing with respect to Gabriel graph property: $Q_{GA} = \min_{(i,j) \in E, k \in V} \frac{|X_k - c_{ij}|}{r_{ij}}$.

5   Experimental Evaluation

In this section, we describe the experiment we conducted on 10 graphs to assess the effectiveness and limitations of our approach. The graphs used are depicted in Fig. 3 along with information about each graph. The graphs have been chosen to represent a variety of graph classes such as trees, cycles, grids, bipartite graphs, cubic graphs, and symmetric graphs.

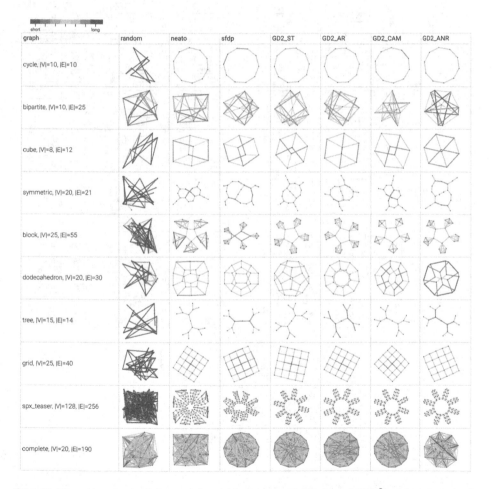

Fig. 3. Drawings from different algorithms: neato, sfdp and $(GD)^2$ with stress (**ST**), aspect ratio (**AR**), crossing angle maximization (**CAM**) and angular resolution (**ANR**) optimization on a set of 10 graphs. Edge color is determined by the discrepancy between actual and ideal edge length (here all ideal edge lengths are 1); informally, short edges are red and long edges are blue. (Color figure online)

Table 2. The values of the nine criteria corresponding to the 10 graphs for the layouts computed by neato, sfdp, random, and 3 runs of $(GD)^2$ initialized with neato, sfdp, and random layouts. Bold values are the best. Green cells show an improvement, yellow cells show a tie, with respect to the initial values.

Crossings	neato	sfdp	rnd	$(GD)^2_n$	$(GD)^2_s$	$(GD)^2_r$
dodec.	6.0	6.0	79.0	6.0	6.0	10.0
cycle	0.0	0.0	11.0	0.0	0.0	0.0
tree	0.0	0.0	31.0	0.0	0.0	0.0
block	23.0	16.0	297.0	23.0	16.0	25.0
compl.	3454	3571	3572	3454	3571	3572
cube	2.0	2.0	18.0	2.0	2.0	2.0
symme.	1.0	0.0	77.0	1.0	0.0	0.0
bipar.	40.0	52.0	40.0	40.0	40.0	40.0
grid	0.0	0.0	190.0	0.0	0.0	0.0
spx t.	73.0	71.0	7254.0	73.0	71.0	76.0

Ideal edge length	neato	sfdp	rnd	$(GD)^2_n$	$(GD)^2_s$	$(GD)^2_r$
dodec.	0.14	0.15	0.53	0.1	0.15	0.08
cycle	0.0	0.0	0.42	0.0	0.0	0.0
tree	0.03	0.13	0.31	0.03	0.04	0.09
block	0.31	0.43	0.5	0.25	0.33	0.31
compl.	0.42	0.41	0.45	0.41	0.41	0.41
cube	0.08	0.12	0.29	0.03	0.0	0.12
symme.	0.08	0.19	0.46	0.07	0.05	0.04
bipar.	0.31	0.26	0.44	0.16	0.13	0.1
grid	0.01	0.09	0.41	0.0	0.0	0.01
spx t.	0.4	0.32	0.45	0.3	0.2	0.32

Stress	neato	sfdp	rnd	$(GD)^2_n$	$(GD)^2_s$	$(GD)^2_r$
dodec.	21.4	17.58	111.05	17.45	17.58	17.6
cycle	0.77	0.77	30.24	0.77	0.77	0.77
tree	2.11	2.7	98.49	2.11	2.62	5.5
block	26.79	28.22	203.31	12.72	23.71	11.2
compl.	33.54	31.58	37.87	31.49	31.49	31.47
cube	2.75	2.71	11.69	2.66	2.69	2.65
symme.	9.88	5.38	180.48	9.88	3.36	3.97
bipar.	9.25	8.5	12.48	8.52	8.5	9.6
grid	6.77	7.38	221.66	6.77	6.78	6.77
spx t.	674.8	418.4	9794	227.1	235.3	227.2

Angular resolution	neato	sfdp	rnd	$(GD)^2_n$	$(GD)^2_s$	$(GD)^2_r$
dodec.	0.39	0.39	0.01	0.6	0.39	0.6
cycle	0.8	0.8	0.05	0.8	0.8	0.8
tree	0.61	0.56	0.04	0.78	0.83	0.88
block	0.05	0.01	0.0	0.36	0.02	0.29
compl.	0.0	0.01	0.0	0.0	0.01	0.0
cube	0.28	0.3	0.01	0.46	0.44	0.4
symme.	0.66	0.6	0.03	0.68	0.76	0.77
bipar.	0.01	0.03	0.01	0.02	0.04	0.11
grid	0.52	0.54	0.0	0.52	0.54	0.52
spx t.	0.02	0.0	0.0	0.03	0.0	0.0

Neighbor preservation	neato	sfdp	rnd	$(GD)^2_n$	$(GD)^2_s$	$(GD)^2_r$
dodec.	0.32	0.3	0.1	0.5	0.3	0.5
cycle	1.0	1.0	0.08	1.0	1.0	1.0
tree	1.0	1.0	0.02	1.0	1.0	1.0
block	0.57	0.93	0.12	0.83	0.93	1.0
compl.	1.0	1.0	1.0	1.0	1.0	1.0
cube	0.5	0.5	0.12	0.5	0.5	0.5
symme.	0.75	0.95	0.05	0.75	1.0	1.0
bipar.	0.47	0.47	0.43	0.47	0.47	0.43
grid	1.0	1.0	0.05	1.0	1.0	1.0
spx t.	0.36	0.44	0.03	0.49	0.46	0.53

Gabriel graph property	neato	sfdp	rnd	$(GD)^2_n$	$(GD)^2_s$	$(GD)^2_r$
dodec.	0.16	0.64	0.07	0.32	0.64	0.32
cycle	1.0	1.0	0.29	1.0	1.0	1.0
tree	1.0	1.0	0.05	1.0	1.0	1.0
block	0.16	0.03	0.04	0.57	0.14	0.59
compl.	0.0	0.01	0.02	0.04	0.01	0.07
cube	0.43	0.51	0.01	0.75	0.8	0.71
symme.	0.54	1.0	0.15	0.7	1.0	1.0
bipar.	0.08	0.11	0.25	0.48	0.64	0.74
grid	1.0	1.0	0.03	1.0	1.0	1.0
spx t.	0.04	0.0	0.02	0.06	0.08	0.08

Vertex resolution	neato	sfdp	rnd	$(GD)^2_n$	$(GD)^2_s$	$(GD)^2_r$
dodec.	0.52	0.54	0.07	0.7	0.81	0.68
cycle	0.98	0.98	0.32	0.98	0.98	0.98
tree	0.68	0.57	0.23	0.69	0.68	0.68
block	0.66	0.38	0.1	0.72	0.59	0.51
compl.	0.8	1.0	0.18	0.84	1.0	0.91
cube	0.66	0.82	0.11	0.66	0.82	0.67
symme.	0.35	0.43	0.06	0.38	0.51	0.6
bipar.	0.83	0.87	0.21	0.83	0.87	0.35
grid	0.87	0.8	0.08	0.88	0.88	0.88
spx t.	0.47	0.48	0.05	0.47	0.48	0.32

Aspect ratio	neato	sfdp	rnd	$(GD)^2_n$	$(GD)^2_s$	$(GD)^2_r$
dodec.	0.92	0.91	0.88	0.96	0.96	0.96
cycle	0.96	0.95	0.67	0.96	0.95	0.96
tree	0.73	0.67	0.88	0.86	0.76	0.88
block	0.9	0.74	0.7	0.96	0.9	0.96
compl.	0.89	0.97	0.91	0.98	0.98	0.98
cube	0.76	0.79	0.57	0.87	0.79	0.88
symme.	0.58	0.67	0.89	0.6	0.67	0.89
bipar.	0.82	0.9	0.91	0.82	0.9	0.91
grid	1.0	1.0	0.82	1.0	1.0	1.0
spx t.	0.98	0.86	0.88	0.99	0.99	0.99

Crossing angle	neato	sfdp	rnd	$(GD)^2_n$	$(GD)^2_s$	$(GD)^2_r$
dodec.	0.06	0.12	0.24	0.06	0.09	0.15
cycle	0.0	0.0	0.19	0.0	0.0	0.0
tree	0.0	0.0	0.23	0.0	0.0	0.0
block	0.11	0.1	0.24	0.05	0.06	0.09
compl.	0.25	0.24	0.24	0.24	0.24	0.24
cube	0.03	0.03	0.21	0.03	0.03	0.04
symme.	0.03	0.0	0.24	0.03	0.0	0.0
bipar.	0.16	0.17	0.23	0.16	0.17	0.19
grid	0.0	0.0	0.23	0.0	0.0	0.0
spx t.	0.16	0.22	0.25	0.16	0.15	0.21

In our experiment we compare $(GD)^2$ with neato [21] and sfdp [21], which are classical implementations of a stress-minimization layout and scalable force-directed layout. In particular, we focus on 9 readability criteria: stress (**ST**), vertex resolution (VR), ideal edge lengths (IL), neighbor preservation (NP), crossing angle (CA), angular resolution (ANR), aspect ratio (AR), Gabriel graph properties (GG), and crossings (CR). We provide the values of the nine criteria corresponding to the 10 graphs for the layouts computed by by neato, sfdp, random, and 3 runs of $(GD)^2$ initialized with neato, sfdp, and random layouts in Table 2. The best result is shown with bold font, green cells indicate improvement, yellow cells represent ties, with respect to the initial values (scores for different criteria obtained using neato, sfdp, and random initialization). From the experimental results we see that $(GD)^2$ improves the random layout in 90% of the tests. $(GD)^2$ also improves or ties initial layouts from neato and sfdp, but the improvements are not as strong or as frequent, most notably for the CR, NP, and CA criteria.

In this experiment, we focused on optimizing a single metric. In some applications, it is desirable to optimize multiple criteria. We can use a similar technique i.e., take a weighted sum of the metrics and optimize the sum of scores. In the prototype (http://hdc.cs.arizona.edu/~mwli/graph-drawing/), there is a slider for each criterion, making it possible to combine different criteria.

6 Limitations

Although $(GD)^2$ is a flexible framework that can optimize a wide range of criteria, it cannot handle the class of constraints where the node coordinates are related by some inequalities, i.e., the framework does not support hard constraints. Similarly, this framework does not naturally support shape-based drawing constraints such as those in [17, 18, 36]. $(GD)^2$ takes under a minute for the small graphs considered in this paper. We have not experimented with larger graphs as the implementation has not been optimized for speed.

7 Conclusions and Future Work

We introduced the graph drawing framework $(GD)^2$ and showed how this approach can be used to optimize different graph drawing criteria and combinations thereof. The framework is flexible and natural directions for future work include adding further drawing criteria and better ways to combine them. To compute the layout of large graphs, a multi-level algorithmic model might be needed. It would also be useful to have a way to compute appropriate weights for the different criteria.

Acknowledgments. This work was supported in part by NSF grants CCF-1740858, CCF-1712119, and DMS-1839274.

References

1. Abadi, M., et al.: TensorFlow: a system for large-scale machine learning. In: 12th USENIX Symposium on Operating Systems Design and Implementation (OSDI 2016), pp. 265–283 (2016)

2. Ábrego, B.M., Fernández-Merchant, S., Salazar, G.: The rectilinear crossing number of k_n: closing in (or are we?). In: Pach, J. (ed.) Thirty Essays on Geometric Graph Theory. Springer, New York (2012). https://doi.org/10.1007/978-1-4614-0110-0_2
3. Ahmed, R., De Luca, F., Devkota, S., Kobourov, S., Li, M.: Graph drawing via gradient descent, $(GD)^2$. arXiv preprint arXiv:2008.05584 (2020)
4. Argyriou, E.N., Bekos, M.A., Symvonis, A.: Maximizing the total resolution of graphs. In: Brandes, U., Cornelsen, S. (eds.) GD 2010. LNCS, vol. 6502, pp. 62–67. Springer, Heidelberg (2011). https://doi.org/10.1007/978-3-642-18469-7_6
5. Bekos, M.A., et al.: A heuristic approach towards drawings of graphs with high crossing resolution. In: Biedl, T., Kerren, A. (eds.) GD 2018. LNCS, vol. 11282, pp. 271–285. Springer, Cham (2018). https://doi.org/10.1007/978-3-030-04414-5_19
6. Berman, M., Rannen Triki, A., Blaschko, M.B.: The Lovász-softmax loss: a tractable surrogate for the optimization of the intersection-over-union measure in neural networks. In: Proceedings of the IEEE Conference on Computer Vision and Pattern Recognition, pp. 4413–4421 (2018)
7. Bostock, M., Ogievetsky, V., Heer, J.: D3: data-driven documents. IEEE Trans. Vis. Comput. Graph. **17**(12), 2301–2309 (2011)
8. Buchheim, C., Chimani, M., Gutwenger, C., Jünger, M., Mutzel, P.: Crossings and planarization. In: Handbook of Graph Drawing and Visualization, pp. 43–85 (2013)
9. Chen, K.T., Dwyer, T., Marriott, K., Bach, B.: DoughNets: visualising networks using torus wrapping. In: Proceedings of the 2020 CHI Conference on Human Factors in Computing Systems, pp. 1–11 (2020)
10. Chrobak, M., Goodrich, M.T., Tamassia, R.: Convex drawings of graphs in two and three dimensions. In: Proceedings of the 12th Annual Symposium on Computational Geometry, pp. 319–328 (1996)
11. Davidson, R., Harel, D.: Drawing graphs nicely using simulated annealing. ACM Trans. Graph. (TOG) **15**(4), 301–331 (1996)
12. Demel, A., Dürrschnabel, D., Mchedlidze, T., Radermacher, M., Wulf, L.: A Greedy heuristic for crossing-angle maximization. In: Biedl, T., Kerren, A. (eds.) GD 2018. LNCS, vol. 11282, pp. 286–299. Springer, Cham (2018). https://doi.org/10.1007/978-3-030-04414-5_20
13. Brandenburg, F.J., Duncan, C.A., Gansner, E., Kobourov, S.G.: Graph-drawing contest report. In: Pach, J. (ed.) GD 2004. LNCS, vol. 3383, pp. 512–516. Springer, Heidelberg (2005). https://doi.org/10.1007/978-3-540-31843-9_56
14. Devkota, S., Ahmed, R., De Luca, F., Isaacs, K.E., Kobourov, S.: Stress-plus-X (SPX) graph layout. In: Archambault, D., Tóth, C.D. (eds.) GD 2019. LNCS, vol. 11904, pp. 291–304. Springer, Cham (2019). https://doi.org/10.1007/978-3-030-35802-0_23
15. Didimo, W., Liotta, G.: The Crossing-angle Resolution in Graph Drawing. In: Pach, J. (ed.) Thirty essays on geometric graph theory. Springer, New York (2014). https://doi.org/10.1007/978-1-4614-0110-0
16. Duncan, C.A., Goodrich, M.T., Kobourov, S.G.: Balanced aspect ratio trees and their use for drawing very large graphs. In: Whitesides, S.H. (ed.) GD 1998. LNCS, vol. 1547, pp. 111–124. Springer, Heidelberg (1998). https://doi.org/10.1007/3-540-37623-2_9
17. Dwyer, T.: Scalable, versatile and simple constrained graph layout. Comput. Graph. Forum **28**, 991–998 (2009)
18. Dwyer, T., Koren, Y., Marriott, K.: IPSep-CoLa: an incremental procedure for separation constraint layout of graphs. IEEE Trans. Vis. Comput. Graph. **12**, 821–8 (2006)

19. Eades, P., Hong, S.-H., Klein, K., Nguyen, A.: Shape-based quality metrics for large graph visualization. In: Di Giacomo, E., Lubiw, A. (eds.) GD 2015. LNCS, vol. 9411, pp. 502–514. Springer, Cham (2015). https://doi.org/10.1007/978-3-319-27261-0_41

20. Eades, P., Huang, W., Hong, S.H.: A force-directed method for large crossing angle graph drawing. arXiv preprint arXiv:1012.4559 (2010)

21. Ellson, J., Gansner, E., Koutsofios, L., North, S.C., Woodhull, G.: Graphviz—open source graph drawing tools. In: Mutzel, P., Jünger, M., Leipert, S. (eds.) GD 2001. LNCS, vol. 2265, pp. 483–484. Springer, Heidelberg (2002). https://doi.org/10.1007/3-540-45848-4_57

22. Gansner, E.R., Koren, Y., North, S.: Graph drawing by stress majorization. In: Pach, J. (ed.) GD 2004. LNCS, vol. 3383, pp. 239–250. Springer, Heidelberg (2005). https://doi.org/10.1007/978-3-540-31843-9_25

23. Griewank, A., Walther, A.: Evaluating derivatives: principles and techniques of algorithmic differentiation, vol. 105. SIAM (2008)

24. Huang, W., Eades, P., Hong, S.H.: Larger crossing angles make graphs easier to read. J. Vis. Lang. Comput. **25**(4), 452–465 (2014)

25. Huang, W., Eades, P., Hong, S.H., Lin, C.C.: Improving multiple aesthetics produces better graph drawings. J. Vis. Lang. Comput. **24**(4), 262–272 (2013)

26. Kamada, T., Kawai, S.: An algorithm for drawing general undirected graphs. Inf. Process. Lett. **31**(1), 7–15 (1989)

27. Kruiger, J.F., Rauber, P.E., Martins, R.M., Kerren, A., Kobourov, S., Telea, A.C.: Graph layouts by t-SNE. Comput. Graph. Forum **36**(3), 283–294 (2017)

28. Kruskal, J.B.: Multidimensional scaling by optimizing goodness of fit to a nonmetric hypothesis. Psychometrika **29**(1), 1–27 (1964)

29. Paszke, A., et al.: PyTorch: an imperative style, high-performance deep learning library. In: Advances in Neural Information Processing Systems, pp. 8024–8035 (2019)

30. Purchase, H.: Which aesthetic has the greatest effect on human understanding? In: DiBattista, G. (ed.) GD 1997. LNCS, vol. 1353, pp. 248–261. Springer, Heidelberg (1997). https://doi.org/10.1007/3-540-63938-1_67

31. Radermacher, M., Reichard, K., Rutter, I., Wagner, D.: A geometric heuristic for rectilinear crossing minimization. In: The 20th Workshop on Algorithm Engineering and Experiments, pp. 129–138 (2018)

32. Schulz, A.: Drawing 3-polytopes with good vertex resolution. J. Graph Algorithms Appl. **15**(1), 33–52 (2011)

33. Shabbeer, A., Ozcaglar, C., Gonzalez, M., Bennett, K.P.: Optimal embedding of heterogeneous graph data with edge crossing constraints. In: NIPS Workshop on Challenges of Data Visualization (2010)

34. Shepard, R.N.: The analysis of proximities: multidimensional scaling with an unknown distance function. Psychometrika **27**(2), 125–140 (1962)

35. Smilkov, D., et al.: Tensorflow.js: machine learning for the web and beyond. In: Proceedings of Machine Learning and Systems 2019, pp. 309–321 (2019)

36. Wang, Y., et al.: Revisiting stress majorization as a unified framework for interactive constrained graph visualization. IEEE Trans. Vis. Comput. Graph. **24**(1), 489–499 (2017)

37. Ware, C., Purchase, H., Colpoys, L., McGill, M.: Cognitive measurements of graph aesthetics. Inf. Vis. **1**(2), 103–110 (2002)

38. Zheng, J.X., Pawar, S., Goodman, D.F.: Graph drawing by stochastic gradient descent. IEEE Trans. Vis. Comput. Graph. **25**(9), 2738–2748 (2018)

Stochastic Gradient Descent Works Really Well for Stress Minimization

Katharina Börsig, Ulrik Brandes$^{(\boxtimes)}$ (iD), and Barna Pasztor

ETH Zürich, Zürich, Switzerland
ubrandes@ethz.ch

Abstract. Stress minimization is among the best studied force-directed graph layout methods because it reliably yields high-quality layouts. It thus comes as a surprise that a novel approach based on stochastic gradient descent (Zheng, Pawar and Goodman, TVCG 2019) is claimed to improve on state-of-the-art approaches based on majorization. We present experimental evidence that the new approach does not actually yield better layouts, but that it is still to be preferred because it is simpler and robust against poor initialization.

Keywords: Multidimensional scaling · Stress minimization · Stochastic gradient descent · Experiments.

1 Introduction

The class of force-directed graph drawing algorithms is large both in terms of objectives and optimization algorithms [1,13]. Experimental [4] and anecdotal evidence suggest that a most desirable objective is the stress function of distance-based multidimensional scaling [14]. Given a simple undirected graph $G = (V, E)$, the layout $x = (\mathbb{R}^2)^V$ of a straight-line drawing is considered suitable, if the weighted deviation

$$stress(x) = \sum_{i<j} d_{ij}^{-2}(\|x_i - x_j\| - d_{ij})^2 \qquad (1)$$

of Euclidean distances $\|x_i - x_j\|$ in the layout from shortest-path distances d_{ij} in the graph is small.

The stress function has been varied in numerous ways to accommodate additional objectives or constraints [2,5,8,9,16]. Since stress minimization is computationally intractable, similarly many approaches have been proposed to save computation time [11,15,17]. These methods are generally designed to improve an initial layout iteratively and thus yield local minima of the stress function that cannot be improved further by moving single vertices.

Here we are interested in assessing a recent proposal by Zheng, Pawar, and Goodman [18] that is based on stochastic gradient descent and claimed to outperform majorization approaches [10].

© Springer Nature Switzerland AG 2020
D. Auber and P. Valtr (Eds.): GD 2020, LNCS 12590, pp. 18–25, 2020.
https://doi.org/10.1007/978-3-030-68766-3_2

Our own computational experiments suggest that the new approach does not lead to better layouts, but that it is still preferable due to its simplicity and, crucially, indifference to initialization. We do not address actual running times because any comparison would be relative to the choice of speed-up techniques and the overall similarity of the computation suggests that the same algorithm engineering techniques could be used in either approach.

The remainder is organized as follows. In Sect. 2, we briefly describe the proposal of Zheng et al. in the context of previous approaches. The results of our experiments are presented and discussed in Sect. 3, and we conclude with some general implications in Sect. 4.

2   Stress Minimization

We very briefly review some major developments in the use of multidimensional scaling in graph drawing. This is not to provide the details of each method but to contrast the approach based on stochastic gradient descent with previous approaches.

Gradient Descent. While first uses of multidimensional scaling for graph drawing date back to the 1960s, it was popularized by Kamada and Kawai [12], who also introduced a localized version of the gradient descent approach used until then. Since a necessary condition for a local minimum of the stress function is that all partial derivatives are zero, they iteratively pick a vertex for which the vector of partial derivatives with respect to its two coordinates has maximum length. Then a two-dimensional Newton-Raphson method is applied to the stress function with all other vertices fixed. Their layout is thus obtained by iteratively moving one vertex at a time toward a position where the different stress terms cancel each other out.

Majorization. Ganser, Koren, and North [10] proposed to use majorization [7] instead. Here, the complex stress function is replaced with a convex function that is larger for each layout but the current, for which it is equal. Minimizing this function leads to a new layout that is guaranteed to have lower stress, and the process is iterated until it converges to a local minimum.

The process can also be localized to move only a single vertex such that the majorizing function is reduced. This yields an intuitive algorithm because the update

$$x_i \leftarrow \frac{1}{\sum\limits_{j \neq i} d_{ij}^{-2}} \sum_{j \neq i} d_{ij}^{-2} \cdot \frac{x_j + d_{ij}(x_i - x_j)}{\|x_i - x_j\|}$$

places vertex i directly into a position that balances out the influences of all other vertices. One iteration consists of an update of each vertex.

Because of its simplicity and guaranteed convergence, this approach is considered the state of the art.

Stochastic Dradient Descent. In this method the gradient is replaced by an unbiased estimator. For additive objective functions such as the stress function in Eq. (1), the estimator may simply be a single term of the sum. Since stress has one term for every pair of vertices, the contribution of this term can be reduced by moving the two vertices either closer together or farther apart.

A single update thus moves both vertices along the vector δ to extend or shrink the line segment $\overline{x_i x_j}$ to match the target length d_{ij} more closely,

$$
\begin{aligned}
x_i &\leftarrow x_i - \tfrac{\mu(t)}{2} \cdot \delta \\
x_j &\leftarrow x_j + \tfrac{\mu(t)}{2} \cdot \delta
\end{aligned}
\qquad \text{where} \qquad
\delta = \frac{\|x_i - x_j\| - d_{ij}}{\|x_i - x_j\|} \cdot (x_i - x_j),
$$

and $\mu(t) = \min\{1, d_{ij}^{-2}\eta(t)\}$ is a weighted step width capped at 1. Since an individual move is almost certainly in conflict with the desired distances of other pairs, the method does not converge in general. Instead, the unweighted step width $\eta(t)$ is made to exhibit an exponential decay over iteration time t, and convergence is thus enforced.

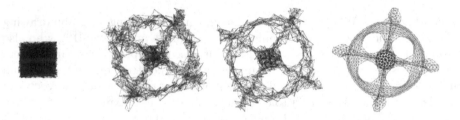

Fig. 1. Example run of stochastic gradient descent on graph dwt_1005 with random initialization and intermediate layouts after 1, 6, and 15 iteration.

One iteration consists of an update of all pairs of vertices in random order. The method is thus similar to localized majorization but instead of aggregating the influence of all other vertices before moving one, those influences are considered separately in random order. The running time of one iteration is in $\Theta(n^2)$ for both stochastic gradient descent and localized majorization, but instead of over a linear number of linear-time vertex movements the computation is spread out over a quadratic number of constant-time dyadic updates.

3   Experiments

Our experiments address the claim [18] that stochastic gradient descent (SGD) outperforms majorization (SMACOF). The graphs used as benchmarks are from the University of Florida sparse matrix collection [6].

On Par, But Not Better. The claim of superior performance is based on experiments in which both approaches are initialized with a random layout as in the example in Fig. 1. It was already concluded from earlier experiments, however, that the performance of majorization depends on the initialization and that random initialization leads to poor local minima [4].

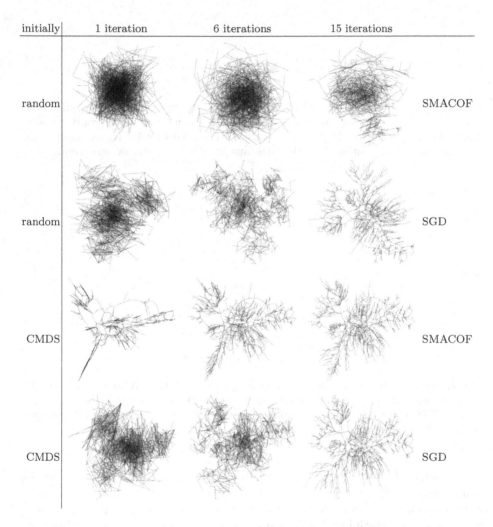

Fig. 2. An example graph (1138_bus) after 1, 6, and 15 iterations.

We therefore ran experiments comparing the reduction in stress when initializing at random or with classical MDS (CMDS). Classical MDS results in layouts that are essentially unique and represent large distances well. Moreover, it can be approximated at comparatively negligible cost using PivotMDS [3]. Two typical examples of the results are shown in Fig. 3, and for a better intuition, we also

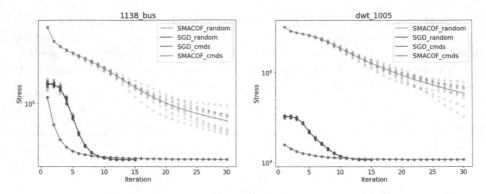

Fig. 3. Stress values for SGD and SMACOF on two example graphs. Random initialization is within a unit square whilst classical MDS is used at an appropriate scale. The plots show results of 10 runs for each algorithm, with circles representing single runs and lines interpolating through the means of all 10 runs. Initial stress omitted.

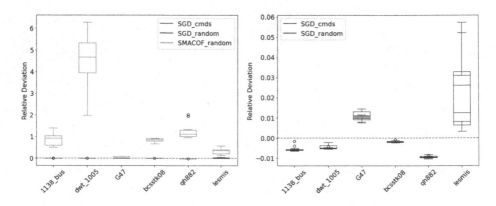

Fig. 4. Stress relative to baseline from SMACOF after CMDS. With 10 runs for each instance, we find that random initialization results in significantly higher stress for SMACOF (left chart). The stress obtained from SGD differs by about ±1% (rescaled on the right).

show some of the corresponding layouts in Fig. 2. While the result on all benchmark graphs confirm that SGD indeed yields much lower stress than majorization when initialized with a random layout, there is no noteworthy difference in the final stress when the initial layout takes care of the global arrangement. Notably, the result of SGD is largely independent of the initialization strategy.

Our experiments on a much larger set of benchmark graphs support these conclusions. The evaluations in Fig. 4 confirm quantitatively that majorization with random initialization is a poor baseline because it results in significantly higher stress compared to majorization after classical scaling. Whether SGD or the latter combination yield lower stress depends on the graph, but relative differences are small, anyway.

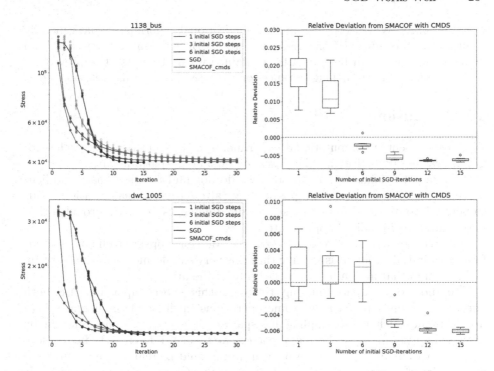

Fig. 5. Stress reduction by SMACOF initialized with CMDS or a few iterations of SGD (left) and the relative deviation of the final stress from the baseline of SMACOF with CMDS (right) on example graphs 1138_bus and dwt_1005. The initial iterations of SGD start from a random initialization in the unit square, and each instance was run 10 times.

Self-initializing. The seeming indifference of SGD to the initial layout prompted a second suite of experiments.

We hypothesized that the initially large displacements in SGD are responsible for the overall quality of the final outcome. If this was the case, then the differences between SGD and SMACOF should disappear when we initialize SMACOF with a small number of SGD iterations.

As illustrated in Fig. 5 this is indeed the case. Even a single step of SGD prevents majorization from sinking into a poor local minimum. After about seven iterations of SGD, majorization yields layouts that are even slightly better than those obtained from initialization with CMDS. We also note that in the next iterations, SMACOF reduces stress faster than SGD, but the number of iterations to the final layout is roughly the same for both. This number becomes smaller than for SMACOF initialized with CMDS, offsetting the higher cost of SGD iterations compared to PivotMDS.

We conclude that a, if not the, major advantage of the approach based on stochastic gradient descent is the reliable untangling of any initial layout during the first few iterations. No separate initialization strategy is required.

Well Designed. We performed a number of additional experiments that generally confirm the recommendations given for stochastic gradient descent [18], and indicate that little can be gained by straightforward attempts at improvement such as an initial focus on long distances or the integration of majorization steps.

4   Conclusions

We have presented computational experiments comparing two approaches for graph drawing by multidimensional scaling of shortest-path distances.

Contrary to claims by the authors, we do not find that stochastic gradient descent, which was recently proposed as an alternative to majorization, leads to better layouts [18]. We find no significant differences in stress, provided that majorization is initialized appropriately.

The true advantage of stochastic gradient descent appears to lie in its indifference to initialization. It is striking that this very simple and uniform algorithm yields results that are on par with the state of the art.

We did not compare running times in this short paper because both approaches largely perform the same operations in different order and speed-up techniques such as subsampling and spatial aggregation abound. Since many of these apply similarly to both approaches, we expect differences to be too subtle for any general claims. Since pairs in a maximal matching can be updated without interference, stochastic gradient descent appears to be more amenable to parallelization, though.

References

1. Brandes, U.: Force-directed graph drawing. In: Kao, M.Y. (ed.) Encyclopedia of Algorithms, pp. 1–6. Springer, New York (2014). https://doi.org/10.1007/978-3-642-27848-8_648-1
2. Brandes, U., Mader, M.: A quantitative comparison of stress-minimization approaches for offline dynamic graph drawing. In: van Kreveld, M., Speckmann, B. (eds.) GD 2011. LNCS, vol. 7034, pp. 99–110. Springer, Heidelberg (2012). https://doi.org/10.1007/978-3-642-25878-7_11
3. Brandes, U., Pich, C.: Eigensolver methods for progressive multidimensional scaling of large data. In: Kaufmann, M., Wagner, D. (eds.) GD 2006. LNCS, vol. 4372, pp. 42–53. Springer, Heidelberg (2007). https://doi.org/10.1007/978-3-540-70904-6_6
4. Brandes, U., Pich, C.: An experimental study on distance-based graph drawing. In: Tollis, I.G., Patrignani, M. (eds.) GD 2008. LNCS, vol. 5417, pp. 218–229. Springer, Heidelberg (2009). https://doi.org/10.1007/978-3-642-00219-9_21
5. Brandes, U., Pich, C.: More flexible radial layout. J. Graph Alg. Appl. **15**(1), 157–173 (2011). https://doi.org/10.7155/jgaa.00221
6. Davis, T.A., Hu, Y.: The University of Florida sparse matrix collection. ACM Trans. Math. Softw. **38**(1), 1 (2011). https://doi.org/10.1145/2049662.2049663
7. De Leeuw, J.: Applications of convex analysis to multidimensional scaling. In: Barra, J.R., Brodeau, F., Romier, G., Van Cutsem, B. (eds.) Recent Developments in Statistics, pp. 133–145. North Holland Publishing Company (1977). http://www.stat.ucla.edu/~deleeuw/janspubs/1977/chapters/deleeuw_C_77.pdf

8. Dwyer, T., Koren, Y., Marriott, K.: Constrained graph layout by stress majorization and gradient projection. Discrete Math. **309**(7), 1895–1908 (2009). https://doi.org/10.1016/j.disc.2007.12.103

9. Gansner, E.R., Hu, Y., North, S.C.: A maxent-stress model for graph layout. IEEE Trans. Vis. Comput. Graph. **19**(6), 927–940 (2013). https://doi.org/10.1109/TVCG.2012.299

10. Gansner, E.R., Koren, Y., North, S.: Graph drawing by stress majorization. In: Pach, J. (ed.) GD 2004. LNCS, vol. 3383, pp. 239–250. Springer, Heidelberg (2005). https://doi.org/10.1007/978-3-540-31843-9_25

11. Ingram, S., Munzner, T., Olano, M.: Glimmer: multilevel MDS on the GPU. IEEE Trans. Vis. Comput. Graph. **15**(2), 249–261 (2009). https://doi.org/10.1109/TVCG.2008.85

12. Kamada, T., Kawai, S.: An algorithm for drawing general undirected graphs. Inf. Process. Lett. **31**, 7–15 (1989). https://doi.org/10.1016/0020-0190(89)90102-6

13. Kobourov, S.G.: Force-directed drawing algorithms. In: Tamassia, R. (ed.) Handbook of Graph Drawing and Visualization, pp. 383–408. CRC Press, Oxford (2013)

14. McGee, V.E.: The multidimensional scaling of 'elastic' distances. Br. J. Math. Stat. Psychol. **19**(2), 181–196 (1966). https://doi.org/10.1111/j.2044-8317.1966.tb00367.x

15. Meyerhenke, H., Nöllenburg, M., Schulz, C.: Drawing large graphs by multilevel maxent-stress optimization. IEEE Trans. Visual Comput. Graphics **24**(5), 1814–1827 (2018). https://doi.org/10.1109/TVCG.2017.2689016

16. Nocaj, A., Ortmann, M., Brandes, U.: Untangling the hairballs of multi-centered, small-world online social media networks. J. Graph Alg. Appl. **19**(2), 595–618 (2015). https://doi.org/10.7155/jgaa.00370

17. Ortmann, M., Klimenta, M., Brandes, U.: A sparse stress model. J. Graph Alg. Appl. **21**(5), 791–821 (2017). https://doi.org/10.7155/jgaa.00440

18. Zheng, J.X., Pawar, S., Goodman, D.F.M.: Graph drawing by stochastic gradient descent. IEEE Trans. Visual Comput. Graphics **25**(9), 2738–2748 (2019). https://doi.org/10.1109/TVCG.2018.2859997

The Local Queue Number of Graphs with Bounded Treewidth

Laura Merker$^{(\boxtimes)}$ and Torsten Ueckerdt

Institute of Theoretical Informatics, Karlsruhe Institute of Technology (KIT),
Karlsruhe, Germany
laura.merker@student.kit.edu, torsten.ueckerdt@kit.edu

Abstract. A queue layout of a graph G consists of a vertex ordering of G and a partition of the edges into so-called queues such that no two edges in the same queue nest, i.e., have their endpoints ordered in an ABBA-pattern. Continuing the research on local ordered covering numbers, we introduce the local queue number of a graph G as the minimum ℓ such that G admits a queue layout with each vertex having incident edges in no more than ℓ queues. Similarly to the local page number [Merker, Ueckerdt, GD'19], the local queue number is closely related to the graph's density and can be arbitrarily far from the classical queue number.

We present tools to bound the local queue number of graphs from above and below, focusing on graphs of treewidth k. Using these, we show that every graph of treewidth k has local queue number at most $k+1$ and that this bound is tight for $k = 2$, while a general lower bound is $\lceil k/2 \rceil + 1$. Our results imply, inter alia, that the maximum local queue number among planar graphs is either 3 or 4.

Keywords: Queue number · Local covering number · Treewidth

1 Introduction

Given a graph, we aim to find a vertex ordering $\prec$ and a partition of the edges into queues, where two edges uv and xy may not be in the same queue if $u \prec x \prec y \prec v$. Since Heath and Rosenberg [14] introduced this concept in 1992, one of the main concerns of studying queue layouts is the investigation of the maximum queue number of the class of planar graphs and the class of graphs with bounded treewidth, see for instance [4,7,8,13,22,25]. Despite recent breakthroughs, there are still large gaps between lower and upper bounds on the maximum queue number of both graph classes. In particular, the maximum queue number of planar graphs is between 4 and 49 due to Alam et al. [4], respectively Dujmović et al. [7], and Wiechert [25] provides a linear lower bound and an exponential upper bound on the maximum queue number of graphs with treewidth k. We continue the research in this direction by proposing a new graph parameter, the *local queue number*, that minimizes the number of queues in which any one vertex has incident edges. Compared to the classical queue number, the investigation

© Springer Nature Switzerland AG 2020
D. Auber and P. Valtr (Eds.): GD 2020, LNCS 12590, pp. 26–39, 2020.
https://doi.org/10.1007/978-3-030-68766-3_3

of the local queue number leads to stronger lower bounds and weaker upper bounds. The latter might offer a way to support conjectured upper bounds on the classical queue number. We remark that analogously to the local queue number considered here, we recently introduced [17] the *local page number* as a weaker version of the classical page number.

All necessary definitions are given in Sect. 1.1, including the formal definition of local queue numbers. In Sect. 1.2, we briefly locate local queue numbers in the general covering number framework, and outline the state of the art on queue numbers and local page numbers of planar graphs and graphs with bounded treewidth. We summarize our results in Sect. 1.3 and point out which results on local page numbers immediately generalize to local queue numbers. We then investigate the local queue number of k-trees in Sect. 2. Finally, we discuss possible applications of the presented tools and propose open problems for further research in Sect. 3.

1.1 Definitions

Consider a graph G with a linear ordering $\prec$ of its vertex set. The sets $V(G)$ and $E(G)$ denote the vertex set, respectively edge set, of G. For subsets $X, Y \subseteq V(G)$, we write $X \prec Y$ and say X is *to the left* of Y and Y is *to the right* of X if $x \prec y$ for all vertices $x \in X, y \in Y$. If the sets consist only of a single vertex, we use x instead of $\{x\}$. Let the *span* of X contain all vertices lying between the leftmost and the rightmost vertex of X, that is $\mathrm{span}(X) = \{v \in V(G) : \exists x, x' \in X \text{ with } x \preccurlyeq v \preccurlyeq x'\}$. For a subgraph H of G and a vertex $v \notin V(H)$, we say v is *below* H if $v \in \mathrm{span}(V(H))$ and we say v is *outside* H otherwise.

Fig. 1. Left to right: 1-queue layout, 3-rainbow, 1-page book embedding, 3-twist.

Two edges $uv, xy \in E(G)$ *nest* if $u \prec x \prec y \prec v$ or $x \prec u \prec v \prec y$, and they *cross* if $u \prec x \prec v \prec y$ or $x \prec u \prec y \prec v$. A set of k pairwise nesting (crossing) edges is called a *k-rainbow* (*k-twist*). A *queue* (*page*) is an edge set in which no two edges nest (cross), see Fig. 1. A *k-queue layout* (*k-page book embedding*) of G consists of a vertex ordering $\prec$ and a partition of the edges of G into k queues (pages). Finally, the *queue number* $\mathrm{qn}(G)$ (*page number* $\mathrm{pn}(G)$, also known as *stack number* or *book thickness*) of a graph G is the smallest k such that there is a k-queue layout (k-page book embedding) for G. Both concepts are called *ordered covering numbers* as a partition of edges can also be considered as covering the graph with queues or pages, respectively.

We now define local variants of the parameters defined above. For this, we allow partitions of arbitrary size but minimize the number of parts at every vertex. An *ℓ-local queue layout* (*ℓ-local book embedding*) is one in which every vertex has incident edges in at most ℓ queues (pages). The *local queue number* $\mathrm{qn}_\ell(G)$ (*local page number* $\mathrm{pn}_\ell(G)$) is the smallest ℓ for which there is an ℓ-local queue layout for G. Note that we have $\mathrm{qn}_\ell(G) \leqslant \mathrm{qn}(G)$ and $\mathrm{pn}_\ell(G) \leqslant \mathrm{pn}(G)$ as layouts of size ℓ are also ℓ-local.

Finally, a *k-tree* is a $(k+1)$-clique or is obtained from a smaller k-tree by choosing a clique C of size k and adding a new vertex u which is adjacent to all vertices of C. Fixing an arbitrary construction ordering, the vertex u is called a *child* of C, and C is called the *parent clique* of u. We also say u is a child of each vertex of C. A child is called *nesting* with respect to a vertex ordering if it is placed below its parent clique, and *non-nesting* otherwise. Note that k-trees are exactly the maximal graphs with treewidth k. As local queue and page numbers are monotone, it suffices for us to investigate k-trees, instead of arbitrary graphs of treewidth k.

1.2   Related Work and Motivation

The notion of local ordered covering numbers unifies the concepts of local covering numbers and ordered covering numbers. The first was introduced by Knauer and Ueckerdt [16], while existing research on the latter focuses on queue numbers and page numbers, which were established by Bernhart and Kainen [5] and Heath and Rosenberg [14], respectively.

We first give a brief overview of global and local covering numbers as introduced in [16]. Consider a class of graphs $\mathcal{G}$, called *guest class*, and an input graph H. We say the graph H is *covered* by some covering graphs $G_1, \ldots, G_t \in \mathcal{G}$ if G_i is a subgraph of H for each i and every edge of H is contained in some covering graph, i.e. if $G_1 \cup \cdots \cup G_t = H$. The set of covering graphs is called an *injective $\mathcal{G}$-cover* of H. The *global covering number* is the minimum number of covering graphs needed to cover a graph H, that is the size of the smallest injective $\mathcal{G}$-cover of H. For the *local covering number*, we use $\mathcal{G}$-covers of arbitrary size and minimize the number of covering graphs at every vertex. For this, we say a $\mathcal{G}$-cover for a graph H is ℓ-*local* if every vertex is contained in at most ℓ covering graphs. Now, the *local covering number* of a graph H with guest class $\mathcal{G}$ is defined as the smallest ℓ such that there is an ℓ-local injective $\mathcal{G}$-cover of H.

Many known graph parameters are covering numbers. For instance, the thickness and outerthickness are global covering numbers for the guest classes of planar and outerplanar graphs, respectively [11,19]. In addition, all kinds of arboricity are global covering numbers for the guest class of the respective forests [2,3,10,20]. The local covering number was considered for the guest classes of complete bipartite graphs [9], complete graphs [24], and different forests [16].

We continue by summarizing known results on the queue number and local page number of planar graphs and graphs with bounded treewidth. While every 1-queue graph is planar [14], the maximum queue number among all outerplanar graphs is 2 [13] and among all planar graphs it is between 4 and 49 [4,7]. The lower bound of 4 is obtained by a planar 3-tree. Alam et al. [4] also show that every planar 3-tree admits a 5-queue layout. Trees have queue number 1 using a

BFS-ordering [14], and BFS-orderings proved also useful for queue layouts of planar graphs [7], outerplanar graphs [13], and graphs with bounded treewidth [25]. Rengarajan and Veni Madhavan [22] prove that every 2-tree admits a 3-queue layout, while Wiechert [25] proves that this bound is tight. More general, there is a graph with treewidth k and queue number at least $k + 1$ for each $k > 1$, while the best known upper bound is $2^k - 1$ [25].

The local version of page numbers was introduced and investigated in [17]. The local page number of any graph is always near its maximum average degree, while the classical page number can be arbitrarily far off: For any $d \geqslant 3$, there are n-vertex graphs with local page number at most $d + 2$ but page number $\Omega(\sqrt{d}n^{1/2-1/d})$. The maximum local page number for k-trees is at least k and at most $k + 1$, and for planar graphs it is either 3 or 4.

Our main motivation for defining local ordered covering numbers is to combine the well-studied notions of ordered covering numbers and local covering numbers and thereby continue research on both concepts. The questions we ask for the new graph parameters naturally arise from those asked for the known concepts. Studying ordered graphs and covering numbers is additionally motivated by applications in very-large-scale integration (VLSI) circuit design and bioinformatics [1,6,15,23]. In addition, covers appear in network design [21], while queue layouts are closely related to 3-dimensional graph drawing [26] and parallel multiplications of sparse matrices [12].

1.3   Contribution

We first observe that there are graphs whose local queue number is arbitrarily far from its queue number. In addition, the local queue number is tied to the *maximum average degree*, which is defined as $\mathrm{mad}(G) = \max\{2|E(H)|/|V(H)|: H \subseteq G, H \neq \emptyset\}$. Both results are derived from the analogous results for local page numbers [17], which is why we omit the proofs here. They can be found in the long version of this paper [18]. Theorem 2 also implies that the local queue number is tied to the local page number, which is conjectured for the classical page number and classical queue number.

Theorem 1. *For any $d \geqslant 3$ and infinitely many n, there exist n-vertex graphs with local queue number at most $d + 2$ but queue number $\Omega(\sqrt{d}n^{1/2-1/d})$.*

Theorem 2. *For any graph G, we have*

$$\frac{\mathrm{mad}(G)}{4} \leqslant \mathrm{qn}_\ell(G) \leqslant \frac{\mathrm{mad}(G)}{2} + 2.$$

While the best upper bound for the queue number of k-trees is $2^k - 1$ due to Wiechert [25], Theorem 2 already provides a linear upper bound for the local queue number of k-trees, which can be slightly improved.

Theorem 3. *Every graph with treewidth k admits a $(k + 1)$-local queue layout.*

Suspecting that the bound in Theorem 3 might be tight, we focus on lower bounds for the local queue number of k-trees in Sect. 2. Our main contribution is a tool that allows to focus on the construction of cliques with non-nesting children. We use this to prove that Theorem 3 is tight for $k = 2$ and that there are k-trees whose local queue number is at least $\lceil k/2 \rceil + 1$ for $k > 1$.

Theorem 4. *There is a graph with treewidth 2 and local queue number 3.*

Theorem 5. *For every $k > 1$, there is a graph G with treewidth k and local queue number at least $\lceil k/2 \rceil + 1$.*

As the maximum average degree of planar graphs is strictly smaller than 6 and 2-trees are planar, Theorems 2 and 4 bound the maximum local queue number of the class of planar graphs.

Corollary 6. *Every planar graph admits a 4-local queue layout and there is a planar graph whose local queue number is at least 3.*

2   The Local Queue Number of k-Trees

We first provide a straight-forward construction for the upper bound of $k+1$ for the local queue number of k-trees, which proves Theorem 3.

Proof (of Theorem 3). We partition the edges of a k-tree G into stars, each forming a queue. Consider an arbitrary construction ordering of G and let $v_1, \ldots, v_n$ denote the vertices of G in this ordering. For each vertex $v_i \in V(G)$, $i = 1, \ldots, n$, we define a queue Q_i that contains all edges from v_i to its children, that is $Q_i = \{v_i v_j \in E(G) : i < j\}$. Choosing an arbitrary vertex ordering yields a queue layout since edges of a star cannot nest. The layout is $(k + 1)$-local since every vertex has at most k neighbors with smaller index.   □

As our main tool for constructing k-trees with large local queue number, we introduce a sequence of two-player games that are adaptions of a game introduced by Wiechert [25]. Taking turns, Alice constructs a k-tree, which is laid out by Bob. The rules for Alice stay the same in all games, whereas the rules for Bob include only the first n conditions in the n-th game (see below). We always assume that Bob has an optimal strategy and prepare Alice to react on all possible moves of Bob. That is, when we write *Alice wins*, then we mean that she wins regardless of the layout Bob chooses.

The graph which is laid out in the r-th round of any game is denoted by G_r, the layout Bob creates by $(\prec_r, \mathcal{Q}_r)$. In the beginning, there is an initial clique C_{init} whose edges are assigned to arbitrary queues. In particular, we have $G_0 = C_{\text{init}}$. The notation we introduce for the games is summarized in Fig. 2. In the r-th round, Alice chooses a k-clique C_r from the current graph G_{r-1} and an integer m_r. Now, m_r new vertices $x_1^r, \ldots, x_{m_r}^r$ are introduced and become adjacent to the vertices of C_r. The clique C_r is the *parent clique* of the new vertices and edges. Vertices with the same parent clique are called *twin vertices* and two edges that share a vertex in the parent clique and are introduced in the same round are called *twin edges*. Then, in the n-th game, Bob inserts the new vertices into the current vertex ordering and assigns the new edges to queues satisfying the first n of the following conditions:

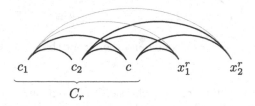

Fig. 2. Notation for the Games (i) to (v). C_r is the parent clique that Alice chooses in the r-th round and she chooses to add $m_r = 2$ children. The two children x_1^r and x_2^r are twin vertices. The two orange (thin) edges are twin edges. (Color figure online)

   (i) The layout $(\prec_r, \mathcal{Q}_r)$ is an ℓ-local queue layout of G_r.
  (ii) In the first round, all new vertices are placed to the right of C_{init}. Without loss of generality, we have $C_{\text{init}} \prec x_1^1 \prec \cdots \prec x_{m_1}^1$.
 (iii) The new vertices $x_1^r, \ldots, x_{m_r}^r$ are inserted consecutively, i.e. y is not in the span of $x_1^r, \ldots, x_{m_r}^r$ for all vertices $y \in V(G_{r-1})$ from the previous rounds.
 (iv) Each two twin edges are assigned to the same queue.
  (v) The new vertices are placed to the right of their parent clique. Without loss of generality, we have $C_r \prec x_1^r \prec \cdots \prec x_{m_r}^r$. The edges between a vertex x_i^r, $i \in \{1, \ldots, m_r\}$, and its parent clique C_r are assigned to pairwise different queues. In particular, if $\ell = k$, then Bob cannot introduce new queues at vertex x_i^r in the following rounds.

Alice wins the n-th game if Bob cannot extend the layout without violating one of the first n conditions. In particular, if Alice wins the first game, this implies the existence of a k-tree with local queue number $\ell + 1$. However, the first game is the hardest for Alice, whereas the games become easier when Bob's moves are more restricted. During the proofs, we decide what Alice does but cannot control Bob's moves. We say that Bob *has to* act in a certain way if Alice wins otherwise.

We now set out to show how Alice wins Game (i) for $k = \ell = 2$. We first present a 2-tree with which Alice wins Game (v) and then show how to augment it until arriving at a 2-tree with which she wins the first game.

Lemma 7. *There is a graph with which Alice wins Game (v) for $k = \ell = 2$.*

Proof. Consider the 2-tree presented in Fig. 3. The edge $\{1, 2\}$ is the initial clique C_{init} and Alice introduces in five rounds the vertices one-by-one in the order indicated by their number. We first argue why we may assume that Bob does not assign any new edge to the same queue as its parent clique.

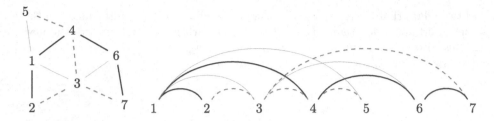

Fig. 3. 2-tree for Game (v) (left) and the layout chosen by Bob (right)

Assume that in some round r, Bob chooses to assign an edge e between the parent clique C_r and its child x_1^r to the same queue Q as C_r. Condition (v) ensures that the other edge e' between C_r and x_1^r is assigned to a different queue Q'. Thus, both endpoints of e' have incident edges in the same two queues Q and Q'. In particular, Bob cannot introduce new queues for edges adjacent to e'. Condition (v) further states that Bob cannot introduce new queues at any vertex that does not belong to the initial clique. It follows that Bob can use only the two queues Q and Q' for any subgraph that starts with e' as initial clique. Now, Alice wins by constructing the 2-tree with queue number 3 provided by Wiechert [25] using e' as the initial clique.

Hence, Bob has no choice when assigning the edges to queues. The vertex ordering is determined by the rules except for the placement of vertex 6 which may be placed between the vertices 4 and 5. In this case, however, the edges $\{1,5\}$ and $\{3,6\}$ form a rainbow. Thus, Bob chooses the layout shown in Fig. 3, which again has a rainbow (edges $\{3,7\}$ and $\{4,5\}$) and Alice wins Game (v). □

We give the reductions from Game (v) to Game (ii) in a more general way, so we can reuse them for subsequent lemmas.

Lemma 8. *Let $k > 1$ and $\ell \leqslant k$. If Alice wins Game (v), then she also wins Game (ii).*

Proof. We prove that Alice wins the n-th game provided a strategy with which she wins the $(n+1)$-st game. We do this by adapting the strategy such that the $(n+1)$-st condition is satisfied in the n-th game. Alternatively, we prove that Alice wins the n-th game if Bob does not act as claimed in the $(n+1)$-st condition. Thus, we may assume that the $(n+1)$-st condition holds and apply the given strategy for the $(n+1)$-st game.

Game (v) ⤳ Game (iv). We assume that Alice has a strategy to win Game (v) and present how Alice adapts her moves to ensure Condition v in Game (iv). We exploit the vertex placement and queue assignment of twin vertices and twin edges to prove that Bob creates a rainbow unless he places new vertices to the right of their parent clique. We thereby observe that each two non-twin edges introduced in the same round are assigned to different queues. We proceed by induction on the number of rounds r. The vertex placement in the first round is established by Condition (ii). In each of the succeeding rounds, let Alice increase the number m_r of added vertices by 2.

Consider a set of twin vertices $X = \{x_0^p, x_1^p, \ldots, x_{m_p+1}^p\}$ that were added to a clique C_p in a former round p. To simplify the notation, we write x_i instead of x_i^p and m instead of m_p. The vertices of C_p are denoted by $c_1 \prec \cdots \prec c_k$. By induction, we have $C_p \prec X$. Observe that this already implies that the edges at each x_i are assigned to pairwise different queues since $c_h x_1$ and $c_j x_0$ nest for $1 \leqslant h < j \leqslant k$ and twin edges are assigned to the same queue due to Condition (iv) (see Fig. 4). In particular, Bob has to choose one of the existing queues for new edges incident to x_i.

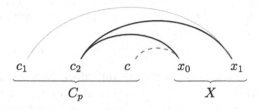

Fig. 4. The twin edges $c_2 x_0$ and $c_2 x_1$ are in the same queue. All edges between C_p and X have a twin edge forming a rainbow with one of them.

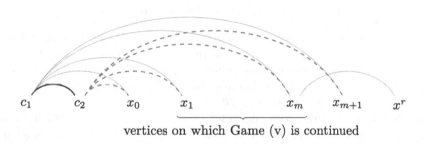

vertices on which Game (v) is continued

Fig. 5. Vertices $x_0, \ldots, x_{m+1}$ and their parent clique C_p. A child x^r of x_i, $i \in \{1, \ldots, m\}$, can only be placed to the right of x_{m+1}.

Alice now continues applying her strategy for Game (v) using the vertices $x_1, \ldots, x_m$. Assume that, in some round r, she chooses a clique C_r that contains x_i for some $i \in \{1, \ldots, m\}$ (see Fig. 5). Bob has to insert a child x^r of C_r into the vertex ordering and has to choose a queue Q for $x_i x^r$ which already contains an edge $c x_i$ with $c \in V(C_p)$. If Bob places x^r between c and x_{m+1}, then $c x_{m+1}$ and $x_i x^r$ form a rainbow in the same queue. If Bob places x^r to the left of c, then $c x_0$ and $x_i x^r$ nest and are assigned to the same queue. Hence, he has to place the new vertices to the right of x_{m+1} and therefore satisfies Condition (v). Thus, Alice may apply her strategy for Game (v) to win.

Game (iv) $\rightsquigarrow$ **Game** (iii). We consider Game (iii) and show how Alice ensures Condition (iv), i.e. that each two twin edges are assigned to the same queue. For this, recall that each vertex has incident edges in at most ℓ different queues. Thus, Bob has at most ℓ^k possibilities to assign the k edges of a new vertex to queues. Alice multiplies the number m_r of added vertices by ℓ^k in each round and thus finds m_r twin vertices whose twin edges are assigned to the same queues. She continues the game on those with the same strategy with which she wins Game (iv) and ignores the others.

Game (iii)$\rightsquigarrow$**Game** (ii). Condition (iii) forces Bob to insert twin vertices consecutively into the vertex ordering. Let m_r be the number of vertices that Alice adds to the current graph in the r-th round of Game (ii). To simulate Condition (iii) in Game (ii), Alice chooses to add $(m_r+1)\cdot|V(G_{r-1})|$ vertices instead of only m_r. By pigeonhole principle, Bob places at least m_r new vertices consecutively. Alice now uses her strategy from Game (iii) to win Game (ii). $\square$

Finally, the following lemma justifies to introduce Condition (ii) if $k = \ell = 2$. That is, we use Lemma 9 to win Game (i) provided a strategy to win Game (ii). We choose $s = 2m_1$, where m_1 is the number of vertices that are introduced in the first round of Game (ii). The clique given by Lemma 9 serves as initial clique C_{init}. Without loss of generality, C_{init} has m_1 children to the right and therefore satisfies Condition (ii).

Lemma 9. *For any $s > 0$, there is a 2-tree G such that for every 2-local queue layout there is an edge with at least s non-nesting children.*

Proof. An *m-ary 2-tree of depth t* is constructed as follows. We start with an edge whose depth, and also the depth of its endpoints, is defined to be 0. For $0 < i \leqslant t$, depth-i edges are introduced inductively by adding m children to each depth-$(i-1)$ edge. The depth of the new children is i. G is an $(s+4)$-ary 2-tree of depth 6 and is partly shown in Fig. 6.

For the sake of contradiction, consider a 2-local queue layout of G such that every edge of depth less than 6 has at least five nesting children. We shall find a rainbow in one of the queues.

Let vw denote the depth-0 edge and let w' be a nesting child of vw. The initial edge vw is assigned to some queue Q_{black}. We now assume that vw' is assigned to a different queue Q_{orange} and handle the other case later. Consider a nesting child w'' of vw'. Next, we have five depth-3 children of vw'' which are placed below their parent edge by assumption. Since the layout is 2-local, there are only four possible combinations how the edges of depth 3 incident to v and w'' can be assigned to queues. By pigeonhole principle, there are two vertices x and y with vx and vy assigned to the same queue Q, where $Q = Q_{\text{black}}$ or $Q = Q_{\text{orange}}$, and $xw'', yw'' \in Q_{\text{blue}}$ for some queue Q_{blue}. Note that $Q_{\text{blue}} \neq Q_{\text{black}}, Q_{\text{orange}}$ as otherwise this would create a rainbow in the respective queue. Without loss of generality, we have $x \prec y$.

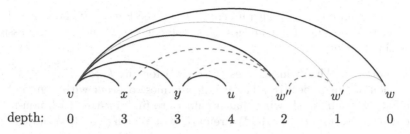

depth: 0   3   3   4   2   1   0

Fig. 6. 2-tree with nesting children. The edges in black (thick), orange (thin), and blue (dashed) correspond to the respective queues. The edges vx and vy could also be orange, yu is black, orange, or blue and creates a rainbow in either case. (Color figure online)

Finally, consider a nesting child u of the edge yw''. Since the layout is 2-local, we have $yu \in Q$ or $yu \in Q_{\text{blue}}$, that is yu is assigned to Q_{black}, Q_{orange}, or Q_{blue}. In all three cases, there is a rainbow in the respective queue.

To end the proof, consider the case that vw' is assigned to Q_{black}. If $vx, vy \in Q_{\text{black}}$, then the argumentation above still works. Otherwise, use vx instead of vw'.

We conclude that in every 2-local queue layout, one of the edges has fewer nesting children than we used in the construction, in particular fewer than five. Therefore, all further children are non-nesting.

$\square$

Lemmas 7 to 9 together show that Alice wins Game (i) for $k = \ell = 2$. That is, there is a 2-tree with local queue number 3, which proves Theorem 4.

Since Bob may use more queues for larger k, the approach for Lemma 7 using a k-tree with queue number $k + 1$ works only for $k = 2$. However, we introduce two new conditions that restrict Bob's moves further and offer Alice a way to win Game (v) for any $k > 1$ and $\ell \leqslant k$.

For Games (vi) and (vii), we change the initial setup. Instead of starting with only one clique, we start with two cliques and proceed on both cliques in parallel. We thereby get two copies of the same graph (a *left graph* and a *right graph*), where each vertex, edge, and clique has a corresponding copy in the other graph. In the r-th round, Alice now chooses a k-clique in the left graph and its copy in the right graph, and adds m_r new vertices to each. Conditions (i) to (v) apply to the left graph and to the right graph independently. In addition, Bob has to satisfy the following conditions (only the first for Game (vi) and both for Game (vii)):

(vi) Each two cliques C and C' that are copies of each other are laid out alternatingly, that is $c_1 \prec c'_1 \prec c_2 \prec c'_2 \prec \cdots \prec c_k \prec c'_k$ for vertex sets $V(C) = \{c_1, \ldots, c_k\}$ and $V(C') = \{c'_1, \ldots, c'_k\}$. In particular, new vertices and their copies are placed in the order of their parent cliques to the right of both parent cliques. Every edge is assigned to the same queue as its copy.

(vii) Consider an edge v_1v_2 with its copy w_1w_2 (see Fig. 7). If there is a child x of v_1 to the right of both edges, then the edges v_1x and v_1v_2 are assigned to different queues.

We show how Alice wins Games (v) to (vi) for any $k > 1$ and $\ell \leqslant k$ in the long version of this paper [18]. The two new games together with Lemma 8 lead to the following lemma showing that it suffices to find a k-tree with non-nesting children to prove lower bounds. We remark that placing children outside their parent clique would be a natural way to avoid nesting edges. Note that Condition (ii) is justified by the requirement of Lemma 10.

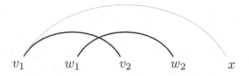

v_1 w_1 v_2 w_2 x

Fig. 7. Two sister edges and a child x to the right of both

Lemma 10. *Let $k > 1$ and $\ell \leqslant k$. Assume that for any $s > 0$ there is a k-tree such that for every ℓ-local queue layout, there is a k-clique with at least s non-nesting children. Then, there is a k-tree with local queue number at least $\ell + 1$.*

Observe that we can embed k'-trees in k-trees with $k' < k$. A queue layout of the k-tree, however, can result in additional restrictions to the queue layout of the embedded k'-tree. In particular, if every ℓ-local queue layout of some k-tree contains a k'-tree having a k'-clique with non-nesting children, we may apply Lemma 10 to this clique. The resulting k'-tree with local queue number at least $\ell + 1$ can then be augmented to a k-tree with local queue number at least $\ell + 1$. This leads to the following strengthening of Lemma 10.

Lemma 11. *Let $1 < k' \leqslant k$ and $\ell \leqslant k'$. Assume that for any $s > 0$ there is a k-tree G such that for every ℓ-local queue layout, G contains a k'-clique with at least s non-nesting children. Then, there is a k-tree with local queue number at least $\ell + 1$.*

Note that adding $2s$ children to a k-clique yields a k-tree that contains a $\lceil k/2 \rceil$-clique with at least s non-nesting children. Thus, Lemma 11 with $\ell = k' = \lceil k/2 \rceil$ proves Theorem 5.

3   Conclusions

Based on the notions of queue numbers and local covering numbers, we introduced the *local queue number* as a novel graph parameter. We presented a tool to deal with k-trees which led to the construction of a 2-tree with local queue number 3. This strengthens the lower bound of 3 for the queue number of 2-trees due to Wiechert [25]. It remains open whether there are k-trees with local queue number $k + 1$ for $k > 2$. Given Lemma 10, this is equivalent to the existence of k-trees that have a k-clique with non-nesting children for any k-local queue layout. That is, if every k-tree admits a k-local queue layout, then every k-tree also admits a k-local queue layout such that all children are placed below their parent clique. As such a vertex placement produces many nesting edges, this does not seem to be a promising strategy.

Question 12. What is the maximum local queue number of treewidth-k graphs?

There is a third parameter that is closely related to queue numbers and local queue numbers. For the *union queue number* $\mathrm{qn_u}(G)$ of a graph G, we define a *union queue* to be a vertex-disjoint union of queues and then minimize the number of union queues that are necessary to cover all edges of G. As we minimize the size of the cover, one could consider the union queue number close to the queue number. Surprisingly, the union queue number is tied to the local queue number and we have a linear upper bound on the union queue number of k-trees. We refer to [17] for an analogous proof for local and union page numbers. This observation has interesting consequences for queue layouts of k-trees. If the best known upper bound of $2^k - 1$ [25] is tight, then there are queue layouts consisting of linearly many union queues but at least exponentially many queues.

On the other hand, Lemma 10 might be extendable for ℓ-queue layouts with $\ell > k$. Condition (v) is crucial for this as it forbids Bob to introduce new queues. If Bob may use more than k queues, however, the given proof fails. Note that the requirement of a k-clique with non-nesting children is satisfied by a construction presented by Wiechert [25, Lemma 10].

Finally, the presented 2-tree also serves as a witness that there are planar graphs with local queue number at least 3. However, it is open whether this can be improved to 4.

Question 13. Is there a planar graph with local queue number 4?

References

1. Aggarwal, A., Klawe, M., Shor, P.: Multilayer grid embeddings for VLSI. Algorithmica **6**(1), 129–151 (1991). https://doi.org/10.1007/BF01759038
2. Akiyama, J., Exoo, G., Harary, F.: Covering and packing in graphs IV: linear arboricity. Networks **11**(1), 69–72 (1981). https://doi.org/10.1002/net.3230110108
3. Akiyama, J., Kano, M.: Path factors of a graph. Graph theory and its Applications, pp. 11–22 (1984)
4. Alam, J.M., Bekos, M.A., Gronemann, M., Kaufmann, M., Pupyrev, S.: Queue layouts of planar 3-trees. Algorithmica **82**(9), 2564–2585 (2020). https://doi.org/10.1007/s00453-020-00697-4

5. Bernhart, F., Kainen, P.C.: The book thickness of a graph. J. Comb. Theory Ser. B **27**(3), 320–331 (1979). https://doi.org/10.1016/0095-8956(79)90021-2
6. Clote, P., Dobrev, S., Dotu, I., Kranakis, E., Krizanc, D., Urrutia, J.: On the page number of RNA secondary structures with pseudoknots. J. Math. Biol. **65**(6), 1337–1357 (2012). https://doi.org/10.1007/s00285-011-0493-6
7. Dujmović, V., Joret, G., Micek, P., Morin, P., Ueckerdt, T., Wood, D.R.: Planar graphs have bounded queue-number. In: 60th IEEE Annual Symposium on Foundations of Computer Science (2019)
8. Dujmović, V., Morin, P., Wood, D.R.: Layout of graphs with bounded tree-width. SIAM J. Comput. **34**(3), 553–579 (2005). https://doi.org/10.1137/S0097539702416141
9. Fishburn, P.C., Hammer, P.L.: Bipartite dimensions and bipartite degrees of graphs. Discrete Math. **160**(1), 127–148 (1996). https://doi.org/10.1016/0012-365X(95)00154-O
10. Gonçalves, D.: Caterpillar arboricity of planar graphs. Discrete Math. **307**(16), 2112–2121 (2007). https://doi.org/10.1016/j.disc.2005.12.055. euroComb '03 - Graphs and Algorithms
11. Guy, R.K., Nowakowski, R.J.: The outerthickness & outercoarseness of graphs I. The complete graph & the n-cube. In: Bodendiek, R., Henn, R. (eds.) Topics in Combinatorics and Graph Theory: Essays in Honour of Gerhard Ringel, pp. 297–310. Physica-Verlag HD, Heidelberg (1990). https://doi.org/10.1007/978-3-642-46908-4_34
12. Heath, L.S., Ribbens, C.J., Pemmaraju, S.V.: Processor-efficient sparse matrix-vector multiplication. Comput. Math. Appl. **48**(3), 589–608 (2004). https://doi.org/10.1016/j.camwa.2003.06.009
13. Heath, L.S., Leighton, F., Rosenberg, A.: Comparing queues and stacks as machines for laying out graphs. SIAM J. Discrete Math. **5**(3), 398–412 (1992). https://doi.org/10.1137/0405031
14. Heath, L.S., Rosenberg, A.: Laying out graphs using queues. SIAM J. Comput. **21**(5), 927–958 (1992). https://doi.org/10.1137/0221055
15. Joseph, D., Meidanis, J., Tiwari, P.: Determining DNA sequence similarity using maximum independent set algorithms for interval graphs. In: Nurmi, O., Ukkonen, E. (eds.) SWAT 1992. LNCS, vol. 621, pp. 326–337. Springer, Heidelberg (1992). https://doi.org/10.1007/3-540-55706-7_29
16. Knauer, K., Ueckerdt, T.: Three ways to cover a graph. Discrete Math. **339**(2), 745–758 (2016). https://doi.org/10.1016/j.disc.2015.10.023
17. Merker, L., Ueckerdt, T.: Local and union page numbers. In: Archambault, D., Tóth, C.D. (eds.) GD 2019. LNCS, vol. 11904, pp. 447–459. Springer, Cham (2019). https://doi.org/10.1007/978-3-030-35802-0_34
18. Merker, L., Ueckerdt, T.: The local queue number of graphs with bounded treewidth (2020). https://arxiv.org/abs/2008.05392
19. Mutzel, P., Odenthal, T., Scharbrodt, M.: The thickness of graphs: a survey. Graphs Comb. **14**(1), 59–73 (1998). https://doi.org/10.1007/PL00007219
20. Nash-Williams, C.S.A.: Decomposition of finite graphs into forests. J. Lond. Math. Soc. **s1−39**(1), 12–12 (1964). https://doi.org/10.1112/jlms/s1-39.1.12
21. Ramanathan, S., Lloyd, E.L.: Scheduling algorithms for multi-hop radio networks. SIGCOMM Comput. Commun. Rev. **22**(4), 211–222 (1992). https://doi.org/10.1145/144191.144283
22. Rengarajan, S., Veni Madhavan, C.E.: Stack and queue number of 2-trees. In: Du, D.-Z., Li, M. (eds.) COCOON 1995. LNCS, vol. 959, pp. 203–212. Springer, Heidelberg (1995). https://doi.org/10.1007/BFb0030834

23. Rosenberg, A.: The diogenes approach to testable fault-tolerant arrays of processors. IEEE Trans. Comput. **C-32**(10), 902–910 (1983). https://doi.org/10.1109/TC.1983.1676134
24. Skums, P.V., Suzdal, S.V., Tyshkevich, R.I.: Edge intersection graphs of linear 3-uniform hypergraphs. Discrete Math. **309**(11), 3500–3517 (2009). https://doi.org/10.1016/j.disc.2007.12.082. 7th International Colloquium on Graph Theory
25. Wiechert, V.: On the queue-number of graphs with bounded tree-width. Electr. J. Comb. **24**(1), P1.65 (2017). http://www.combinatorics.org/ojs/index.php/eljc/article/view/v24i1p65
26. Wood, D.R.: queue layouts, tree-width, and three-dimensional graph drawing. In: Agrawal, M., Seth, A. (eds.) FSTTCS 2002. LNCS, vol. 2556, pp. 348–359. Springer, Heidelberg (2002). https://doi.org/10.1007/3-540-36206-1_31

Parameterized Algorithms
for Queue Layouts

Sujoy Bhore[1,2], Robert Ganian[1], Fabrizio Montecchiani[3(✉)],
and Martin Nöllenburg[1]

[1] Algorithms and Complexity Group, TU Wien, Vienna, Austria
{sujoy,rganian,noellenburg}@ac.tuwien.ac.at
[2] Université libre de Bruxelles (ULB), Bruxelles, Belgium
[3] Dipartimento di Ingegneria, Università degli Studi di Perugia, Perugia, Italy
fabrizio.montecchiani@unipg.it

Abstract. An *h-queue layout* of a graph G consists of a *linear order* of
its vertices and a partition of its edges into h *queues*, such that no two
independent edges of the same queue nest. The minimum h such that
G admits an h-queue layout is the *queue number* of G. We present two
fixed-parameter tractable algorithms that exploit structural properties
of graphs to compute optimal queue layouts. As our first result, we show
that deciding whether a graph G has queue number 1 and computing
a corresponding layout is fixed-parameter tractable when parameterized
by the treedepth of G. Our second result then uses a more restrictive
parameter, the vertex cover number, to solve the problem for arbitrary h.

Keywords: Queue number · Parameterized complexity · Treedepth ·
Vertex cover number · Kernelization

1 Introduction

An *h-queue layout* of a graph G is a linear layout of G consisting of a *linear
order* of its vertices and a partition of its edges into *queues*, such that no two
independent edges of the same queue nest [22]; see Fig. 1 for an illustration. The
queue number qn(G) of a graph G is the minimum number of queues in any
queue layout of G. While such linear layouts represent an abstraction of various
problems such as, for instance, sorting and scheduling [3,28], they also play a
central role in three-dimensional graph drawing. It is known that a graph class
has bounded queue number if and only if every graph in this class has a three-
dimensional crossing-free straight-line grid drawing in linear volume [10,14]. We
refer the reader to [16,25] for further references and applications. Moreover, it
is worth recalling that *stack layouts* [24,30] (or *book embeddings*), which allow
nesting edges but forbid edge crossings, form the "dual" concept of queue layouts.

Research of FM partially supported by Dip. Ingegneria Univ. Perugia, RICBA19FM:
"Modelli, algoritmi e sistemi per la visualizzazione di grafi e reti". RG acknowledges
support from the Austrian Science Fund (FWF) grant P 31336, SB and MN acknowledge support from FWF grant P 31119.

D. Auber and P. Valtr (Eds.): GD 2020, LNCS 12590, pp. 40–54, 2020.
https://doi.org/10.1007/978-3-030-68766-3_4

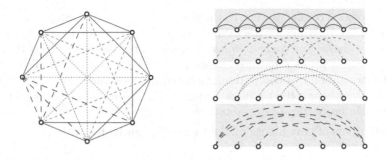

Fig. 1. A 4-queue layout of K_8.

A rich body of literature is concerned with the study of upper bounds for the queue number of several planar and non-planar graph families (see, e.g., [1,2,9,12–14,21,29] and also [15] for additional references). For instance, a graph of treewidth w has queue number at most $\mathcal{O}(2^w)$ [29], while every proper minor-closed class of graphs (including planar graphs) has constant queue number [13].

Of particular interest to us is the corresponding recognition problem, which we denote by QUEUE NUMBER: Given a graph G and a positive integer h, decide whether G admits an h-queue layout. In 1992, in a seminal paper, Heath and Rosenberg proved that 1-QUEUE NUMBER, i.e., the restriction of QUEUE NUMBER to $h = 1$, is NP-complete [22]. In particular, they characterized the graphs that admit queue layouts with only one queue as the arched leveled-planar graphs, and showed that the recognition of these graphs is NP-complete [22].

Since QUEUE NUMBER is NP-complete even for a single queue, it is natural to ask under which conditions the problem can be solved efficiently. For instance, it is known that if the linear order of the vertices is given (and the aim is thus to simply partition the edges of the graph into queues), then the problem becomes solvable in polynomial time [21]. We follow up on recent work made for the stack number [5] and initiate the study of the parameterized complexity of QUEUE NUMBER by asking under which parameterizations the problem is fixed-parameter tractable. In other words, we are interested in whether (1-)QUEUE NUMBER can be solved in time $f(k) \cdot n^{\mathcal{O}(1)}$ for some computable function f of the considered structural parameter k of the n-vertex input graph G.

As our main result, we show 1-QUEUE NUMBER is fixed-parameter tractable parameterized by the *treedepth* of the input graph (Sect. 3). We remark that treedepth is a fundamental graph parameter with close ties to the theory of graph sparsity (see, e.g., [23]). The main technique used by the algorithm is iterative pruning, where we recursively identify irrelevant parts of the input and remove these until we obtain a bounded-size equivalent instance (a *kernel*) solvable by brute force. While the iterative pruning technique has already been used in a few other algorithms that exploit treedepth [18–20], the unique challenge here lay in establishing that the removal of seemingly irrelevant parts of the graph cannot change NO-instances to YES-instances. The proof of this claim, formalized in Lemma 1, uses a new type of block decomposition of 1-queue layouts.

For our second result, we turn to the general QUEUE NUMBER problem. Here, we establish fixed-parameter tractability when parameterized by a larger parameter, namely the *vertex cover number* (Sect. 4). This result is also achieved by kernelization and forms a natural counterpart to the recently established fixed-parameter tractability of computing the stack number under the same parameterization [5], although the technical arguments and steps of the proof differ due to the specific properties of queue layouts.

Note: Full proofs of statements marked with (*) can be found in [4].

2 Preliminaries

We can assume that our input graphs are connected, as the queue number of a graph is the maximum queue number over all its connected components. Given a graph $G = (V, E)$ and a vertex $v \in V$, let $N(v)$ be the set of neighbors of v in G. Also, for $r \in \mathbb{N}$, we denote by $[r]$ the set $\{1, \dots, r\}$. An *h-queue layout* of G is a pair $\langle \prec, \sigma \rangle$, where $\prec$ is a linear order of V, and $\sigma \colon E \to [h]$ is a function that maps each edge of E to one of h queues. In an h-queue layout $\langle \prec, \sigma \rangle$ of G, it is required that no two independent edges in the same queue *nest*, that is, for no pair of edges $uv, wx \in E$ with four distinct end-vertices and $\sigma(uv) = \sigma(wx)$, the vertices are ordered as $u \prec w \prec x \prec v$. Given two distinct vertices u and v of G, u is to the *left* of v if $u \prec v$, else u is to the *right* of v. Note that a 1-queue layout of G is simply defined by a linear order $\prec$ of V and $\sigma \equiv 1$.

We assume familiarity with basic notions in parameterized complexity [8,11].

Treedepth. Treedepth is a parameter closely related to treewidth, and the structure of graphs of bounded treedepth is well understood [23]. We formalize a few notions needed to define treedepth, see also Fig. 2 for an illustration. A *rooted forest* $\mathcal{F}$ is a disjoint union of rooted trees. For a vertex x in a tree T of $\mathcal{F}$, the *height* (or *depth*) of x in $\mathcal{F}$ is the number of vertices in the path from the root of T to x. The *height of a rooted forest* is the maximum height of a vertex of the forest. Let $V(T)$ be the vertex set of any tree $T \in \mathcal{F}$.

Definition 1 (Treedepth). *Let the* closure *of a rooted forest $\mathcal{F}$ be the graph* $clos(\mathcal{F}) = (V_c, E_c)$ *with the vertex set* $V_c = \bigcup_{T \in \mathcal{F}} V(T)$ *and the edge set* $E_c = \{xy \mid x \text{ is an ancestor of } y \text{ in some } T \in \mathcal{F}\}$. *A* treedepth decomposition *of a graph G is a rooted forest $\mathcal{F}$ such that $G \subseteq clos(\mathcal{F})$. The* treedepth $td(G)$ *of a graph G is the minimum height of any treedepth decomposition of G.*

An optimal treedepth decomposition can be computed by an FPT algorithm.

Proposition 1 [27]. *Given an n-vertex graph G and an integer k, it is possible to decide whether G has treedepth at most k, and if so, to compute an optimal treedepth decomposition of G in time $2^{\mathcal{O}(k^2)} \cdot n$.*

Proposition 2 [23]. *Let G be a graph and $td(G) \leq k$. Then G has no path of length 2^k.*

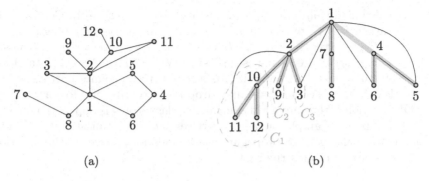

Fig. 2. (a) A graph G and (b) a treedepth decomposition $\mathcal{F}$ of G of height 4. In particular, $P_2 = \{1,2\}$, $A_2 = \{C_1, C_2, C_3\}$, and $m_2 = 3$.

Vertex Cover Number. A *vertex cover* C of a graph $G = (V, E)$ is a subset $C \subseteq V$ such that each edge in E has at least one incident vertex in C. The *vertex cover number* of G, denoted by $\tau(G)$, is the size of a minimum vertex cover of G. Observe that $td(G) \leq \tau(G) + 1$: it suffices to build $\mathcal{F}$ as a single path with vertex set C and with leaves $V \setminus C$. Computing an optimal vertex cover of G is FPT.

Proposition 3 [6]. *Given an n-vertex graph G and a constant τ, it is possible to decide whether G has vertex cover number at most τ, and if so, to compute a vertex cover C of size τ of G in time $\mathcal{O}(2^\tau + \tau \cdot n)$.*

3   Parameterization by Treedepth

In this section, we establish our main result: the fixed-parameter tractability of 1-QUEUE NUMBER parameterized by treedepth. We formalize the statement below.

Theorem 1. *Let G be a graph with n vertices and constant treedepth k. We can decide in $\mathcal{O}(n)$ time whether G has queue number one, and, if this is the case, we can also output a 1-queue layout of G.*

3.1   Algorithm Description

Since we assume G to be connected, any treedepth decomposition of G consists of a single tree T. Now, suppose that a treedepth decomposition T of G of depth k is given. For a vertex t of T, let P_t be the set of ancestors of t including t, let A_t be the set of connected components of $G - P_t$ which contain a child of t, and m_t be the maximum number of vertices in a component in A_t; see also Fig. 2(b).

Observation 1. *For every component $C \in A_t$ and for every vertex $v \in C$, it holds that $N(v) \subseteq C \cup P_t$. Moreover, $|C \cup P_t| \leq m_t + k$.*

Now, we define the following equivalence over components in A_t. Components $B, C \in A_t$ satisfy $B \sim C$ if and only if there exists a bijective renaming function $\eta_{B,C} : B \to C$ over (the vertices of) B, C such that each vertex $b_i \in B$ has a counterpart $\eta_{B,C}(b_i) = c_i \in C$ that satisfies: (i) $N(b_i) \cap P_t = N(c_i) \cap P_t$ and (ii) b_i is adjacent to $b_j \in B$ if and only if c_i is adjacent to its counterpart c_j. When B, C are clear from the context, we may drop the subscript of η for brevity.

By Observation 1, the number of equivalence classes of $\sim$ is upper-bounded by the number of possible graphs on $k + m_t$ vertices, which is at most $2^{(k+m_t)^2}$. The next observation allows us to propagate the bounds formalized by the notation above from children towards the root.

Observation 2. *If for a vertex t of T there exist integers a, b such that each child q of t satisfies $|A_q| \leq a$ and $m_q \leq b$, then $m_t \leq (a \cdot b) + 1$.*

The main component of our treedepth algorithm is Lemma 1, stated below. Intuitively, applying Lemma 1 bottom-up on T (together with Observation 2) allows us to iteratively remove subtrees from T while preserving the (non-)existence of a hypothetical solution—in particular, we will be able to prune subtrees of parents with a very large number of children until we reach an equivalent instance where each vertex has a bounded number of children. To formalize the meaning of "very large", we define the following function for $i \geq 2$:

$$\#children(k, i) = \left(\left((2^{(k+1)} + 1)^{size(k,i)^2} + 1 \right) \cdot (size(k,i) + k)! \right) \cdot 2^{(k+size(k,i))^2},$$

where $size(k, i)$ is a recursively defined function that captures the size bound given by Observation 2 as follows:

- $size(k, i) = (size(k, i - 1) \cdot \#children(k, i - 1)) + 1$ for $i \geq 2$, and
- $size(k, 1) = \#children(k, 1) = 0$.

Lemma 1. *Assume G has a vertex t at depth i in T such that $|A_t| \geq \#children(k, i)$, but $m_t \leq size(k, i)$ and every descendant q of t in T satisfies that $|A_q| \leq \#children(k, i-1)$. Then there exists a component B of A_t such that $G - B$ has queue number one if and only if G has queue number one. Moreover, B can be computed in time $size(k, i)! \cdot \#children(k, i)^2$.*

The proof of the lemma is deferred to Sect. 3.2. Before proceeding, we show how Lemma 1 is used to obtain Theorem 1.

Proof (of Theorem 1). We start by applying Proposition 1 to compute a treedepth decomposition T of G of depth at most k. Consider now vertices at depth $k - 1$ in T, i.e., vertices whose children are all leaves in T, and set $i = 2$. Observe that every vertex v at this depth satisfies $m_v \leq size(k, 2)$ since $size(k, 2) = 1$ and $m_v = 1$. If $|A_v| \geq \#children(k, 2)$, we apply Lemma 1 to obtain an equivalent graph with fewer vertices and restart on that graph. Otherwise, every vertex v at depth $k - 1$ satisfies $|A_v| < \#children(k, 2)$.

We now inductively repeat the above argument for every depth less than $k - 1$. In particular, assume that for some depth $1 \leq d \leq k - 1$ every vertex v

at depth d satisfies $|A_v| < \#children(k, i)$, where $i = k - d + 1$. Then we can set $d' := d - 1$, $i' := i + 1$, and recall from Observation 2 that every vertex v' at depth d' satisfies $m_v \leq size(k, i')$. Hence, if v' has too many subtrees—in particular, if $|A_{v'}| \geq \#children(k, i')$—we will once again apply Lemma 1 to obtain an equivalent smaller instance, in which case we restart the algorithm. Repeating this procedure for d' will eventually stop, and at that point it will hold that $|A_{v'}| < \#children(k, i')$ for every v' at depth d', in turn allowing us to continue with the induction.

The above procedure will halt only once the root r of T satisfies $|A_r| < \#children(k, k)$ and $m_r \leq size(k, k)$. At that point, we have a kernel G' [8,11]— an equivalent graph that has size bounded by a function of k, notably by $f(k) = \#children(k, k) \cdot size(k, k) + 1$. To prove Theorem 1, it suffices to decide whether G' admits a 1-queue layout by a brute-force algorithm that runs in time $\mathcal{O}(f(k)! \cdot f(k)^2)$. Since Lemma 1 is applied $\mathcal{O}(n)$ times and the runtime of the associated algorithm is $\mathcal{O}(size(k, k) \cdot \#children(k, k)^2)$, the total runtime is upper-bounded by a function of k times n. Finally, we note that while it would be possible to provide a term upper-bounding the dependency on k of the running time, it is clear that such a term must necessarily be non-elementary—indeed, the recursive definition of the two functions $\#children(k, k)$ and $size(k, k)$ results in a tower of exponents of height k. $\qquad\square$

3.2   Proof of Lemma 1

Since we have

$$|A_t| \geq \left(\left((2^{(k+1)} + 1)^{size(k,i)^2} + 1\right) \cdot (size(k, i) + k)!\right) \cdot 2^{(k + size(k,i))^2} = \#children(k, i)$$

and the number of equivalence classes of $\sim$ is upper-bounded by $2^{(k+m_t)^2} \leq 2^{(k+size(k,i))^2}$, there must exist an equivalence class, denoted $A_t^{\sim} \subseteq A_t$, containing at least $\left((2^{(k+1)} + 1)^{size(k,i)^2} + 1\right) \cdot (size(k, i) + k)!$ connected components in A_t which are pairwise equivalent w.r.t. $\sim$. Moreover, this equivalence class can be computed in time at most $size(k, i)! \cdot \#children(k, i)^2$ by simply brute-forcing over all potential renaming functions η between arbitrarily chosen $\#children(k, i)$-many components in A_t to construct the set of all equivalence classes of these components. Let B be an arbitrarily selected component in $A_t^{\sim}$. First, observe that if G is a YES-instance then so is $G-B$, as deleting vertices and edges cannot increase the queue number. On the other hand, assume there is a 1-queue layout of $G - B$ with linear order $\prec$. Our aim for the rest of the proof is to obtain a linear order $\prec'$ of G that extends $\prec$ and yields a valid 1-queue layout of G.

A Refined Equivalence. Let $\equiv_{\prec}$ be an equivalence over components in $A_t^{\sim}$ defined as follows. $C \equiv_{\prec} D$ if and only if the following holds: the linear order $\prec$ restricted to $P_t \cup \eta_{C,D}(C)$ is the same as $\prec$ restricted to $P_t \cup C$. In other words, $\equiv_{\prec}$ is a refinement of $\sim$ restricted to $A_t^{\sim}$ which groups components based on the order in which their vertices appear (also taking into account which subinterval they appear in w.r.t. P_t). Note that $\equiv_{\prec}$ has at most $(m_t + k)! \leq (size(k, i) + k)!$

many equivalence classes, and hence by the virtue of $A_t^\sim$ having size at least $((2^{(k+1)} + 1)^{size(k,i)^2} + 1) \cdot (size(k,i) + k)!$, there must exist an equivalence class U of $\equiv_\prec$ containing at least $(2^{(k+1)} + 1)^{size(k,i)^2} + 1$ components of $A_t^\sim$.

We adopt the following terminology for U: we will denote the components in U as $C_1, C_2, \ldots, C_u$, where $u = |U|$, we will identify the vertices in a component C_i by using the lower index i, and for each such vertex v, say $v = v_i \in C_i$, use v_j to denote its counterpart $\eta_{C_i, C_j}(v_i)$.

Identifying Delimiting Components. Consider two adjacent vertices $v_i, w_i \in C_i$. We say that component C_j is *vw-separate* from C_i if edges $v_i w_i$ and $v_j w_j$ neither nest nor cross each other. On the other hand, C_j is *vw-interleaving* (respectively, *vw-nesting*) with C_i if $v_i w_i$ and $v_j w_j$ cross each other (respectively, if one of $v_i w_i$ and $v_j w_j$ nests the other). By the definition of $\equiv_\prec$ and U, these three cases are exhaustive. Moreover, if $v_i w_i$ is an edge then so is $v_j w_j$ and hence C_j cannot be *vw-nesting* with C_i. Our next aim will be to find two components—we will call them *delimiting components*—that are not *vw-separate* for *any* edge vw. To this end, for some two adjacent vertices v_i, w_i of C_i, denote by D_1 the component whose counterpart to v_i (say v_1) is placed leftmost in $\prec$ among all components in U. We now define a sequence of components as follows: D_ℓ is the unique component that is (i) *vw-separate* from $D_{\ell-1}$ and whose vertex v_ℓ is placed (ii) to the right of $v_{\ell-1}$, and (iii) v_ℓ is placed leftmost among all components satisfying properties (i) and (ii). Let d be the maximum integer such that D_d exists.

Lemma 2 (*). $d \le 2^{k+1} + 1$.

Moreover, each component C_q in U can be uniquely assigned to one component D_ℓ as defined above (w.r.t. the chosen edge vw) as follows: If $C_q = D_\ell$ for some ℓ, then C_q is assigned to itself; otherwise, D_ℓ is the component whose vertex v_ℓ is to the left of and simultaneously closest to the corresponding vertex v_q in C_q among all components $D_1, \ldots, D_d$.

Lemma 3 (*). *Let C_q and C_p be two components assigned to the same component D_ℓ w.r.t. the edge vw. Then C_q and C_p are vw-interleaving.*

We are now ready to construct our delimiting components. Recall that at this point, $|U| \ge (2^{(k+1)} + 1)^{size(k,i)^2} + 1$ while the maximum number of edges inside a component in U is upper-bounded by $m_t^2 \le size(k,i)^2$. Hence by the pigeon-hole principle and by applying the bound provided in Lemma 2 for each edge inside the components of U, there must exist two components in U, say C_x and C_y, which for each edge vw are assigned to the same component D_ℓ^{vw}. By Lemma 3 it now follows that they are *vw-interleaving* for every edge vw.

Using Delimiting Components. Before we use C_x and C_y to insert B, we can show that the way they interleave with each other is "consistent" in $\prec$.

Lemma 4 (*). *Assume, w.l.o.g., that some vertex v_x is to the left of v_y. Then for each vertex w_x it holds that w_x is to the left of w_y.*

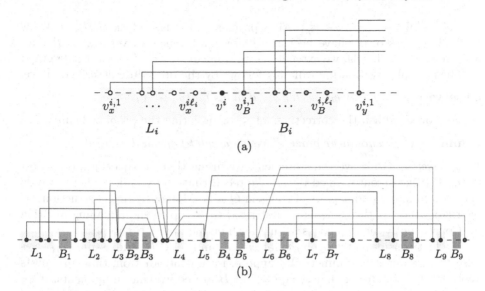

Fig. 3. Reinsertion of B_i: (a) A schematic illustration, and (b) an example where blue and red vertices belong to C_x and C_y, respectively.

We remark that it is not the case that C_x must be vw-interleaving with C_y if vw is not an edge – this is, in fact, a major complication that we will need to overcome to complete the proof. W.l.o.g. and recalling Lemma 4, we will hereinafter assume that every vertex $v_x \in C_x$ is placed to the left of its counterpart $v_y \in C_y$. The following definition allows us to partition the vertices of C_x into subsequences that should not be interleaved with vertices of B.

Definition 2 (Block). *A block $L = \{v_x^1, v_x^2, \ldots, v_x^h\}$ of C_x is a maximal set of vertices of C_x such that: (1) there is no vertex v_y^i (the counterpart in C_y of v_x^i), with $1 \leq i \leq h$, between two vertices of L in $\prec$; (2) there are no two vertices of L such that one has a neighbor to its left and one has a neighbor to its right.*

We observe that, as an immediate consequence of Definition 2, no two vertices of L are adjacent (an edge uv in L would imply that u has a neighbor to its right and v has a neighbor to its left, or vice versa).

For each block $L = \{v_x^1, v_x^2, \ldots, v_x^h\}$ of C_x, there is a corresponding set of vertices $\{v_B^1, v_B^2, \ldots, v_B^h\}$ of B, i.e., the set containing the counterparts of L in B. We will obtain a linear order of G by processing the blocks of C_x one by one as encountered in a left-to-right sweep of $\prec$, and for each block L, we will extend $\prec$ by suitably inserting the corresponding vertices of B.

Consider the i-th encountered block $L_i = \{v_x^{i,1}, v_x^{i,2}, \ldots, v_x^{i,\ell_i}\}$ of C_x, refer to Fig. 3 for an illustration. Note that, because C_x and C_y are equivalent components, it holds $v_y^{i,1} \prec v_y^{i,2} \cdots \prec v_y^{i,\ell_i}$ (even though such vertices might not be consecutive). Also, let v^i be the first vertex to the left of $v_y^{i,1}$ in $\prec$ (possibly $v^i = v_x^{i,\ell_i}$). We insert all vertices in the corresponding block B_i of B such that:

$v_i \prec v_B^{i,1} \prec v_B^{i,2} \prec \ldots v_B^{i,\ell_i} \prec v_y^{i,1}$. After processing the last block of C_x, we know that all vertices of C_x have been considered and hence all vertices of B have been reinserted, that is, we extended $\prec$ to a linear order $\prec'$ of the whole graph G. The next observation immediately follows by the procedure described above.

Observation 3. *For every vertex v_x, it holds that $v_x \prec' v_B \prec' v_y$.*

We now establish the correctness of $\prec'$, completing the proof of Lemma 1.

Lemma 5 (*). *The linear order $\prec'$ yields a valid 1-queue layout of G.*

Proof (sketch). To prove the statement, we argue that no two edges of G nest in the 1-queue layout defined by $\prec'$. We recall that $\prec'$ extends $\prec$, hence we do not need to argue about pairs of edges in $G - B$. Moreover, by construction, $\prec'$ restricted to C_x is the same as $\prec'$ restricted to B (up to the renaming function η). Consequently, no two edges having both endpoints in B can nest.

We first consider any edge $v_B w$ for $w \in P_t$ and $v_B \in B$, and assume $v_B \prec' w$ (else the argument is symmetric). Suppose, for a contradiction, that $v_B w$ nests another edge ab. Recall that since C_x and B are equivalent components, if v_B is to the left of w, the same holds for v_x. By Observation 3, we know $v_x \prec' v_B \prec' w$, which implies that ab is nested by $v_x w$ as well, a contradiction with the correctness of $\prec$. Similarly, if $v_B w$ is nested by an edge ab, then we know $v_B \prec' v_y \prec' w$, which implies that ab nests $v_y w$ as well, again a contradiction.

We now consider any edge $v_B w_B$, with $v_B \prec' w_B$, and we assume for a contradiction that $v_B w_B$ nests an edge ab. Since Definition 2 ensures that a block cannot contain a pair of adjacent vertices, we know that v_x and w_x belong to different blocks, say L_i and L_j (with $i < j$) respectively. Therefore, we can rename the vertices as $v_x = v_x^{i,i'}$ and $w_x = v_x^{j,j'}$, and similarly $v_B = v_B^{i,i'}$ and $w_B = v_B^{j,j'}$; refer to Fig. 4(a) for an illustration. By Observation 3, it holds $v_x^{i,i'} \prec' v_B^{i,i'} \prec' v_y^{i,i'}$ and $v_x^{j,j'} \prec' v_B^{j,j'} \prec' v_y^{j,j'}$. Moreover, the correctness of $\prec$ implies that $v_B^{i,i'} \prec' a \prec' v_y^{i,i'}$ (since $v_y^{i,i'} v_y^{j,j'}$ cannot nest ab) and $v_x^{j,j'} \prec' b \prec' v_B^{j,j'}$ (since $v_x^{i,i'} v_x^{j,j'}$ cannot nest ab). Because a is between $v_B^{i,i'}$ and $v_y^{i,i'}$, either there exists another vertex $v_y^{i,1}$ (the counterpart to the first vertex in block L_i, where possibly $v_y^{i,1} = a$) such that $v_B^{i,i'} \prec' v_y^{i,1} \preceq' a \prec' v_y^{i,i'}$, or $a = v_y^{i,i'}$.

Suppose first $a \neq v_y^{i,1}$ and $a \neq v_y^{i,i'}$. Observe that $v_x^{i,1}$ has at least one neighbor in C_x (because C_x is connected), and that $v_x^{j,j'}$ is to the right of $v_x^{i,i'}$, hence, by Definition 2, $v_x^{i,1}$ also has a neighbor to its right, say v_x^{l,j^*}. Because no two edges nest in $\prec$, it must be: (i) $v_x^{i,1} \prec' v_x^{i,i'}$, (ii) $v_x^{l,j^*} \prec' b$, and (iii) $v_y^{l,j^*} \prec' b$ (possibly $v_y^{l,j^*} = b$). Altogether, this implies that $v_x^{j,j'}$ and v_x^{l,j^*} are in the same block (i.e., $l = j$) and hence $v_B^{j,j'} \prec' v_y^{j,j^*} \prec' b$, which contradicts $b \prec' v_B^{j,j'}$. If instead $a = v_y^{i,1}$ or $a = v_y^{i,i'}$, then b is either a vertex of C_y or a vertex of P_t. If $b \in C_y$, the argument is similar, as we can set $b = v_y^{j,j^*}$ and observe that $v_B^{j,j'}$ should be to the left of v_y^{j,j^*}, see Fig. 4(b). If $b \in P_t$, we would have $v_x^{j,j'} \prec' b \prec' v_y^{j,j'}$, which contradicts the fact that C_x and C_y are equivalent components, see Fig. 4(c). $\qquad\square$

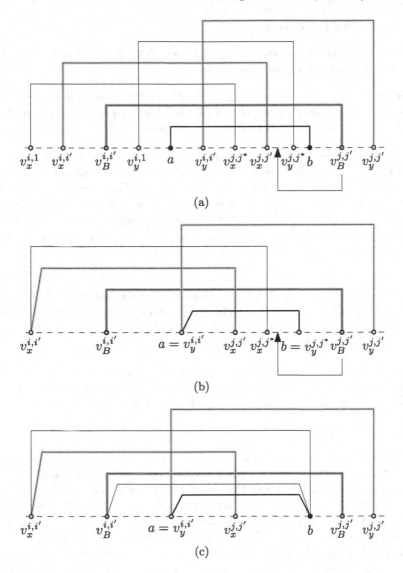

Fig. 4. Illustration for the proof of Lemma 5: $v_B^{i,i'} v_B^{j,j'}$ nests an edge ab.

4   Parameterization by Vertex Cover Number

We now turn to the general QUEUE NUMBER problem and show that it is fixed-parameter tractable when parameterized by the vertex cover number by proving:

Theorem 2. *Let G be a graph with n vertices and vertex cover number $\tau = \tau(G)$. A queue layout of G with the minimum number of queues can be computed in $\mathcal{O}(2^{\tau^{\mathcal{O}(\tau)}} + \tau \log \tau \cdot n)$ time.*

4.1   Algorithm Description

Before describing the algorithm behind Theorem 2, we make an easy observation (which matches an analogous observation in [5]).

Lemma 6. *Every n-vertex graph $G = (V, E)$ with a vertex cover C of size τ admits a τ-queue layout. Moreover, if G and C are given as input, such a τ-queue layout can be computed in $\mathcal{O}(n + \tau \cdot n)$ time.*

Proof. Denote by $c_1, \ldots, c_\tau$ the τ vertices of C and let $\prec$ be any linear order of G such that $c_i \prec c_{i+1}$, for $i = 1, 2, \ldots, \tau - 1$. A queue assignment σ of G on h queues can be obtained as follows. Let $U = V \setminus C$. For each $i \in [\tau]$ all edges uc_i with $u \in U \cup \{c_1, \ldots, c_{i-1}\}$ are assigned to queue i. Now, consider the edges assigned to any queue $i \in [\tau]$. By construction, they are all incident to vertex c_i, and thus no two of them nest each other. Therefore, the pair $\langle \prec, \sigma \rangle$ is a τ-queue layout of G and can be computed in $\mathcal{O}(n + \tau \cdot n)$ time.

□

Let C be a vertex cover of size τ of graph G. For any subset U of C, a vertex $v \in V \setminus C$ is of *type U* if $N(v) = U$. This defines an equivalence relation on $V \setminus C$ and in particular partitions $V \setminus C$ into at most $\sum_{i=1}^{\tau} \binom{\tau}{i} = 2^{\tau-1} < 2^\tau$ distinct types. Denote by V_U the set of vertices of type U.

Lemma 7. *Let $h \in \mathbb{N}$ and $v \in V_U$ such that $|V_U| \geq 2 \cdot h^\tau + 2$. Then G admits an h-queue layout if and only if $G' = G - \{v\}$ does. Moreover, an h-queue layout of G' can be extended to an h-queue layout of G in linear time.*

The proof of Lemma 7 is deferred to Sect. 4.2.

Proof (of Theorem 2). By Proposition 3, we can determine the vertex cover number τ of G and compute a vertex cover C of size τ in time $\mathcal{O}(2^\tau + \tau \cdot n)$. With Lemma 7 in hand, we can then apply a binary search on the number of queues $h \leq \tau$ as follows. If $h > \tau$, by Lemma 6 we can immediately conclude that G admits a τ-queue layout and compute one in $\mathcal{O}(n + \tau \cdot n)$ time. Hence we shall assume that $h \leq \tau$. We construct a kernel G^* from G of size $h^{\mathcal{O}(\tau)}$ as follows. We first classify each vertex of G based on its type. We then remove an arbitrary vertex from each set V_U with $|V_U| > 2 \cdot h^\tau + 1$ until $|V_U| \leq 2 \cdot h^\tau + 1$. Thus, constructing G^* can be done in $\mathcal{O}(2^\tau + \tau \cdot n)$ time, since 2^τ is the number of types and $\tau \cdot n$ is the maximum number of edges of G. From Lemma 7 we conclude that G admits an h-queue layout if and only if G^* does.

Given a linear order $\prec^*$ of G^*, a queue assignment σ^* such that $\langle \prec^*, \sigma^* \rangle$ is an h-queue layout of G^* exists if and only if σ^* contains no h-rainbow [21], i.e., h independent edges that pairwise nest, which can be easily checked (and computed if it exists) in $h^{\mathcal{O}(\tau)}$ time [21]. Consequently, determining whether G^* admits an h-queue layout can be done by first guessing all linear orders, and then for each of them by testing for the existence of an h-rainbow. Since we have 2^τ types, and each of the at most $2 \cdot h^\tau + 1$ elements of the same type are equivalent in the queue layout (that is, the position of two elements of the same type can be exchanged in $\prec^*$ without affecting σ^*), the number of linear orders can be upper bounded by $(2^\tau)^{\mathcal{O}(h^\tau)} = 2^{\tau^{\mathcal{O}(\tau)}}$. Thus, whether h queues suffice for G^* can be determined in $2^{\tau^{\mathcal{O}(\tau)}} \cdot h^{\mathcal{O}(\tau)} = 2^{\tau^{\mathcal{O}(\tau)}}$ time. An h-queue layout of G^* (if any) can be extended to one of G by iteratively applying the constructive procedure of Lemma 7, in $\mathcal{O}(\tau \cdot n)$ time. Finally, by applying a binary search on h we obtain an overall time complexity of $\mathcal{O}(2^{\tau^{\mathcal{O}(\tau)}} + \tau \log \tau \cdot n)$, as desired.   □

4.2 Proof of Lemma 7

One direction follows easily, since removing a vertex from an h-queue layout still gives an h-queue-layout of the resulting graph. So let $\langle \prec, \sigma \rangle$ be an h-queue layout of G'. We prove that an h-queue layout of G can be constructed by inserting v immediately to the right of a suitable vertex u in V_U and by assigning the edges of v to the same queues as the corresponding edges of u.

We say that two vertices $u_1, u_2 \in V_U$ are *queue equivalent*, if for each vertex $w \in U$, the edges $u_1 w$ and $u_2 w$ are both assigned to the same queue according to σ. Each vertex in V_U has degree exactly $|U|$, hence this relation partitions the vertices of V_U into at most $h^{|U|} \leq h^\tau$ sets. Let $V_U^* = V_U \setminus \{v\}$. Since $|V_U^*| \geq 2 \cdot h^\tau + 1$, at least three vertices of this set, which we denote by u_1, u_2, and u_3, are queue equivalent. Consider now the graph induced by the edges of these three vertices that are assigned to a particular queue. By the above argument, such a graph is a $K_{l,3}$, for some $l > 0$. However, $K_{3,3}$ does not admit a 1-page queue layout, because any graph with queue number 1 is planar [22]. As a consequence, $l \leq 2$, that is, each $u_i \in V_U^*$ has at most two edges on each queue. Denote such two edges by $u_i w$ and $u_i z$ and assume, w.l.o.g., that $u_1 \prec u_2 \prec u_3$ and $w \prec z$. We now claim that $w \prec u_1 \prec u_2 \prec u_3 \prec z$, else two edges would nest. We can distinguish a few cases based on the position of u_1 (recall that $u_1 \prec u_2 \prec u_3$), refer to Fig. 5 for an illustration.

– **Case A:** $w \prec z \prec u_1$, then the nesting edges are $z u_1$ and $w u_2$.
– **Case B:** $u_1 \prec w \prec z$, then we distinguish three more subcases.

• **Case B.1:** $u_2 \prec w$, then the nesting edges are $u_1 z$ and $u_2 w$.
• **Case B.2:** $w \prec u_2 \prec z$, then the nesting edges are $u_1 z$ and $w u_2$.
• **Case B.3:** $z \prec u_2$, then the nesting edges are $z u_2$ and $w u_3$.

– **Case C:** $w \prec u_1 \prec z$, if $w \prec u_2 \prec u_3 \prec z$ the claim follows. Else, we have two more subcases based again on the position of u_2.

• **Case C.1:** $w \prec z \prec u_2$, then the nesting edges are $w u_2$ and $u_1 z$.
• **Case C.2:** $w \prec u_2 \prec z \prec u_3$, then the nesting edges are $w u_3$ and $u_1 z$.

It follows that we can extend $\prec$ by introducing v as the first vertex to the right of u_1 and, for each edge vw such that $w \in U$, we can assign vw to the same queue as $u_1 w$. This operation does not introduce any nesting. Namely, if vw is assigned to a queue containing only one edge of u_1, the graph induced by the edges in this queue is a star with center w and no two edges can nest. If vw is assigned to a queue containing two edges of u_1, say $u_1 w$ and $u_1 z$, then we know that all vertices of V_U are between w and z in $\prec$ and again no two edges nest.

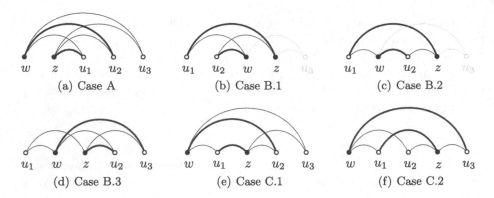

Fig. 5. Illustration for the proof of Lemma 7.

5   Conclusions and Open Problems

We proved that h-QUEUE NUMBER is fixed-parameter tractable parameterized by treedepth for $h = 1$, and by the vertex cover number for arbitrary $h \geq 1$. Several interesting questions arise from our research, among them:

1. A first natural question is to understand whether Theorem 1 can be extended to the general case ($h \geq 1$). In particular, our arguments establishing the existence of interleaving components already fail for $h = 2$.
2. Extending Theorem 1 to graphs of bounded treewidth is also an interesting problem; here the main issue is to be able to forget information about vertices in a partial order, thus an approach based on testing arched leveled-planarity might be more suitable.
3. Finally, we mention the possibility of studying the parameterized complexity of *mixed* linear layouts, using both queues and stacks, see [7,17,22,26].

References

1. Bannister, M.J., Devanny, W.E., Dujmović, V., Eppstein, D., Wood, D.R.: Track layouts, layered path decompositions, and leveled planarity. Algorithmica **81**(4), 1561–1583 (2018). https://doi.org/10.1007/s00453-018-0487-5
2. Bekos, M.A., et al.: Planar graphs of bounded degree have bounded queue number. SIAM J. Comput. **48**(5), 1487–1502 (2019). https://doi.org/10.1137/19M125340X
3. Bhatt, S.N., Chung, F.R.K., Leighton, F.T., Rosenberg, A.L.: Scheduling tree-dags using FIFO queues: A control-memory trade-off. J. Parallel Distrib. Comput. **33**(1), 55–68 (1996). https://doi.org/10.1006/jpdc.1996.0024
4. Bhore, S., Ganian, R., Montecchiani, F., Nöllenburg, M.: Parameterized algorithms for queue layouts. CoRR abs/2008.08288 (2020)
5. Bhore, S., Ganian, R., Montecchiani, F., Nöllenburg, M.: Parameterized algorithms for book embedding problems. J. Graph Algorithms Appl. (2020). https://doi.org/10.7155/jgaa.00526

6. Chen, J., Kanj, I.A., Xia, G.: Improved upper bounds for vertex cover. Theor. Comput. Sci. **411**(40–42), 3736–3756 (2010). https://doi.org/10.1016/j.tcs.2010. 06.026

7. de Col, P., Klute, F., Nöllenburg, M.: Mixed linear layouts: complexity, heuristics, and experiments. In: Archambault, D., Tóth, C.D. (eds.) GD 2019. LNCS, vol. 11904, pp. 460–467. Springer, Cham (2019). https://doi.org/10.1007/978-3-030-35802-0_35

8. Cygan, M., et al.: Parameterized Algorithms. Springer, Cham (2015). https://doi. org/10.1007/978-3-319-21275-3

9. Di Battista, G., Frati, F., Pach, J.: On the queue number of planar graphs. SIAM J. Comput. **42**(6), 2243–2285 (2013). https://doi.org/10.1137/130908051

10. Di Giacomo, E., Liotta, G., Meijer, H.: Computing straight-line 3d grid drawings of graphs in linear volume. Comput. Geom. **32**(1), 26–58 (2005). https://doi.org/ 10.1016/j.comgeo.2004.11.003

11. Downey, R.G., Fellows, M.R.: Fundamentals of Parameterized Complexity. TCS. Springer, London (2013). https://doi.org/10.1007/978-1-4471-5559-1

12. Dujmović, V.: Graph layouts via layered separators. J. Comb. Theory, Ser. B **110**, 79–89 (2015). https://doi.org/10.1016/j.jctb.2014.07.005

13. Dujmović, V., Joret, G., Micek, P., Morin, P., Ueckerdt, T., Wood, D.R.: Planar graphs have bounded queue-number. In: Foundations of Computer Science (FOCS'19), pp. 862–875. IEEE (2019). https://doi.org/10.1109/FOCS.2019.00056

14. Dujmović, V., Morin, P., Wood, D.R.: Layout of graphs with bounded tree-width. SIAM J. Comput. **34**(3), 553–579 (2005). https://doi.org/10.1137/ S0097539702416141

15. Dujmović, V., Morin, P., Wood, D.R.: Layered separators in minor-closed graph classes with applications. J. Comb. Theory, Ser. B **127**, 111–147 (2017). https:// doi.org/10.1016/j.jctb.2017.05.006

16. Dujmović, V., Wood, D.R.: On linear layouts of graphs. Discrete Math. Theor. Comput. Sci. **6**(2), 339–358 (2004)

17. Dujmović, V., Wood, D.R.: Stacks, queues and tracks: layouts of graph subdivisions. Discrete Math. Theor. Comput. Sci. **7**(1), 155–202 (2005)

18. Ganian, R., Ordyniak, S.: The complexity landscape of decompositional parameters for ILP. Artif. Intell. **257**, 61–71 (2018). https://doi.org/10.1016/j.artint.2017.12. 006

19. Ganian, R., Peitl, T., Slivovsky, F., Szeider, S.: Fixed-parameter tractability of dependency QBF with structural parameters. In: Principles of Knowledge Representation and Reasoning (KR'20) (2020, to appear)

20. Gutin, G.Z., Jones, M., Wahlström, M.: The mixed Chinese postman problem parameterized by pathwidth and treedepth. SIAM J. Discrete Math. **30**(4), 2177–2205 (2016). https://doi.org/10.1137/15M1034337

21. Heath, L.S., Leighton, F.T., Rosenberg, A.L.: Comparing queues and stacks as mechanisms for laying out graphs. SIAM J. Discrete Math. **5**(3), 398–412 (1992). https://doi.org/10.1137/0405031

22. Heath, L.S., Rosenberg, A.L.: Laying out graphs using queues. SIAM J. Comput. **21**(5), 927–958 (1992). https://doi.org/10.1137/0221055

23. Nešetřil, J., Ossona de Mendez, P.: Sparsity. AC, vol. 28. Springer, Heidelberg (2012). https://doi.org/10.1007/978-3-642-27875-4

24. Ollmann, T.: On the book thicknesses of various graphs. In: Southeastern Conference on Combinatorics, Graph Theory and Computing. Congressus Numerantium, vol. VIII, p. 459 (1973)

25. Pemmaraju, S.V.: Exploring the powers of stacks and queues via graph layouts. Ph.D. thesis, Virginia Tech (1992)
26. Pupyrev, S.: Mixed linear layouts of planar graphs. In: Frati, F., Ma, K.-L. (eds.) GD 2017. LNCS, vol. 10692, pp. 197–209. Springer, Cham (2018). https://doi.org/10.1007/978-3-319-73915-1_17
27. Reidl, F., Rossmanith, P., Villaamil, F.S., Sikdar, S.: A faster parameterized algorithm for treedepth. In: Esparza, J., Fraigniaud, P., Husfeldt, T., Koutsoupias, E. (eds.) ICALP 2014, Part I. LNCS, vol. 8572, pp. 931–942. Springer, Heidelberg (2014). https://doi.org/10.1007/978-3-662-43948-7_77
28. Tarjan, R.E.: Sorting using networks of queues and stacks. J. ACM **19**(2), 341–346 (1972). https://doi.org/10.1145/321694.321704
29. Wiechert, V.: On the queue-number of graphs with bounded tree-width. Electr. J. Comb. **24**(1), 65 (2017). https://doi.org/10.37236/6429
30. Yannakakis, M.: Embedding planar graphs in four pages. J. Comput. Syst. Sci. **38**(1), 36–67 (1989). https://doi.org/10.1016/0022-0000(89)90032-9

Lazy Queue Layouts of Posets

Jawaherul Md. Alam[1], Michael A. Bekos[2]([⊠]), Martin Gronemann[3],
Michael Kaufmann[2], and Sergey Pupyrev[4]

[1] Amazon Inc., Tempe, AZ, USA
jawaherul@gmail.com
[2] Institut für Informatik, Universität Tübingen, Tübingen, Germany
{bekos,mk}@informatik.uni-tuebingen.de
[3] Theoretical Computer Science, Osnabrück University, Osnabrück, Germany
martin.gronemann@uni-osnabrueck.de
[4] Facebook, Inc., Menlo Park, CA, USA
spupyrev@gmail.com

Abstract. We investigate the queue number of posets in terms of their width, that is, the maximum number of pairwise incomparable elements. A long-standing conjecture of Heath and Pemmaraju asserts that every poset of width w has queue number at most w. The conjecture has been confirmed for posets of width $w = 2$ via so-called *lazy* linear extension.

We extend and thoroughly analyze lazy linear extensions for posets of width $w > 2$. Our analysis implies an upper bound of $(w-1)^2+1$ on the queue number of width-w posets, which is tight for the strategy and yields an improvement over the previously best-known bound. Further, we provide an example of a poset that requires at least $w+1$ queues in every linear extension, thereby disproving the conjecture for posets of width $w > 2$.

Keywords: Queue layouts · Posets · Linear extensions

1 Introduction

A *queue layout* of a graph consists of a total order $\prec$ of its vertices and a partition of its edges into *queues* such that no two edges in a single queue *nest*, that is, there are no edges (u,v) and (x,y) in a queue with $u \prec x \prec y \prec v$. If the input graph is directed, then the total order has to be compatible with its edge directions, i.e., it has to be a topological ordering of it [13,14]. The minimum number of queues needed in a queue layout of a graph is commonly referred to as its *queue number*.

There is a rich literature exploring bounds on the queue number of different classes of graphs [1,11,15,17–19]. A remarkable work by Dujmović et al. [8] proves that the queue number of (undirected) planar graphs is constant, thus improving upon previous (poly-)logarithmic bounds [4,6,7] and resolving an old conjecture by Heath, Leighton and Rosenberg [11]. For a survey, we refer to [9].

© Springer Nature Switzerland AG 2020
D. Auber and P. Valtr (Eds.): GD 2020, LNCS 12590, pp. 55–68, 2020.
https://doi.org/10.1007/978-3-030-68766-3_5

In this paper, we investigate bounds on the queue number of posets. Recall that a *poset* $\langle P, < \rangle$ is a finite set of elements P equipped with a partial order $<$; refer to Sect. 2 for formal definitions. The queue number of $\langle P, < \rangle$ is the queue number of the acyclic digraph $G(P, <)$ associated with the poset that contains all non-transitive relations among the elements of P. This digraph is known as the *cover graph* and can be visualized using a Hasse diagram; see Fig. 1.

The study of the queue number of posets was initiated in 1997 by Heath and Pemmaraju [12], who provided bounds on the queue number of a poset expressed in terms of its *width*, that is, the maximum number of pairwise incomparable elements with respect to $<$. In particular, they observed that the queue number of a poset of width w cannot exceed w^2 and posed the following conjecture.

Conjecture 1 (Heath and Pemmaraju [12]**).** Every poset of width w has queue number at most w.

Heath and Pemmaraju [12] made a step towards settling the conjecture by providing a linear upper bound of $4w - 1$ on the queue number of planar posets of width w. This bound was recently improved to $3w - 2$ by Knauer, Micek, and Ueckerdt [16], who also gave a planar poset whose queue number is exactly w, thus establishing a lower bound. Furthermore, they investigated (non-planar) posets of width 2, and proved that their queue number is at most 2. Therefore, Conjecture 1 holds when $w = 2$.*

Our Contribution. We present improvements upon the aforementioned results, thus continuing the study of the queue number of posets expressed in terms of their width, which is one of the open problems by Dujmović et al. [8].

(i) For a fixed total order of a graph, the queue number is the size of a maximum *rainbow*, that is, a set of pairwise nested edges [11]. Thus to determine the queue number of a poset $\langle P, < \rangle$ one has to compute a *linear extension* (that is, a total order complying with $<$), which minimizes the size of a maximum rainbow. In in [2], we present a poset and a linear extension of it which yields a rainbow of size w^2. Knauer et al. [16] studied a special type of linear extensions, called *lazy*, for posets of width-2 to show that their queue number is at most 2. Thus, it is tempting to generalize and analyze lazy linear extensions for posets of width $w > 2$. We provide such an analysis and show that the maximum size of a rainbow in a lazy linear extension of a width-w poset is at most $w^2 - w$ (Theorem 1 in Sect. 3). Furthermore, we show that the bound is worst-case optimal for lazy linear extensions (for details refer to [2]).

(ii) The above bound already provides an improvement over the existing upper bound on the queue number of posets. However, a carefully chosen lazy linear extension, which we call *most recently used* (MRU), further improves the bound to $(w-1)^2 + 1$ (Theorem 2 in Sect. 4). Therefore, the queue number of a width-w poset is at most $(w-1)^2 + 1$. Again we show this bound to be worst-case optimal for MRU extensions (for details refer to [2]).

---

* Knauer et al. [16] also claim to reduce the queue number of posets of width w from w^2 to $w^2 - 2\lfloor w/2 \rfloor$. However, as we discuss in [2], their argument is incomplete.

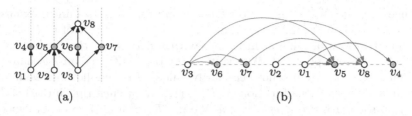

        (a)   (b)

Fig. 1. (a) The Hasse diagram of a width-4 poset; gray elements are pairwise incomparable; the chains of a certain decomposition are shown by vertical lines. (b) A 2-queue layout with a 2-rainbow formed by edges (v_2, v_5) and (v_6, v_8).

(iii) We demonstrate a non-planar poset of width 3 whose queue number is 4 (Theorem 3). We generalize this example to posets of width $w > 3$ (Theorem 4), thus disproving Conjecture 1. These two proofs are mostly deferred to [2].

2   Preliminaries

A *partial order* over a finite set of elements P is a binary relation $<$ that is irreflexive and transitive. A set P together with a partial order, $<$, is a *partially ordered set* (or simply a *poset*) and is denoted by $\langle P, < \rangle$. Two elements x and y with $x < y$ or $y < x$ are called *comparable*; otherwise x and y are *incomparable*. A subset of pairwise comparable (incomparable) elements of a poset is called a *chain* (*antichain*, respectively). The *width* of a poset is defined as the cardinality of a largest antichain. For two elements x and y of P with $x < y$, we say that x *is covered* by y if there is no element $z \in P$ such that $x < z < y$. A poset $\langle P, < \rangle$ is naturally associated with an acyclic digraph $G(P, <)$, called the *cover graph*, whose vertex-set V consists of the elements of P, and there exists an edge from u to v if u is covered by v; see Fig. 1a. By definition, $G(P, <)$ has no transitive edges.

A *linear extension* L of a poset $\langle P, < \rangle$ is a total order of P, which complies with $<$, that is, for every two elements x and y in P with $x < y$, x precedes y in L. Given a linear extension L of a poset, we write $x \prec y$ to denote that x precedes y in L; if in addition x and y may coincide, we write $x \preceq y$. We use $[x_1, x_2, \ldots, x_k]$ to denote $x_i \prec x_{i+1}$ for all $1 \leq i < k$; such a subsequence of L is also called a *pattern*. Let $F = \{(x_i, y_i); \ i = 1, 2, \ldots, k\}$ be a set of $k \geq 2$ *independent* (that is, having no common endpoints) edges of $G(P, <)$. It follows that $x_i \prec y_i$ for all $1 \leq i \leq k$. If $[x_1, \ldots, x_k, y_k, \ldots, y_1]$ holds in L, then the edges of F form a k-*rainbow* (see Fig. 1b). Edge (x_i, y_i) *nests* edge (x_j, y_j), if $1 \leq i < j \leq k$.

A *queue layout* of an acyclic digraph G consists of a total order of its vertices that is compatible with the edge directions of G and of a partition of its edges into *queues*, such that no two edges in a queue are nested. The *queue number* of G is the minimum number of queues required by its queue layouts. The *queue number of a poset* $\langle P, < \rangle$ is the queue number of its cover graph $G(P, <)$. Equivalently, the queue number of $\langle P, < \rangle$ is at most k if and only if it admits a linear extension L such that no $(k+1)$-rainbow is formed by some of the edges of $G(P, <)$ [15]. If certain edges form a rainbow in L, we say that L *contains* the rainbow.

The elements of a poset $\langle P, < \rangle$ of width w can be partitioned into w chains [5]. Note that such a partition is not necessarily unique. In the following, we fix this partition, and treat it as a function $\mathcal{C} : P \to \{1, \dots, w\}$ such that if $\mathcal{C}(u) = \mathcal{C}(v)$ and $u \neq v$, then either $u < v$ or $v < u$. We use $\mathcal{R}$, $\mathcal{B}$, and $\mathcal{G}$ to denote specific chains from a chain decomposition. A set of edges of the cover graph $G(P, <)$ of the poset that form a rainbow in a linear extension is called an *incoming $\mathcal{R}$-rainbow* $T_{\mathcal{R}}$ of size s if it consists of s edges $(u_1, r_1), \dots, (u_s, r_s)$ such that $r_i \in \mathcal{R}$ for all $1 \leq i \leq s$ and $\mathcal{C}(u_i) \neq \mathcal{C}(u_j)$ for all $1 \leq i, j \leq s$ with $i \neq j$. If $s = w$, $T_{\mathcal{R}}$ is called *complete* and is denoted by $T_{\mathcal{R}}^*$. An edge e of $T_{\mathcal{R}}$ with both endpoints in $\mathcal{R}$ is called an $\mathcal{R}$-*self edge*. For example, $T_{\mathcal{R}}^* \setminus \{e\}$ is an incoming $\mathcal{R}$-rainbow of size $w-1$ without the $\mathcal{R}$-self edge e. Similar notation is used for chains $\mathcal{B}$ and $\mathcal{G}$.

3   Lazy Linear Extensions

First let us recall two properties of linear extensions, whose proofs immediately follow from the fact that a cover graph of a poset contains no transitive edges.

Proposition 1. *A linear extension of a poset $\langle P, < \rangle$ does not contain pattern* $[r_1 \dots r_2 \dots r_3]$, *where $\mathcal{C}(r_1) = \mathcal{C}(r_2) = \mathcal{C}(r_3)$ and (r_1, r_3) is an edge of $G(P, <)$.*

Proposition 2. *A linear extension of a poset $\langle P, < \rangle$ does not contain pattern* $[r_1 \dots r_2 \dots b_2 \dots b_1]$, *where $\mathcal{C}(r_1) = \mathcal{C}(r_2)$, $\mathcal{C}(b_1) = \mathcal{C}(b_2)$, and (r_1, b_1) and (r_2, b_2) are edges of $G(P, <)$.*

Proposition 2 implies that for any linear extension of a poset, the maximum size of a rainbow is at most w^2 [12]. In [2] we show that for every $w \geq 2$, there exists a width-w poset and a linear extension of it containing a w^2-rainbow. Hence, a linear extension has be to chosen carefully, if one seeks for a bound on the queue number of posets that is strictly less than w^2.

In this section, we present and analyze such an extension, which we call *lazy*. Assume that a poset is given with a decomposition into w chains. Intuitively, a lazy linear extension is constructed incrementally starting from a minimal element of the poset. In every iteration, the next element is chosen from the same chain, if possible. Formally, for $i = 1, \dots, n-1$, assume that we have computed a lazy linear extension L for i vertices of $G(P, <)$ and let v_i be last vertex in L (if any). To determine the next vertex v_{i+1} of L, we compute the following set consisting of all source-vertices of the subgraph of $G(P, <)$ induced by $V \setminus L$:

$$S = \{v \in V \setminus L : \nexists (u, v) \in E \text{ with } u \in V \setminus L\} \tag{1}$$

If there is a vertex u in S with $\mathcal{C}(u) = \mathcal{C}(v_i)$, we set $v_{i+1} = u$; otherwise v_{i+1} is freely chosen from S. For the example of Fig. 1a, observe that $v_1 \prec v_4 \prec v_2 \prec v_3 \prec v_6 \prec v_7 \prec v_5 \prec v_8$ is a lazy linear extension.

Lemma 1. *If a lazy linear extension L of poset $\langle P, < \rangle$ contains the pattern $[r_1 \ldots b \ldots r_2]$, where $\mathcal{C}(r_1) = \mathcal{C}(r_2) \neq \mathcal{C}(b)$, then there exists some $x \in P$ with $\mathcal{C}(x) \neq \mathcal{C}(r_1)$ between r_1 and r_2 in L, such that $x < r_2$.*

Proof. Since the pattern is $[r_1 \ldots b \ldots r_2]$, $G(P, <)$ contains an edge from a vertex x with $\mathcal{C}(x) \neq \mathcal{C}(r_1)$ to a vertex $y \in \mathcal{C}(r_1)$ that is between r_1 and r_2 in L (notice that x may or may not coincide with b). Since the edge belongs to $G(P, <)$, it follows that $x < y \leq r_2$. □

Lemma 2. *A lazy linear extension of poset $\langle P, < \rangle$ does not contain pattern*

where $(u_1, r_1), \ldots, (u_{w-1}, r_{w-1})$ form an incoming $\mathcal{C}(r)$-rainbow of size $w - 1$, such that $\mathcal{C}(r) \neq \mathcal{C}(u_i)$ for all $1 \leq i \leq w - 1$ and $\mathcal{C}(r) \neq \mathcal{C}(b)$.

Proof. Assume to the contrary that there is a lazy linear extension L containing the pattern. Since $[r \ldots b \ldots r_{w-1}]$ holds in L, by Lemma 1, there is x with $\mathcal{C}(x) \neq \mathcal{C}(r_{w-1})$ between r and r_{w-1} in L such that $x < r_{w-1}$. Since $\mathcal{C}(x) \neq \mathcal{C}(r_{w-1})$, there is $1 \leq j \leq w - 1$ such that $\mathcal{C}(x) = \mathcal{C}(u_j)$, which implies $u_j < x$. Thus:

Since $u_j < x < r_{w-1} \leq r_j$, there is a path from u_j to r_j in $G(P, <)$. Thus, edge (u_j, r_j) is transitive; a contradiction. □

Theorem 1. *The maximum size of a rainbow formed by the edges of $G(P, <)$ in a lazy linear extension of a poset $\langle P, < \rangle$ of width w is at most $w^2 - w$.*

Proof. Assume to the contrary that there is a lazy linear extension L that contains a $(w^2 - w + 1)$-rainbow T. By Proposition 2 and the pigeonhole principle, T contains at least one complete incoming rainbow of size w; denote it by $T^*_{\mathcal{R}}$ and the corresponding chain by $\mathcal{R}$. By Proposition 1, the $\mathcal{R}$-self edge of $T^*_{\mathcal{R}}$ is innermost in $T^*_{\mathcal{R}}$. Thus, if $(u_1, r_1), \ldots, (u_w, r_w)$ are the edges of $T^*_{\mathcal{R}}$ and $u_w \in \mathcal{R}$, then without loss of generality, we may assume that the following holds in L.

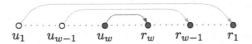

We next show that (u_w, r_w) is the innermost and (u_{w-1}, r_{w-1}) is the second innermost edge in T. Assume to the contrary that there exists an edge (x, y) in T that does not belong to $T^*_{\mathcal{R}}$ (that is, $\mathcal{C}(y) \neq \mathcal{R}$) and which is nested by

(u_{w-1}, r_{w-1}). Regardless of whether (x, y) nests (u_w, r_w) or not, we deduce the following.

Together with $u_w \in \mathcal{R}$ and $y \notin \mathcal{R}$, we apply Lemma 2, which yields a contradiction. Since (u_w, r_w) and (u_{w-1}, r_{w-1}) are the two innermost edges of T, it follows that T does not contain another complete incoming rainbow of size w.

Hence, each of the remaining $w - 1$ incoming rainbows has size exactly $w - 1$. Consider vertex u_{w-1} and let without loss of generality $\mathcal{C}(u_{w-1}) = \mathcal{B}$. By Proposition 1, $\mathcal{B} \neq \mathcal{R}$. We claim that the incoming $\mathcal{B}$-rainbow $T_{\mathcal{B}}$ does not contain the $\mathcal{B}$-self edge. Assuming the contrary, this $\mathcal{B}$-self edge nests (u_{w-1}, r_{w-1}) because (u_w, r_w) and (u_{w-1}, r_{w-1}) are the two innermost edges of T. Since $\mathcal{C}(u_{w-1}) = \mathcal{B}$, we obtain a contradiction by Proposition 1. Thus, $T_{\mathcal{B}}$ is a $\mathcal{B}$-rainbow of size $w - 1$ containing no $\mathcal{B}$-self edge. All edges of $T_{\mathcal{B}}$ nest (u_{w-1}, r_{w-1}), which yields the forbidden pattern of Lemma 2 formed by vertices of $T_{\mathcal{B}}$, $u_{w-1} \in \mathcal{B}$, and $r_{w-1} \in \mathcal{R}$; a contradiction. □

In [2] we show that our analysis is tight, i.e., there are posets of width w and corresponding lazy linear extensions containing $(w^2 - w)$-rainbows.

4   MRU Extensions

We now define a special type of lazy linear extensions for a width-w poset $\langle P, < \rangle$, which we call *most recently used*, or simply *MRU*. For $i = 1, \ldots, n - 1$, assume that we have computed a linear extension L for i vertices of $G(P, <)$, which are denoted by $v_1, \ldots, v_i$. To determine the next vertex of L, we compute set S of Eq. (1). Among all vertices in S, we select one from the most recently used chain (if any). Formally, we select a vertex $u \in S$ such that $\mathcal{C}(u) = \mathcal{C}(v_j)$ for the largest $1 \leq j \leq i$. If such vertex does not exist, we choose v_{i+1} arbitrarily from S. For the example of Fig. 1a, observe that $v_1 \prec v_4 \prec v_2 \prec v_3 \prec v_6 \prec v_5 \prec v_7 \prec v_8$ is an MRU extension.

For a linear extension L of poset $\langle P, < \rangle$, and two elements x and y in P, let $\mathcal{C}[x, y]$ be the subset of chains whose elements appear between x and y (inclusively) in L, that is, $\mathcal{C}[x, y] = \{\mathcal{C}(z) : x \preceq z \preceq y\}$.

Lemma 3. *Let L be an MRU extension of a width-w poset $\langle P, < \rangle$ containing pattern $[r_1 \ldots r_2 \ldots b]$, such that $\mathcal{C}(r_1) = \mathcal{C}(r_2) \neq \mathcal{C}(b)$ and there is no element in L between r_1 and r_2 from chain $\mathcal{C}(r_1)$. If $\mathcal{C}[r_1, r_2] = \mathcal{C}[r_1, b]$, then $r_2 < b$.*

Proof. Assume to the contrary that there is some b for which $r_2 < b$ does not hold. Without loss of generality, let b be the first (after r_2) of those elements in L. Since $\mathcal{C}[r_1, r_2] = \mathcal{C}[r_1, b]$, there are elements between r_1 and r_2 in L from chain $\mathcal{C}(b)$. Let b_1 be the last such element in L. Hence, $r_1 \prec b_1 \prec r_2 \prec b$. Consider the incremental construction of L. Since there is no element between r_1 and r_2 in L from chain $\mathcal{C}(r_1)$, the chain of b was "more recent" than the one of r_2, when r_2 was chosen as the next element. Thus, there is an edge (x, b) in $G(P, <)$ with $r_2 \prec x$ in L. Since b is the first element that is not comparable to r_2, then $r_2 < x$ holds. Hence, $r_2 < b$; a contradiction to our assumption that $r_2 < b$ does not hold.   □

Corollary 1. *Let L be an MRU extension of a width-w poset $\langle P, < \rangle$ containing pattern $[r_1 \ldots r_2]$, such that $\mathcal{C}(r_1) = \mathcal{C}(r_2)$ and there is no element in L between r_1 and r_2 from chain $\mathcal{C}(r_1)$. If $|\mathcal{C}[r_1, r_2]| = w$, then r_2 is comparable to all subsequent elements in L.*

Next we describe a forbidden pattern which is central in our proofs.

Lemma 4. *An MRU extension L of a width-w poset $\langle P, < \rangle$ does not contain the following pattern, even if $u_k = b_1$*

- $\mathcal{C}(u_i) \neq \mathcal{C}(u_j)$ for $1 \leq i, j \leq w$ with $i \neq j$,
- $(u_1, r_1), \ldots, (u_k, r_k)$ form an incoming $\mathcal{R}$-rainbow of size k for some $1 \leq k \leq w$,
- between b_1 and b_2 in L, there is an element from $\mathcal{R}$ but no elements from $\mathcal{B} = \mathcal{C}(b_1) = \mathcal{C}(b_2)$.

Proof. Since there are no elements between b_1 and b_2 in L from $\mathcal{B}$ and since $\mathcal{C}(u_i) \neq \mathcal{C}(u_j)$ for $1 \leq i, j \leq w$ with $i \neq j$, one of $u_1, \ldots, u_k$ belongs to $\mathcal{B}$. Let u_i be this element with $1 \leq i \leq k$, that is, $\mathcal{C}(u_i) = \mathcal{B}$. Since $(u_1, r_1), \ldots, (u_k, r_k)$ form an incoming $\mathcal{R}$-rainbow, (u_i, r_i) is an edge of $G(P, <)$. Notice that $[u_i \ldots b_1 \ldots b_2 \ldots r_i]$ holds in L and that $u_i = b_1$ may hold if $i = k$.

Our proof is by induction on $|\mathcal{C}| - |\mathcal{C}[b_1, b_2]|$, which ranges between 0 and $w-2$. In the base case $|\mathcal{C}| - |\mathcal{C}[b_1, b_2]| = 0$, that is, $|\mathcal{C}[b_1, b_2]| = w$. By Corollary 1, b_2 is comparable to all subsequent elements in L. In particular, $b_2 < r_i$, which implies that (u_i, r_i) is transitive in $G(P, <)$, since $u_i \leq b_1 < b_2 < r_i$; a contradiction.

Assume $|\mathcal{C}| - |\mathcal{C}[b_1, b_2]| > 0$. Let r_0 be the first vertex from $\mathcal{R}$ after b_2 in L, that is, $r_0 \preceq r_k$. If there are no elements between b_2 and r_0 from $\mathcal{C} \setminus \mathcal{C}[b_1, b_2]$ (that is, $\mathcal{C}[b_1, b_2] = \mathcal{C}[b_2, r_0]$), then by Lemma 3 it follows that $b_2 < r_0$, which implies $u_i \leq b_1 < b_2 < r_0 \leq r_i$. Thus, edge (u_i, r_i) is transitive in $G(P, <)$; a contradiction. Therefore, we may assume that there are elements between b_2 and r_0 in L from $\mathcal{C} \setminus \mathcal{C}[b_1, b_2]$. Let g_1 be the first such element; denote $\mathcal{C}(g_1) = \mathcal{G}$. Since between b_1 and b_2 in L there is an element from $\mathcal{R}$ (that is, $\mathcal{R} \in \mathcal{C}[b_1, b_2]$), $\mathcal{G} \neq \mathcal{R}$ holds. Similarly, $\mathcal{G} \neq \mathcal{B}$. Let (u_ℓ, r_ℓ) be the edge of the incoming $\mathcal{R}$-rainbow with $\mathcal{C}(u_\ell) = \mathcal{G}$; notice that such an edge exists as $\mathcal{G} \in \mathcal{C} \setminus \mathcal{C}[b_1, b_2]$. Since r_0 is the first element from $\mathcal{R}$ after b_2 in L, $r_0 \preceq r_\ell$. Thus, $[u_\ell \ldots b_1 \ldots b_2 \ldots g_1 \ldots r_0 \ldots r_\ell]$

holds in L such that $\mathcal{C}(u_\ell) = \mathcal{G} \notin \{\mathcal{R}, \mathcal{B}\}$. Let g_2 be the last element between u_ℓ and b_1 from $\mathcal{G}$, that is, $u_\ell \preceq g_2 \prec b_1$ in L. Now, consider the pattern:

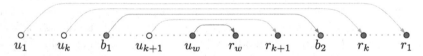

which satisfies the conditions of the lemma, since between g_2 and g_1 in L there is an element of $\mathcal{R}$ (namely, the one between b_1 and b_2 in L) and no elements of $\mathcal{G}$ (by the choice of g_1 and g_2). Further, $|\mathcal{C}| - |\mathcal{C}[g_2, g_1]| < |\mathcal{C}| - |\mathcal{C}[b_1, b_2]|$, since $\{\mathcal{G}\} = \mathcal{C}[g_2, g_1] \setminus \mathcal{C}[b_1, b_2]$. By the inductive hypothesis, the aforementioned pattern is not contained in L. Thus, also the initial one is not contained.   □

In the next five lemmas we study configurations that cannot appear in a rainbow formed by the edges of $G(P, <)$ in an MRU extension.

Lemma 5. *Let $\mathcal{R}$ and $\mathcal{B}$ be different chains of a width-w poset. Then a rainbow in an MRU extension of the poset does not contain all edges from*

$$T_{\mathcal{R}}^* \ \cup \ \{(b_1, b_2)\},$$

where $b_1, b_2 \in \mathcal{B}$ and $T_{\mathcal{R}}^*$ is a complete incoming $\mathcal{R}$-rainbow.

Proof. Assume to the contrary that a rainbow T contains an incoming $\mathcal{R}$-rainbow formed by edges $(u_1, r_1), \ldots, (u_w, r_w)$ and an edge (b_1, b_2) with $b_1, b_2 \in \mathcal{B}$. As in the proof of Theorem 1, we can show that (u_{w-1}, r_{w-1}) and (u_w, r_w) are the two innermost edges of T, and $\mathcal{C}(u_w) = \mathcal{R}$. Assume without loss of generality that $u_k \prec b_1 \prec u_{k+1}$ in L for some $1 \le k \le w - 1$, which implies that $r_{k+1} \prec b_2 \prec r_k$. Thus, the following holds in L.

By Proposition 1, there are no elements from $\mathcal{B}$ between b_1 and b_2. Hence, the conditions of Lemma 4 hold for the pattern; a contradiction.   □

Lemma 6. *Let $\mathcal{R}$ and $\mathcal{B}$ be different chains of a width-w poset. Then a rainbow in an MRU extension of the poset does not contain all edges from*

$$T_{\mathcal{R}}^* \setminus \{(r_1, r_2)\} \ \cup \ T_{\mathcal{B}}^* \setminus \{(b_1, b_2)\},$$

where $r_1, r_2 \in \mathcal{R}$, $b_1, b_2 \in \mathcal{B}$, and $T_{\mathcal{R}}^*, T_{\mathcal{B}}^*$ are complete incoming $\mathcal{R}$-rainbow and $\mathcal{B}$-rainbow, respectively.

Proof. Let $T_{\mathcal{R}}$ be an incoming $\mathcal{R}$-rainbow of size $w - 1$ without the $\mathcal{R}$-self edge; define $T_{\mathcal{B}}$ symmetrically. Assume to the contrary that a rainbow T in an MRU extension L contains both $T_{\mathcal{R}}$ and $T_{\mathcal{B}}$. Let (u_{w-1}, r_{w-1}) and (v_{w-1}, b_{w-1}) be the innermost edges of $T_\mathcal{R}$ and $T_\mathcal{B}$ in T, respectively. Without loss of generality, assume that (v_{w-1}, b_{w-1}) nests (u_{w-1}, r_{w-1}). This implies the following in L:

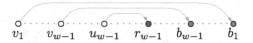

By Lemma 2 applied to $T_\mathcal{B}$, there are no elements from $\mathcal{B}$ between v_{w-1} and r_{w-1} in L. Consider edge (u_i, r_i) of $T_\mathcal{R}$ such that $u_i \in \mathcal{B}$. Element u_i ensures that there are some elements preceding v_{w-1} in L that belong to $\mathcal{B}$. Let b_ℓ be the last such element in L, that is, $b_\ell \preceq v_{w-1}$. Symmetrically, let b_r be the first element from $\mathcal{B}$ following r_{w-1} in L, that is, $r_{w-1} \prec b_r \preceq b_{w-1}$, and we have:

$$\overset{\bullet}{b_\ell} \quad \overset{\circ}{v_{w-1}} \quad \overset{\circ}{u_{w-1}} \quad \overset{\bullet}{r_{w-1}} \quad \overset{\bullet}{b_r} \quad \overset{\bullet}{b_{w-1}}$$

By the choice of b_ℓ and b_r, we further know that between b_ℓ and b_r there are no elements from $\mathcal{B}$, but there is an element from $\mathcal{R}$, namely r_{w-1}. Let $(u_1, r_1), \ldots, (u_k, r_k)$ be the edges of $T_\mathcal{R}$ that nest both b_ℓ and b_r in L. Assuming that $u_w = r_{w-1}$, we conclude that the following holds in L:

$$\overset{\circ}{u_1} \quad \overset{\circ}{u_k} \quad \overset{\bullet}{b_l} \quad \overset{\circ}{u_{k+1}} \quad \overset{\circ}{u_{w-1}} \quad \overset{\circ}{u_w} \quad \overset{\bullet}{b_r} \quad \overset{\bullet}{r_k} \quad \overset{\bullet}{r_1}$$

Since between b_ℓ and b_r there are no elements from $\mathcal{B}$, but there is an element from $\mathcal{R}$, we have the forbidden pattern of Lemma 4; a contradiction. □

Lemma 7. *Let $\mathcal{R}, \mathcal{B}, \mathcal{G}$ be pairwise different chains of a width-w poset. Then a rainbow in an MRU extension of the poset does not contain all edges from*

$$T_\mathcal{R}^* \setminus \{(g_1, r)\} \quad \cup \quad T_\mathcal{B}^* \setminus \{(g_2, b)\},$$

where $g_1, g_2 \in \mathcal{G}$, $r \in \mathcal{R}$, $b \in \mathcal{B}$, and $T_\mathcal{R}^*, T_\mathcal{B}^*$ are complete incoming $\mathcal{R}$-rainbow and $\mathcal{B}$-rainbow, respectively.

Proof. Assume to the contrary that a rainbow T contains both $T_\mathcal{R}$ and $T_\mathcal{B}$ as in the statement of the lemma. Let $(u_1, r_1), \ldots, (u_{w-1}, r_{w-1})$ be the edges of $T_\mathcal{R}$ and $(v_1, b_1), \ldots, (v_{w-1}, b_{w-1})$ be the edges of $T_\mathcal{B}$, where (u_{w-1}, r_{w-1}) and (v_{w-1}, b_{w-1}) are the $\mathcal{R}$- and $\mathcal{B}$-self edges, respectively. By Proposition 1, (u_{w-1}, r_{w-1}) and (v_{w-1}, b_{w-1}) are innermost edges in $T_\mathcal{R}$ and $T_\mathcal{B}$. Without loss of generality, assume that (v_{w-1}, b_{w-1}) nests (u_{w-1}, r_{w-1}), and that v_{w-1} appears between vertices u_k and u_{k+1} of $T_\mathcal{R}$, which implies that $r_{k+1} \prec b_{w-1} \prec r_k$. Hence, the following holds in L:

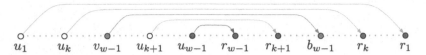

By Proposition 1, there is no vertex of $\mathcal{B}$ between v_{w-1} and b_{w-1} in L. If there is a vertex from $\mathcal{G}$ between v_{w-1} and b_{w-1} in L, then we have the forbidden pattern of Lemma 4, since $\mathcal{C}(u_i) \neq \mathcal{G}$ for all $1 \leq i \leq w-1$.

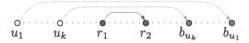

Otherwise, by Lemma 1, there is some $x \notin \mathcal{B}$ between v_{w-1} and b_{w-1} in L, such that $x < b_{w-1}$. As mentioned above, $x \notin \mathcal{G}$ either. Thus, the incoming $\mathcal{B}$-rainbow contains edge (v_i, b_i), which nests (v_{w-1}, b_{w-1}), such that $\mathcal{C}(v_i) = \mathcal{C}(x)$. Since $v_i < x < b_{w-1} < b_i$, the edge (v_i, b_i) is transitive; a contradiction. □

Lemma 8. *Let $\mathcal{R}, \mathcal{B}, \mathcal{G}$ be pairwise different chains of a width-w poset. Then a rainbow in an MRU extension of the poset does not contain all edges from*

$$T_\mathcal{B}^* \setminus \{(b_1, b_2)\} \ \cup \ T_\mathcal{R}^* \setminus \{(m_r, r)\} \ \cup \ T_\mathcal{G}^* \setminus \{(m_g, g)\},$$

where $b_1, b_2 \in \mathcal{B}$, $m_r \in V \setminus \mathcal{R}$, $r \in \mathcal{R}$, $m_g \in V \setminus \mathcal{G}$, $g \in \mathcal{G}$, and $T_\mathcal{B}^*, T_\mathcal{R}^*, T_\mathcal{G}^*$ are complete incoming $\mathcal{B}$-rainbow, $\mathcal{R}$-rainbow $\mathcal{G}$-rainbow, respectively.

Proof. Assume to the contrary that a rainbow T contains three incoming rainbows, $T_\mathcal{B}$, $T_\mathcal{R}$, and $T_\mathcal{G}$, as in the statement of the lemma. Without loss of generality, assume that the $\mathcal{G}$-self edge (g_1, g_2) is nested by the $\mathcal{R}$-self edge, (r_1, r_2); that is, $r_1 \prec g_1 \prec g_2 \prec r_2$. Denote the edges of $T_\mathcal{B}$ by (u_i, b_{u_i}) for $1 \leq i \leq w-1$, and assume that the following holds in L for some $k \leq w-1$.

Suppose there exists a vertex $x \in \mathcal{B}$ such that $r_1 \prec x \prec r_2$; then r_1 and r_2 together with x and edges of $T_\mathcal{B}$ form the forbidden pattern of Lemma 4. Thus, there are no vertices from $\mathcal{B}$ between r_1 and r_2 in L, and (u_k, b_{u_k}) is the innermost edge of $T_\mathcal{B}$ in T. Therefore, we can find two consecutive vertices in chain $\mathcal{B}$, b' and b'', such that $b' \prec r_1 \prec r_2 \prec b'' \preceq b_{u_k}$. Here b' exists because by Lemma 7 at least one of the two edges, $(b, r), (b, g)$, is in T as part of $T_\mathcal{R}$, $T_\mathcal{G}$, respectively. Further, by Lemma 2, the interval between u_k and b_{u_k} does not contain pattern $[u_k \ldots b \ldots x \ldots b_{u_k}]$, where $b \in \mathcal{B}, x \notin \mathcal{B}$. Thus, $b' \prec u_k$ and the interval of L between b'' and b_{u_k} contains vertices only from $\mathcal{B}$ ($b'' = b_{u_k}$ is possible).

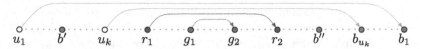

Now if there exists a vertex from $\mathcal{C}(m_r)$ between b' and b'', then $[b' \ldots r_1 \ldots b'']$ together with the edges of $T_{\mathcal{R}}$ form the forbidden pattern of Lemma 4. Thus, there are no vertices from $\mathcal{C}(m_r)$ between b' and b''.

Finally, consider vertices r_1 and r_2 that are consecutive in $\mathcal{R}$. By Lemma 1 and the fact that $r_1 \prec g_1 \prec r_2$, there is $x \notin \mathcal{C}(m_r)$ between r_1 and r_2 such that $x < r_2$. Since $x \notin \mathcal{C}(m_r)$, rainbow $T_{\mathcal{R}}$ contains edge (y, r_y) for some $r_y \in \mathcal{R}$ such that $\mathcal{C}(y) = \mathcal{C}(x)$. Edge (y, r_y) is transitive, as $y < x < r_2 < r_y$; a contradiction.   □

Lemma 9. *Let* $\mathcal{R}, \mathcal{B}, \mathcal{G}$ *be pairwise different chains of a width-w poset. Then a rainbow in an MRU extension of the poset does not contain all edges from*

$$T_{\mathcal{B}}^* \setminus \{(m_b, b)\} \ \cup \ T_{\mathcal{R}}^* \setminus \{(m_r, r)\} \ \cup \ T_{\mathcal{G}}^* \setminus \{(m_g, g)\},$$

where $m_b \in V \setminus \mathcal{B}$, $b \in \mathcal{B}$, $m_r \in V \setminus \mathcal{R}$, $r \in \mathcal{R}$, $m_g \in V \setminus \mathcal{G}$, $g \in \mathcal{G}$, *and* $T_{\mathcal{B}}^*, T_{\mathcal{R}}^*, T_{\mathcal{G}}^*$ *are complete incoming $\mathcal{B}$-rainbow, $\mathcal{R}$-rainbow $\mathcal{G}$-rainbow, respectively.*

Proof. Assume to the contrary that a rainbow T contains three incoming rainbows $T_{\mathcal{B}}$, $T_{\mathcal{R}}$, and $T_{\mathcal{G}}$, as in the statement of the lemma for some MRU extension L of the poset. By Lemma 7, $\mathcal{C}(m_b)$, $\mathcal{C}(m_r)$, and $\mathcal{C}(m_g)$ are pairwise distinct chains.

Without loss of generality, assume that the $\mathcal{R}$-self edge, (r_1, r_2), nests the $\mathcal{B}$-self edge, (b_1, b_2), which in turn nests the $\mathcal{G}$-self edge, (g_1, g_2). Namely, $r_1 \prec b_1 \prec g_1 \prec g_2 \prec b_2 \prec r_2$. Denote the edges of $T_{\mathcal{B}}$ by (u_i, b_{u_i}) for $1 \le i \le w - 1$, and assume that

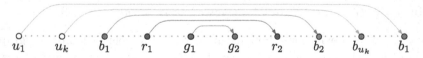

holds in L for some $k \le w - 1$. If there is a vertex from $\mathcal{C}(m_b)$ between r_1 and r_2 in L, then the forbidden pattern of Lemma 4 is formed by $[r_1 \ldots b_1 \ldots r_2]$ and edges of $T_{\mathcal{B}}$. Otherwise by Lemma 1, there is some $x \notin \mathcal{C}(m_b)$ between b_1 and b_2 such that $x < b_2$. Since $|T_{\mathcal{B}}| = w - 1$, $T_{\mathcal{B}}$ contains edge (y, b_y) for some $b_y \in \mathcal{B}$ such that $\mathcal{C}(y) = \mathcal{C}(x)$. Since $y < x < b_2 < b_y$, (y, b_y) is transitive; a contradiction.   □

Now we state the main result of the section.

Theorem 2. *The maximum size of a rainbow formed by the edges of $G(P, <)$ in an MRU extension of a poset $\langle P, < \rangle$ of width w is at most $(w-1)^2 + 1$.*

Proof. When $w = 2$, the theorem holds for any lazy linear extension by Theorem 1 and thus for MRU. Hence, we focus on the case $w \geq 3$. Assume to the contrary that an MRU extension contains a rainbow T of size $(w-1)^2+1$. Let $T_\mathcal{B}$, $T_\mathcal{R}$, $T_\mathcal{G}$ be the largest incoming rainbows in T corresponding to chains $\mathcal{B}$, $\mathcal{R}$, and $\mathcal{G}$, respectively. Assume without loss of generality that $|T_\mathcal{B}| \geq |T_\mathcal{R}| \geq |T_\mathcal{G}|$. By the pigeonhole principle, we have $|T_\mathcal{B}| \geq |T_\mathcal{R}| \geq w - 1$. We claim that $|T_\mathcal{B}| = w - 1$. Indeed, if $|T_\mathcal{B}| = w$, then by Lemma 5, $T_\mathcal{R}$ does not contain the $\mathcal{R}$-self edge. Thus, T contains $T_\mathcal{B}^*$ and $T_\mathcal{R}^* \setminus \{(r_1, r_2)\}$ with $r_1, r_2 \in \mathcal{R}$; a contradiction by Lemma 6.

Thus, $|T_\mathcal{B}| = |T_\mathcal{R}| = |T_\mathcal{G}| = w - 1$ follows, and we distinguish cases based on the number of self edges in $T_\mathcal{B}$, $T_\mathcal{R}$, and $T_\mathcal{G}$. If *each* of them contain its self edge, then we have the forbidden configuration of Lemma 9. If *two* of $T_\mathcal{B}$, $T_\mathcal{R}$, and $T_\mathcal{G}$ contain a self edge, then we have the forbidden configuration of Lemma 8. Finally, if *at most one* of $T_\mathcal{B}$, $T_\mathcal{R}$, and $T_\mathcal{G}$ contains a self edge, say $T_\mathcal{B}$, then $T_\mathcal{R}$ and $T_\mathcal{G}$ form the forbidden configuration of Lemma 6. This concludes the proof.   □

In [2] we show that our analysis is tight, i.e., there are posets of width w and corresponding MRU extensions containing $((w - 1)^2 + 1)$-rainbows.

5   A Counterexample to Conjecture 1

Here we sketc.h our approach to disprove Conjecture 1. We describe a poset in terms of its cover graph $G(p, q)$; see Fig. 2. For $p \geq q - 3$, graph $G(p, q)$ consists of $2p + q$ vertices $a_1, \ldots, a_p$, $b_1, \ldots, b_q$, and $c_1, \ldots, c_p$ that form three chains of lengths p, q, and p, respectively. For all $1 \leq i \leq p$ and for all $1 \leq j \leq q$, the edges (a_i, a_{i+1}), (b_j, b_{j+1}) and (c_i, c_{i+1}) form the intra-chain edges of $G(p, q)$. Graph $G(p, q)$ also contains the following inter-chain edges: (i) (a_i, c_{i+3}) and (c_i, a_{i+3}) for all $1 \leq i + 3 \leq p$, and (ii) (a_i, b_i) and (c_i, b_i) for all $1 \leq i \leq q$.s We denote by $\widetilde{G}(p, q)$ the graph obtained by adding (b_1, a_p) and (b_1, c_p) to $G(p, q)$.

Theorem 3. $\widetilde{G}(31, 22)$ *requires 4 queues in every linear extension.*

Sketch. In [2] we provide lower bounds on the queue number for simple subgraphs of $\widetilde{G}(p, q)$ and then for more complicated ones for appropriate values of p and q. We distinguish two cases depending on the length of edge (b_1, c_p) in a linear extension L of $\widetilde{G}(p, q)$. Either the edge is "short" (that is, b_1 is close to c_p in L) or "long". In the first case, the existence of a 4-rainbow is derived from the properties of the subgraphs. In the latter case, edge (b_1, c_p) nests a large subgraph of $\widetilde{G}(p, q)$, which needs 3 queues.   □

To prove that Conjecture 1 does not hold for $w > 3$, we employ an auxiliary lemma implicitly used in [16]; see [2] for details.

Theorem 4. *For every $w \geq 3$, there is a width-w poset with queue number $w + 1$.*

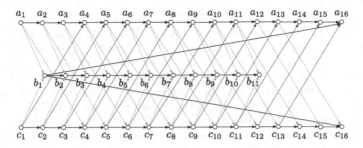

Fig. 2. Illustration of graph $\widetilde{G}(p,q)$ with $p = 16$ and $q = 11$.

6   Conclusions

In this paper, we explored the relationship between the queue number and the width of posets. We disproved Conjecture 1 and we focused on two natural types of linear extensions, lazy and MRU. That led to an improvement of the upper bound on the queue number of posets. A natural future direction is reduce the gap between the lower bound, $w + 1$, and the upper bound, $(w - 1)^2 + 1$, on the queue number of posets of width $w > 2$. In particular, we do not know whether the queue number of width-3 posets is four or five, and whether a subquadratic upper bound is possible. It is also intriguing to ask whether Conjecture 1 holds for *planar* width-w posets whose best-known upper bound is currently $3w - 2$ [16].

Another related open problem is on the *stack number* of directed acyclic graphs (DAGs). The stack number is defined analogously to the queue number except that no two edges in a single stack *cross*. Heath et al. [13,14] asked whether the *stack number* of upward planar DAGs is bounded by a constant. While the question has been settled for some subclasses of planar digraphs [10], the general problem remains unsolved. This is in contrast with the stack number of undirected planar graphs, which has been shown recently to be exactly four [3].

References

1. Alam, J.M., Bekos, M.A., Gronemann, M., Kaufmann, M., Pupyrev, S.: Queue layouts of planar 3-trees. Algorithmica (2020), https://doi.org/10.1007/s00453-020-00697-4
2. Bekos, M.A., Gronemann, M., Kaufmann, M., Pupyrev, S.: Lazy queue layouts of posets. CoRR abs/2008.10336v2 (2020). http://arxiv.org/abs/2008.10336v2
3. Bekos, M.A., Kaufmann, M., Klute, F., Pupyrev, S., Raftopoulou, C.N., Ueckerdt, T.: Four pages are indeed necessary for planar graphs. J. of Comp. Geom. **11**(1), 332–353 (2020). https://jocg.org/index.php/jocg/article/view/504
4. Di Battista, G., Frati, F., Pach, J.: On the queue number of planar graphs. SIAM J. Comput. **42**(6), 2243–2285 (2013). https://doi.org/10.1137/130908051
5. Dilworth, R.P.: A decomposition theorem for partially ordered sets. Ann. Math. **51**(1), 161–166 (1950)
6. Dujmović, V.: Graph layouts via layered separators. J. Comb. Theory Ser. B **110**, 79–89 (2015)

7. Dujmović, V., Frati, F.: Stack and queue layouts via layered separators. J. Graph Algorithms Appl. **22**(1), 89–99 (2018). https://doi.org/10.7155/jgaa.00454
8. Dujmović, V., Joret, G., Micek, P., Morin, P., Ueckerdt, T., Wood, D.R.: Planar graphs have bounded queue-number. In: Zuckerman, D. (ed.) FOCS, pp. 862–875. IEEE Computer Society (2019). https://doi.org/10.1109/FOCS.2019.00056
9. Dujmović, V., Wood, D.R.: On linear layouts of graphs. Discret. Math. Theor. Comput. Sci. **6**(2), 339–358 (2004). http://dmtcs.episciences.org/317
10. Frati, F., Fulek, R., Ruiz-Vargas, A.J.: On the page number of upward planar directed acyclic graphs. J. Graph Algorithms Appl. **17**(3), 221–244 (2013). https://doi.org/10.7155/jgaa.00292
11. Heath, L.S., Leighton, F.T., Rosenberg, A.L.: Comparing queues and stacks as mechanisms for laying out graphs. SIAM J. Discrete Math. **5**(3), 398–412 (1992). https://doi.org/10.1137/0405031
12. Heath, L.S., Pemmaraju, S.V.: Stack and queue layouts of posets. SIAM J. Discret. Math. **10**(4), 599–625 (1997). https://doi.org/10.1137/S0895480193252380
13. Heath, L.S., Pemmaraju, S.V.: Stack and queue layouts of directed acyclic graphs: part II. SIAM J. Comput. **28**(5), 1588–1626 (1999). https://doi.org/10.1137/S0097539795291550
14. Heath, L.S., Pemmaraju, S.V., Trenk, A.N.: Stack and queue layouts of directed acyclic graphs: part I. SIAM J. Comput. **28**(4), 1510–1539 (1999). https://doi.org/10.1137/S0097539795280287
15. Heath, L.S., Rosenberg, A.L.: Laying out graphs using queues. SIAM J. Comput. **21**(5), 927–958 (1992). https://doi.org/10.1137/0221055
16. Knauer, K., Micek, P., Ueckerdt, T.: The queue-number of posets of bounded width or height. In: Biedl, T., Kerren, A. (eds.) GD 2018. LNCS, vol. 11282, pp. 200–212. Springer, Cham (2018). https://doi.org/10.1007/978-3-030-04414-5_14
17. Pupyrev, S.: Mixed linear layouts of planar graphs. In: Frati, F., Ma, K.-L. (eds.) GD 2017. LNCS, vol. 10692, pp. 197–209. Springer, Cham (2018). https://doi.org/10.1007/978-3-319-73915-1_17
18. Rengarajan, S., Veni Madhavan, C.E.: Stack and queue number of 2-trees. In: Du, D.-Z., Li, M. (eds.) COCOON 1995. LNCS, vol. 959, pp. 203–212. Springer, Heidelberg (1995). https://doi.org/10.1007/BFb0030834
19. Wiechert, V.: On the queue-number of graphs with bounded tree-width. Electr. J. Comb. **24**(1), 65 (2017). http://www.combinatorics.org/ojs/index.php/eljc/article/view/v24i1p65

Drawing Tree-Like Graphs, Visualisation, and Special Drawings of Elementary Graphs

Improved Upper and Lower Bounds for LR Drawings of Binary Trees

Timothy M. Chan$^{(\boxtimes)}$ and Zhengcheng Huang

Department of Computer Science, University of Illinois at Urbana-Champaign,
Champaign, USA
{tmc,zh3}@illinois.edu

Abstract. In SODA'99, Chan introduced a simple type of planar straight-line upward order-preserving drawings of binary trees, known as *LR drawings*: such a drawing is obtained by picking a root-to-leaf path, drawing the path as a straight line, and recursively drawing the subtrees along the paths. Chan proved that any binary tree with n nodes admits an LR drawing with $O(n^{0.48})$ width. In SODA'17, Frati, Patrignani, and Roselli proved that there exist families of n-node binary trees for which any LR drawing has $\Omega(n^{0.418})$ width. In this paper, we improve Chan's upper bound to $O(n^{0.437})$ and Frati *et al.*'s lower bound to $\Omega(n^{0.429})$.

1 Introduction

Drawings of trees on a grid with small area have been extensively studied in the graph drawing literature [1–4, 6–9, 13–18, 20–27] (see also the book [10] and a recent survey [12]).

In this paper, we focus on one simple type of drawings of binary trees called *LR drawings*, which was introduced by Chan in SODA'99 [4] (and named in a later paper by Frati, Patrignani, and Roselli [14]): For a given binary tree T, we place the root somewhere on the top side of the bounding box, recursively draw its left subtree L and its right subtree R, and combine the two drawings by applying one of two rules. In the *left rule*, we connect the root of T to the root of R by a vertical line segment, place the bounding box of L's drawing one unit to the left of the vertical line segment, and place the bounding box of R's drawing underneath. In the *right rule*, we connect the root of T to the root of L by a vertical line segment, place the bounding box of R's drawing one unit to the right of the vertical line segment, and place the bounding box of L's drawing underneath. See Fig. 1(a). LR drawings are precisely those that can be obtained by recursive applications of these two rules.

(For historical context, we should mention that a similar notion of *hv drawings* were proposed before in some of the early papers on tree drawings [7–9], and were also defined recursively using two rules; the key differences are that in hv drawings, the root is always placed at the upper left corner, and the order of the left and right subtrees may not be preserved.)

© Springer Nature Switzerland AG 2020
D. Auber and P. Valtr (Eds.): GD 2020, LNCS 12590, pp. 71–84, 2020.
https://doi.org/10.1007/978-3-030-68766-3_6

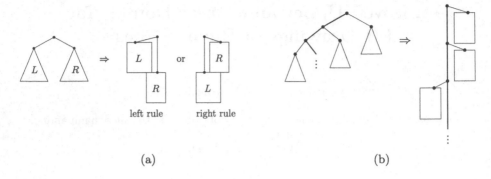

Fig. 1. LR drawing

Alternatively, LR drawings have the following equivalent definition: for a given binary tree T, we pick a root-to-leaf path π, draw π on a vertical line, and recursively draw all subtrees of π (i.e., subtrees rooted at siblings of the nodes along π), placing the bounding boxes of left subtrees of π one unit to the left of the vertical line, and the bounding boxes of the right subtrees of π one unit to the right of the vertical line. See Fig. 1(b).

It is easy to see that LR drawings satisfy the following desirable properties:

1. *Planar*: edges do not cross in the drawing.
2. *Straight-line*: edges are drawn as straight line segments.
3. *Strictly upward*: a parent has a strictly larger y-coordinate than each child.
4. *Order-preserving*: the edge from a parent to its left child is to the left of the edge from the parent to its right child in the drawing.

Indeed, the original motivation for LR drawings is in finding "good" planar, straight-line, strictly upward, order-preserving drawings of binary trees [4]. Goodness here is measured in terms of the *area* of a drawing, defined as the width (the number of grid columns) times the height (the number of grid rows), assuming that nodes are placed on an integer grid. The goal is to prove worst-case bounds on the minimum area needed for such drawings as a function of the number of nodes n. As $\Omega(n)$ height is clearly necessary in the worst case for strictly upward drawings (and LR drawings have $O(n)$ height), the goal becomes bounding the width. Chan's original paper gave several methods to produce LR drawings of arbitrary binary trees, the first method guaranteeing $O(n^{0.695})$ width, a second method with $O(\sqrt{n})$ width, and a final method (described in the appendix of his paper) with $O(n^{0.48})$ width.

More recently, in SODA'17, Frati, Patrignani, and Roselli [14] proved the first nontrivial lower bound, showing that there exist binary trees for which any LR drawing requires $\Omega(n^{0.418})$ width. This raises an intriguing question: can the gap between upper and lower bounds be closed, and the precise value of the exponent be determined?

It should be mentioned that other methods were subsequently found for planar, straight-line, strictly upward, order-preserving drawings of binary trees with smaller width ($2^{O(\sqrt{\log n})}$ in Chan's original paper, and eventually, $O(\log n)$ in a paper by Garg and Rusu [16]). Nevertheless, the question on LR drawings is still interesting and natural, as it is fundamentally about combinatorics of trees, or more specifically, decompositions of trees via path separators (instead of the more usual vertex or edge separators). Indeed, by the alternative definition, the minimum LR-drawing width $W^*(T)$ of a binary tree T can be described by the following self-contained formula, without reference to geometry:

$$W^*(T) = \min_{\pi} \max_{\alpha,\beta} (W^*(\alpha) + W^*(\beta) + 1),$$

where the minimum is over all root-to-leaf paths π in T, and the maximum is over all left subtrees α of π and all right subtrees β of π.

The LR drawing problem was also mentioned in Di Battista and Frati's recent survey [12] (as "Open problem 10").[1] LR drawing techniques have been applied to solve other problems, for example, on octagonal,[2] planar, straight-line, strictly upward, order-preserving drawings of binary trees [4], orthogonal,[3] planar, straight-line, non-upward, order-preserving drawings of binary trees [13], and planar straight-line drawings of outerplanar graphs [11,19], although in each of these applications, better methods not relying on LR drawings were eventually found [3,5,14].

In this paper, we make progress in narrowing the gap on the width bounds for LR drawings of binary trees: we improve Chan's upper bound from $O(n^{0.48})$ to $O(n^{0.437})$, and improve Frati et al.'s lower bound from $\Omega(n^{0.418})$ to $\Omega(n^{0.429})$.

2   Upper Bound

In this section, we present an algorithm for LR drawings that achieves width $O(n^{0.438})$. A small improvement to $O(n^{0.437})$ will be given in the next section. Our algorithm builds upon Chan's approach [4, Appendix A] but uses new ideas to substantially improve his $O(n^{0.48})$ upper bound. Throughout the paper, let $|T|$ denote the size (i.e., the number of nodes) in a tree T.

2.1   The Algorithm

Given a binary tree T with n nodes, we describe a recursive algorithm to produce an LR drawing of T and show by induction that its width is at most cn^p, for some constants p and c to be set later.

For n smaller than a sufficiently large constant, we can draw T arbitrarily. Otherwise, we maintain a path $\pi = \langle v_0, \ldots, v_t \rangle$. A subtree of π refers to a subtree rooted at a sibling of a node in π (it does not include the two subtrees at v_t).

---

[1] Technically, that survey asks about a different but related function: $W^{**}(n) = \max_T \min_\pi \max_{\alpha,\beta}(W^{**}(|\alpha|) + W^{**}(|\beta|) + 1)$, where the outer maximum is over all n-node binary trees T. This function may be larger than $\max_{T:|T|=n} W^*(T)$.

[2] All edges have slope from $\{0, \pm1, \pm\infty\}$.

[3] All edges are horizontal or vertical.

Let α and β denote the largest left subtree and right subtree of π, respectively. We maintain the invariant that

$$|\alpha|^p + |\beta|^p \leq (1 - \delta)n^p$$

for some sufficiently small constant $\delta > 0$. Initially, $t = 0$ and v_0 is the root of T. If v_t is a leaf, then we draw the subtrees of π recursively and combine them by aligning π vertically; the width is bounded by $c|\alpha|^p + c|\beta|^p + 1$, which by the invariant (and the induction hypothesis) is at most $c(1 - \delta)n^p + 1 < cn^p$ for a sufficiently large c (depending on p and δ). From now on, assume that v_t is not a leaf. Let L and R be the left and right subtree of the current node v_t, respectively. For some choice of constants $\delta > 0$ and h, we consider four cases (which cover all possibilities, as we will show in the next subsection).

CASE 1: $|\alpha|^p + |R|^p \leq (1 - \delta)n^p$. Set v_{t+1} to be the left child of v_t. Increment t and repeat.

CASE 2: $|\beta|^p + |L|^p \leq (1 - \delta)n^p$. Set v_{t+1} to be the right child of v_t. Increment t and repeat.

In either of the above two cases, the invariant is clearly preserved.

We may now assume that $|\alpha|^p + |R|^p > (1 - \delta)n^p$ and $|\beta|^p + |L|^p > (1 - \delta)n^p$. In conjunction with the invariant, we know that $|\beta| < |R|$ and $|\alpha| < |L|$.

For the next two cases, we introduce notation for the left and right subtrees of π (see Fig. 2). Let $\alpha_1^{(0)} = \alpha$ (the largest left subtree of π). The parent of $\alpha_1^{(0)}$ divides π into two segments. Let $\alpha_1^{(1)}$ and $\alpha_2^{(1)}$ denote the largest left subtree of the top and bottom segment, respectively. Extend the definition analogously: For each i, the parents of the $2^i - 1$ subtrees in $\{\alpha_j^{(\ell)} \mid 0 \leq \ell < i, \, 1 \leq j \leq 2^\ell\}$ divide π into 2^i segments. In the downward order, let $\alpha_1^{(i)}, \ldots, \alpha_{2^i}^{(i)}$ denote the largest left subtrees of these segments. The above labeling of subtrees resembles a "ruler pattern" (like in [14]). We define the right subtrees $\beta_1^{(i)}, \ldots, \beta_{2^i}^{(i)}$ similarly (we do not care how the left subtrees and the right subtrees of π interleave).

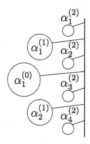

Fig. 2. Notation for left subtrees

CASE 3: There exists $i \leq h$ such that $\sum_{j=1}^{2^i} |\alpha_j^{(i)}|^p + \max\{|L|, |R|\}^p \leq (1 - \delta)n^p$. We generate an LR drawing of T using a procedure called the *i-right-twist*:

We bend π at the parents of all subtrees in $\{\alpha_j^{(\ell)} \mid 0 \leq \ell < i,\ 1 \leq j \leq 2^\ell\}$ (all these subtrees are thus pulled downward in the drawing), as illustrated in Fig. 3. We recursively draw R. We draw most of the subtrees of π recursively as well, but with the following exceptions: for the subtrees in $\{\alpha_j^{(\ell)} \mid 0 \leq \ell < i,\ 1 \leq j \leq 2^\ell\}$, we make their leftmost paths vertically aligned and recursively draw the subtrees of these paths. Similarly, for L, we make its leftmost path vertically aligned and recursively draw the subtrees of the path. Since every subtree of π has size at most $\max\{|\alpha|, |\beta|\} < \max\{|L|, |R|\}$, it is easy to check (using the induction hypothesis) that the resulting LR drawing has width at most $\sum_{j=1}^{2^i} c|\alpha_j^{(i)}|^p + c\max\{|L|, |R|\}^p + 2^h$; this is at most $(1-\delta)cn^p + 2^h < cn^p$, for a sufficiently large c (depending on p, δ, and h).

CASE 4: There exists $i \leq h$ such that $\sum_{j=1}^{2^i} |\beta_j^{(i)}|^p + \max\{|L|, |R|\}^p \leq (1-\delta)n^p$. This is similar to Case 3, by using the *i-left-twist*.

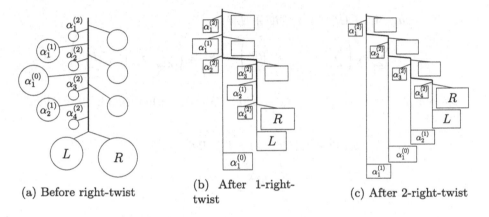

(a) Before right-twist

(b) After 1-right-twist

(c) After 2-right-twist

Fig. 3. Right twist

Remark. The twisting procedures in Cases 3 and 4, and the introduction of the "ruler pattern", are the main new ideas, compared to Chan's previous algorithm [4].

2.2 Analysis

To complete the induction proof, it suffices to show that these four cases cover all possibilities.

Lemma 1. *For $p = 0.438$ and a sufficiently small constant $\delta > 0$ and a sufficiently large constant h,*

$$
\min \left\{
\begin{array}{l}
|\alpha|^p + |R|^p, \\
|\beta|^p + |L|^p, \\
\displaystyle \min_{i=1}^{h} \left(\sum_{j=1}^{2^i} |\alpha_j^{(i)}|^p + \max\{|L|, |R|\}^p \right), \\
\displaystyle \min_{i=1}^{h} \left(\sum_{j=1}^{2^i} |\beta_j^{(i)}|^p + \max\{|L|, |R|\}^p \right)
\end{array}
\right\} \leq (1 - \delta)n^p.
$$

Proof. Assume for the sake of contradiction that the lemma is false. Without loss of generality, assume $|R| \geq |L|$. Let $a_0, \ldots, a_h, b_0, \ldots, b_h$ be positive real numbers with $\sum_{i=0}^{h} a_i + \sum_{i=0}^{h} b_i = 1$, whose values are to be determined later. Let

$$
X := a_0(|\alpha|^p + |R|^p) + b_0(|\beta|^p + |L|^p)
$$

$$
+ \sum_{i=1}^{h} a_i \left(\sum_{j=1}^{2^i} |\alpha_j^{(i)}|^p + |R|^p \right) + \sum_{i=1}^{h} b_i \left(\sum_{j=1}^{2^i} |\beta_j^{(i)}|^p + |R|^p \right).
$$

By our assumption, $X > (1 - \delta)n^p$. On the other hand, by Hölder's inequality,[4]

$$
X = a_0|\alpha|^p + b_0|\beta|^p + b_0|L|^p + (1 - b_0)|R|^p
$$

$$
+ \sum_{i=1}^{h} a_i \sum_{j=1}^{2^i} |\alpha_j^{(i)}|^p + \sum_{i=1}^{h} b_i \sum_{j=1}^{2^i} |\beta_j^{(i)}|^p
$$

$$
\leq \lambda^{1-p} \left(|\alpha| + |\beta| + |L| + |R| + \sum_{j=1}^{h} \sum_{j=1}^{2^i} |\alpha_j^{(i)}| + \sum_{j=1}^{h} \sum_{j=1}^{2^i} |\beta_j^{(i)}| \right)^p
$$

$$
\leq \lambda^{1-p} n^p,
$$

where

$$
\lambda := a_0^{\frac{1}{1-p}} + 2b_0^{\frac{1}{1-p}} + (1 - b_0)^{\frac{1}{1-p}} + \sum_{i=1}^{h} 2^i a_i^{\frac{1}{1-p}} + \sum_{i=1}^{h} 2^i b_i^{\frac{1}{1-p}}.
$$

Thus, we have $\lambda^{1-p} > 1 - \delta$. However, we show that this is not true for some choice of parameters.

---

[4] Hölder's inequality states that $\sum_i |x_i y_i| \leq \left(\sum_i |x_i|^s \right)^{1/s} \left(\sum_i |y_i|^t \right)^{1/t}$ for any $s, t > 1$ with $\frac{1}{s} + \frac{1}{t} = 1$. In our applications, it is more convenient to set $s = \frac{1}{p}$, $t = \frac{1}{1-p}$, $x_i = X_i^p$, and $y_i = c_i$, and rephrase the inequality as: $\sum_i c_i X_i^p \leq \left(\sum_i c_i^{1/(1-p)} \right)^{1-p} \left(\sum_i X_i \right)^p$ for any $0 < p < 1$ and $c_i, X_i \geq 0$.

We first set $a_i = b_i = (2^{-\frac{1-p}{p}})^i a_0$ for $1 \le i \le h$ (by calculus, this choice is actually the best for minimizing λ). Let $\rho = 1 + 2\sum_{i=1}^{h}(2^{-\frac{1-p}{p}})^i$ and $b_0 = 1 - \rho a_0$. Then we indeed have $\sum_{i=0}^{h} a_i + \sum_{i=0}^{h} b_i = 1$, and the above expression simplifies to $\lambda = \rho a_0^{\frac{1}{1-p}} + 2(1 - \rho a_0)^{\frac{1}{1-p}} + (\rho a_0)^{\frac{1}{1-p}}$. For $p = 0.438$, the limit of ρ (as $h \to \infty$) is $2/(1 - 2^{-\frac{1-p}{p}}) - 1 \approx 2.395068$. We can plug in $a_0 = 0.247$ and verify (using a calculator) that the limit of λ is less than 0.9984, which leads to a contradiction for a sufficiently small δ and a sufficiently large h.

3 Slightly Improved Upper Bound

In this section, we describe a refinement of our algorithm to further improve the width upper bound to $O(n^{0.437})$. Although the improvement is tiny, the main purpose is to show that our algorithm is not optimal.

The change lies in the procedure of i-right-twist in Case 3, specifically, how L is drawn. Instead of vertically aligning the leftmost path in L, we choose a different path, exploiting the already "used" width from the drawing of $\alpha_{2^i}^{(i)}$ that is available to the left of the root of L. We define a new path $\pi' = \langle u_0, u_1, \dots \rangle$ in L as follows. Initially, set u_0 to the root of L. For $k = 0, 1, \dots$ (until u_k is a leaf), if the left subtree of u_k has size at most $|\alpha_{2^i}^{(i)}|$, then set u_{k+1} to be the right child of u_k; otherwise, set u_{k+1} to be the left child of u_k (see Fig. 4). This way, every left subtree of π' has size at most $|\alpha_{2^i}^{(i)}|$, and every right subtree of π' has size less than $|L| - |\alpha_{2^i}^{(i)}|$. (Note that $|\alpha_{2^i}^{(i)}| < |L|$.) We draw L by vertically aligning the path π', and recursively drawing the left and right subtrees of π'. The overall LR drawing of T has width at most $\sum_{j=1}^{2^i} c|\alpha_j^{(i)}|^p + c\max\{|L| - |\alpha_{2^i}^{(i)}|, |R|\}^p + 2^h$. (Parts of the drawing may have width bounded instead by $\sum_{j=1}^{2^i-1} c|\alpha_j^{(i)}|^p + c\max\{|L|, |R|\}^p + 2^h$, but this is no worse than the above bound since $|L|^p \le |\alpha_{2^i}^{(i)}|^p + (|L| - |\alpha_{2^i}^{(i)}|)^p$.) Thus, we can relax the condition in Case 3 to $\sum_{j=1}^{2^i} c|\alpha_j^{(i)}|^p + c\max\{|L| - |\alpha_{2^i}^{(i)}|, |R|\}^p \le (1 - \delta)n^p$.

With a similar modification to the i-left-twist procedure, we can relax the condition in Case 4 to $\sum_{j=1}^{2^i} c|\beta_j^{(i)}|^p + c\max\{|L|, |R| - |\beta_{2^i}^{(i)}|\}^p \le (1 - \delta)n^p$.

It suffices to prove the following variant of Lemma 1:

Lemma 2. For $p = 0.437$ and a sufficiently small constant $\delta > 0$ and a sufficiently large constant h,

$$\min \left\{ \begin{array}{l} |\alpha|^p + |R|^p, \\ |\beta|^p + |L|^p, \\ \displaystyle\min_{i=1}^{h} \left(\sum_{j=1}^{2^i} |\alpha_j^{(i)}|^p + \max\{|L| - |\alpha_{2^i}^{(i)}|, |R|\}^p \right), \\ \displaystyle\min_{i=1}^{h} \left(\sum_{j=1}^{2^i} |\beta_j^{(i)}|^p + \max\{|L|, |R| - |\beta_{2^i}^{(i)}|\}^p \right) \end{array} \right\} \le (1 - \delta)n^p.$$

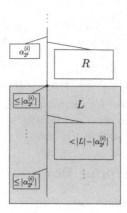

Fig. 4. Choosing a path π' inside L

Proof. Assume for the sake of contradiction that the lemma is false. Without loss of generality, assume $|R| \geq |L|$. Note that $|\beta_{2^i}^{(i)}|$ decreases with i. Let i^* be the smallest integer with $1 \leq i^* \leq h$ such that $|R| - |\beta_{2^{i^*}}^{(i^*)}| \geq |L|$ (if such an integer does not exist, set $i^* = h + 1$).

Let $a_0, \ldots, a_h, b_0, \ldots, b_h$ be positive real numbers with $\sum_{i=0}^{h} a_i + \sum_{i=0}^{h} b_i = 1$, whose values are to be determined later. Let

$$
X := \sum_{i=0}^{h} a_i \left(\sum_{j=1}^{2^i} |\alpha_j^{(i)}|^p + |R|^p \right) + \sum_{i=0}^{i^*-1} b_i \left(\sum_{j=1}^{2^i} |\beta_j^{(i)}|^p + |L|^p \right)
$$

$$
+ \sum_{i=i^*}^{h} b_i \left(\sum_{j=1}^{2^i} |\beta_j^{(i)}|^p + (|R| - |\beta_{2^i}^{(i)}|)^p \right).
$$

By our assumption, $X > (1-\delta)n^p$. On the other hand, by Hölder's inequality, for any $0 < \gamma < 1$,

$$
|\beta_{2^i}^{(i)}|^p + (|R| - |\beta_{2^i}^{(i)}|)^p = (1-\gamma)|\beta_{2^i}^{(i)}|^p + \gamma|\beta_{2^i}^{(i)}|^p + (|R| - |\beta_{2^i}^{(i)}|)^p
$$

$$
\leq (1-\gamma)|\beta_{2^i}^{(i)}|^p + (\gamma^{\frac{1}{1-p}} + 1)^{1-p}|R|^p,
$$

and by Hölder's inequality again,

$$X \le \sum_{i=0}^{h} a_i \sum_{j=1}^{2^i} |\alpha_j^{(i)}|^p + \sum_{i=0}^{i^*-1} b_i \sum_{j=1}^{2^i} |\beta_j^{(i)}|^p + \left(\sum_{i=0}^{i^*-1} b_i\right) |L|^p$$
$$+ \sum_{i=i^*}^{h} b_i \left(\sum_{j=1}^{2^i-1} |\beta_j^{(i)}|^p + (1-\gamma)|\beta_{2^i}^{(i)}|^p\right)$$
$$+ \left(\sum_{i=0}^{h} a_i + (1+\gamma^{\frac{1}{1-p}})^{1-p} \sum_{i=i^*}^{h} b_i\right) |R|^p$$
$$\le \lambda^{1-p} n^p,$$

where

$$\lambda := \sum_{i=0}^{h} 2^i a_i^{\frac{1}{1-p}} + \sum_{i=0}^{i^*-1} 2^i b_i^{\frac{1}{1-p}} + \left(\sum_{i=0}^{i^*-1} b_i\right)^{\frac{1}{1-p}} + \sum_{i=i^*}^{h} (2^i - 1 + (1-\gamma)^{\frac{1}{1-p}}) b_i^{\frac{1}{1-p}}$$
$$+ \left(\sum_{i=0}^{h} a_i + (1+\gamma^{\frac{1}{1-p}})^{1-p} \sum_{i=i^*}^{h} b_i\right)^{\frac{1}{1-p}}.$$

An optimal choice of parameters is now messier to describe, but will not be necessary. We can reuse our earlier choice with $a_0 = 0.247$, $a_i = b_i = (2^{-\frac{1-p}{p}})^i a_0$ for $1 \le i \le h$, and $b_0 = 1 - \sum_{i=0}^{h} a_i - \sum_{i=1}^{h} b_i$. For $p = 0.437$, $\gamma = 0.1$, and $h = 7$, we can verify (with a short computer program) that for each possible $i^* \in \{1, \ldots, 8\}$, λ evaluates to strictly less than 1, a contradiction.

Theorem 1. *For any binary tree with n nodes, there exists an LR drawing with $O(n^{0.437})$ width.*

Remark. It is not difficult to implement the algorithm to construct the drawing in $O(n)$ time.

4   Lower Bound

We now prove an $\Omega(n^{0.429})$ lower bound on the width of LR drawings. Our proof is largely based on Frati, Patrignani, and Roselli's [14]; we show that a simple variation of their proof is sufficient to improve their $\Omega(n^{0.418})$ lower bound.

4.1   Tree Construction

For any given positive integer n, we describe a recursive construction of a binary tree T_n with n nodes and show by induction that any LR drawing of T_n has width at least cn^p, where p and $c > 0$ are constants to be determined later.

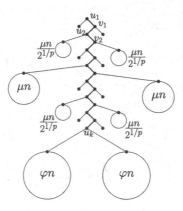

Fig. 5. Tree construction for the lower bound

For n smaller than a sufficiently large constant, we can construct T_n arbitrarily. Otherwise, let h, φ, and μ be parameters, to be chosen later. We construct a tree T_n containing a path $\pi = \langle u_1, v_1, u_2, v_2, \ldots, u_{k-1}, v_{k-1}, u_k \rangle$, where $k = 2^h$ and u_1 is the root. The left and right subtree of u_k, which we denote by L and R, are recursively constructed trees each with $\lceil \varphi n \rceil$ nodes.

We will add left subtrees $\alpha_1, \ldots, \alpha_{k-1}$ to $u_1, \ldots, u_{k-1}$ and right subtrees $\beta_1, \ldots, \beta_{k-1}$ to $v_1, \ldots, v_{k-1}$. Specifically, the subtrees $\alpha_{k/2}$ and $\beta_{k/2}$, which are said to be at *level* 0, are recursively constructed trees each with $\lceil \mu n \rceil$ nodes. The subtrees $\alpha_{k/4}, \alpha_{3k/4}$ and $\beta_{k/4}, \beta_{3k/4}$, which are at *level* 1, are recursively constructed trees each with $\lceil 2^{-1/p} \mu n \rceil$ nodes. Extend the process analogously: For each $i \leq h-2$, the 2^i left subtrees $\alpha_{k/2^{i+1}}, \alpha_{3k/2^{i+1}}, \alpha_{5k/2^{i+1}}, \ldots$ and 2^i right subtrees $\beta_{k/2^{i+1}}, \beta_{3k/2^{i+1}}, \beta_{5k/2^{i+1}}, \ldots$, which are at *level* i, are recursively constructed trees each with $\lceil (2^{-1/p})^i \mu n \rceil$ nodes. As shown in Fig. 5, these subtrees of π form a "ruler pattern" (which somewhat resembles the ruler pattern from our upper bound proof, coincidentally or not).

We set $h = \lfloor p \log(\mu n / c_0) \rfloor$ for a sufficiently large constant c_0, and choose parameters φ and μ to satisfy

$$\varphi + \frac{\mu}{1 - 2^{-\frac{1-p}{p}}} = \frac{1}{2}. \tag{1}$$

Then the total size of the left subtrees at level $0, \ldots, h-2$ is

$$\sum_{i=0}^{h-2} 2^i \left\lceil (2^{-1/p})^i \mu n \right\rceil = \sum_{i=0}^{h-2} (2^{-\frac{1-p}{p}})^i \mu n + O(k)$$

$$= \frac{\mu n}{1 - 2^{-\frac{1-p}{p}}} - \Theta((2^{-\frac{1-p}{p}})^h \mu n) + O(k)$$

$$= \left(\tfrac{1}{2} - \varphi\right) n - \Theta(c_0 k).$$

The same bound holds for the right subtrees at level $0, \ldots, h-2$. Thus, we can distribute $\Theta(c_0)$ nodes to each of the $\Theta(k)$ subtrees at the last level $h-1$ so that $|T_n|$ is exactly n.

4.2   Analysis

We begin with a simple property arising from the ruler pattern:

Lemma 3. *For any set $J \subseteq \{1, \ldots, k-1\}$ of consecutive integers, the largest subtree α_j (or β_j) with $j \in J$ has size at least $\left(\frac{|J|-1}{k}\right)^{1/p} \mu n$.*

Proof. We may assume $|J| \geq 2$ (for otherwise the inequality is trivial). Say $k/2^{i+1} \leq |J| < k/2^i$. The subtrees α_j at level at most i are precisely those with indices j divisible by $k/2^{i+1}$; there exists one such index with $j \in J$. The size of α_j and β_j is at least $(2^{-1/p})^i \mu n \geq (|J|/k)^{1/p} \mu n$.   □

Assume inductively that any LR drawing of $T_{n'}$ has width at least $c(n')^p$, for all $n' < n$. Let $T_n(u_j)$ denote the subtree of T_n rooted at node u_j. We will prove the following claim, for c sufficiently small:

Claim. For $j \in \{1, \ldots, k\}$, every LR drawing Γ of $T_n(u_j)$ has width at least

$$\tfrac{k-j+1}{k} c(\mu n)^p + c(\varphi n)^p.$$

Proof. We do another proof by induction, on j (within the outer induction proof). Let $\pi(\Gamma)$ denote the root-to-leaf path in $T(u_j)$ that is vertically aligned in Γ. Let $\pi_{j \to k}$ denote the path $\langle u_j, \ldots, u_k \rangle$. Consider the last node w that is common to both paths $\pi(\Gamma)$ and $\pi_{j \to k}$.

CASE 1: $w = u_k$. Let α and β be the largest subtree among $\alpha_j, \ldots, \alpha_{k-1}$ and $\beta_j, \ldots, \beta_{k-1}$, respectively (in the special case $j = k$, let $\alpha = \beta = \emptyset$). By Lemma 3,

$$|\alpha|^p, |\beta|^p \geq \tfrac{k-j-1}{k}(\mu n)^p.$$

If $\pi(\Gamma)$ contains the left child of u_k, then the drawings of α and R are separated by the vertical line through $\pi(\Gamma)$, and so (by the outer induction hypothesis) the overall drawing Γ has width at least

$$c|\alpha|^p + c|R|^p + 1 \geq \tfrac{k-j-1}{k} c(\mu n)^p + c(\varphi n)^p + 1 \geq \tfrac{k-j+1}{k} c(\mu n)^p + c(\varphi n)^p,$$

for a sufficiently small c (since $\frac{1}{k}(\mu n)^p = O(1)$). If $\pi(\Gamma)$ contains the right child of u_k, then β_i and L are vertically separated, and the argument is similar.

CASE 2: $w = u_m$ for some $j \leq m < k$. Let α be the largest subtree among $\alpha_j, \ldots, \alpha_{m-1}$ (in the special case $m = j$, let $\alpha = \emptyset$). By Lemma 3,

$$|\alpha|^p \geq \tfrac{m-j-1}{k}(\mu n)^p.$$

Since $\pi(\Gamma)$ contains the left child of u_m, we know that the drawings of α and $T_n(u_{m+1})$ are separated by the vertical line through $\pi(\Gamma)$, and so by the induction hypotheses, the overall drawing Γ has width at least

$$c|\alpha|^p + \tfrac{k-m}{k}c(\mu n)^p + c(\varphi n)^p + 1 \geq \tfrac{k-j-1}{k}c(\mu n)^p + c(\varphi n)^p + 1$$
$$\geq \tfrac{k-j+1}{k}c(\mu n)^p + c(\varphi n)^p,$$

for a sufficiently small c.

CASE 3: $w = v_m$ for some $m < k$. This is similar to Case 2.   □

Applying Claim 4.2 with $j = 1$, we see that any LR drawing of T_n has width at least $c(\mu n)^p + c(\varphi n)^p$, which is at least cn^p, completing the induction proof, provided that

$$\varphi^p + \mu^p \geq 1. \tag{2}$$

For $p = 0.429$, we can choose $\mu = 0.122$ and $\varphi \approx 0.297513$ and verify (using a calculator) that both (1) and (2) are satisfied.

Theorem 2. *For every positive integer n, there is a binary tree with n nodes such that any LR drawing requires $\Omega(n^{0.429})$ width.*

Remark. The maximum value of p that guarantees the existence of μ and φ satisfying (1) and (2) has a concise description: it is given by $p = 1/(1 + x)$, where x is the solution to the equation

$$1 - 2^{-x} = (2^{1/x} - 1)^x.$$

Our lower-bound proof is very similar to Frati, Patrignani, and Roselli's [14], but there are two main differences: First, their tree construction was parameterized by a different parameter h instead of n; they upper-bounded the size n by an exponential function on h and lower-bounded the width by another exponential function on h. Second, and more crucially, they chose $\varphi = \mu$ (in our terminology). Besides convenience, we suspect that their choice was due to the above parameterization issue. With this extra, unnecessary constraint $\varphi = \mu$, the best choice of p was only around 0.418.

5   Final Remarks

The main open problem is to narrow the remaining small gap in the exponents of the upper and lower bound (between 0.437 and 0.429). The fact that both the upper and lower bound proofs use similar "ruler patterns" suggests that we are on the right track (even though looking for further tiny improvements in the upper-bound proof by complicating the analysis, along the lines of Sect. 3, doesn't seem very worthwhile).

Frati *et al.* [14] have computed the exact optimal width for small values of n, and according to their experimental data for all $n \leq 455$, a function of the form $W^*(n) = an^b - c$ with the least-squares fit is $W^*(n) \approx 1.54n^{0.443} - 0.55$. Our results reveal that the true exponent is actually smaller.

Another open problem is to bound the related function $W^{**}(n)$ mentioned in footnote 1 of the introduction; our new upper-bound proof does not work for this problem, but Chan's $O(n^{0.48})$ upper bound [4] still holds.

References

1. Bachmaier, C., Brandenburg, F., Brunner, W., Hofmeier, A., Matzeder, M., Unfried, T.: Tree drawings on the hexagonal grid. In: Proceedings of 16th International Symposium on Graph Drawing (GD), pp. 372–383 (2008). https://doi.org/10.1007/978-3-642-00219-9_36
2. Biedl, T.: Ideal drawings of rooted trees with approximately optimal width. J. Graph Algorithms Appl. **21**, 631–648 (2017). https://doi.org/10.7155/jgaa.00432
3. Biedl, T.: Upward order-preserving 8-grid-drawings of binary trees. In: Proceedings of 29th Canadian Conference on Computational Geometry (CCCG), pp. 232–237 (2017)
4. Chan, T.M.: A near-linear area bound for drawing binary trees. Algorithmica **34**(1), 1–13 (2002). https://doi.org/10.1007/s00453-002-0937-x
5. Chan, T.M.: Tree drawings revisited. Discrete Comput. Geom. **63**(4), 799–820 (2019). https://doi.org/10.1007/s00454-019-00106-w
6. Chan, T.M., Goodrich, M.T., Kosaraju, S.R., Tamassia, R.: Optimizing area and aspect ration in straight-line orthogonal tree drawings. Comput. Geom. **23**(2), 153–162 (2002). https://doi.org/10.1016/S0925-7721(01)00066-9
7. Crescenzi, P., Di Battista, G., Piperno, A.: A note on optimal area algorithms for upward drawings of binary trees. Comput. Geom. **2**, 187–200 (1992). https://doi.org/10.1016/0925-7721(92)90021-J
8. Crescenzi, P., Penna, P.: Minimum-area h-v drawings of complete binary trees. In: DiBattista, G. (ed.) GD 1997. LNCS, vol. 1353, pp. 371–382. Springer, Heidelberg (1997). https://doi.org/10.1007/3-540-63938-1_82
9. Crescenzi, P., Penna, P.: Strictly-upward drawings of ordered search trees. Theor. Comput. Sci. **203**(1), 51–67 (1998). https://doi.org/10.1016/S0304-3975(97)00287-9
10. Di Battista, G., Eades, P., Tamassia, R., Tollis, I.G.: Graph Drawing. Prentice Hall, Upper Saddle River (1999)
11. Di Battista, G., Frati, F.: Small area drawings of outerplanar graphs. Algorithmica **54**(1), 25–53 (2009). https://doi.org/10.1007/s00453-007-9117-3
12. Di Battista, G., Frati, F.: Drawing trees, outerplanar graphs, series-parallel graphs, and planar graphs in small area. In: Pach, J. (ed.) Geometric Graph Theory, pp. 121–165. Springer, Heidelberg (2013). https://doi.org/10.1007/978-1-4614-0110-0_9
13. Frati, F.: Straight-line orthogonal drawings of binary and ternary trees. In: Proceedings of 15th International Symposium on Graph Drawing (GD), pp. 76–87 (2007). https://doi.org/10.1007/978-3-540-77537-9_11
14. Frati, F., Patrignani, M., Roselli, V.: LR-drawings of ordered rooted binary trees and near-linear area drawings of outerplanar graphs. J. Comput. Syst. Sci. **107**, 28–53 (2020). https://doi.org/10.1016/j.jcss.2019.08.001
15. Garg, A., Goodrich, M.T., Tamassia, R.: Planar upward tree drawings with optimal area. Int. J. Comput. Geometry Appl. **6**(3), 333–356 (1996). https://doi.org/10.1142/S0218195996000228

16. Garg, A., Rusu, A.: Area-efficient order-preserving planar straight-line drawings of ordered trees. Int. J. Comput. Geometry Appl. **13**(6), 487–505 (2003). https://doi.org/10.1142/S021819590300130X

17. Garg, A., Rusu, A.: Straight-line drawings of general trees with linear area and arbitrary aspect ratio. In: Proceedings of 3rd International Conference on Computational Science and its Applications (ICCSA), Part III, pp. 876–885 (2003). https://doi.org/10.1007/3-540-44842-X_89

18. Garg, A., Rusu, A.: Straight-line drawings of binary trees with linear area and arbitrary aspect ratio. J. Graph Algorithms Appl. **8**(2), 135–160 (2004)

19. Garg, A., Rusu, A.: Area-efficient planar straight-line drawings of outerplanar graphs. Discrete Appl. Math. **155**(9), 1116–1140 (2007)

20. Lee, S.: Upward octagonal drawings of ternary trees. Master's thesis, University of Waterloo (2016). (Supervised by T. Biedl and T. M. Chan.) https://uwspace.uwaterloo.ca/handle/10012/10832

21. Leiserson, C.E.: Area-efficient graph layouts (for VLSI). In: Proceedings of 21st IEEE Symposium on Foundations of Computer Science (FOCS), pp. 270–281 (1980). https://doi.org/10.1109/SFCS.1980.13

22. Reingold, E.M., Tilford, J.S.: Tidier drawings of trees. IEEE Trans. Softw. Eng. **7**(2), 223–228 (1981). https://doi.org/10.1109/TSE.1981.234519

23. Shiloach, Y.: Linear and Planar Arrangement of Graphs. Ph.D. thesis, Weizmann Institute of Science (1976). https://lib-phds1.weizmann.ac.il/Dissertations/shiloach_yossi.pdf

24. Shin, C., Kim, S.K., Chwa, K.: Area-efficient algorithms for straight-line tree drawings. Comput. Geom. **15**(4), 175–202 (2000). https://doi.org/10.1016/S0925-7721(99)00053-X

25. Shin, C., Kim, S.K., Kim, S., Chwa, K.: Algorithms for drawing binary trees in the plane. Inf. Process. Lett. **66**(3), 133–139 (1998). https://doi.org/10.1016/S0020-0190(98)00049-0

26. Trevisan, L.: A note on minimum-area upward drawing of complete and Fibonacci trees. Inf. Process. Lett. **57**(5), 231–236 (1996). https://doi.org/10.1016/0020-0190(96)81422-0

27. Valiant, L.G.: Universality considerations in VLSI circuits. IEEE Trans. Comput. **30**(2), 135–140 (1981). https://doi.org/10.1109/TC.1981.6312176

On the Edge-Length Ratio of 2-Trees

Václav Blažej[1]([✉]) [ID], Jiří Fiala[2] [ID], and Giuseppe Liotta[3] [ID]

[1] Faculty of Information Technology, Czech Technical University in Prague,
Prague, Czech Republic
vaclav.blazej@fit.cvut.cz
[2] Faculty of Mathematics and Physics, Charles University, Prague, Czech Republic
fiala@kam.mff.cuni.cz
[3] Dipartimento di Ingegneria, Università degli Studi di Perugia, Perugia, Italy
giuseppe.liotta@unipg.it

Abstract. We study planar straight-line drawings of graphs that minimize the ratio between the length of the longest and the shortest edge. We answer a question of Lazard et al. [Theor. Comput. Sci. **770** (2019), 88–94] and, for any given constant r, we provide a 2-tree which does not admit a planar straight-line drawing with a ratio bounded by r. When the ratio is restricted to adjacent edges only, we prove that any 2-tree admits a planar straight-line drawing whose edge-length ratio is at most $4 + \varepsilon$ for any arbitrarily small $\varepsilon > 0$, hence the upper bound on the local edge-length ratio of partial 2-trees is 4.

Keywords: Planar straight-line drawing · Edge-length ratio · 2-tree

1 Introduction

Straight-line drawings of planar graphs are thoroughly studied both for their theoretical interest and their applications in a variety of disciplines (see, e.g., [6,12]). Different quality measures for planar straight-line drawings have been considered in the literature, including area, angular resolution, slope number, average edge length, and total edge length (see, e.g., [8,9,11]).

This paper studies the problem of computing planar straight-line drawings of graphs where the length ratio of the longest to the shortest edge is as small as possible. We recall that the problem of deciding whether a graph admits a planar straight-line drawing with specified edge lengths is NP-complete even

The research was initiated during workshop Homonolo 2018. Research partially supported by MIUR, the Italian Ministry of Education, University and Research, under Grant 20174LF3T8 AHeAD: efficient Algorithms for HArnessing networked Data. V. Blažej acknowledges the support of the OP VVV MEYS funded project CZ.02.1.01/0.0/0.0/16_019/0000765 "Research Center for Informatics". This work was supported by the Grant Agency of the Czech Technical University in Prague, grant No. SGS20/208/OHK3/3T/18. The work of J. Fiala was supported by the grant 19-17314J of the GA ČR.

© Springer Nature Switzerland AG 2020
D. Auber and P. Valtr (Eds.): GD 2020, LNCS 12590, pp. 85–98, 2020.
https://doi.org/10.1007/978-3-030-68766-3_7

when restricted to 3-connected planar graphs [7] and the completeness persists in the case when all given lengths are equal [4]. In addition, deciding whether a degree-4 tree has a planar drawing such that all edges have the same length and the vertices are at integer grid points is NP-complete [1].

In the attempt of relaxing the edge length conditions which make the problem hard, Hoffmann et al. [9] propose to minimize the ratio between the longest and the shortest edges among all straight-line drawings of a graph. While the problem remains hard for general graphs (through approximation of unit disk graphs [5]), Lazard et al. prove [10] that any outerplanar graph admits a planar straight-line drawing such that the length ratio of the longest to the shortest edges is strictly less than 2. This result is tight in the sense that for any $\varepsilon > 0$ there are outerplanar graphs that cannot be drawn with an edge-length ratio smaller than $2 - \varepsilon$. Lazard et al. also ask whether their construction could be extended to the class of series-parallel graphs.

We answer this question in the negative sense, by showing that a subclass of series-parallel graphs, called 2-trees, does not allow any planar straight-line drawing of bounded edge-length ratio. In fact, a corollary of our main result is the existence of an $\Omega(\log n)$ lower bound for the edge-length ratio of planar straight-line drawings of n-vertex 2-trees. Motivated by this negative result, we consider a local measure of edge-length ratio and prove that when the ratio is restricted only to the adjacent edges, any series-parallel graph admits a planar straight-line drawing with local edge-length ratio at most $4+\varepsilon$, for any arbitrarily small $\varepsilon > 0$. The proof of this upper bound is constructive, and it gives rise to a linear-time algorithm assuming a real RAM model of computation.

It is worth noticing that Borrazzo and Frati have shown that any 2-tree on n vertices can be drawn with edge-length ratio $O(n^{0.695})$ [3]. This, together with our $\Omega(\log n)$ result, defines a non-trivial gap between the upper and lower bound on the edge-length ratio of planar straight-line drawings of partial 2-trees. We recall that Borrazzo and Frati also show an $\Omega(n)$ lower bound on the edge-length ratio of general planar graphs [3].

The rest of the paper is organized as follows. Preliminaries are in Sect. 2; the $\Omega(\log n)$ lower bound is proved in Sect. 3; Sect. 4 presents a constructive argument for an upper bound on the local edge-length ratio of partial 2-trees. Conclusions and open problems can be found in Sect. 5. Omitted proof can be found in the appendix of the full version of the paper which is available online [2].

2   Preliminaries

We use capital letters $A, B, \ldots$, for the points in the Euclidean plane. For points A and B, let $|AB|$ denote the Euclidean distance between A and B. The symbol $\triangle ABC$ denotes the triangle determined by three distinct non-colinear points A, B, and C. The symbol $\angle BAC$ stands for the angle at vertex A of the triangle $\triangle ABC$.

For a polygon Q, we denote its perimeter by $P(Q)$ and its area by $A(Q)$.

We consider finite nonempty planar graphs and their planar straight-line drawings. Once a straight-line drawing of a graph G is given, with a slight abuse of notation we use the same symbol for a vertex U and the point U representing the vertex U in the drawing; the same symbol UV for an edge and the corresponding segment; as well as $\triangle UVW$ for an induced cycle of length three and the corresponding triangle.

When we consider graphs as combinatorial objects, we often use lowercase symbols u or e for the vertices and edges.

The *edge-length ratio* of a planar straight-line drawing of a graph G is the ratio between the length of the longest and the shortest edge of the drawing.

Definition 1. *The* edge-length ratio $\rho(G)$ *of a planar graph G is the infimum edge-length ratio taken over all planar straight-line drawings of G.*

The class of 2-trees is defined recursively: an edge is a 2-tree. If e is an edge of a 2-tree, then the graph, formed by adding a new vertex u adjacent to both endpoints of e, is also a 2-tree. In such a situation we say that u has been added as a *simplicial* vertex to e. A *partial 2-tree* is a subgraph of a 2-tree.

3   Edge-Length Ratio of 2-Trees

We recall that 2-trees are planar graphs. The main result of this section is the following.

Theorem 1. *For any $r \geq 1$, there exists a 2-tree G with edge-length ratio $\rho(G) \geq r$.*

To prove Theorem 1, for a given r we argue that a sufficiently large 2-tree, drawn with the longest edge having length r, contains a triangle with area at most $\frac{1}{4}$ (Corollary 1). Then, inside this triangle of small area we build a sequence of triangles with perimeters decreasing by at least 1 at every two steps (Lemmas 5 and 6), which results in a triangle with an edge of length less than 1.

We consider a special subclass $\mathcal{G} = \{G_0, G_1, \dots\}$ of 2-trees with labeled vertices and edges constructed as follows: G_0 is the complete graph K_3 whose vertices and edges are given the label 0. The graph G_{i+1} is obtained by adding five simplicial vertices to each edge of label i of G_i. Each newly created vertex and edge gets label $i+1$. See Fig. 1 for an example where the black vertices and edges have label 0, the blue ones have label 1, and the red ones have label 2.

A *separating triangle of level i* in a straight-line drawing of a 2-tree G is an unordered triple $\{U, V, W\}$ of mutually adjacent vertices such that the vertex W of label i was added as a simplicial vertex to the edge UV in the recursive construction of G and the triangle $\triangle UVW$ contains in its interior at least two other vertices with label i which are simplicial to the edge UV. For example, in Fig. 1a) vertices $\{U, V, W\}$ form a separating triangle of level 1.

Lemma 1. *For any $k > i \geq 1$, for any planar straight-line drawing of the graph G_k, and for any edge e of G_k labeled by i, there exists a separating triangle of level $i+1$ containing the endpoints of e.*

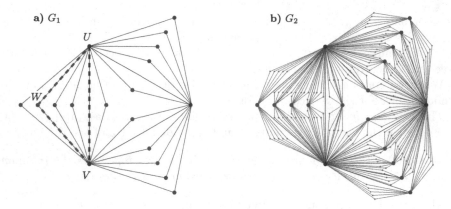

Fig. 1. The 2-trees G_1 and G_2. Black color corresponds to label 0, blue to 1, and red to 2. Separating triangle Δ_1 is emphasized by a dashed line in G_1. (Color figure online)

Proof. If a common edge of two triangles is traversed in the same direction when following their boundaries in the clockwise manner, then these triangles are nested, i.e. the interior of one contains the other one. Since we have five vertices simplicial to e, out of the corresponding five triangles in at least three e traversed in the same direction when following their boundaries in the clockwise manner. Thus at least three triangles are nested and the outermost of these is the desired separating triangle.

(For the clarity of presentation we have assumed a straight-line drawing, where the graph-theoretic term triangle coincides with the geometric one. This assumption could indeed be neglected when we consider a triangle in a planar drawing as the Jordan curve formed from the drawing of a 3-cycle.)

We proceed to show that any drawing of G_k contains a triangle of sufficiently small area. To this aim, we construct a sequence of nested triangles such that each triangle's area is half of the previous triangle's area. We denote as Δ_i a separating triangle of level i in an embedding of G_k, where $i \leq k$.

Lemma 2. *For any $k \geq 1$, any planar straight-line drawing of G_k contains a sequence of triangles $\Delta_1, \Delta_2, \ldots, \Delta_k$, where for any $i \in \{1, \ldots, k\}$ the triangle Δ_i is a separating triangle of level i, and for each $i > 1$, in addition, Δ_i is in the interior of Δ_{i-1} and $A(\Delta_i) \leq \frac{1}{2}A(\Delta_{i-1})$.*

Proof. We prove the lemma by induction on i. For $i = 1$ we apply Lemma 1 on any edge e of label 0 in G_k to get the triangle Δ_1.

When $i \in \{2, \ldots, k\}$, we assume by inductive hypothesis that the graph G_k contains a sequence of triangles $\Delta_1, \Delta_2, \ldots, \Delta_{i-1}$ satisfying the constraints. Let U be one of the two vertices of label $i - 1$ in the interior of Δ_{i-1} and let e and f be the two edges of label $i - 1$ incident to U.

We apply Lemma 1 on both of e and f to obtain two separating triangles of level i inside Δ_{i-1}, see Fig. 2. Since the drawing was planar, the two triangles are non-overlapping. We choose the triangle with the smaller area to be Δ_i to assure that $A(\Delta_i) \le \frac{1}{2}A(\Delta_{i-1})$.

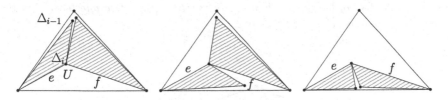

Fig. 2. Two separating triangles created in the interior of Δ_{i-1}

Corollary 1. *For any $r > 1$ and $k \ge 2 + 2\log_2 r$, every planar straight-line drawing of G_k with edge lengths at most r contains a separating triangle of area at most $\frac{1}{4}$.*

Proof. If all edges have length at most r, the area of Δ_1 is bounded by $\frac{\sqrt{3}}{4}r^2$. By Lemma 2, any drawing of G_k contains a sequence of nested separating triangles whose last element Δ_k has area at most $\frac{1}{2^{k-1}}\frac{\sqrt{3}r^2}{4} \le \frac{1}{4}$.

Before we proceed to the next step in our construction, we need some elementary facts from the trigonometry.

We call *thin* any triangle with edges of length at least 1 and area at most $\frac{1}{4}$. Any thin triangle has height at most $\frac{1}{2}$ and hence it has one obtuse angle of size at least $\frac{2\pi}{3}$ and two acute angles, each of size at most $\frac{\pi}{6}$.

Lemma 3. *Let $\triangle ABC$ be a thin triangle, where the longest edge is AB and let $D \in \triangle ABC$ be such that $|CD| \ge 1$. Then one of the angles $\angle ACD$ or $\angle BCD$ is obtuse.*

Proof. Assume by contradiction that both $\angle ACD$ and $\angle BCD$ are acute. Without loss of generality we may also assume that $\angle ACD \ge \angle BCD$. Since $\angle ACD + \angle BCD = \angle ACB \ge \frac{2\pi}{3}$, it follows that $\angle ACD \ge \frac{\pi}{3}$.

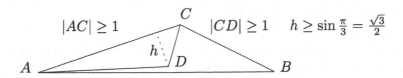

Fig. 3. To the argument that $\angle ACD$ cannot be acute.

Then the triangle $\triangle ACD$ has height at least $\frac{\sqrt{3}}{2}$, see Fig. 3. Thus it has area at least $\frac{\sqrt{3}}{4}$, a contradiction with the fact that the surrounding thin triangle $\triangle ABC$ has area at most $\frac{1}{4}$.

Now we focus our attention on the perimeters of the considered triangles.

Lemma 4. *Let $\triangle ABC$ be a thin triangle, where the longest edge is AB. Denote by Q the polygon, created by cutting off an isosceles triangle $\triangle BDE$ with both edges BD and BE of length 1. Then the perimeter of any triangle located in the polygon Q is at most $P(\triangle ABC) - 1$.*

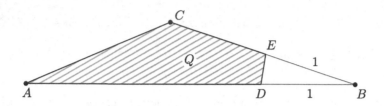

Fig. 4. Cutting-off the triangle $\triangle BDE$.

See Fig. 4 for an example of cutting off an isosceles triangle.

Proof. Assume for a contradiction that some triangle T has perimeter $P(T) > P(\triangle ABC) - 1$. Since T and Q are nested convex objects, we have that that $P(Q) \geq P(T) > P(\triangle ABC) - 1$. Then the length of the edge DE is greater than 1 and hence the angle $\angle DBE \geq \frac{\pi}{3}$, a contradiction with the property that the acute angles of a thin triangle are at most $\frac{\pi}{6}$.

We now return to our construction and show that a separating triangle with a small area is guaranteed to contain a separating triangle of a significantly smaller perimeter. In the following two lemmas we distinguish two complementary cases, namely whether the edge of level $i-1$ of a separating triangle of level i is incident to its obtuse angle or not.

Lemma 5. *Let G_k have a planar straight-line drawing with edge lengths at least 1 and let $\triangle UVW$ be a thin separating triangle of level i, where $i \leq k-1$. Assume that the edge UV is of level $i-1$ and that it is incident to the obtuse angle of $\triangle UVW$. Then $\triangle UVW$ contains a thin separating triangle T of level $i+1$ whose perimeter satisfies $P(T) \leq P(\triangle UVW) - 1$.*

Proof. Let X and Y be the two vertices of level i simplicial to the edge UV inside the triangle $\triangle UVW$. As the embedding of G_k is non-crossing straight-line, we may assume without loss of generality that the vertex X is inside $\triangle UVY$.

As all triangles in our further consideration are inside the thin triangle $\triangle UVW$, they have area at most $\frac{1}{4}$. By the definition of thin triangle they are also thin, as otherwise we would get in G_k an edge shorter than 1, which violates the assumptions of the Lemma. We distinguish several cases depending on the position of the obtuse angle of the considered triangles, see Fig. 5.

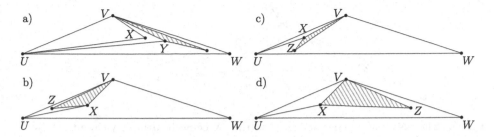

Fig. 5. Case analysis for Lemma 5.

a) The obtuse angle of $\triangle UVX$ is at V. By Lemma 1 we find a separating triangle T incident with the edge VY. Since T takes place within the angle $\angle WVX$, it is at distance at least 1 from U. Hence we may apply Lemma 4 to cut away the isosceles triangle in the neighborhood of vertex U, to argue that the perimeter of T is at most $P(\triangle UVW) - 1$.

b) The obtuse angle of $\triangle UVX$ is at X and the separating triangle $\triangle VXZ$ incident with VX obtained by by Lemma 1 is inside $\triangle UVX$. As $\triangle UXV$ is thin, we get that $\angle WVX \geq \frac{\pi}{2}$. Hence all points of $\triangle VXZ$ are at distance at least 1 from W. We cut away the vertex W and obtain $P(\triangle VXZ) \leq P(\triangle UVW) - 1$.

c) The angle $\angle UXV$ is obtuse, the separating triangle $\triangle VXZ$ is outside $\triangle UVX$ and the angle $\angle VXZ$ is obtuse. We apply Lemma 3 to get that $\angle WVZ$ is obtuse—the case of $\angle UVZ$ being obtuse is excluded as this angle is composed from acute angles of two thin triangles: $\triangle UVX$ and $\triangle VXZ$. Then we cut away the vertex W as in the previous case and obtain the claimed result.

d) The angle $\angle UXV$ is obtuse, the separating triangle $\triangle VXZ$ is outside $\triangle UVX$ and the angle $\angle VXZ$ is acute. Now cut away the vertex U (as $|UX| \geq 1$), and get $P(\triangle VXZ) \leq P(\triangle UVW) - 1$.

Note that *only* when Case a) occurred, we used the existence of two vertices of label i within the separating triangle $\triangle UVW$. If Y was not present, we would have to discuss the case that the obtuse angle of $\triangle UVX$ is at V and both separating triangles of level $i+1$ are inside $\triangle UVX$. For such a case it is possible to find a configuration where Lemma 4 cannot be immediately applied, see Fig. 6.

Lemma 6. *Let G_k have a planar straight-line drawing with edge length at least 1 and let $\triangle UVW$ be a thin separating triangle of level $i \leq k - 2$. Assume that the edge UV is of level $i - 1$ and that it is not incident to the obtuse angle of $\triangle UVW$. Then $\triangle UVW$ contains a thin separating triangle T of level at most $i + 2$ whose perimeter satisfies $P(T) \leq P(\triangle UVW) - 1$.*

Proof. Similarly to the previous lemma, let X be one of the two vertices of level i simplicial to the edge UV inside the triangle $\triangle UVW$, see Fig. 7. By Lemma 1 we construct a separating triangle $\triangle UXZ$ incident with the edge UX.

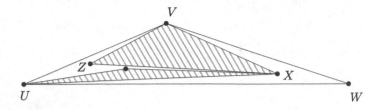

Fig. 6. The case that avoids cutting. Note that X could be arbitrary close to W and Z to U.

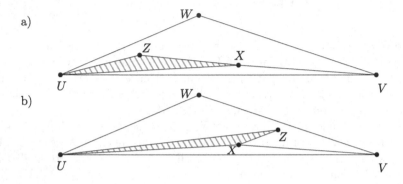

Fig. 7. Case analysis for Lemma 6.

a) If the angle $\angle UXZ$ is acute, then we cut away V and apply Lemma 4 to obtain $P(\triangle UXZ) \leq P(\triangle UVW) - 1$.

b) If the angle $\angle UXZ$ is obtuse, then we apply Lemma 5 for the triangle $\triangle UXZ$ to find a suitable separating triangle T of level $i + 2$ within $\triangle UXZ$.

Corollary 2. *For any $r > 1$, $k \geq 1$, $l \geq 0$ and any planar straight-line drawing of G_{k+l} with edge length at least 1 it holds: If the drawing contains a thin separating triangle of level $k \geq 1$, then it has a triangle of perimeter at most $2r + \frac{1}{4} - \lfloor \frac{l}{2} \rfloor$.*

Proof. Denote by Δ_0 the thin triangle of level k in the drawing of G_{k+l}. Since all edges have length at most r, any thin triangle it could be drawn inside a rectangle $r \times \frac{1}{8}$, hence it has perimeter at most $2r + \frac{1}{4}$.

We involve Lemmas 5 and 6, to find in the drawing of G_{k+l} a sequence of nested separating triangles of length at least $l + 1$ with decreasing perimeters.

We argue that the sequence can be chosen such that for any $i \in \{1, 2, \ldots, \lfloor \frac{l}{2} \rfloor\}$: $P(\Delta_{2i}) \leq P(\Delta_{2i-2}) - 1 \leq P(\Delta_0) - i$. We distinguish two cases whether the edge of level $2i - 3$ in Δ_{2i-2} is incident to the obtuse angle of Δ_{2i-2} or not:

- In the first case we apply Lemma 5 to get $P(\Delta_{2i-1}) \leq P(\Delta_{2i-2}) - 1$. As Δ_{2i} is inside Δ_{2i-1}, we get $P(\Delta_{2i}) \leq P(\Delta_{2i-2}) - 1$.
- Otherwise we apply Lemma 6 to derive $P(\Delta_{2i}) \leq P(\Delta_{2i-2}) - 1$ directly.

Now we combine the two parts together to prove Theorem 1.

Proof (of Theorem 1). For given r we choose $k = \lceil 2 + 2\log_2 r \rceil$ and consider the graph G_{k+4r}. Assume for a contradiction that G_{k+4r} allows a drawing of edge-length ratio at most r. Up to an appropriate scaling, we assume that the longest edge of such drawing has length r and hence the shortest has length at least 1.

In the drawing of the graph G_{k+4r} consider a sequence of separating triangles $\Delta_1, \ldots, \Delta_{k+4r}$ where $\Delta_1, \ldots, \Delta_k$ are chosen as shown in Corollary 1.

By Corollary 1, the triangle Δ_k is thin, so we can extend the sequence with $\Delta_k, \ldots, \Delta_{k+4r}$ according to Corollary 2.

By Corollary 2, $P(\Delta_{k+4r}) \leq 2r + \frac{1}{4} - 2r = \frac{1}{4}$, a contradiction to the assumption that all triangles of G_{k+4r} have sides of length at least one.

Note that the graph G_{k+4r} has $O^*\big((10^4)^r\big)$ vertices and edges, as in each iteration we add 10 edges of level i per every edge of level $i-1$. The dependency between the edge-length ratio and the number of vertices could be rephrased as follows:

Corollary 3. *The edge-length ratio over the class of n-vertex 2-trees is $\Omega(\log n)$.*

We recall that Borrazzo and Frati prove that every partial 2-tree with n vertices admits a planar straight-line drawing whose edge-length ratio is in $O(n^{0.695})$ [3, Corollary 1].

4 Local Edge-Length Ratio of 2-Trees

The aesthetic criterion studied in the previous section took into account any pair of edges. By our construction of nested triangles, it might happen that two edges attaining the maximum length ratio are far in the graph distance (in the Euclidean distance they are close as the triangles are nested). This observation leads us to the question, whether 2-trees allow drawings where the length ratio of any two adjacent edges could be bounded by a constant. For this purpose we define the local variant of the edge-length ratio as follows:

The *local edge-length ratio* of a planar straight-line drawing of a graph G is the maximum ratio between the lengths of two adjacent edges (sharing a common vertex) of the drawing.

Definition 2. *The* local edge-length ratio $\rho_l(G)$ *of a planar graph G is the infimum local edge-length ratio taken over all planar straight-line drawings of G.*

$$\rho_l(G) = \inf_{drawing\ of\ G} \max_{UV,VW \in E_G} \frac{|UV|}{|VW|}$$

Observe that the local edge-length ratio $\rho_l(G)$ is by definition bounded by the global edge-length ratio $\rho(G)$. In particular, every outerplanar graph G allows a drawing witnessing $\rho_l(G) \leq 2$ [10]. We extend this positive result to the class of all 2-trees with a slightly increased bound on the ratio.

Theorem 2. *The local edge-length ratio of any n-vertex 2-tree G is $\rho_l(G) \leq 4$. Also, for any arbitrarily small positive constant ε, a planar straight-line drawing of G with local edge-length ratio at most $4 + \varepsilon$ can be computed in $O(n)$ time assuming the real RAM model of computation.*

The proof of Theorem 2 is based on a construction that provides a straight-line drawing of local edge-length ratio $4 + \varepsilon$ for any given 2-tree G and any $\varepsilon > 0$.

We use a breadth first search (BFS) and and decompose V_G into layers based on the distance from the initial edge e of the recursive definition of the 2-tree. Each such layer $L_i = \{u : \text{dist}(u, e) = i\}$ is a forest, see Fig. 8a).

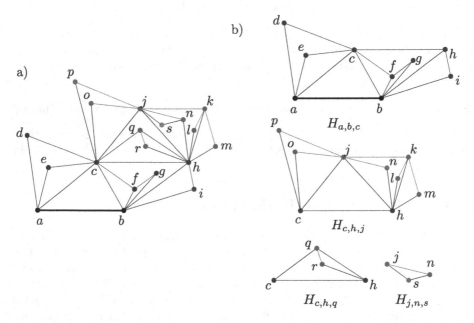

Fig. 8. a) A decomposition of a 2-tree G into layers: black L_0 (the initial edge), blue L_1, red L_2, and green L_3; b) The tree components of G (Color figure online)

Moreover, for every component C of $L_i, i \geq 1$ we may due to the definition of a 2-tree identify a unique vertex $w \in C$ and two its neighbors $u, v \in L_{i-1}$, as w is the first vertex of C inserted into G, and in the time of its insertion it was simplicial to the edge uv. We call the subgraph of G induced by $C \cup \{u, v\}$ a *tree-component* rooted in u, v and denote it by $H_{u,v,w}$, see Fig. 8b). Observe that each tree-component of itself is a 2-tree. Moreover the vertices of $H_{u,v,w}$ distinct from u, v and w can be partitioned into two disjoint sets: those adjacent to u and those adjacent to v.

Note that BFS can be executed in $O(n)$ time for a planar graph with n vertices. This procedure can be extended in a straightforward way to determine the tree-components in $O(n)$ time—on each vertex we spend additional constant time to identify the tree component it belongs.

For a line segment AB, let $\overline{AB} = AB \setminus \{A, B\}$ denote for the segment AB without its endpoints.

Definition 3. *Let UV be an edge of a planar straight-line drawing of G on at least three vertices. The* vacant region *for UV is the intersection of all open half-planes determined by all pairs of vertices such that these half-planes contain $\overline{UV}$.*

For example, Fig. 9 shows the vacant region for an edge UV in a planar straight-line drawing of a 2-tree. Note that, by the definition, the vacant region for UV is an open convex set with U and V on the boundary.

Fig. 9. The filled gray region is the vacant region of the edge UV.

We proceed to the main technical step of our construction.

Lemma 7. *Let $H_{X,Y,Z}$ be a tree-component of a 2-tree G. For any $\delta > 0$, any open convex set S and any two points on the boundary of S, the graph $H_{X,Y,Z}$ can be drawn with the local edge-length ratio at most $2 + \delta$ such that vertices X, Y are placed on the chosen two points, the rest of the drawing of $H_{X,Y,Z}$ is inside S, and XY is the longest edge of the drawing.*

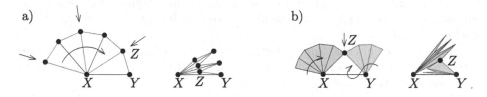

Fig. 10. Folding a path in the tree component $H_{X,Y,Z}$. The arrows indicate the vertex movement. (Color figure online)

Instead of the proof of Lemma 7, that is omitted due to space restrictions, we provide a very brief idea of the construction. Observe that in the case that in the case when the tree component $H_{X,Y,Z}$ is a fan centered at X, then it can be folded like an umbrella into the vacant region of XY as depicted in Fig. 10a).

In the folded drawing the red edges have the same length upto an additive factor δ, while the blue are twice longer (again upto $+\delta$).

Analogously, if the vertices adjacent to X in $H_{X,Y,Z}$ induce a path and the same for the neighbors of Y, then these two paths can be folded from both sides of XY inside its vacant region, see Fig. 10b). By much more technically involved argument it can be shown that the whole branch of a tree can be folded into the area near the first edge of the branch.

Proof (of Theorem 2). When $G = K_2$, then it has $\rho_l(K_2) = 1$, by Definition 2 (note that U and W need not to be distinct.) Otherwise we proceed by induction on the number of tree components of G.

For any $\varepsilon \in (0, 1)$, let $\delta = \frac{\varepsilon}{3}$. The induction hypothesis we aim to prove is:

Claim. Any 2-tree G allows a drawing with local edge-length ratio at most $4+\varepsilon$, where each tree component $H_{X,Y,Z}$ is drawn with local edge length ratio at most $2 + \delta$ and XY is the longest edge of the drawing of $H_{X,Y,Z}$.

For the base of the induction G consists of a single tree component $H_{X,Y,Z}$, where XY is the initial edge of construction of G as a 2-tree. We choose any open convex set S and two points X, Y on its boundary and apply Lemma 7.

For the induction step assume that $H_{X,Y,Z}$ is a tree component of G, where Z belongs to the highest possible level. The graph $G' = G \setminus (H_{X,Y,Z} \setminus \{X, Y\})$ (i.e. when we remove from G the component of L_i containing the vertex Z) is a 2-tree, since we may create G' as a 2-tree by the same order of insertions as is used for G, only restricted to the vertices of G'. By induction hypothesis G' allows a drawing with local edge-length ratio at most $4 + \varepsilon$.

In this drawing we identify the vacant region S for XY and involve Lemma 7 to extend the drawing of G' to the entire G. The only vertices common to G' and $H_{X,Y,Z}$ are X and Y, hence we shall argue that edges incident with X or Y have edge-length ratio at most $4 + \varepsilon$, as inside $H_{X,Y,Z}$ the ratio is at most $2 + \delta < 4 + \varepsilon$ by Lemma 7.

By the construction of the 2-tree, the edge XY may belong to several tree components rooted in X, Y, where it is the longest edge, but only to a single tree-component rotted in the vertices of the preceding level. Consequently, the edge-length ratio of any two edges incident with X or with Y is at most $(2+\delta)^2 = 4 + 2\delta + \delta^2 < 4 + 3\delta = 4 + \varepsilon$.

Finally, we remark that computing the coordinates of the vertices can be executed in constant time per vertex, assuming a real RAM model of computation. It follows that the drawing of G can be computed in $O(n)$ time.

Since any graph of treewidth at most 2, in particular all series-parallel graphs, can be augmented to a 2-tree, Theorem 2 directly implies the following.

Corollary 4. *For any graph G of treewidth at most 2, it holds that $\rho_l(G) \leq 4$.*

5   Conclusions and Open Problems

This paper studied the edge-length ratio of planar straight-line drawing of partial 2-trees. It proved an $\Omega(\log n)$ lower bound on such edge-length ratio and it proved that every partial 2-tree admits a planar straight-line drawing such that the local edge-length ratio is at most $4 + \varepsilon$ for any arbitrarily small positive ε. Several questions are naturally related with our results. We conclude the paper by listing some of those that, in our opinion, are among the most interesting ones.

1. Corollary 3 of this paper gives a logarithmic lower bound while Corollary 1 of [3] gives a sub-linear upper bound on the edge-length ratio of planar straight-line drawings of partial 2-trees. We find it interesting to close the gap between the upper and lower bound.
2. Theorem 2 gives an upper bound of 4 on the local edge-length ratio of partial 2-trees. It would be interesting to establish whether such an upper bound is tight. Also, studying the local edge-length ratio of other families of planar graphs is an interesting topic.
3. The construction in Theorem 2 creates drawings where the majority of angles are very close to 0 or π radians. Hence, it would make sense to study the interplay between (local or global) edge-length ratio and angular resolution in planar straight-line drawings.

Acknowledgement. We thank all three reviewers for their positive comments and careful review, which helped improve our contribution.

References

1. Bhatt, S.N., Cosmadakis, S.S.: The complexity of minimizing wire lengths in VLSI layouts. Inf. Process. Lett. **25**(4), 263–267 (1987). https://doi.org/10.1016/0020-0190(87)90173-6
2. Blažej, V., Fiala, J., Liotta, G.: On the edge-length ratio of 2-trees. CoRR abs/1909.11152 (2019). http://arxiv.org/abs/1909.11152
3. Borrazzo, M., Frati, F.: On the planar edge-length ratio of planar graphs. JoCG **11**(1), 137–155 (2020). https://journals.carleton.ca/jocg/index.php/jocg/article/view/470
4. Cabello, S., Demaine, E.D., Rote, G.: Planar embeddings of graphs with specified edge lengths. J. Graph Algorithms Appl. **11**(1), 259–276 (2007). https://doi.org/10.7155/jgaa.00145
5. Chen, J., Jiang, A., Kanj, I.A., Xia, G., Zhang, F.: Separability and topology control of quasi unit disk graphs. Wirel. Netw. **17**(1), 53–67 (2011). https://doi.org/10.1007/s11276-010-0264-0
6. Di Battista, G., Eades, P., Tamassia, R., Tollis, I.G.: Graph Drawing: Algorithms for the Visualization of Graphs. Prentice-Hall, Upper Saddle River (1999)
7. Eades, P., Wormald, N.C.: Fixed edge-length graph drawing is NP-hard. Discrete Appl. Math. **28**(2), 111–134 (1990). https://doi.org/10.1016/0166-218X(90)90110-X
8. Giacomo, E.D., Liotta, G., Tamassia, R.: Graph drawing. In: Goodman, J.E., O'Rourke, J., Tóth, C.D. (eds.) Handbook of Discrete and Computational Geometry, pp. 247–284. CRC Press LLC, Boca Raton (2017)

9. Hoffmann, M., van Kreveld, M.J., Kusters, V., Rote, G.: Quality ratios of measures for graph drawing styles. In: Proceedings of the 26th Canadian Conference on Computational Geometry, CCCG 2014, Halifax, Nova Scotia, Canada, 2014. Carleton University, Ottawa, Canada (2014). http://www.cccg.ca/proceedings/2014/papers/paper05.pdf
10. Lazard, S., Lenhart, W.J., Liotta, G.: On the edge-length ratio of outerplanar graphs. Theor. Comput. Sci. **770**, 88–94 (2019). https://doi.org/10.1016/j.tcs.2018.10.002
11. Nishizeki, T., Rahman, M.S.: Planar graph drawing. In: Lecture Notes Series on Computing, vol. 12. World Scientific (2004). https://doi.org/10.1142/5648
12. Tamassia, R. (ed.): Handbook on Graph Drawing and Visualization. Chapman and Hall/CRC, Boca Raton (2013). https://www.crcpress.com/Handbook-of-Graph-Drawing-and-Visualization/Tamassia/9781584884125

HOTVis: Higher-Order Time-Aware Visualisation of Dynamic Graphs

Vincenzo Perri[1]([✉])(iD) and Ingo Scholtes[1,2](iD)

[1] Data Analytics Group, Department of Informatics (IfI), University of Zurich,
Zürich, Switzerland
perri@ifi.uzh.ch
[2] Chair of Data Analytics, Faculty of Mathematics and Natural Sciences,
University of Wuppertal, Wuppertal, Germany

Abstract. Network visualisation techniques are important tools for the exploratory analysis of complex systems. While these methods are regularly applied to visualise data on complex networks, we increasingly have access to time series data that can be modelled as temporal networks or dynamic graphs. In dynamic graphs, the temporal ordering of time-stamped edges determines the *causal topology* of a system, i.e., which nodes can, directly *and indirectly*, influence each other via a so-called *causal path*. This causal topology is crucial to understand dynamical processes, assess the role of nodes, or detect clusters. However, we lack graph drawing techniques that incorporate this information into static visualisations. Addressing this gap, we present a novel dynamic graph visualisation algorithm that utilises higher-order graphical models of causal paths in time series data to compute time-aware static graph visualisations. These visualisations combine the simplicity and interpretability of static graphs with a time-aware layout algorithm that highlights patterns in the causal topology that result from the temporal dynamics of edges.

1 Introduction

Network visualisation techniques are a cornerstone in the exploratory analysis of data on complex systems. They help us recognize patterns in relational data on complex networks, such as, e.g., clusters or groups of well-connected nodes, hierarchical and core-periphery structures, or highly important nodes [7,18,24]. However, apart from knowing which elements are connected, we increasingly have information on when and in which chronological order connections occurred. Sources of time-stamped data include social interactions, click stream data in the web, financial transactions, passenger itineraries in transportation networks, or gene regulatory interactions [16]. Despite these applications, visualising *time-stamped network data* is still a challenge [2,16]. Common approaches

Electronic supplementary material The online version of this chapter (https:// doi.org/10.1007/978-3-030-68766-3_8) contains supplementary material, which is available to authorized users.

D. Auber and P. Valtr (Eds.): GD 2020, LNCS 12590, pp. 99–114, 2020.
https://doi.org/10.1007/978-3-030-68766-3_8

use sequences of snapshots, where each snapshot is a graph of the connections active in a time interval or at a point in time, to animate the evolution of the topology. Such animations can help us gain a high-level understanding of temporal activities in dynamic graphs. However, they are complex and cognitively demanding, which makes it hard to recognise patterns that determine how nodes influence each other over time. Moreover the application of graph drawing algorithms to temporal snapshot necessitates a coarse graining of time. This introduces a major issue: we lose information on the chronological ordering of links that determines so-called *time-respecting* or *causal paths* [16,19]. In a nutshell, for two time-stamped edges $(a, b; t_1)$ and $(b, c; t_2)$ that occur at times t_1 and t_2, a *causal path* $\overrightarrow{abc}$ from node a via node b to node c can only exist if edge (a, b) occurs before (b, c), i.e. if $t_1 < t_2$. If the ordering of edges is reversed, such a causal path does not exist, i.e. node a cannot influence node c via b, neither directly nor indirectly. This simple example highlights how the temporal ordering of edges gives rise to *causal topologies*. While two edges (a, b) and (b, c) in a static graph imply that a (transitive) path $\overrightarrow{abc}$ exists, the temporal ordering of edges in dynamic graphs can invalidate this assumption. This has important implications for the modelling of epidemic processes, random walk and diffusion dynamics, centrality measures used to rank nodes, or clustering techniques. It calls for a new class of higher-order network modelling, analysis, and visualisation techniques [21]. In a recent review on state-of-the-art temporal network analysis [16], Holme highlights a lack of visualisation techniques that (i) go beyond cognitively demanding animations, and (ii) consider the complex topology of causal paths in high-resolution time series data: *"[...] temporal networks lack the intuitive visual component of static networks. Probably this is a fundamental property that cannot be completely altered, but there should be better visualization methods than we have now. Highest on our wish list is a method that both simplifies some structures and keeps (at least some) of the time-respecting paths (maybe at the cost of not having time on the abscissa)."* [16], p. 23.

Addressing this gap, we develop a visualisation algorithm that incorporates information on causal paths in dynamic graphs into simple (static) visualisations. Our contributions are: (i) we highlight a lack of time-aware graph visualisation techniques that respects the *causal topology* resulting from the ordering of edges in high-resolution data on dynamic graphs; (ii) we develop a dynamic graph drawing algorithm that generalises force-directed layouts to high-dimensional De Bruijn graph models of causal paths [5,30]; (iii) we assess the quality of our visualisations in synthetic and empirical time series data, and show that they help to detect temporal clusters invisible in static visualisations, and identify important vertices with high temporal centrality; (iv) we provide an Open Source `python` implementation of our algorithm [6]. Focusing on time-aware static visualisations that highlight temporal patterns neglected by existing techniques, we take a new approach to dynamic graph drawing. Considering recent works on learning optimal higher-order graph models from rich time series data [30], our work opens perspectives to combine machine learning and visual data mining.

2   Preliminaries and Related Work

We define a dynamic graph as a tuple $G^{(t)} = (V, E^{(t)})$, where V is a set of vertices and $E^{(t)}$ is a set of time-stamped edges $E^{(t)} \subseteq V \times V \times \mathbb{N}$. We assume that $(v, w; t) \in E^{(t)}$ denotes that a directed edge between source vertex v and target vertex w occurred *instantaneously* at discrete time $t \in \mathbb{N}$. We say that a (static) graph $G = (V, E)$ is the *time-aggregated* graph corresponding to a dynamic graph $G^{(t)}$ iff $(v, w) \in E \leftrightarrow \exists t \in \mathbb{N} : (v, w; t) \in E^{(t)}$. We further assume that the edge weights $w : E \rightarrow \mathbb{N}$ of such a time-aggregated graph capture the number of times edges have been active in the corresponding dynamic graph, i.e. we define $w(v, w) := |\{t \in \mathbb{N} : (v, w; t) \in E^{(t)}\}|$. A simple example for a dynamic graph with eight time-stamped edges and five nodes is shown in Fig. 1 (a). A key concept in the study of dynamic graphs is that of a *time-respecting* or *causal path* [16, 19]. For a dynamic graph $G^{(t)} = (V, E^{(t)})$ we call a sequence $(v_0, v_1; t_0), (v_1, v_2; t_1), \dots, (v_{l-1}, v_l; t_l)$ of time-stamped edges a *causal path* $p = \overrightarrow{v_0 v_1 v_2 \dots v_l}$ of length l from vertex v_0 to v_l iff (i) $(v_i, v_{i+1}; t_i) \in E^{(t)}$, and (ii) $0 < t_{i+1} - t_i \leq \delta$ holds for $i \in \{0, 1, \dots, l-1\}$. We thus define the length of causal paths as the number of edges that they traverse, which implies that time-stamped edges are trivial causal paths of length one. In this definition, the condition $0 < t_{i+1} - t_i$ ensures that the sequence of time-stamped edges respects the "arrow of time", while the condition $t_{i+1} - t_i \leq \delta$ ensures that to form a causal path two time-stamped edges must occur within time δ of each other.

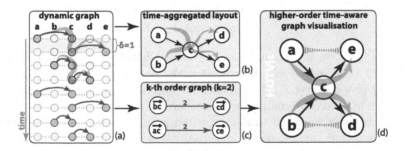

Fig. 1. Information on *causal paths* (coloured arrows) contained in dynamic graphs (a) is discarded by standard time-aggregated visualisations (b). HOTVis uses higher-order graph models of causal paths (c) to produce time-aware, static visualisations (d) that highlight the causal topology of dynamic graphs.

We note that the existence of a causal path $\overrightarrow{v_0 \dots v_l}$ is a necessary condition for a vertex v_0 in a dynamic graph to *causally influence* another vertex v_l. We further observe that each causal path in $G^{(t)}$ necessarily implies that the same path exists in the time-aggregated graph G. Conversely, the existence of a path in graph G corresponding to the dynamic graph $G^{(t)}$ does *not* imply that the corresponding causal path exists in $G^{(t)}$. The example in Fig. 1 (a) illustrates how the chronological order of edges can break transitivity in a dynamic graph.

Here, the timing and ordering of time-stamped edges implies that only two of the four theoretically possible *causal paths* of length two exist. Hence, despite the presence of corresponding paths in the static topology, vertices a and b cannot indirectly influence d and e via causal paths $\overrightarrow{acd}$ and $\overrightarrow{bce}$ respectively.

To address the issue that time-aggregated graph representations discard information on causal paths, we utilize higher-order graph models that capture how the chronological ordering of edges influences causal paths [21]. For a given dynamic graph $G^{(t)}$ and order $k \geq 1$ we define a *higher-order graph* $G^{(k)}$ as tuple $G^{(k)} = (V^{(k)}, E^{(k)})$ of higher-order vertices $V^{(k)} \subseteq V^k$ and higher-order edges $E^{(k)} \subseteq V^{(k)} \times V^{(k)}$.

Each higher-order vertex $v =: \overrightarrow{v_0 v_1 \ldots v_k} \in V^{(k)}$ is an ordered tuple of k vertices $v_i \in V$ in the dynamic graph $G^{(t)}$ that also satisfies the conditions of a causal path. Higher-order edges are constructed using the iterative line graph construction of high-dimensional De Bruijn graphs [5]. The construction of a De Bruijn graph of order k restricts edges to connect higher-order vertices that overlap in $k - 1$ vertices, i.e. we require:

$$(\overrightarrow{v_0 v_1 \ldots v_{k-1} v_k}, \overrightarrow{w_0 w_1 \ldots w_{k-1} w_k}) \in E^{(k)} \Rightarrow v_i = w_{i-1}(i = 1, \ldots, k)$$

Utilising the modelling framework introduced in [30] we use (weighted) higher-order edges of a k-th order graph $G^{(k)}$ to represent the frequency of causal paths of length k in a dynamic graph, i.e we define weights $w : E^{(k)} \to \mathbb{N}$ as

$$w(\overrightarrow{v_0 \ldots v_{k-1}}, \overrightarrow{v_1 \ldots v_k}) := \{|(t_0, \ldots, t_{k-1}) : (e; t_i) \in E^{(t)}$$
$$\text{from causal path } \overrightarrow{v_0 \ldots v_k})\}|$$

Figure 1(c) shows an example for a (trivial) higher-order graph model of order $k = 2$ that represents the causal paths $\overrightarrow{bcd}$ and $\overrightarrow{ace}$ of length two in the dynamic graph in Fig. 1 (a). Higher-order graphs generalise time-aggregated graph representations of dynamic graphs, where for $k = 1$ we have $V^{(1)} = V$ and $E^{(1)} = E$. Hence, a weighted time-aggregated graph is a first-order model of a dynamic graph that counts edges, i.e. causal paths of length one. For $k > 1$, we obtain higher-order models that capture both the topology and the chronological ordering of time-stamped edges in a dynamic graph, where the second-order model is the simplest model that is sensitive to the timing and ordering of edges.

Related Work. Having illustrated the effects that are due to the arrow of time, we review works on dynamic graph drawing. We only present methods relevant to our work, referring the reader to [2] for a detailed review.

A natural approach to visualise time series data on graphs are animated visualisations that show the temporal evolution of vertices and/or edges. To generate such animations, we need to create a sequence of graphs, where each graph is a *static snapshot* of the vertices and edges at one point in time. An independent visualisation of such snapshots by means of standard graph layout algorithms is likely to result in animations that make it difficult to associate structures in subsequent frames, a problem often framed as maintaining the user's "mental map" [1,29]. A large number of works focuses on optimising graph layouts across multiple snapshots [8,10,14,20,22], or in generating smooth transitions [11,23] that minimise the cognitive effort required to trace time-varying vertices, edges, or clusters through subsequent snapshots.

Despite these improvements, identifying patterns in animations remains challenging. Also, their use is limited since animations cannot be embedded in scholarly articles, books, or posters. Addressing these issues, a second line of research focuses on methods to visualise dynamic graphs in terms of *timeline representations*, which map the time dimension of dynamic graphs to a (static) spatial dimension. Examples includes directed acyclic time-unfolded graph representations of dynamic graphs [19,28], *time arc* or *time radar trees* [4,15], sequences of layered adjacencies [36], stacked 3D representations where consecutive time slices are arranged along a third dimension [10], and circular representations [9]. However, timelines are limited to a small number of time stamps, which hinders their application to data with high temporal resolution (e.g. seconds or even milliseconds). The application of static graph drawing algorithms to such data requires a coarse-graining of time into *time slices*, such that each time slice gives rise to a graph snapshot that can be visualised using, e.g., force-directed layout algorithms. As pointed out in [32], this coarse-graining of time leads to a loss of information on temporal patterns. Some recent works have thus explored

---

Algorithm 1. HOTVis: Higher-order time-aware layout with max. order K

1: **procedure** HOTVis($G^{(t)}, K, N, \delta, \alpha_2, \ldots, \alpha_K$)
2:      $A, \text{Pos} = dict(), \text{Temp} = t_0$
3:      **for** $k \in \text{range}(1, K)$ **do**
4:          ▷ superimpose attractive forces
5:          $G^{(k)} = \text{HigherOrderGraph}(G^{(t)}, \delta, k)$
6:          **for** $(\overrightarrow{v_0 \ldots v_{k-1}}, \overrightarrow{v_1 \ldots v_k}) \in E^{(k)}$ **do**
7:              **if** $(v_0, v_k) \in A$ **then**
8:                  $A[v_0, v_k] = A[v_0, v_k] + \alpha_k \cdot w(\overrightarrow{v_0 \ldots v_{k-1}}, \overrightarrow{v_1 \ldots v_k})$
9:              **else**
10:                  $A[v_0, v_k] = \alpha_k \cdot w(\overrightarrow{v_0 \ldots v_{k-1}}, \overrightarrow{v_1 \ldots v_k})$
11:      **for** $i \in \text{range}(N)$ **do**
12:          ▷ apply many-body simulation [12]
13:          **for** $v \in V$ **do**
14:              $\Theta = 0$
15:              **for** $w \in V, w \neq v$ **do**
16:                  $\Delta = \text{Pos}[w] - \text{P}[v]$
17:                  $\Theta = \Theta - \Delta/|\Delta| \cdot k^2/|\Delta|$
18:              **for** $(v, w) \in A$ **do**
19:                  $\Delta = \text{Pos}[w] - \text{Pos}[v]$
20:                  $\Theta = \Theta + \Delta/|\Delta| \cdot A[v, w] \cdot |\Delta|^2/k$
21:              $P[v] = P[v] + \Theta/|\Theta| \cdot \min(|\Theta|, \text{Temp})$
22:          $\text{Temp} = \text{cool}(\text{Temp})$
         return Pos

---

dynamic graph visualisations that are not based on time slices, e.g. using a continuous space-time cube [32] or using visualisations that highlight higher-order dependencies at the level of individual nodes [34]. To the best of our knowledge none of the existing methods has explicitly addressed static representations of dynamic graphs that retain information on which nodes can influence each other via causal paths, which is the motivation for our work.

3   Higher-Order Time-Aware Network Visualisation

To address the research gap outlined in Sect. 2, we propose an algorithm to generate higher-order time-aware visualisations (HOTVis). It captures the influence of the temporal dimension of a graph on its causal topology, i.e. which vertices can influence each other via causal paths, generalising the force-directed layout algorithm introduced in [12] to high-dimensional graphs.

Force-directed layouts optimally position vertices in a Euclidean space by means of a many-body simulation. Attractive forces along edges move connected nodes close to each other while a repulsive force between all nodes separates them. Simulating these forces until an equilibrium state is reached leads to graph layouts that highlight topological structures and symmetries [7]. HOTVis generalises the attractive forces of force-directed layouts so that they capture the topology of causal paths in time-stamped data. In particular, our algorithm *superimposes* attractive forces that act between the endpoints of edges in multiple higher-order graphs up to a configurable maximum order K. Figure 1(d) illustrates this idea based on the edges in the second-order model shown in Fig. 1(c). The additional attractive forces between vertex pairs a, e and b, d (coloured lines in Fig. 1(d)) change the positioning of vertices such that those vertices that can causally influence each other are positioned in proximity.

The pseudocode of HOTVis is shown in Algorithm 1. It takes a dynamic graph $G^{(t)}$, a maximum time difference δ used to define causal paths, a maximum order K, a number of iterations N, and parameters α_k controlling the influence of paths of length k on the layout. The algorithm works in two phases. The first phase generates higher-order graphs $G^{(k)}$ up to order K (lines 4–6). For each edge $(\overrightarrow{v_0 \ldots v_{k-1}}, \overrightarrow{v_1 \ldots v_k})$ in $G^{(k)}$, an attractive force is added between vertices v_0 and v_k that can influence each other via causal path $\overrightarrow{v_0 v_1 \ldots v_k}$ (lines 7–10). Its strength depends on (i) the frequency of causal paths, and (ii) a parameter α_k that controls the influence of causal paths of length k on the generated time-aware layout. For $\alpha_2 = \ldots = \alpha_K = 0$ we obtain a standard force-directed (first-order) layout in which time is ignored. For $\alpha_k > 0$ and $k > 1$ vertex positions are additionally influenced by the ordering of time-stamped edges. The second phase of HOTVis (lines 11–22), uses the many-body simulation proposed in [12] to simulate repulsive and superimposed attractive forces between nodes. The algorithm returns a dictionary of vertex positions that produces a static, time-aware visualisation. The computational complexity of the algorithm is given by the sum of the computational complexities of two phases, the first consisting in the generation of k−th order graph models for $k = 2, \ldots, K$, the second in the layouting of the nodes. The complexity of the first phase has a non-trivial dependency on the temporal distribution of time-stamped edges of the dynamical graph and is further discussed in the full version [25]. The computational complexity of the second phase corresponds to that of the algorithm proposed in [12].

Illustration in Synthetic Example. A demonstration of a time-aware visualisation in synthetic data on a temporal network with $K = 2$ is shown in Fig. 2. The data was generated using a stochastic model creating time-stamped edge sequences with temporal correlations that lead to an over-representation of causal paths of length two that (indirectly) connect pairs of vertices in three clusters (coloured vertices in Fig. 2). Different from what one would expect based on the time-aggregated topology, the chronological ordering of time-stamped edges leads to an under-representation of causal paths between vertices with *different colours*. Hence, we obtain *temporal clusters in the causal topology*, where vertices in the same cluster can indirectly influence each other via causal paths more than vertices in different clusters. Details of the model are included in the full version [25] and in the Zenodo package [26]. *Temporal clusters* in the causal topology are visible only in a second-order time-aware visualisation that superimposes attractive forces calculated in both the first and the second-order graph (middle panel Fig. 2). These clusters are not visible in the static graph layout shown in Fig. 2 (left panel). To demonstrate that clusters are solely due to the chronological ordering of time-stamped edges and the resulting causal paths, we shuffle the timestamps of edges and reapply our time-aware visualisation algorithm. The resulting layout in Fig. 2 (right panel) shows that the shuffling of timestamps destroys the cluster structure, confirming that HOTVis visualises patterns due to the ordering of edges.

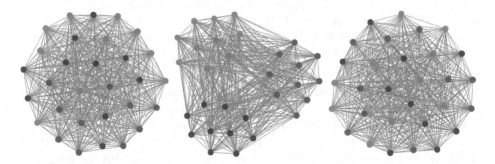

Fig. 2. Application of HOTVis to synthetic temporal network with three temporal clusters (coloured nodes). A second-order time-aware layout (middle) highlights clusters not visible in a static visualisation (left). A time-aware visualisation of the data with randomised time stamps (right) confirms that clusters are due to the ordering of edges.

4   Experimental Evaluation

Having illustrated `HOTVis` in a synthetic example, we now evaluate whether it produces better visual representations of empirical time-stamped network data. We compare `HOTVis` (at different orders) against a baseline visualisation that is generated using the Fruchterman-Reingold algorithm [12]. To quantitatively assess the "quality" of visualisations generated by `HOTVis`, we define measures that capture how well the causal topology is represented. For the following definitions, let $G^{(t)}$ be a dynamic graph that gives rise to a multi-set S of N causal paths $S = \{p_0, \ldots, p_N\}$. We also assume that Algorithm 1 assigns each vertex $v \in V$ to a position $\pi_v := Pos[v] \in \mathbb{R}^2$.

Edge Crossing (ξ). This standard measure counts the number of pairs of edges that cross each other in the visualisation. It is widely used in the evaluation of graph drawing algorithm. It rests on the idea that a large number of edge crossings ξ makes it difficult to identify which vertices are connected by edges, i.e. "high-quality" drawings minimise ξ. We efficiently calculate edge crossing based on the orientation predicate [31].

Causal Path Dispersion (σ). The *causal path dispersion* σ captures whether the sets of vertices traversed by causal paths are less spatially dispersed than expected based on the spatial distribution of vertices. It intuitively captures whether vertices that can influence each other directly and indirectly are positioned in close proximity. For this, we consider a multi-set S of causal paths p with cardinality $N := |S|$ that traverse a graph with vertices V. We define σ as

$$\sigma = \frac{\sum_{p \in S} \sum_{u_i \in p} \|Pos[u_i] - B(p)\| \cdot |V|}{N \cdot \sum_{v_i \in V} \|Pos[v_i] - B(V)\|} \in [0, 1],$$

where $B : 2^V \to \mathbb{R}^2$ is a function that returns the barycentre of vertex positions $P[v]$ for vertex multi-set V. For $\sigma \approx 1$ the spatial distribution of nodes traversed by causal paths is the same as for nodes traversed by random paths in the network topology. Values of $\sigma < 1$ highlight that vertices connected via causal paths occupy a smaller area than expected at random.

Closeness Eccentricity (Δ). In force-directed layouts the distance of a vertex from the barycentre of the visualisation is correlated with the vertex' *closeness centrality*, defined as the inverse of the sum of shortest path lengths between the vertex and all other vertices [3]. It is thus interesting to study whether vertex positions in our time-aware visualisation are correlated with the *temporal closeness centrality* of a vertex v. For a set S of causal paths p, we define this as

$$CC(v) := \sum_{w \neq v \in V} \frac{\sum_{p \in S} \delta_w(p) \delta_v(p)}{\sum_{p \in S, w \in p} \text{dist}(v, w; p)}.$$

Here, $\text{dist}(v, w; p)$ denotes the (topological) distance between vertices v and w via causal path p and $\delta_v(p)$ is one if path p traverses vertex v and zero otherwise.

With $T_\gamma = \{u_1, u_2, \ldots, u_n\}$ being the set of n nodes whose temporal closeness centrality is in the γ upper percentile, we define closeness eccentricity $\Delta(\gamma)$ as:

$$\Delta(\gamma) := \frac{\sum_{u_i \in T_\gamma} ||Pos[u_i] - B(V)|| \cdot |V|}{|T_\gamma| \cdot \sum_{v \in V} ||Pos[v] - B(V)||} \in [0, 1]$$

$\Delta(\gamma)$ captures whether the n vertices with highest temporal closeness centrality are closer to ($\Delta < 1$) or farther away ($\Delta > 1$) from the barycentre of the visualisation than we would expect at random.

4.1 Experimental Results

We now report the results of our experimental evaluation of HOTVis in (i) the synthetically generated data with temporal clusters introduced above, and (ii) five time-stamped data sets on real complex networks. The five data sets fall into two classes, highlighting different types of data in which our algorithm can be used. The first class captures paths or trajectories in a networked system. Here we utilise two data sets (i) flights, which captures $280k$ passenger itineraries between 175 US airports recorded in 2001, and (ii) tube, which contains 4.2 million passenger trips in the London metro in 2014. Details on those data are available on Zenodo [26]. The second class consists of time-stamped data on social interactions, in which we can calculate *causal paths* as defined in Sect. 2. Here we used three data sets from the Sociopatterns collaboration, namely (i) hospital, which captures 32,424 time-stamped proximity events between 75 patients, medical and administrative staff in a hospital recorded over a period of five days [35], (ii) workplace, which consists of 9,827 face-to-face interactions between 92 company employees recorded in an office building over a period of ten days [13] and (iii) school which contains 77,602 proximity events between 242 individuals (232 children and 10 teachers) [33].

We now tests whether we can "learn" consistent patterns in the causal topology of the time-stamped network data from the time-aware visualisation generated by HOTVis. To assess the consistency of the patterns identified by our algorithm we use a cross-validation approach: we generate time-aware visualisations with different maximum orders K in a training sample, and then assess their quality in a validation set that we withheld from our algorithm. Thanks to the evaluation on unseen data, we can use our measures to compare the generalisability of the patterns displayed at each maximum order K. A benefit of our method is that the optimal maximum order needed to visualise the topology of causal paths can be determined using the statistical model selection techniques described in [27,30]. It provides a principled method to balance the complexity and explanatory power of the higher-order model used for our visualisation, learning a model that avoids both over- and underfitting. In the following, we evaluate our algorithm for all maximum orders $K = 1, \ldots, K_{opt} + 2$, where K_{opt} is the *optimal* order returned by the method described in [30]. All results were obtained by applying the cross-validation approach described above in 100 layout calculations for each data set and for each maximum order K. To focus on the effect of the *topology* of causal paths and rule out distortions due to skewed distributions of weights w (cf. [17]), we set the weight of all paths to a constant value of one. Moreover, to ensure that all orders have an equal influence on the overall layout, we set the parameter α_k in a higher-order model with order k to $\alpha_k := m_k^{-1}$, where m_k is the number of unique paths of length k. This ensures that the forces in each k-th order model are scaled according to the density of edges. We note that the choice of those parameters is motivated by simplicity and ease of reproducibility. In particular, it does not require sophisticated parameter tuning, which could be used to optimise the visualisation for a specific data set.

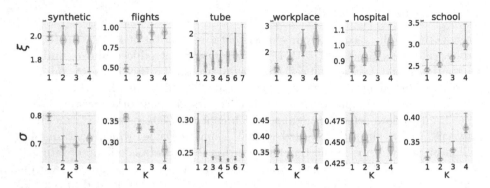

Fig. 3. Edge crossing ξ (top) and causal path dispersion σ (bottom) for a synthetic dynamic graph with three clusters (column synthetic), empirical data on flight itineraries (column flights), metro trips (tube), and time-stamped interactions between workers in an office environment (workplace), patients and hospital staff (hospital), and children and teachers in a primary school (school). All values are averages of 100 cross-validation experiments, where a time-aware layout with maximum order K (x-axis) was computed for a 70% random training set of causal paths, calculating quality measures (y-axis) in the layout for a test set of remaining 30% of causal paths.

The results of our evaluation are shown in Fig. 3. For the edge crossing ξ, in the synthetic data we find no significant change with increasing K, while the empirical data sets show significant increases as the maximum order grows. A general growth of edge crossings with increasing K is expected since, apart from the topology of *edges*, the time-aware visualisation considers the topology of causal paths. The causal path dispersion σ decreases considerably for orders $K > 1$ in all data sets, highlighting that our algorithm positions those nodes close to each other that strongly influence each other via causal paths. For a suitably chosen order K, we further observe that relatively large decreases of causal path dispersion σ (e.g. a decrease of 15% for order $K = 5$ in tube) are associated with relatively mild or insignificant increases of edge crossing ξ (e.g. no significant change for order K in tube). For those orders K, our method provides a good trade-off between a visualisation that best represents the topology of causal paths and a visualisation that best represents the static topology. On the one hand, this supports our hypothesis that HOTVis better represents the causal topology of temporal networks compared to time-aggregated (first-order) visualisations. On the other hand, this raises the issue of finding the "optimal" order K of a higher-order graph model, which can be addressed using the statistical model selection technique presented in [30]. In agreement with the results of our cross-validation, this technique yields an optimal order $K_{opt} = 5$ for tube and $K_{opt} = 2$ for the other data sets. This indicates that we can use statistical model selection to learn the optimal maximum order parameter K to be used in HOTVis.

In Fig. 4 we illustrate HOTVis in the school data set for $K = 1$ (left) and the optimal order $K = 2$ (right). Node colours indicate the membership of students in different classes. Importantly, this group structure in the data is not expressed in the topology of links (see Fig. 4 left). Consequently, a time-neglecting first-order layout places nodes in a single group, which leads to a cluttered visualisation that makes it difficult to visually detect the ground truth group structure. A second-order layout generated by HOTVis (see Fig. 4 right) better highlights group structures that are expressed in the topology of causal paths, thus leading to *temporal cluster* patterns that cannot be seen in the static topology. This example demonstrates that the mechanism by which HOTVis visualises temporal clusters—as illustrated in the synthetic example in Fig. 2— can successfully visualize group structures in empirical social networks.

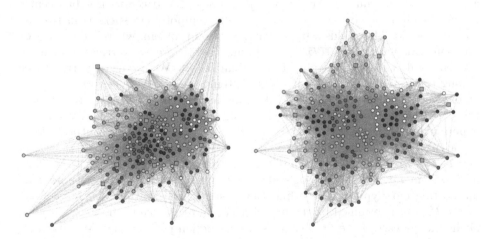

Fig. 4. Comparison between a time-neglecting (left) and time-aware (right) layout for the school data set. Node colours represent the class each student belongs to, while square markers identify teachers. The time-aware visualisation generated by HOTVis (right) positions nodes that influence each other through causal paths close to each other. HOTVis highlights ground truth groups of students in school classes. We used the default parameter $\alpha_k = m_k^{-1}$ as described in Sect. 4.

Temporal Closeness. We finally show results for closeness eccentricity Δ. We specifically test whether, similar to static force-directed layout algorithms,

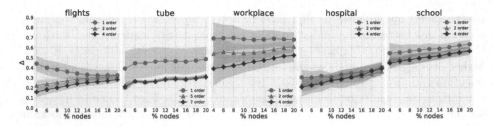

Fig. 5. Closeness eccentricity Δ (y-axis) for a varying top percentage n of vertices with highest temporal closeness (x-axis). Results are presented for $K=1$ (blue dots) $K=K_{opt}$ (green triangles), $K=K_{opt+2}$ (black pluses). Hulls indicate the 2σ interval. (Color figure online)

HOTVis places nodes with high temporal closeness in the centre of the visualisation. The results for the four empirical data sets are shown in Fig. 5 (again for 100 cross-validation experiments). In all data sets, higher values of K correspond to lower values of Δ. For tube and flights, values significantly differ with a 2σ confidence interval. This indicates that HOTVis represents the temporal closeness of vertices better than a first-order layout. To quantify our ability to visually identify vertices with high temporal centrality, we additionally ran a prediction experiment: We use the proximity of vertices to the barycentre of the visualisation to identify vertices whose temporal closeness is in the top 10% percentile. This yields a binary classification problem, where we use vertex positions calculated by HOTVis in a training set to predict vertices with highest temporal closeness centrality in a validation set. We predict a vertex to be among the top 10% vertices with highest temporal closeness centrality, if it is among the top 10% vertices with smallest distance to the barycentre. To compare the performance of different orders K we use ROC curves (100 cross-validation experiments). The results in Fig. 6 show that for $K > 1$ *HOTVis* outperforms a static (first-order) visualisation in all data sets. AUC scores for $K = 1$, $K = K_{opt}$ and $K = K_{opt} + 2$ are: 0.82, 0.95, 0.95 in flights; 0.87, 0.92, 0.92 in tube; 0.71, 0.77, 0.77 in workplace; 0.93, 0.96, 0.94 in hospital; 0.80, 0.87, 0.85 in school. Supporting our hypothesis, we find that visualisations with order $K > K_{opt}$, where K_{opt} is determined by the model selection technique from [30], only yield negligible increases (or even decreases) in prediction performance. We illustrate this in the flights dataset for $K = 1$ and $K = K_{opt} = 2$. The two layouts in Fig. 7 strongly differ in the positioning of the top 10% vertices with highest temporal closeness centrality (in red). Different from a time-neglecting layout, in HOTVis the majority of vertices classified as most central due to their proximity to the barycentre are within the top 10% values of temporal closeness centrality.

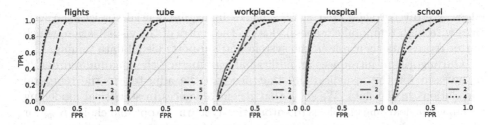

Fig. 6. ROC curves illustrate the ability to predict nodes with top 10% closeness centrality for $K = 1$ (blue dashed), $K=K_{opt}$ (green solid), $K=K_{opt+2}$ (black dotted). (Color figure online)

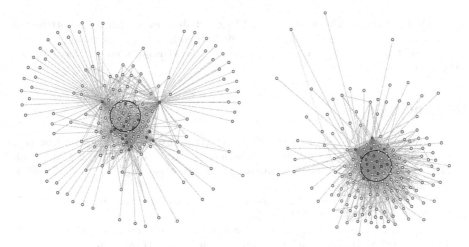

Fig. 7. Comparison between time-neglecting (left) and time-aware (right) layout for `flights`. Vertices whose temporal closeness centrality is among top 10% of values are highlighted in red. Black circles delineate the area containing 10% of vertices closest to the barycentre. The time-aware visualisation generated by `HOTVis` (right) places vertices with high temporal closeness centrality close to the barycentre, enabling us to identify temporally important vertices. We used $\alpha_k = m_k^{-1}$ as described in Sect. 4. (Color figure online)

5   Conclusion and Outlook

Despite advances in dynamic graph drawing, the visualisation of high-resolution time-stamped network data is still a challenge. Existing methods suffer from a limited ability to highlight patterns in the *causal topology of dynamic graphs*, which is determined by the interplay between its topology (i.e. *which* edges exist) and the temporal dynamics of edges (i.e. *when* time-stamped edges occur). We address this issue through `HOTVis`, a dynamic graph drawing algorithm that uses higher-order graph to produce static, time-aware visualisations. Experiments in synthetic and empirical data support our hypothesis that the resulting visualisations better highlight temporal clusters due to the chronological ordering of edges. We further show that HOTVis places highly influential vertices (i.e. vertices with high temporal closeness) close to the centre of the visualisation, which better represents their role in the system and supports visual data mining.

Our algorithm introduces an additional parameter—the maximum order K to be used for the visualisation—that needs to be adjusted to the temporal correlations in the data. We show that recent advances in statistical modelling and machine learning enable us to automatically learn the optimal choice K_{opt} for this parameter, thus turning it into a practicable method to visualise patterns in temporal data. Our work highlights a largely unexplored potential for new visual data mining techniques that combine graph drawing, higher-order network models [21], and machine learning that we seek to explore in the future.

Acknowledgements. Vincenzo Perri and Ingo Scholtes acknowledge support by the Swiss National Science Foundation, grant 176938. Ingo Scholtes acknowledges support by the project bergisch.smart.mobility, funded by the German state of North Rhine-Westphalia.

References

1. Archambault, D., Purchase, H.C.: Mental map preservation helps user orientation in dynamic graphs. In: Didimo, W., Patrignani, M. (eds.) GD 2012. LNCS, vol. 7704, pp. 475–486. Springer, Heidelberg (2013). https://doi.org/10.1007/978-3-642-36763-2_42
2. Beck, F., Burch, M., Diehl, S., Weiskopf, D.: A taxonomy and survey of dynamic graph visualization **36**(1), 133–159 (2017)
3. Brandes, U., Kenis, P., Wagner, D.: Communicating centrality in policy network drawings. IEEE Trans. Visual. Comput. Graph. **9**(2), 241–253 (2003)
4. Burch, M., et al.: Radial edge splatting for visualizing dynamic directed graphs, pp. 603–612 (2012)
5. De Bruijn, N.G.: A combinatorial problem. Koninklijke Nederlandse Akademie v. Wetenschappen **49**(49), 758–764 (1946)
6. Pathpy developers: pathpy software package (2020). https://github.com/pathpy/pathpy
7. Di Battista, G., Eades, P., Tamassia, R., Tollis, I.G.: Algorithms for drawing graphs: an annotated bibliography. Comput. Geom. **4**(5), 235–282 (1994)
8. Diehl, S., Görg, C., Kerren, A.: Preserving the mental map using foresighted layout. In: Ebert, D.S., Favre, J.M., Peikert, R. (eds.) Data Visualization 2001, pp. 175–184. Springer, Vienna (2001). https://doi.org/10.1007/978-3-7091-6215-6_19
9. van den Elzen, S., Holten, D., Blaas, J., van Wijk, J.J.: Dynamic network visualization with extended massive sequence views. IEEE Trans. Visual. Comput. Graph. **20**(8), 1087–1099 (2013)
10. Erten, C., Kobourov, S.G., Le, V., Navabi, A.: Simultaneous graph drawing: layout algorithms and visualization schemes. In: Liotta, G. (ed.) GD 2003. LNCS, vol. 2912, pp. 437–449. Springer, Heidelberg (2004). https://doi.org/10.1007/978-3-540-24595-7_41
11. Friedrich, C., Eades, P.: The marey graph animation tool demo. In: Marks, J. (ed.) GD 2000. LNCS, vol. 1984, pp. 396–406. Springer, Heidelberg (2001). https://doi.org/10.1007/3-540-44541-2_37

12. Fruchterman, T.M., Reingold, E.M.: Graph drawing by force-directed placement. Softw. Pract. Exp. **21**(11), 1129–1164 (1991)
13. Génois, M., et al.: Data on face-to-face contacts in an office building suggest a low-cost vaccination strategy based on community linkers. Netw. Sci. **3**(3), 326–347 (2015)
14. Görg, C., Birke, P., Pohl, M., Diehl, S.: Dynamic graph drawing of sequences of orthogonal and hierarchical graphs. In: Pach, J. (ed.) GD 2004. LNCS, vol. 3383, pp. 228–238. Springer, Heidelberg (2005). https://doi.org/10.1007/978-3-540-31843-9_24
15. Greilich, M., Burch, M., Diehl, S.: Visualizing the evolution of compound digraphs with timearctrees. In: Computer Graphics Forum, vol. 28, pp. 975–982. Wiley Online Library (2009)
16. Holme, P.: Modern temporal network theory: a colloquium. Eur. Phys. J. B **88**(9), 1–30 (2015). https://doi.org/10.1140/epjb/e2015-60657-4
17. Jacomy, M., Venturini, T., Heymann, S., Bastian, M.: Forceatlas2, a continuous graph layout algorithm for handy network visualization designed for the gephi software. PloS one **9**(6), e98679 (2014)
18. Kaufmann, M., Wagner, D.: Drawing Graphs: Methods and Models, vol. 2025. Springer, Cham (2003)
19. Kempe, D., Kleinberg, J., Kumar, A.: Connectivity and inference problems for temporal networks. J. Comput. Syst. Sci. **64**, 820–842 (2002)
20. Kumar, G., Garland, M.: Visual exploration of complex time-varying graphs. IEEE Trans. Visual. Comput. Graph. **12**(5), 805–812 (2006)
21. Lambiotte, R., Rosvall, M., Scholtes, I.: From networks to optimal higher-order models of complex systems. Nat. Phys. **15**(4), 313–320 (2019)
22. Loubier, E., Dousset, B.: Temporal and relational data representation by graph morphing. Saf. Reliab. Manag. Risk (ESREL 2008), Hammamet **14**(02), 2008–2016 (2008)
23. Nesbitt, K.V., Friedrich, C.: Applying gestalt principles to animated visualizations of network data. In: Proceedings Sixth International Conference on Information Visualisation, pp. 737–743. IEEE (2002)
24. Noguchi, C., Kawamoto, T.: Evaluating network partitions through visualization. arXiv e-prints arXiv:1906.00699, June 2019
25. Perri, V., Scholtes, I.: Hotvis: Higher-order time-aware visualisation of dynamic graphs (2020). https://arxiv.org/abs/1908.05976
26. Perri, V., Scholtes, I.: Hotvis: Higher-order time-aware visualisation of dynamic graphs (supplementary code and data) (2020). https://doi.org/10.5281/zenodo.3994152
27. Petrović, L.V., Scholtes, I.: Learning the markov order of paths in a network. arXiv preprint arXiv:2007.02861 (2020)
28. Pfitzner, R., Scholtes, I., Garas, A., Tessone, C.J., Schweitzer, F.: Betweenness preference: quantifying correlations in the topological dynamics of temporal networks. Phys. Rev. Lett. **110**, 198701 (2013)
29. Purchase, H.C., Hoggan, E., Görg, C.: How important is the "mental map" – an empirical investigation of a dynamic graph layout algorithm. In: Kaufmann, M., Wagner, D. (eds.) GD 2006. LNCS, vol. 4372, pp. 184–195. Springer, Heidelberg (2007). https://doi.org/10.1007/978-3-540-70904-6_19
30. Scholtes, I.: When is a network a network? Multi-order graphical model selection in pathways and temporal networks, pp. 1037–1046. ACM (2017)
31. Shewchuk, J.R.: Adaptive precision floating-point arithmetic and fast robust geometric predicates. Discrete Comput. Geom. **18**(3), 305–363 (1997)

32. Simonetto, P., Archambault, D., Kobourov, S.: Drawing dynamic graphs without timeslices. In: Frati, F., Ma, K.-L. (eds.) GD 2017. LNCS, vol. 10692, pp. 394–409. Springer, Cham (2018). https://doi.org/10.1007/978-3-319-73915-1_31
33. Stehlé, J., Voirin, N., Barrat, A., Cattuto, C., Isella, L., Pinton, J.F., Quaggiotto, M., Van den Broeck, W., Régis, C., Lina, B., et al.: High-resolution measurements of face-to-face contact patterns in a primary school. PloS one 6(8), e23176 (2011)
34. Tao, J., Xu, J., Wang, C., Chawla, N.V.: Honvis: Visualizing and exploring higher-order networks. In: 2017 IEEE Pacific Visualization Symposium (PacificVis), pp. 1–10. IEEE (2017)
35. Vanhems, P., et al.: Estimating potential infection transmission routes in hospital wards using wearable proximity sensors. PloS one 8(9), e73970 (2013)
36. Vehlow, C., Burch, M., Schmauder, H., Weiskopf, D.: Radial layered matrix visualization of dynamic graphs. In: 2013 17th International Conference on Information Visualisation, pp. 51–58. IEEE (2013)

VAIM: Visual Analytics for Influence Maximization

Alessio Arleo[1]($\boxtimes$) (iD), Walter Didimo[2](iD), Giuseppe Liotta[2](iD), Silvia Miksch[1](iD), and Fabrizio Montecchiani[2](iD)

[1] TU Wien, Vienna, Austria
{alessio.arleo,silvia.miksch}@tuwien.ac.at
[2] Università degli Studi di Perugia, Perugia, Italy
{walter.didimo,giuseppe.liotta,fabrizio.montecchiani}@unipg.it

Abstract. In social networks, individuals' decisions are strongly influenced by recommendations from their friends and acquaintances. The *influence maximization* (IM) problem asks to select a *seed set* of users that maximizes the influence spread, i.e., the expected number of users influenced through a stochastic diffusion process triggered by the seeds. In this paper, we present VAIM, a visual analytics system that supports users in analyzing the information diffusion process determined by different IM algorithms. By using VAIM one can: (i) simulate the information spread for a given seed set on a large network, (ii) analyze and compare the effectiveness of different seed sets, and (iii) modify the seed sets to improve the corresponding influence spread.

Keywords: Influence maximization · Information diffusion · Visual analytics

1 Introduction

People in social networks influence each other in both direct and indirect ways, through a mechanism often known as the *word-of-mouth effect* (see, e.g., [12,13]). For this reason social networks are becoming the favorite venue where companies advertise their products/services and where politicians run their campaigns. The *influence maximization* (IM) problem asks to select a *seed set* of users that maximizes the influence spread, i.e., the expected number of users positively influenced by an information diffusion process triggered by the seeds and that spreads through the network according to some stochastic model. We refer the reader to the works by Guille et al. [10] and by Li et al. [16] for surveys about influence maximization and information diffusion in social networks.

Research of WD, GL and FM partially supported by: (*i*) MIUR, grant 20174LF3T8 "AHeAD: efficient Algorithms for HArnessing networked Data", (*ii*) Dip. di Ingegneria - Università degli Studi di Perugia, grant RICBA19FM: "Modelli, algoritmi e sistemi per la visualizzazione di grafi e reti". Research of AA and SM partially supported by TU Wien "Smart CT" research cluster.

© Springer Nature Switzerland AG 2020
D. Auber and P. Valtr (Eds.): GD 2020, LNCS 12590, pp. 115–123, 2020.
https://doi.org/10.1007/978-3-030-68766-3_9

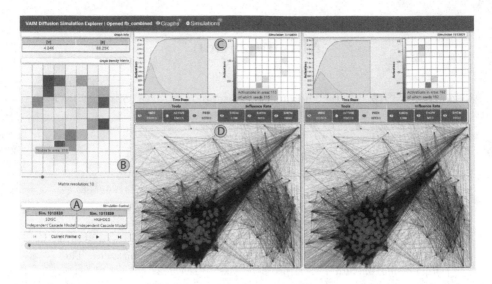

Fig. 1. VAIM's visual interface, at $t = 0$ of the case study in Sect. 3. Its components are marked as follows: A) Simulation control, B) Density Matrix view, C) Diffusion Matrix view, D) Node-link view. (Color figure online)

Analyzing and engineering an IM algorithm is a demanding task; as reported by Arora et al. [4], there is no single state-of-the-art technique for IM. Under the most common diffusion models, finding the optimal seed set in a network is known to be an NP-hard problem [12]. Besides the problem hardness, being the information diffusion process stochastic, even the evaluation of influence spread of any seed set is computationally complex [8], which makes the design of scalable and effective IM algorithms a great challenge that motivated a large and still increasing body of literature [16]. In this context, we want to exploit the power of information visualization to support expert users in analyzing, evaluating, and comparing IM algorithms. Our main contributions are as follows.

(*i*) We present VAIM, a system that provides facilities to simulate an information diffusion process over a given network and problem-oriented visual analytics (VA) tools to explore the related data (Sect. 2). VAIM has a modular architecture that currently includes some of the most popular IM algorithms and information diffusion models. An interface with multiple coordinated views makes it possible to visually compare and analyze the performance of a diffusion model over potentially very large networks and for different choices of the seed sets (i.e., for different IM algorithms). The user can interactively modify the seed set and iterate the process until a satisfying spread is achieved.

(*ii*) The effectiveness of VAIM is evaluated through a case study (Sect. 3). We show how tacking advantage of VAIM for (a) comparing different seed selection algorithms on the same network, and (b) improving the seed selection by either a manual or a system-assisted modification of the initial seed set.

Related Work. There are several visualization systems designed to analyze information diffusion processes in social networks. TwitInfo [18,19] aggregates tweets in the spatial, temporal, and event dimensions supporting the exploration of event propagation processes. Whisper [5] exploits a flower-like visualization for real-time monitoring of the diffusion of a given topic, highlighting the spatio-temporal information of the process over the world. OpinionFlow [24] uses Sankey graphs and density maps to visually summarize opinion diffusion processes. FluxFlow [25] adopts a timeline visualization to analyze anomalous information diffusion spreading. D-Map [6] collects data from Sina Weibo and offers a map-based ego-centric visualization to reveal dynamic patterns of how people are involved and influenced in a diffusion process. SocialWave [21] uses abstract visualizations to explore and analyze spatio-temporal diffusion of information. More approaches are elaborated in Chen et al. [7]. All these approaches are designed to reveal different facets of information diffusion processes and they often rely on geographical and other user-related information. On the other hand, they neither support the user in analyzing the impact of the seeds (which in fact may be unknown) and of the network structure in terms of influence spread, nor offer simulation tools to experiment different diffusion models.

Long and Wong [17] introduce Visual-VM, a visualization system for viral marketing. Similar to VAIM, Visual-VM allows users to simulate stochastic diffusion processes and to visually analyze their output. However, Visual-VM offers a simple visual interface, which strongly relies on geographical information to lay out the network. The networks analyzed with VAIM may come from diverse scenarios and may not contain geographical information about users.

Finally, Vallet et al. [22,23] present a visualization framework to compare different diffusion models based on a common set of graph rewriting rules. Different from VAIM, the work of Vallet et al. does not focus on comparing different IM algorithms and it is mainly tailored to networks of small or medium size.

Background and Notation. We model a social network as a directed graph $G = (V, E)$. A *diffusion model* M captures the stochastic diffusion process among the vertices of G. During the process, a vertex $v \in V$ can be either *active* or *inactive*. The *influence spread* of a seed set S, denoted by $\sigma_{G,M}(S)$, is the expected number of active vertices once the diffusion process (over the graph G and under the model M) terminates. More formally, the IM problem asks for a set $S^* \subseteq V$ of at most $0 < k \leq |V|$ seeds that maximizes the influence spread, i.e., $S^* = arg\,max\{\sigma_{G,M}(S)|S \subseteq V \wedge |S| \leq k\}$. One of the most commonly used diffusion models is the *Independent Cascade* (IC) [16]. Other models (such as the Linear Threshold model) make use of additional parameters but do not differ significantly in terms of the underlying iterative framework. In the IC model, a diffusion instance unfolds through an iterative process: In step 0, only the seed vertices are active; in step $j > 0$, each vertex u activated at step $j - 1$ will activate each of its inactive neighbors v with probability $0 \leq p(u, v) \leq 1$. The process halts when no more vertices can be activated. Unfortunately, the IM problem is NP-hard under the IC model, as well as under other models [12]. For a broader discussion refer to [10,16].

2   VAIM Design

The design of VAIM relies on the "Data-Users-Tasks" model proposed in [20].

Data. To estimate the influence spread of a seed set, we rely on a simulation-based approach. To obtain statistically relevant data, the simulation is repeated multiple times. Each single repetition is a time-dependent process taking as input a graph and a set of seeds. Hence, the data model of VAIM includes the input network, and set-typed temporal data represented by the active set of vertices and edges at every timestamp of the simulated diffusion process.

Users. VAIM targets a single class of expert users. Those users are knowledgeable in their own application domain and in the use of visual analytics tools. Also, they are interested not only in the resulting influence spread, but also on how the structure of the network influences the diffusion process.

Tasks. VAIM is designed to support the following user tasks:

T1 Simulate. It should be possible to simulate a diffusion process on a given network, with the seeds from an IM algorithm, under a given diffusion model.

T2 Evaluate. The user should be allowed to visually analyze both the quality of spread of a seed set and the impact of the network structure on the diffusion process, such as areas with a higher rate of active nodes, isolated areas, etc. The user can fast forward, rewind, and pause the process animation.

T3 Compare. It should be possible to visually compare the performance of different seed sets computed by different IM algorithms.

T4 Feedback. The user should be facilitated in modifying the seed set and iterate the simulate-evaluate-compare process.

2.1   Visualization Design

The visualization design adopts an overview+detail approach. The interface is organized as a dashboard with multiple coordinated views (see also Fig. 1). The chosen colour schemes and palettes are colorblind friendly [11].

– Simulation control (Fig. 1-A). Here the user can set different parameters about the diffusion process, such as the stochastic model and the number of iterations (Task T1).

– Density matrix view (Fig. 1-B). The main purpose of this view is to provide an overview of the network structure in a scalable manner. This is achieved with a simplified matrix visualization, which is obtained by firstly computing a node-link layout of the whole (potentially very large) network with some fast algorithm, such as centralized or distributed force-directed techniques (e.g., [2, 3,14]), and then by slicing the plane into cells. The color intensity of each cell reflects the number of nodes inside. The size of the matrix can be increased or decreased through a simple slider. Hovering with the mouse on a cell, the number of nodes in that cell is reported.

– Diffusion matrix view (Fig. 1-C). It allows users to visually compare multiple simulations over the same network. A legend below it shows the considered IM algorithms. Each simulation is conveyed using a distinct matrix visualization whose cells' colors vary in a *YlOrRd* scale (yellow to orange to red) and reflect the number of active nodes in the corresponding area of the network. Notably, the density and diffusion matrices have the same set of cells, so to facilitate comparisons and associations among them. Similarly as for the density matrix view, the computation and the rendering of this view must be fast enough to allow the visualization of multiple simulations over large networks. At the left side of each diffusion matrix, the **process trend chart** is a plot with two curves showing, for each iteration, the number of new nodes activated in that iteration and the cumulative number of nodes activated up to that iteration. A red vertical segment indicates the currently selected iteration. VAIM can animate the diffusion process over time. Other facilities allow users to highlight those cells containing some seeds, or whose influence rate is low ($< 30\%$), medium ($[30\%, 60\%]$), or high ($> 60\%$). Clicking on a cell, its influence rate is shown and a list of nodes that can be either removed or promoted as seeds is suggested, based on node degrees and influence rate.

– Node-link view (Fig. 1-D). Below each diffusion matrix, there is a panel in which a detailed node-link diagram of a portion of the network can be visualized. This portion can be freely chosen by the user through a brushing selection of any group of $k \times h$ cells in the density matrix. The combination of the node-link view with the two matrix views described above is particularly useful for very large networks, for which detailed visualizations are feasible only for small portions. In the diagram, blue nodes represent seeds while dark red nodes and edges represent the active elements at the considered time instant (Fig. 2(a)). The user can hide all edges or leave only the active ones.

The three views together are designed to support Tasks T2, T3 and T4.

3   Evaluation and Discussion

We discuss an evaluation of VAIM on the following case study (a second one is discussed in [1]). The input is the fb-combined social network, extracted from Facebook [15], having 4,039 nodes and 88,234 edges (https://snap.stanford.edu/data/). We simulated an IC diffusion process (T1), using two seed sets of 400 nodes each, computed by two popular IM algorithms: HIGHDEG [12,13] and SDISC [9], based on degree centrality and discount, respectively. We compared

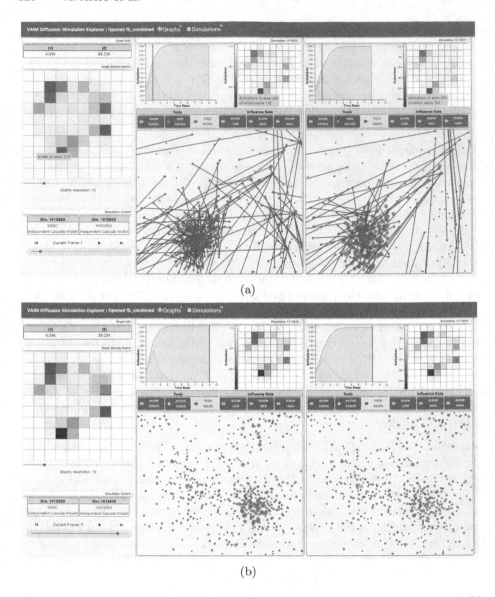

Fig. 2. Snapshot of VAIM after (a) the first iteration of the diffusion process, and (b) at the end of the diffusion process.

and evaluated (T2 and T3) the performance of the two diffusion processes. Figure 1 shows a snapshot of the interface at the beginning of each diffusion process. The process trend charts reveal that SDISC leads to a higher number of active nodes in fewer iterations. By exploring the diffusion matrices we can observe a different distribution of the seeds selected by the two IM algorithms. For example, focusing on the densest cell of the network (which can be easily spotted in the density matrix), we see that HIGHDEG (the right-side simulation) concentrates a higher number of seeds than SDISC (the left-side simulation) in that cell (182 seeds of HIGHDEG vs 115 seeds of SDISC), while putting relatively fewer seeds in sparser cells. Also, within the densest cell, SDISC distributes the seeds more uniformly than HIGHDEG. Figure 2(a) shows the processes at the next iteration, and still focuses on the densest cell. Despite the smaller number of seeds, SDISC yields a higher number of newly active nodes (red nodes) in that cell (268 of SDISC vs 243 of HIGHDEG). Also, the greater number of red edges (those used by the diffusion process) exiting the cell, reveals a higher influence of the nodes of this cell towards nodes outside it. Figure 2(b) shows the end of the processes. Using the influence rate function, we observe that the cells selected from the density matrix have a smaller number of active nodes with HIGHDEG than with SDISC. Looking at the node-link view for these cells (edges are hidden), this seems to be caused by the very small number of seeds that HIGHDEG placed in this portion of the network. The above discussion helps understanding how the seeding strategy adopted by SDISC leads to better performance, which corroborates the results of an experimental analysis performed on a collaboration graph presented in [9].

In order to improve the information spread of SDISC (T4), VAIM suggested 20 nodes (with smallest degree in the cell with highest influence rate) to be removed from the original seed set and 20 nodes (with highest degree in the cell with lowest influence rate) to be promoted as seeds. We modified the seed set accordingly and we simulated again the diffusion process. The new process lead to 2% more of active nodes.

4   Conclusion and Future Work

We discussed the use of visual analytics to support the analysis and fine tuning of IM strategies. We plan to extend the system with features such as edge bundling to mitigate edge clutter in the node-link view. We will also implement new diffusion models, together with ad-hoc views to explore the additional parameters of these models. Considering networks with node and edge attributes (e.g., geolocations) is also an interesting direction. Finally, we want to further evaluate VAIM with more case studies and experiments, in particular to test its scalability (both in terms of simulation and visualization) to very large networks.

References

1. Arleo, A., Didimo, W., Liotta, G., Miksch, S., Montecchiani, F.: Vaim: Visual analytics for influence maximization. arXiv:2008.08821v1 [cs.SI] (2020). https://arxiv.org/abs/2008.08821
2. Arleo, A., Didimo, W., Liotta, G., Montecchiani, F.: Large graph visualizations using a distributed computing platform. Inf. Sci. **381**, 124–141 (2017). https://doi.org/10.1016/j.ins.2016.11.012

3. Arleo, A., Didimo, W., Liotta, G., Montecchiani, F.: A distributed multilevel force-directed algorithm. IEEE Trans. Parallel Distrib. Syst. **30**(4), 754–765 (2019). https://doi.org/10.1109/TPDS.2018.2869805
4. Arora, A., Galhotra, S., Ranu, S.: Debunking the myths of influence maximization: an in-depth benchmarking study. In: SIGMOD Conference, pp. 651–666. ACM (2017)
5. Cao, N., Lin, Y., Sun, X., Lazer, D., Liu, S., Qu, H.: Whisper: tracing the spatiotemporal process of information diffusion in real time. IEEE Trans. Vis. Comput. Graph. **18**(12), 2649–2658 (2012)
6. Chen, S., et al.: D-Map: visual analysis of ego-centric information diffusion patterns in social media. In: VAST, pp. 41–50. IEEE Computer Society (2016)
7. Chen, S., Lin, L., Yuan, X.: Social media visual analytics. Comput. Graph. Forum **36**(3), 563–587 (2017)
8. Chen, W., Wang, C., Wang, Y.: Scalable influence maximization for prevalent viral marketing in large-scale social networks. In: KDD, pp. 1029–1038. ACM (2010)
9. Chen, W., Wang, Y., Yang, S.: Efficient influence maximization in social networks. In: Proceedings of the 15th ACM SIGKDD International Conference on Knowledge Discovery and Data Mining, KDD 2009, pp. 199–208. Association for Computing Machinery, New York (2009). https://doi.org/10.1145/1557019.1557047
10. Guille, A., Hacid, H., Favre, C., Zighed, D.A.: Information diffusion in online social networks: a survey. SIGMOD Rec. **42**(2), 17–28 (2013)
11. Harrower, M., Brewer, C.A.: Colorbrewer.org: an online tool for selecting colour schemes for maps. Cartographic J. **40**(1), 27–37 (2003)
12. Kempe, D., Kleinberg, J.M., Tardos, É.: Maximizing the spread of influence through a social network. In: KDD, pp. 137–146. ACM (2003)
13. Kempe, D., Kleinberg, J.M., Tardos, É.: Maximizing the spread of influence through a social network. Theory Comput. **11**, 105–147 (2015). https://doi.org/10.4086/toc.2015.v011a004
14. Kobourov, S.G.: Force-directed drawing algorithms. In: Tamassia, R. (ed.) Handbook on Graph Drawing and Visualization, pp. 383–408. Chapman and Hall/CRC, Boca Raton (2013)
15. Leskovec, J., Mcauley, J.J.: Learning to discover social circles in ego networks. In: Advances in Neural Information Processing Systems, pp. 539–547 (2012)
16. Li, Y., Fan, J., Wang, Y., Tan, K.: Influence maximization on social graphs: a survey. IEEE Trans. Knowl. Data Eng. **30**(10), 1852–1872 (2018)
17. Long, C., Wong, R.C.: Visual-VM: a social network visualization tool for viral marketing. In: ICDM Workshops, pp. 1223–1226. IEEE Computer Society (2014)
18. Marcus, A., Bernstein, M.S., Badar, O., Karger, D.R., Madden, S., Miller, R.C.: Processing and visualizing the data in tweets. SIGMOD Rec. **40**(4), 21–27 (2011)
19. Marcus, A., Bernstein, M.S., Badar, O., Karger, D.R., Madden, S., Miller, R.C.: TwitInfo: aggregating and visualizing microblogs for event exploration. In: CHI, pp. 227–236. ACM (2011)
20. Miksch, S., Aigner, W.: A matter of time: applying a data-users-tasks design triangle to visual analytics of time-oriented data. Comput. Graph. **38**, 286–290 (2014)
21. Sun, G., Tang, T., Peng, T., Liang, R., Wu, Y.: Socialwave: visual analysis of spatio-temporal diffusion of information on social media. ACM TIST **9**(2), 15:1–15:23 (2018)

22. Vallet, J., Kirchner, H., Pinaud, B., Melançon, G.: A visual analytics approach to compare propagation models in social networks. In: Rensink, A., Zambon, E. (eds.) Proceedings Graphs as Models, GaM@ETAPS 2015, London, UK, 11–12 April 2015. EPTCS, vol. 181, pp. 65–79 (2015). https://doi.org/10.4204/EPTCS. 181.5

23. Vallet, J., Pinaud, B., Melançon, G.: Studying propagation dynamics in networks through rule-based modeling. In: Chen, M., Ebert, D.S., North, C. (eds.) 2014 IEEE Conference on Visual Analytics Science and Technology, VAST 2014, Paris, France, 25–31 October 2014, pp. 281–282. IEEE Computer Society (2014). https:// doi.org/10.1109/VAST.2014.7042530

24. Wu, Y., Liu, S., Yan, K., Liu, M., Wu, F.: OpinionFlow: visual analysis of opinion diffusion on social media. IEEE Trans. Vis. Comput. Graph. **20**(12), 1763–1772 (2014)

25. Zhao, J., Cao, N., Wen, Z., Song, Y., Lin, Y., Collins, C.: #FluxFlow: visual analysis of anomalous information spreading on social media. IEEE Trans. Vis. Comput. Graph. **20**(12), 1773–1782 (2014)

Odd Wheels Are Not Odd-distance Graphs

Gábor Damásdi[(✉)] [iD]

MTA-ELTE Lendület Combinatorial Geometry Research Group, Budapest, Hungary
damasdigabor@caesar.elte.hu

Abstract. An odd wheel graph is a graph formed by connecting a new vertex to all vertices of an odd cycle. We answer a question of Rosenfeld and Le by showing that odd wheels cannot be drawn in the plane such that the lengths of the edges are odd integers.

Keywords: Geometric graphs · Odd-distance graphs · Forbidden subgraphs

1 Introduction

A *geometric graph* is a graph drawn in the plane so that the vertices are represented by distinct points and the edges are represented by possibly intersecting straight line segments connecting the corresponding points . A *unit-distance graph* is a geometric graph where all edges are represented by segments of length 1. The study of unit-distance graphs started with the question of Edward Nelson, who raised the problem of determining the minimum number of colors that are needed to color the points of the plane so that no two points unit distance apart are assigned the same color. This number is known as the chromatic number of the plane. Until recently the best lower bound was 4, but Aubrey de Grey [6] constructed a unit-distance graph which cannot be colored with 4 colors. The best upper bound is 7. For more details on unit-distance graphs see for example [13].

Erdős [4] raised the problem to determine the maximal number of edges in a unit-distance graph with n vertices and this question became known as the Erdős Unit Distance Problem.

Later Erdős and Rosenfeld [1] asked the same two questions for odd distances. Namely, let G^{odd} be the graph whose vertex set is the plane and two vertices are connected if their distance is an odd integer. They asked to determine the chromatic number of G^{odd}, and to determine how many distances among n points in the plane can be odd integers.

Analogously we define *odd-distance graphs* to be the geometric graphs having an embedding in the Euclidean plane in which all edges are of odd integer length. In other words, the odd-distance graphs are the finite subgraphs of G^{odd}. There are odd-distance graphs whose chromatic number is five [1,6] but contrary to

D. Auber and P. Valtr (Eds.): GD 2020, LNCS 12590, pp. 124–134, 2020.
https://doi.org/10.1007/978-3-030-68766-3_10

the unit distance case we do not have any upper bound. The chromatic number of G^{odd} might be infinite. In the case when we require the color classes to be measurable sets, it has been shown that the chromatic number is indeed infinite [2,14].

Four points in the plane with pairwise odd integer distances do not exist, hence K_4 is not an odd-distance graph. From Turán's theorem we know that the complete tripartite graph $K_{n,n,n}$ has the maximal number of edges among K_4-free graphs. Piepemeyer [11] showed that $K_{n,n,n}$, and therefore any 3-colorable graph, is an odd-distance graph. This settles the second question of Erdős and Rosenfeld.

Let W_n be the wheel graph formed by connecting a new vertex to all vertices of a cycle on n vertices.[1] The wheel graph W_{2k} is 3-colorable, hence it is an odd-distance graph.

Rosenfeld and Le [12] showed that having K_4, which is also W_3, as a subgraph is not the only obstruction for being an odd-distance graph, since W_5 is also not an odd-distance graph. This led them to the following question: Is it true that W_{2k+1} is not a subgraph of G^{odd} for any k? We answer this for the affirmative.

Theorem 1. *The odd wheels are not odd-distance graphs.*

In Sect. 2 we consider drawings of wheel graphs in general, not assuming that the edge lengths are odd numbers. We develop a number of useful lemmas, that might prove useful for related questions. For example Harborth's conjecture asks whether all planar graphs admit a planar drawing with integer edge lengths. Since the maximal planar graphs are the triangulations, they contain many wheels. Hence, understanding the possible drawings of wheels is vital for solving the conjecture. Then, in Sect. 3, we prove Theorem 1.

2   Wheels with Integer Edge Lengths

Embeddings of Wheel Graphs

Every set of $n + 1$ ordered points of the plane determines an *embedding* of the wheel graph W_n. Throughout this paper we will always assume that the center of the wheel is embedded at the origin O and the other points are $A_1, A_2, \ldots, A_n$, following the order of the vertices in the defining cycle of the wheel (see Fig. 1). In the following notations the index is understood cyclically, i.e. the index $n+1$ is equivalent to the index 1. For example every embedding determines n triangles: OA_iA_{i+1} for $i \in \{1, \ldots, n\}$. These will be referred to as the triangles of the embedding. We will use the following notations:

$$r_i = |OA_i|, \quad r_{i,i+1} = |A_iA_{i+1}|$$

$$\theta_{i,i+1} = \angle A_iOA_{i+1}$$

---

[1] There is some discrepancy in the literature, since some authors prefer to denote by W_n the wheel graph on n vertices.

That is, the i-th triangle has sides of length r_i, r_{i+1} and $r_{i,i+1}$, and its inner angle is $\theta_{i,i+1}$. Note that the $\theta_{i,i+1}$-s are directed angles. We do not assume planarity or even general position of the points. For example crossings are allowed, and O does not need to be in the interior of the cycle (see Fig. 1).

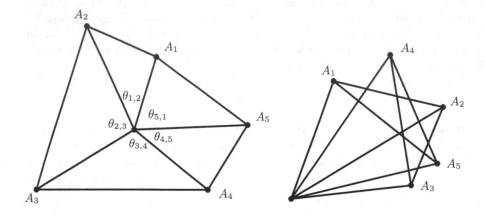

Fig. 1. Two embeddings of the wheel graph W_5.

Geometry of a Triangle

Let us recall some classical results from elementary geometry. Let $T(a, b, c)$ denote a triangle with sides a, b, c and angles α, β, γ. By the law of cosines:

$$\cos(\alpha) = \frac{b^2 + c^2 - a^2}{2bc} \tag{1}$$

$$\sin(\alpha) = \sqrt{1 - \cos(\alpha)^2} = \frac{\sqrt{4b^2c^2 - (b^2 + c^2 - a^2)^2}}{2bc} \tag{2}$$

Let A denote the area of $T(a, b, c)$.

$$A = \frac{bc\sin(\alpha)}{2} = \frac{\sqrt{4b^2c^2 - (b^2 + c^2 - a^2)^2}}{4} \tag{3}$$

Using these formulas we will introduce two notions, the characteristic of a triangle and the residual of an angle. Strictly speaking we will only need residuals for the proof of Theorem 1, but there is a strong connection to the characteristics of triangles so they are worth mentioning.

Characteristic of a Triangle

From (3) we can see that if a, b and c are integers, then we can write the area of $T(a, b, c)$ as $r\sqrt{D}$ for some rational number r and a square-free integer D. If the area is 0, then $r = 0$ and D can be any square-free integer. If the area is non-zero, then D must be the square-free part of $4b^2c^2 - (b^2 + c^2 - a^2)^2$. In this case the number D is called the *characteristic* of the triangle.

We say that a point set in the plane is *integral* if the pairwise distances are integers. The characteristic of triangles is a useful tool in the study and algorithmic generation of integral point sets (see for example [8]). The following statement is folklore, for a proof see [9].

Theorem 2. *The triangles spanned by each three non collinear points in a plane integral point set have the same characteristic.*

Consider an embedding of a wheel graph where the edges have integer lengths. The rest of the distances might be non-integer, so the n triangles of the embedding can have different characteristics. (See for example Fig. 2). When we started the study of embeddings of wheel graphs, we hoped to show that there cannot be too many characteristics appearing in a embedding. It turned out that there can be arbitrarily many, but we can still gain some information by considering them. Later in this section we will show the following statement.

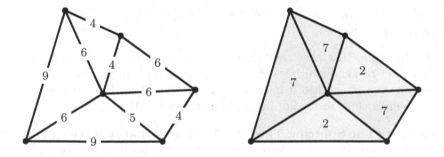

Fig. 2. Characteristics of the triangles in an embedding.

Lemma 1. *Consider an embedding of a wheel graph with integer edge lengths. Then the angles among the $\theta_{i,i+1}$-s corresponding to the triangles of a given characteristic add up to an integer multiple of π.*

Residual of an Angle

Considering (1) and (3) we can see that the characteristic of a triangle transfers to the angles in the following sense. If a triangle that have integer sides have characteristic D, then the sine of its angles have the form $q\sqrt{D}$ for some rational number q, and the cosines of the angles are rational. Hence we will say that an *angle θ has residual D* if D is square-free, $\sin(\theta) = q\sqrt{D}$ for some rational number q, and furthermore $\cos(\theta)$ is rational.

Most angles in general do not have any residual. Integer multiples of π have residual D for any square-free integer D, but other angles have at most one residual. If the residual is unique, it will be called *the residual of the angle*. For example the residual of $\frac{\pi}{2}$ is 1, the residual of $\frac{\pi}{3}$ is 3, but $\frac{\pi}{6}$ does not have any residual. Just as the characteristic of triangles, the residual of the angles is a useful tool, in [5] it was used to find trisectible angles in triangles that have integer sides.

The trigonometric addition formulas $\sin(\theta+\phi) = \sin(\theta)\cos(\phi)+\sin(\phi)\cos(\theta)$ and $\cos(\theta + \phi) = \cos(\theta)\cos(\phi) - \sin(\theta)\sin(\phi)$ immediately imply that the set of angles that have residual D are closed under addition. Also, for any ϕ the following angles have the same set of residuals: $\phi, -\phi, \pi + \phi, \pi - \phi$.

Angles Whose Squared Trigonometric Functions Are Rational

Conway, Radin and Sadun [3] studied angles whose squared trigonometric functions are rational. They said that θ is a *pure geodetic angle* if the square of its sine is rational and they showed the following theorem.

Theorem 3 (The Splitting Theorem [3]). *If the value of a rational linear combination of pure geodetic angles is a rational multiple of π, then so is the value of its restriction to those angles whose tangents are rational multiples of any given square root.*

Clearly, angles that have residual D are pure geodetic angles and have tangents that are rational multiples of $\sqrt{D}$. Therefore, Theorem 3 applies to them, but we can even strengthen it in some sense. Note that in the next theorem we consider simple sums instead of rational linear combinations.

Theorem 4 (The Splitting Theorem for angles that have residual). *Let us consider some angles that have a residual. If the value of the sum of these angles is a rational multiple of π, then so is the value of its restriction to those angles that have a given residual. Furthermore, these restricted sums must add up to integer multiples of $\frac{\pi}{3}$ or $\frac{\pi}{2}$.*

Proof. The first part is clear from Theorem 3. For the second part we recall Niven's Theorem.

Theorem 5 (Niven's theorem [10]). *Consider the angles in range $0 \le \theta \le \frac{\pi}{2}$. The only values of θ such that both $\frac{\theta}{\pi}$ and $\cos(\theta)$ are rational are $0, \frac{\pi}{3}$ and $\frac{\pi}{2}$.*

Since angles corresponding to a given residual are closed under addition, the sum restricted to residual D gives us an angle that has residual D. But angles that have residual D have rational cosine so we can apply Theorem 5 to the restricted sums. □

Now we are ready to prove Lemma 1. Since triangles that have characteristic D have angles that have residual D, it is enough to show the following residual version.

Lemma 2. *Consider an embedding of a wheel graph with integer edge lengths. Then the angles among the $\theta_{i,i+1}$-s corresponding to a given residual add up to an integer multiple of π.*

Proof. We know that $\sum_{i=1}^{n} \theta_{i,i+1}$ is an integer multiple of 2π. Hence we can apply Theorem 4 for the angles $\theta_{i,i+1}$. Suppose that the angles corresponding to a given residual D add up to θ. From Theorem 4 it is clear that θ is either an integer multiple of $\frac{\pi}{2}$ or an integer multiple of $\frac{\pi}{3}$. Let $\theta' = \theta \mod \pi$. Then $\theta' = 0, \frac{\pi}{3}, \frac{2\pi}{3}$ or $\frac{\pi}{2}$. Note that since θ is the sum of some angles that have residual D, it also has residual D. Hence θ' also has residual D.

Since $\sin(\frac{\pi}{3}) = \sin(\frac{2\pi}{3}) = \frac{\sqrt{3}}{2}$, we have $D = 3$ for $\theta' = \frac{\pi}{3}$ and also for $\theta' = \frac{2\pi}{3}$. Similarly we have $D = 1$ for $\theta' = \frac{\pi}{2}$. Therefore, if we group the terms of $\sum_{i=1}^{n} \theta_{i,i+1}$ based on the residuals, every group will sum up to an integer multiple of π except maybe the ones corresponding to $D = 1$ and $D = 3$. (Some $\theta_{i,i+1}$ might not have a unique residual but those are themselves integer multiples of π, we can just pick an arbitrary residual for them). Since the whole sum should be an integer multiple of π, the exceptional cases together must add up to a integer multiple of π. This can only happen if both of them add up to an integer multiple of π, since $\frac{\pi}{3} + \frac{\pi}{2}$ and $\frac{2\pi}{3} + \frac{\pi}{2}$ are not integer multiples of π. Hence, every sum corresponding to a given residual is a integer multiple of π. □

3   Wheels with Odd Edge Lengths

In the previous section we considered wheels with arbitrary integer edge lengths. Now we are ready to turn our attention to drawings of odd wheels where the edge lengths are odd numbers.

Lemma 3. *If a, b and c are odd numbers and the characteristic of the triangle $T(a, b, c)$ is D, then $D \equiv 3 \mod 8$.*

Proof. From (3) we know that the characteristic of the triangle is the square-free part of $4a^2b^2 - (a^2 + b^2 - c^2)^2$. Since squares of odd numbers are congruent to 1 modulo 8, we have $4a^2b^2 - (a^2 + b^2 - c^2)^2 \equiv 3 \mod 8$. Since the square part of $4a^2b^2 - (a^2 + b^2 - c^2)^2$ is the square of an odd number it is congruent to 1 modulo 8. Hence $D \equiv 3 \mod 8$. □

This means that if we have an embedding of a wheel graph where the edge lengths are odd integers, then each $\theta_{i,i+1}$ have a unique residual that is congruent to 3 modulo 8. The next idea is to classify the angles whose residual is congruent to 3 modulo 8.

Lemma 4. *Suppose $D \equiv 3$ mod 8 and ϕ is an angle that has residual D. Then $\cos(\phi)$ can be written as $\frac{m}{2p}$, where $p \equiv 1$ mod 8 and m is an integer. Furthermore the remainder of m modulo 8 is determined by the angle, and it is either $1, 2, 3, 5, 6$ or 7.*

We will call this remainder the *class* of ϕ.

Proof. By the definition of having a residual $\cos(\phi)$ is rational. Since $D \equiv 3$ mod 8 the value of $\cos(\phi)$ is non-zero. Hence, we can write $\cos(\phi) = \frac{a}{b}$ for some non-zero integers a, b such that $gcd(a, b) = 1$. There are two cases.

First, suppose that a and b are odd. Odd numbers have an inverse in $\mathbb{Z}_8$. So, if b is odd, there is an odd number k such that $bk \equiv 1$ mod 8. Hence we can write $\cos(\phi) = \frac{2ak}{2bk}$. Now ak is odd, therefore $2ak$ is not divisible by 4. Hence $m = 2ak$, $p = bk$ works.

Second, suppose that a or b is even. Since $gcd(a, b) = 1$, one of them is even and the other one is odd. Consider that $\sin(\phi) = \pm\sqrt{1 - \frac{a^2}{b^2}} = \pm\frac{\sqrt{b^2 - a^2}}{b}$. The square-free part of $b^2 - a^2$ is D, and $b^2 - a^2$ is odd, so $b^2 - a^2 \equiv 3$ mod 8. Since the only quadratic residuals modulo 8 are 0, 1 and 4, the only possibility is that $b^2 \equiv 4$ mod 8 and $a^2 \equiv 1$ mod 8. Since $b^2 \equiv 4$ mod 8, $b' = \frac{b}{2}$ is odd and similarly to the previous case there is an odd k such that $kb' \equiv 1$ mod 8. Since $a^2 \equiv 1$ mod 8, a must be odd. So $m = ak$, $p = b'k$ works.

It is also easy to see that an angle cannot fall into two classes, notice that $\frac{m_1}{2p_1} = \frac{m_2}{2p_2}$ implies $m_1 p_2 \equiv m_2 p_1$ mod 8. $\square$

The aim of the next lemma is to answer the following question. Suppose we have two angles one of class m_1 and one of class m_2. If their sum have a class, what could that be?

Lemma 5. *If $\cos(\theta) = \frac{m_1}{2p_1}$, $\cos(\phi) = \frac{m_2}{2p_2}$ and $\cos(\theta + \phi) = \frac{m_3}{2p_2}$ for some integers p_1, p_2, p_3 that are congruent to 1 modulo 8 and integers m_1, m_2, m_3, then*

$$m_1^2 + m_2^2 + m_3^2 - m_1 m_2 m_3 - 4 \equiv 0 \quad \text{mod } 8 \tag{4}$$

Proof. Using the cosine addition formula $\cos(\theta + \phi) = \cos(\theta)\cos(\phi) - \sin(\theta)\sin(\phi)$:

$$\frac{m_3}{2p_3} = \frac{m_1}{2p_1} \cdot \frac{m_2}{2p_2} - \left(\pm\frac{\sqrt{4p_1^2 - m_1^2}}{2p_1}\right) \cdot \left(\pm\frac{\sqrt{4p_2^2 - m_2^2}}{2p_2}\right)$$

$$(2m_3 p_1 p_2 - m_1 m_2 p_3)^2 = p_3^2(4p_1^2 - m_1^2)(4p_2^2 - m_2^2)$$

$$4m_3^2 p_1^2 p_2^2 + m_1^2 m_2^2 p_3^2 - 4m_1 m_2 m_3 p_1 p_2 p_3 = 16p_1^2 p_2^2 p_3^2 - 4p_1^2 m_2^2 p_3^2 - 4m_1^2 p_2^2 p_3^2 + m_1^2 m_2^2 p_3^2$$

$$p_1^2 p_2^2 m_3^2 - p_1 p_2 p_3 m_1 m_2 m_3 - 4p_1^2 p_2^2 p_3^2 + p_1^2 m_2^2 p_3^2 + m_1^2 p_2^2 p_3^2 = 0$$

Using that $p_1 \equiv p_2 \equiv p_3 \equiv 1$ mod 8 we get (4). $\square$

Consider the solutions of (4) in $\mathbb{Z}_8$. Clearly, every triple (m_1, m_2, m_3) that is not a solution of this equation encodes a forbidden change in the class when adding two angles. For example, since $(1, 2, 3)$ is not a solution, adding an angle of class 1 and an angle of class 2 cannot result in an angle of class 3. The equation is symmetric in m_1, m_2 and m_3. We will be later interested in solutions where one of the m_i-s is $1, 3, 5$ or 7. Checking every triple we find that these solutions are the following ones and the re-orderings of these: $(1, 1, 2)$, $(1, 1, 7)$, $(1, 2, 5)$, $(1, 3, 5)$, $(1, 3, 6)$, $(1, 6, 7)$, $(2, 3, 3)$, $(2, 3, 7)$, $(2, 5, 5)$, $(2, 7, 7)$, $(3, 3, 7)$, $(3, 5, 6)$, $(5, 5, 7)$, $(5, 6, 7)$, $(7, 7, 7)$.

Proof of Main Theorem

The idea of the proof is simple, we want to show that $\sum_{i=1}^{n} \theta_{i,i+1}$ is not a multiple of 2π using the fact that each $\theta_{i,i+1}$ is an angle of a triangle whose sides have odd length. This will not work this easily, for example both $\frac{\pi}{3}$ and $\frac{2\pi}{3}$ appears in triangles with odd sides and $\frac{\pi}{3} + \frac{\pi}{3} + \frac{\pi}{3} + \frac{\pi}{3} + \frac{2\pi}{3} = 2\pi$. To reach a contradiction we will also use that the triangles in a wheel embedding share sides with their neighbours.

Proof (of Theorem 1)
Suppose there is a counterexample to Theorem 1. From Lemma 3 we know that each $\theta_{i,i+1}$ has a unique residual. Let $\phi_1, \ldots, \phi_n$ be a reordering of the angles $\theta_{1,2}, \theta_{2,3}, \ldots, \theta_{n,1}$ in such a way that the angles of given residuals are consecutive. In general an arbitrary angle might not have any residual. The advantage of this ordering is that $\sum_{i=1}^{\ell} \phi_i$ has a residual for each $\ell \in \{0, 1, \ldots, n\}$. To see this suppose that the residual of ϕ_ℓ is D. From Lemma 2 we see that the ϕ_i-s before ϕ_ℓ whose residual is not D sum up to an integer multiple of π. Thus, they do not affect the residual of $\sum_{i=1}^{\ell} \phi_i$. Since angles that have residual D are closed under addition, the rest sums up to an angle that has residual D.

We also know that $D \equiv 3 \mod 8$ from Lemma 3. Hence by Lemma 4 we know that $\sum_{i=1}^{\ell} \phi_i$ have a class for each $\ell \in \{0, 1, \ldots, n\}$. Consider how the class changes as ℓ goes from 0 to n.

In each step we increase the angle by some $\theta_{j,j+1}$. We have $\cos(\theta_{j,j+1}) = \frac{r_j^2 + r_{j+1}^2 - r_{j,j+1}^2}{2 r_j r_{j+1}} = \frac{(r_j^2 + r_{j+1}^2 - r_{j,j+1}^2) r_j r_{j+1}}{2 r_j^2 r_{j+1}^2}$. Since r_j, r_{j+1} and $r_{j,j+1}$ are odd numbers, $(r_j^2 + r_{j+1}^2 - r_{j,j+1}^2) \equiv 1 \mod 8$ and $r_j^2 r_{j+1}^2 \equiv 1 \mod 8$. Therefore the class of $\theta_{j,j+1}$ is the remainder of $r_j r_{j+1}$ by eight, which is either $1, 3, 5$ or 7. We will use this fact in the following form. If $r_j r_{j+1} \equiv 1 \mod 4$, then the class of $\theta_{j,j+1}$ is either 1 or 5, and if $r_j r_{j+1} \equiv 3 \mod 4$, then the class of $\theta_{j,j+1}$ is either 3 or 7. Therefore, as we increase ℓ the angle $\sum_{i=1}^{\ell} \phi_i$ changes either by an angle whose class is 1 or 5, or by an angle whose class is 3 or 7 depending on the remainder of $r_j r_{j+1}$ divided by 4.

Now we are ready to use Lemma 5. The solutions of (4) have an underlying structure, that we can use. This is depicted in Fig. 3. We create a graph G whose vertex set is $\{1, 2, 3, 5, 6, 7\}$. For solutions of the form $(1, x, y)$ and $(5, x, y)$ we have connected x and y by a dashed edge. For solutions of the form $(3, x, y)$ and $(7, x, y)$ we have connected them by and solid edge, allowing loop edges. These two sets of edges are disjoint. Note that the dashed edges form a bipartite graph such that the solid edges connect vertices inside the two parts. This will allow us to use a parity argument, as any closed trail in this graph must use an even number of dashed edges.

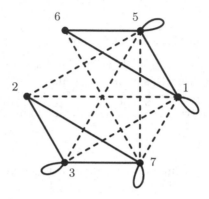

Fig. 3. Possible changes in the class when adding an angle of class 1, 3, 5 or 7.

Now we are ready to finish the proof. Let T be the trail of length $n + 1$ in G whose ℓ-th vertex is the class of $\sum_{i=1}^{\ell-1} \phi_i$. We know that $\sum_{i=1}^{n} \phi_i$ is a integer multiple of 2π, so $\cos(\sum_{i=1}^{n} \phi_i) = \frac{2}{2 \cdot 1}$. Hence the trail should start and end at 2. By Lemma 5 when the class of ϕ_i is 1 or 5, we follow one of the solid edges, if the class is 3 or 7, we follow a dashed edge.

Finally, we show that we followed a dashed edge an odd number of times. Considering the equation $(r_1 r_2)(r_2 r_3) \cdots (r_{n-1} r_n)(r_n r_1) = (\prod r_i)^2 \equiv 1 \mod 4$ we have $r_i r_{i+1} \equiv 3 \mod 4$ for an even number of i-s. Since n is odd, this implies that we have an odd number of i-s when $r_i r_{i+1} \equiv 1 \mod 4$. Hence the trail contains an odd number of dashed edges. Since the dashed edges form a bipartite graph and the solid edges connect vertices inside the two parts the trail cannot end where it started, a contradiction. This shows that a counterexample to Theorem 1 cannot exists. □

4   Final Remarks

We note that some parts of the proof can be replaced by other arguments. For example Lemma 5 also follows from the analysis of Cayley-Menger determinants.

The main goal of understanding odd-distance graphs is to determine the chromatic number of G^{odd}. Odd wheels are the simplest graphs that are not 3-colorable, yet they are not odd-distance graphs. Our proof heavily relies on the fact that a wheel graph contains many triangles. An other nice question of Rosenfeld and Nam Lê Tien [12] is the following. Are there triangle-free graphs that are not odd-distance graphs?

Piepemeyer's construction which shows that $K_{n,n,n}$ is an odd-distance graph comes from an integral point set. Naturally, one might be tempted to look for odd-distance graphs with high chromatic number in a similar way. Take an integral point set and then consider the odd-distance graph given by the edges of odd length. We note that this method cannot lead to success since the chromatic number of these graphs is at most 3. We leave the proof of this statement to the interested readers.

We can also consider the natural analog of Harborth's conjecture. Which planar graphs have a planar drawing where the length of the edges are odd integers?

Take for example a maximal planar graph, in other words a triangulation. If it contains an odd wheel, it is not an odd-distance graph. On the other hand if it does not contain an odd wheel, it is 3-colorable. Hence it is an odd distance graph, but this does not imply that we can find an plane drawing without crossings. Is it true that all 3-colorable planar graphs have an embedding without crossings where the length of the edges are odd integers?

Acknowledgement. We would like to thank SciExperts for providing free access to the software Wolfram Mathematica, and therefore to the database of Ed Pegg Jr. [7] on embeddings of wheels. We also thank Dömötör Pálvölgyi and our anonymous reviewers for valuable suggestions and encouragement.

References

1. Ardal, H., Maňuch, J., Rosenfeld, M., Shelah, S., Stacho, L.: The odd-distance plane graph. Discrete Comput. Geom. **42**(2), 132–141 (2009). https://doi.org/10.1007/s00454-009-9190-2
2. Bukh, B.: Measurable sets with excluded distances. Geom. Funct. Anal. **18**(3), 668–697 (2008). https://doi.org/10.1007/s00039-008-0673-8
3. Conway, J.H., Radin, C., Sadun, L.: On angles whose squared trigonometric functions are rational. Discrete Computat. Geom. **22**(3), 321–332 (1999). https://doi.org/10.1007/PL00009463
4. Erdős, P.: On sets of distances of n points. Amer. Math. Monthly **53**, 248–250 (1946). https://doi.org/10.2307/2305092
5. Gordon, R.A.: Integer-sided triangles with trisectible angles. Math. Mag. **87**(3), 198–211 (2014). https://doi.org/10.4169/math.mag.87.3.198
6. de Grey, A.D.N.J.: The chromatic number of the plane is at least 5. Geombinatorics **28**(1), 18–31 (2018)
7. Jr., E.P.: Wheel graphs with integer edges (2015). https://demonstrations.wolfram.com/WheelGraphsWithIntegerEdges/

134   G. Damásdi

8. Kreisel, T., Kurz, S.: There are integral heptagons, no three points on a line, no four on a circle. Discrete Comput. Geom. **39**(4), 786–790 (2007). https://doi.org/10.1007/s00454-007-9038-6
9. Kurz, S.: On the characteristic of integral point sets in $\mathbb{E}^m$. Australas. J. Combin. **36**, 241–248 (2006)
10. Niven, I.: Irrational numbers. The Carus Mathematical Monographs, No. 11, The Mathematical Association of America. Distributed by John Wiley and Sons Inc., New York, NY (1956)
11. Piepmeyer, L.: The maximum number of odd integral distances between points in the plane. Discrete Comput. Geom. **16**(1), 113–115 (1996). https://doi.org/10.1007/BF02711135
12. Rosenfeld, M., Tien, N.L.: Forbidden subgraphs of the odd-distance graph. J. Graph Theory **75**(4), 323–330 (2014). https://doi.org/10.1002/jgt.21738
13. Soifer, A.: The mathematical coloring book. Springer, New York (2009)
14. Steinhardt, J.: On coloring the odd-distance graph. Electron. J. Combin. 16(1), Note 12, 7 (2009), http://www.combinatorics.org/Volume_16/Abstracts/v16i1n12.html

Polygons with Prescribed Angles in 2D and 3D

Alon Efrat[1], Radoslav Fulek[1(✉)], Stephen Kobourov[1], and Csaba D. Tóth[2,3]

[1] University of Arizona, Tucson, AZ, USA
{alon,rfulek,kobourov}@arizona.edu
[2] California State University Northridge, Los Angeles, CA, USA
csaba.toth@csun.edu
[3] Tufts University, Medford, MA, USA

Abstract. We consider the construction of a polygon P with n vertices whose turning angles at the vertices are given by a sequence $A = (\alpha_0, \ldots, \alpha_{n-1})$, $\alpha_i \in (-\pi, \pi)$, for $i \in \{0, \ldots, n-1\}$. The problem of realizing A by a polygon can be seen as that of constructing a straight-line drawing of a graph with prescribed angles at vertices, and hence, it is a special case of the well studied problem of constructing an *angle graph*. In 2D, we characterize sequences A for which every generic polygon $P \subset \mathbb{R}^2$ realizing A has at least c crossings, for every $c \in \mathbb{N}$, and describe an efficient algorithm that constructs, for a given sequence A, a generic polygon $P \subset \mathbb{R}^2$ that realizes A with the minimum number of crossings. In 3D, we describe an efficient algorithm that tests whether a given sequence A can be realized by a (not necessarily generic) polygon $P \subset \mathbb{R}^3$, and for every realizable sequence the algorithm finds a realization.

Keywords: Crossing number · Polygon · Spherical polygon · Angle graph

1 Introduction

Straight-line realizations of graphs with given metric properties have been one of the earliest applications of graph theory. Rigidity theory, for example, studies realizations of graphs with prescribed edge lengths, but also considers a mixed model where the edges have prescribed lengths or directions [4,13–15,21]. In this paper, we extend research on the so-called *angle graphs*, introduced by Vijayan [27] in the 1980s, which are geometric graphs with prescribed angles between adjacent edges. Angle graphs found applications in mesh flattening [29], and computation of conformal transformations [8,22] with applications in the theory of minimal surfaces and fluid dynamics.

Research on this paper is supported, in part, by NSF grants CCF-1740858, CCF-1712119, and DMS-1839274. The full version is available at http://arxiv.org/abs/2008.10192.

© Springer Nature Switzerland AG 2020
D. Auber and P. Valtr (Eds.): GD 2020, LNCS 12590, pp. 135–147, 2020.
https://doi.org/10.1007/978-3-030-68766-3_11

Viyajan [27] characterized planar angle graphs under various constraints, including the case when the graph is a cycle [27, Theorem 2] and when the graph is 2-connected [27, Theorem 3]. In both cases, the characterization leads to an efficient algorithm to find a planar straight-line drawing or report that none exists. Di Battista and Vismara [6] showed that for 3-connected angle graphs (e.g.., a triangulation), planarity testing reduces to solving a system of linear equations and inequalities in linear time. Garg [10] proved that planarity testing for angle graphs is NP-hard, disproving a conjecture by Viyajan. Bekos et al. [2] showed that the problem remains NP-hard even if all angles are multiples of $\pi/4$.

The problem of computing (straight-line) realizations of angle graphs can be seen as the problem of reconstructing a drawing of a graph from the given partial information. The research problems to decide if the given data uniquely determine the realization or its parameters of interest are already interesting for cycles, and were previously considered in the areas of conformal transformations [22] and visibility graphs [7].

In 2D, we are concerned with realizations of angle cycles as polygons minimizing the number of crossings which, as we shall see, depends only on the sum of the turning angles. It follows from the seminal work of Tutte [26] and Thomassen [25] that every positive instance of a 3-connected planar angle graph admits a crossing-free realization if the prescribed angles yield convex faces. Convexity will also play a crucial role in our proofs.

In 3D, we would like to determine whether a given angle cycle can be realized by a (not necessarily generic) polygon. Somewhat counter-intuitively, self-intersections cannot be always avoided in a polygon realizing the given angle cycle in 3D; we present examples below. Di Battista et al. [5] characterized oriented polygons that can be realized in $\mathbb{R}^3$ without self-intersections with axis-parallel edges of given directions. Patrignani [20] showed that recognizing crossing-free realizibility is NP-hard for graphs of maximum degree 6 in this setting.

Throughout the paper we assume modulo n arithmetic on the indices, and use $\langle .,. \rangle$ scalar product notation.

Angle sequences in 2-space. In the plane, an *angle sequence* A is a sequence $(\alpha_0, \ldots, \alpha_{n-1})$ of real numbers such that $\alpha_i \in (-\pi, \pi)$ for all $i \in \{0, \ldots, n-1\}$. Let $P \subset \mathbb{R}^2$ be an oriented polygon with n vertices $v_0, \ldots, v_{n-1}$ that appear in the given order along P, which is consistent with the given orientation of P. The *turning angle* of P at v_i is the angle in $(-\pi, \pi)$ between the vector $v_i - v_{i-1}$ and $v_{i+1} - v_i$. The sign of the angle is positive if rotating the plane, so that the vector $v_i - v_{i-1}$ points in the positive direction of the x-axis, makes the y-coordinate of $v_{i+1} - v_i$ positive. Otherwise, the angle nonpositive; see Fig. 1.

The oriented polygon P *realizes* the angle sequence A if the turning angle of P at v_i is equal to α_i, for $i = 0, \ldots, n-1$. A polygon P is *generic* if all its self-intersections are transversal (that is, proper crossings), vertices of P are distinct points, and no vertex of P is contained in a relative interior of an edge of P. Following the terminology of Viyajan [27], an *angle sequence* is *consistent* if there exists a generic closed polygon P with n vertices realizing A. For a

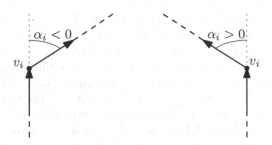

Fig. 1. A negative, or right, (on the left) and a positive, or left, (on the right) turning angle α_i at the vertex v_i of an oriented polygon.

polygon P that realizes an angle sequence $A = (\alpha_0, \ldots, \alpha_{n-1})$ in the plane, the *total curvature* of P is $\mathrm{TC}(P) = \sum_{i=0}^{n-1} \alpha_i$, and the *turning number* (also known as *rotation number*) of P is $\mathrm{tn}(P) = \mathrm{TC}(P)/(2\pi)$; it is known that $\mathrm{tn}(P) \in \mathbb{Z}$ in the plane [24].

The *crossing number*, denoted by $\mathrm{cr}(P)$, of a generic polygon is the number of self-crossings of P. The *crossing number* of a consistent angle sequence A is the minimum integer c, denoted by $\mathrm{cr}(A)$, such that there exists a generic polygon $P \in \mathbb{R}^2$ realizing A with $\mathrm{cr}(P) = c$. Our first main results is the following theorem.

Theorem 1. *For a consistent angle sequence* $A = (\alpha_0, \ldots, \alpha_{n-1})$ *in the plane, we have*

$$
\mathrm{cr}(A) = \begin{cases} 1 & \text{if } \sum_{i=0}^{n-1} \alpha_i = 0, \\ |k| - 1 & \text{if } \sum_{i=0}^{n-1} \alpha_i = 2k\pi \text{ and } k \neq 0. \end{cases}
$$

The proof of Theorem 1 can be easily converted into a weakly linear-time algorithm that constructs, for a given consistent sequence A, a generic polygon $P \subset \mathbb{R}^2$ that realizes A with the minimum number of crossings.

Angle sequences in 3-space and spherical polygonal linkages. In $\mathbb{R}^d$, $d \geq 3$, the sign of a turning angle no longer plays a role: The *turning angle* of an oriented polygon P at v_i is in $(0, \pi)$, and an angle sequence $A = (\alpha_0, \ldots, \alpha_{n-1})$ is in $(0, \pi)^n$. The unit-length direction vectors of the edges of P determine a spherical polygon P' in $\mathbb{S}^{d-1}$. Note that the turning angles of P correspond to the spherical lengths of the segments of P'. It is not hard to see that this observation reduces the problem of realizability of A by a polygon in $\mathbb{R}^d$ to the problem of realizability of A by a spherical polygon in $\mathbb{S}^{d-1}$, in the sense defined below, that additionally contains the origin $\mathbf{0}$ in the interior of its convex hull.

Let $\mathbb{S}^2 \subset \mathbb{R}^3$ denote the unit 2-sphere. A *great circle* $C \subset \mathbb{S}^2$ is the intersection of $\mathbb{S}^2$ with a 2-dimensional hyperplane in $\mathbb{R}^3$ containing $\mathbf{0}$. A *spherical line segment* is a connected subset of a great circle that does not contain a pair of antipodal points of $\mathbb{S}^2$. The *length* of a spherical line segment ab equals the measure of the central angle subtended by ab. A *spherical polygon* $P \subset \mathbb{S}^2$ is a closed curve consisting of finitely many spherical segments; and a spherical polygon $P = (\mathbf{u}_0, \ldots, \mathbf{u}_{n-1})$, $\mathbf{u}_i \in \mathbb{S}^2$, realizes an angle sequence $A = (\alpha_0, \ldots, \alpha_{n-1})$ if the spherical segment $(\mathbf{u}_{i-1}, \mathbf{u}_i)$ has (spherical) length α_i, for every i. As usual, the *turning angle* of P at $\mathbf{u}_i$ is the angle in $[0, \pi]$ between the tangents to $\mathbb{S}^2$ at $\mathbf{u}_i$ that are co-planar with the great circles containing $(\mathbf{u}_i, \mathbf{u}_{i+1})$ and $(\mathbf{u}_i, \mathbf{u}_{i-1})$. Unlike for polygons in $\mathbb{R}^2$ and $\mathbb{R}^3$, we do not put any constraints on turning angles of spherical polygons (i.e., angles 0 and π are allowed).

Regarding realizations of A by spherical polygons, we prove the following.

Theorem 2. *Let $A = (\alpha_0, \ldots, \alpha_{n-1})$, $n \geq 3$, be an angle sequence. There exists a generic polygon $P \subset \mathbb{R}^3$ realizing A if and only if $\sum_{i=0}^{n-1} \alpha_i \geq 2\pi$ and there exists a spherical polygon $P' \subset \mathbb{S}^2$ realizing A. Furthermore, P can be constructed efficiently if P' is given.*

Theorem 3. *There exists a constructive weakly polynomial-time algorithm to test whether a given angle sequence $A = (\alpha_0, \ldots, \alpha_{n-1})$ can be realized by a spherical polygon $P' \subset \mathbb{S}^2$.*

A simple exponential-time algorithm for realizability of angle sequences by spherical polygons follows from a known characterization [3, Theorem 2.5], which also implies that the order of angles in A does not matter for the spherical realizability. The topology of the configuration spaces of spherical polygonal linkages have also been studied [16]. Independently, Streinu et al. [19,23] showed that the configuration space of *noncrossing* spherical linkages is connected if $\sum_{i=0}^{n-1} \alpha_i \leq 2\pi$. However, these results do not seem to help prove Theorem 3.

The combination of Theorems 2 and 3 yields our second main result.

Theorem 4. *There exists a constructive weakly polynomial-time algorithm to test whether a given angle sequence $A = (\alpha_0, \ldots, \alpha_{n-1})$ can be realized by a polygon $P \subset \mathbb{R}^3$.*

Organization. We prove Theorem 1 in Sect. 2 and Theorems 2, 3, and 4 in Sect. 3. We finish with concluding remarks in Sect. 4.

2   Crossing Minimization in the Plane

The first part of the following lemma gives a folklore necessary condition for the consistency of an angle sequence A in the plane. The condition is also sufficient except when $k = 0$. The second part follows from a result of Grünbaum and Shepard [11, Theorem 6], using a decomposition due to Wiener [28]. We provide a proof for the sake of completeness.

Lemma 1. *If an angle sequence $A = (\alpha_0, \ldots, \alpha_{n-1})$ is consistent, then $\sum_{i=0}^{n-1} \alpha_i = 2k\pi$ for some $k \in \mathbb{Z}$. Furthermore, if $k \neq 0$ then $\mathrm{cr}(A) \geq |k| - 1$.*

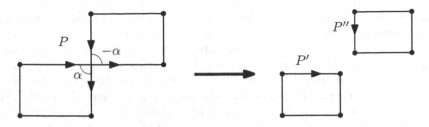

Fig. 2. Splitting an oriented closed polygon P at a self-crossing point into 2 oriented closed polygons P' and P'' such that $\operatorname{tn}(P) = \operatorname{tn}(P') + \operatorname{tn}(P'')$.

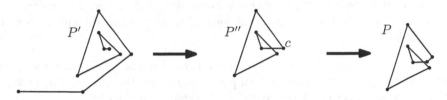

Fig. 3. Constructing a polygon P with $|\operatorname{tn}(P)| - 1$ crossings.

Proof. Let P be a polygon such that $\operatorname{cr}(A) = \operatorname{cr}(P)$. We prove that $\operatorname{cr}(A) \geq |k| - 1 = |\operatorname{tn}(P)| - 1$, by induction on $\operatorname{cr}(P)$.

We consider the base case, where $\operatorname{cr}(P) = 0$. By the Jordan-Schönflies curve theorem, P bounds a compact region homeomorphic to a disk. By a well-known fact, the internal angles at the vertices of P sum up to $(n - 2)\pi$. Since A is consistent, $\sum_{i=0}^{n-1} \alpha_i = 2k\pi$, and thus, $(n - 2)\pi = \sum_{i=0}^{n-1}(\pi - \alpha_i) = (n - 2k)\pi$ or $(n - 2)\pi = \sum_{i=0}^{n-1}(\pi + \alpha_i) = (n + 2k)\pi$, depending on the orientation of the polygon. The claim follows since $|\operatorname{tn}(P)| = |k| = 1$ in this case.

Refer to Fig. 2. In the inductive step, we have $\operatorname{cr}(P) \geq 1$. By splitting P into two closed parts P' and P'' at a self-crossing, we obtain a pair of closed polygons such that $\operatorname{tn}(P) = \operatorname{tn}(P') + \operatorname{tn}(P'')$. We have $\operatorname{cr}(P) \geq 1 + \operatorname{cr}(P') + \operatorname{cr}(P'') \geq 1 + |\operatorname{tn}(P')| - 1 + |\operatorname{tn}(P'')| - 1 \geq |\operatorname{tn}(P)| - 1$. Thus, the induction goes through, since both $\operatorname{cr}(P')$ and $\operatorname{cr}(P'')$ are smaller than $\operatorname{cr}(P)$. □

The following lemma shows that the lower bound in Lemma 1 is tight when $\alpha_i > 0$ for all $i \in \{0, \ldots, n-1\}$.

Lemma 2. *If $A = (\alpha_0, \ldots, \alpha_{n-1})$ is an angle sequence such that $\sum_{i=0}^{n-1} \alpha_i = 2k\pi$, $k \neq 0$, and $\alpha_i > 0$, for all i, then $\operatorname{cr}(A) \leq |k| - 1$.*

Proof. Refer to Fig. 3. In three steps, we construct a polygon P realizing A with $|\operatorname{tn}(P)| - 1$ self-crossings thereby proving $\operatorname{cr}(A) \leq |k| - 1 = |\operatorname{tn}(P)| - 1$. In the first step, we construct an oriented self-crossing-free polygonal line P' with $n+2$ vertices, whose first and last (directed) edges are parallel to the positive x-axis, and whose internal vertices have turning angles $\alpha_0, \ldots, \alpha_{n-1}$ in this order. We construct P' incrementally: The first edge has unit length starting from the origin; and every successive edge lies on a ray emanating from the endpoint of the previous edge. If the ray intersects neither the x-axis nor previous edges, then the next edge has unit length, otherwise its length is chosen to avoid any such intersection. In the second step, we prolong the last edge of P' until it creates the last self-intersection/crossing c and denote by P'' the resulting closed polygon composed of the part of P' from c to c via the prolonged part. By making the differences between the lengths of the edges of P' sufficiently large a prolongation of the last edge of P' has to eventually create at least one desired self-intersection. Hence, P'' is well-defined. Finally, we construct P realizing A from P'' by an appropriate modification of P'' in a small neighborhood of c without creating additional self-crossings. The number of self-crossings of P follows by the winding number of P with respect to the point just a bit north from the end vertex of P', which is k or $-k$.   □

To prove the upper bound in Theorem 1, it remains to consider the case that $A = (\alpha_0, \ldots, \alpha_{n-1})$ contains both positive and negative angles. The crucial notion in the proof is that of an (essential) sign change of A which we define next. Let $\beta_i = \sum_{j=0}^{i} \alpha_j \mod 2\pi$. Let $\mathbf{v}_i \in \mathbb{R}^2$ denote the unit vector $(\cos \beta_i, \sin \beta_i)$. Hence, $\mathbf{v}_i$ is the direction vector of the $(i+1)$-st edge of an oriented polygon P realizing A if the direction vector of the first edge of P is $(1,0) \in \mathbb{R}^2$. As observed by Garg [10], [Section 6], the consistency of A implies that $\mathbf{0}$ is a strictly positive convex combination of vectors $\mathbf{v}_i$, that is, there exist scalars $\lambda_0, \ldots, \lambda_{n-1} > 0$ such that $\sum_{i=0}^{n-1} \lambda \mathbf{v}_i = \mathbf{0}$ and $\sum_{i=0}^{n-1} \lambda_i = 1$.

The *sign change* of A is an index i such that $\alpha_i < 0$ and $\alpha_{i+1} > 0$, or vice versa, $\alpha_i > 0$ and $\alpha_{i+1} < 0$. Let $\mathrm{sc}(A)$ denote the number of sign changes of A. Note that the number of sign changes of A is even. A sign change i of a consistent angle sequence A is *essential* if $\mathbf{0}$ is not a strictly positive convex combination of $\{\mathbf{v}_0, \ldots, \mathbf{v}_{i-1}, \mathbf{v}_{i+1}, \ldots, \mathbf{v}_{n-1}\}$.

Lemma 3. *If* $A = (\alpha_0, \ldots, \alpha_{n-1})$ *is a consistent angle sequence, where* $\sum_{i=0}^{n-1} \alpha_i = 2k\pi$, $k \in \mathbb{Z}$, *and all sign changes are essential, then* $\mathrm{cr}(A) \leq \big| |k| - 1 \big|$.

Proof. We distinguish between two cases depending on whether $\sum_{i=0}^{n-1} \alpha_i = 0$.
Case 1: $\sum_{i=0}^{n-1} \alpha_i = 0$. Since $\sum_{i=0}^{n-1} \alpha_i = 0$, we have $\mathrm{sc}(A) \geq 2$. Since all sign changes are essential, for any two distinct sign changes $i \neq j$, we have $\mathbf{v}_i \neq \mathbf{v}_j$, therefore counting different vectors $\mathbf{v}_i$, where i is a sign change, is equivalent to counting essential sign changes. We show next that $\mathrm{sc}(A) = 2$.

Suppose, to the contrary, that $\mathrm{sc}(A) > 2$. Note that $\mathrm{sc}(A)$ is even, since the number of sign changes in a cyclic sequence of signs is even. Thus, we have $\mathrm{sc}(A) \geq 4$. Note that if $\mathbf{v}_i$ corresponds to an essential sign change i, then there exists an open halfplane bounded by a line through the origin that contains only $\mathbf{v}_i$ in $\{\mathbf{v}_0, \ldots, \mathbf{v}_{n-1}\}$. Thus, if i and i' are distinct essential sign changes, for any other essential sign change j we have that $\mathbf{v}_j$ is contained in a closed convex cone bounded by $-\mathbf{v}_i$ and $-\mathbf{v}_{i'}$ unless $-\mathbf{v}_i = \mathbf{v}_{i'}$. Hence, the only possibility for having 4 essential sign changes i, i', j, and j' is if they satisfy $\mathbf{v}_i = -\mathbf{v}_{i'}$, $\mathbf{v}_j = -\mathbf{v}_{j'}$ and $\mathbf{v}_i \neq \pm\mathbf{v}_j$. Since all i, i', j, and j' are sign changes, there exists a fifth vector $\mathbf{v}_k$, which implies that one of i, i', j, and j' is not essential (contradiction).

Assume w.l.o.g. that j and $n-1$ are the only two essential sign changes. We have that $\mathbf{v}_j \neq -\mathbf{v}_{n-1}$. Indeed, since the sign changes j and $n-1$ are essential, all the other vectors $\mathbf{v}_i$, other than $\mathbf{v}_j$ and $\mathbf{v}_{n-1}$, either must be contained in the same open half-plane defined by a line through $\mathbf{v}_j$ and $-\mathbf{v}_{n-1}$, which is impossible due to the consistency of A, or must be orthogonal to $\mathbf{v}_j$ and $\mathbf{v}_{n-1}$. Then due to the consistency of A, there exists a pair $\{i, i'\}$ such that $\mathbf{v}_i = -\mathbf{v}_{i'}$. However, j and $n-1$ are the only sign changes by assumption, and thus there exists some index ℓ such that $\mathbf{v}_\ell \neq \pm\mathbf{v}_i$ (contradiction).

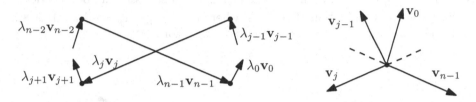

Fig. 4. The case of exactly 2 sign changes $n-1$ and j, both of which are essential, when $\sum_{i=0}^{n-1} \alpha_i = 0$. Both missing parts of the polygon on the left are convex chains.

It follows that $\mathbf{v}_j$ and $\mathbf{v}_{n-1}$ are not collinear, and we have that the remaining $\mathbf{v}_i$'s belong to the closed convex cone bounded by $-\mathbf{v}_j$ and $-\mathbf{v}_{n-1}$; refer to Fig. 4. Thus, we may assume that (i) $\beta_{n-1} = 0$, (ii) the sign changes of A are j and $n-1$, and (iii) $0 < \beta_0 < \ldots < \beta_j$ and $\beta_j > \beta_{j+1} > \ldots > \beta_{n-1} = 0$. Now, realizing A by a generic polygon with exactly 1 crossing between the line segments in the direction of $\mathbf{v}_j$ and $\mathbf{v}_{n-1}$ is a simple exercise.

Case 2: $\sum_{i=0}^{n-1} \alpha_i \neq 0$. We show that, unlike in the first case, none of the sign changes of A can be essential. Indeed, suppose j is an essential sign change, and let $A' = (\alpha'_0, \ldots, \alpha'_{n-2}) = (\alpha_0, \ldots, \alpha_{j-1}, \alpha_j + \alpha_{j+1}, \ldots, \alpha_{n-1})$ and $\beta'_i = \sum_{j=0}^{i} \alpha'_j$ mod 2π. Consider the unit vectors $\mathbf{v}'_0, \ldots, \mathbf{v}'_{n-2}$, where $\mathbf{v}'_i = (\cos\beta'_i, \sin\beta'_i)$. Since j is an essential sign change, there exists a nonzero vector $\mathbf{v}$ such that $\langle \mathbf{v}, \mathbf{v}_j \rangle > 0$ and $\langle \mathbf{v}, \mathbf{v}'_i \rangle \leq 0$ for all i. Hence, by symmetry, we may assume that $0 \leq \beta'_i \leq \pi$, for all i. Since j is a sign change, we have $-\pi < \alpha'_i < \pi$ for all i, consequently $\beta'_j = \sum_{i=0}^{j} \alpha'_i$ mod $2\pi = \sum_{i=0}^{j} \alpha'_i$, which in turn implies, by Lemma 1, that $0 = \beta'_{n-2} = \sum_{i=0}^{n-2} \alpha'_i = \sum_{i=0}^{n-1} \alpha_i$ (contradiction).

We have shown that A has no sign changes. By Lemma 2, we have $\mathrm{cr}(A) \leq |k| - 1$, which concludes the proof. $\qquad\square$

Theorem 1. *For a consistent angle sequence $A = (\alpha_0, \ldots, \alpha_{n-1})$ in the plane, we have*

$$\mathrm{cr}(A) = \begin{cases} 1 & \text{if } \sum_{i=0}^{n-1} \alpha_i = 0, \\ |k| - 1 & \text{if } \sum_{i=0}^{n-1} \alpha_i = 2k\pi \text{ and } k \neq 0. \end{cases}$$

Proof. The claimed lower bound $\mathrm{cr}(A) \geq \left| |k| - 1 \right|$ on the crossing number of A follows by Lemma 1, in the case when $k \neq 0$, and the result of Viyajan [27, Theorem 2] in the case when $k = 0$. It remains to prove the upper bound $\mathrm{cr}(A) \leq \left| |k| - 1 \right|$.

We proceed by induction on n. In the base case, we have $n = 3$. Then P is a triangle, $\sum_{i=0}^{2} \alpha_i = \pm 2\pi$, and $\mathrm{cr}(A) = 0$, as required. In the inductive step, assume $n \geq 4$, and that the claim holds for all shorter angle sequences. Let $A = (\alpha_0, \ldots, \alpha_{n-1})$ be an angle sequence with $\sum_{i=0}^{n-1} \alpha_i = 2k\pi$.

If A has no sign changes or if all sign changes are essential, then Lemma 2 or Lemma 3 completes the proof. Otherwise, we have at least one nonessential sign change s. Let $A' = (\alpha_0', \ldots, \alpha_{n-2}') = (\alpha_0, \ldots, \alpha_{s-1}, \alpha_s + \alpha_{s+1}, \ldots, \alpha_{n-1})$. Note that $\sum_{i=0}^{n-2} \alpha_i' = 2k\pi$. Since the sign change s is nonessential, $\mathbf{0}$ is a strictly positive convex combination of the β_i''s, where $\beta_i' = \sum_{j=0}^{i} \alpha_j' \mod 2\pi$. Indeed, this follows from $\beta_i' = \beta_i$, for $i < s$, and $\beta_i' = \beta_{i+1}$, for $i \geq s$.

Fig. 5. Re-introducing the j-th vertex to a polygon realizing A' in order to obtain a polygon realizing A.

Refer to Fig. 5. Hence, by applying the induction hypothesis we obtain a realization of A' as a generic polygon P' with $\left| |k| - 1 \right|$ crossing. A generic polygon realizing A is then obtained by modifying P in a small neighborhood of one of its vertices without introducing any additional crossing, similarly as in the paper by Guibas et al. [12]. $\qquad\qquad\square$

3   Realizing Angle Sequences in 3-Space

In this section, we describe a polynomial-time algorithm to decide whether an angle sequence $A = (\alpha_0, \ldots \alpha_{n-1})$, where $0 < \alpha_i < \pi$ for all i, can be realized as a polygon in $\mathbb{R}^3$.

We remark that our problem can be expressed as solving a system of polynomial equations, where $3n$ variables describe the coordinates of the n vertices of P, and each of n equations is obtained by the cosine theorem applied for a vertex and two incident edges of P. However, it is not clear to us how to solve such a system efficiently.

By Fenchel's theorem in differential geometry [9], the total curvature of any smooth curve in $\mathbb{R}^d$ is at least 2π. Fenchel's theorem has been adapted to closed polygons [24, Theorem 2.4], and it gives the following a necessary condition for an angle sequence A to have a realization in $\mathbb{R}^d$, for all $d \geq 2$:

$$\sum_{i=0}^{n-1} \alpha_i \geq 2\pi. \tag{1}$$

We show that a slightly stronger condition is both necessary and sufficient, hence it characterizes realizable angle sequences in $\mathbb{R}^3$.

Lemma 4. *Let* $A = (\alpha_0, \ldots, \alpha_{n-1})$, $n \geq 3$, *be an angle sequence. There exists a polygon* $P \subset \mathbb{R}^3$ *realizing* A *if and only if there exists a spherical polygon* $P' \subset \mathbb{S}^2$ *realizing* A *such that* $\mathbf{0} \in \mathrm{relint}(\mathrm{conv}(P'))$ *(relative interior of* $\mathrm{conv}(P')$*). Furthermore,* P *can be constructed efficiently if* P' *is given.*

Proof. Assume that an oriented polygon $P = (v_0, \ldots, v_{n-1})$ realizes A in $\mathbb{R}^3$. Let $\mathbf{u}_i = (v_{i+1} - v_i)/\|v_{i+1} - v_i\| \in \mathbb{S}^2$ be the unit direction vector of the edge $v_i v_{i+1}$ of P according to its orientation. Then $P' = (\mathbf{u}_0, \ldots, \mathbf{u}_{n-1})$ is a spherical polygon that realizes A. Suppose, for the sake of contradiction, that $\mathbf{0}$ is not in the relative interior of $\mathrm{conv}(P')$. Then there is a plane H that separates $\mathbf{0}$ and P', that is, if $\mathbf{n}$ is the normal vector of H, then $\langle \mathbf{n}, \mathbf{u}_i \rangle > 0$ for all $i \in \{0, \ldots, n-1\}$. This implies $\langle \mathbf{n}, (v_{i+1} - v_i) \rangle > 0$ for all i, hence $\langle \mathbf{n}, \sum_{i=1}^{n-1}(v_{i+1} - v_i) \rangle > 0$, which contradicts the fact that $\sum_{i=1}^{n-1}(v_{i+1} - v_i) = \mathbf{0}$, and $\langle \mathbf{n}, \mathbf{0} \rangle = 0$.

Conversely, assume that a spherical polygon P' realizes A, with edge lengths $\alpha_0, \ldots, \alpha_{n-1} > 0$. If all the vertices of P' lie on a common great circle, then $\mathbf{0} \in \mathrm{relint}(\mathrm{conv}(P'))$ implies $\sum_{i=0}^{n-1} \pm \alpha_i = 0 \mod 2\pi$, where the sign is determined by the direction (cw. or ccw.) in which a particular segment of P' traverses the common great circle according to its orientation. As observed by Garg [10, Section 6], the signed angle sequence is consistent in this case due to the assumption that $\mathbf{0} \in \mathrm{relint}(\mathrm{conv}(P'))$. Thus, we obtain a realization of A that is contained in a plane.

Otherwise we may assume that $\mathbf{0} \in \mathrm{int}(\mathrm{conv}(P'))$. By Carathéodory's theorem [17, Thereom 1.2.3], P' has 4 vertices whose convex combination is the origin $\mathbf{0}$. Then we can express $\mathbf{0}$ as a strictly positive convex combination of *all* vertices of P'. The coefficients in the convex combination encode the lengths of the edges of a polygon P realizing A, which concludes the proof in this case.

We now show how to compute strictly positive coefficients in strongly polynomial time. Let $\mathbf{c} = \frac{1}{n}\sum_{i=0}^{n-1} \mathbf{u}_i$ be the centroid of the vertices of P'. If $\mathbf{c} = \mathbf{0}$, we are done. Otherwise we can find a tetrahedron $T = \mathrm{conv}\{\mathbf{u}_{i_0}, \ldots, \mathbf{u}_{i_3}\}$ such that $\mathbf{0} \in T$ and such that the ray from $\mathbf{0}$ in the direction $-\mathbf{c}$ intersects $\mathrm{int}(T)$, by solving an LP feasibility problem in $\mathbb{R}^3$. By computing the intersection of the ray with the faces of T, we find the maximum $\mu > 0$ such that $-\mu\mathbf{c} \in \partial T$ (the boundary of T). We have $-\mu\mathbf{c} = \sum_{j=0}^{3} \lambda_j \mathbf{u}_{i_j}$ and $\sum_{j=0}^{3} \lambda_j = 1$ for suitable coefficients $\lambda_j \geq 0$. Now $\mathbf{0} = \mu\mathbf{c} - \mu\mathbf{c} = \frac{\mu}{n}\sum_{i=0}^{n-1} \mathbf{u}_i + \sum_{j=0}^{3} \lambda_j \mathbf{u}_{i_j}$ is a strictly positive convex combination of the vertices of P'. $\square$

It is easy to find an angle sequence A that satisfies (1) but does not correspond to a spherical polygon P'. Consider, for example, $A = (\pi - \varepsilon, \pi - \varepsilon, \pi - \varepsilon, \varepsilon)$, for some small $\varepsilon > 0$. Points in $\mathbb{S}^2$ at (spherical) distance $\pi - \varepsilon$ are nearly antipodal. Hence, the endpoints of a polygonal chain $(\pi - \varepsilon, \pi - \varepsilon, \pi - \varepsilon)$ are nearly antipodal as well, and cannot be connected by an edge of (spherical) length ε. Thus a spherical polygon cannot realize A.

Algorithms. In the remainder of this section, we show how to find a realization $P \subset \mathbb{R}^3$ or report that none exists, in polynomial time. Our first concern is to decide whether an angle sequence is realizable by a spherical polygon.

Proof. Let $A = (\alpha_0, \ldots, \alpha_{n-1}) \in (0, \pi)^n$ be a given angle sequence. Let $\mathbf{n} = (0, 0, 1) \in \mathbb{S}^2$, that is, $\mathbf{n}$ is the north pole. For $i \in \{0, 1, \ldots, n-1\}$, let $U_i \subseteq \mathbb{S}^2$ be the locus of the end vertices $\mathbf{u}_i$ of all (spherical) polygonal lines $P_i' = (\mathbf{n}, \mathbf{u}_0, \ldots, \mathbf{u}_i)$ with edge lengths $\alpha_0, \ldots, \alpha_{i-1}$. It is clear that A is realizable by a spherical polygon P' if and only if $\mathbf{n} \in U_{n-1}$.

Note that for all $i \in \{0, \ldots, n-1\}$, the set U_i is invariant under rotations about the z-axis, since $\mathbf{n}$ is a fixed point and rotations are isometries. We show how to compute the sets U_i, $i \in \{0, \ldots, n-1\}$, efficiently.

We define a *spherical zone* as a subset of $\mathbb{S}^2$ between two horizontal planes (possibly, a circle, a spherical cap, or a pole). Recall the parameterization of $\mathbb{S}^2$ using spherical coordinates (cf. Fig. 6 (left)): for every $\mathbf{v} \in \mathbb{S}^2$, $\mathbf{v}(\psi, \varphi) = (\sin\psi \sin\varphi, \cos\psi \sin\varphi, \cos\varphi)$, with longitude $\psi \in [0, 2\pi)$ and polar angle $\varphi \in [0, \pi]$, where the *polar angle* φ is the angle between $\mathbf{v}$ and $\mathbf{n}$. Using this parameterization, a spherical zone is a Cartesian product $[0, 2\pi) \times I$ for some circular arc $I \subset [0, \pi]$. In the remainder of the proof, we associate each spherical zone with such a circular arc I.

We define additions and subtraction on polar angles $\alpha, \beta \in [0, \pi]$ by

$$\alpha \oplus \beta = \min\{\alpha + \beta, 2\pi - (\alpha + \beta)\}, \quad \alpha \ominus \beta = \max\{\alpha - \beta, \beta - \alpha\};$$

see Fig. 6 (right). (This may be interpreted as addition mod 2π, restricted to the quotient space defined by the equivalence relation $\varphi \sim 2\pi - \varphi$.)

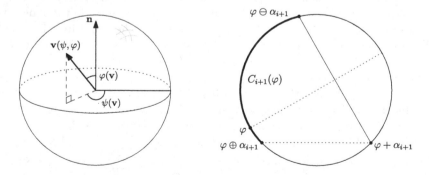

Fig. 6. Parametrization of the unit vectors (left). Circular arc $C_{i+1}(\varphi)$ (right).

We show that U_i is a spherical zone for all $i \in \{0, \ldots, n-1\}$, and show how to compute the intervals $I_i \subset [0, \pi]$ efficiently. First note that U_0 is a circle at (spherical) distance α_0 from $\mathbf{n}$, hence U_0 is a spherical zone with $I_0 = [\alpha_0, \alpha_0]$.

Assume that U_i is a spherical zone associated with $I_i \subset [0, \pi]$. Let $\mathbf{u}_i \in U_i$, where $\mathbf{u}_i = \mathbf{v}(\psi, \varphi)$ with $\psi \in [0, 2\pi)$ and $\varphi \in I_i$. By the definition U_i, there exists a polygonal line $(\mathbf{n}, \mathbf{u}_0, \ldots, \mathbf{u}_i)$ with edge lengths $\alpha_0, \ldots, \alpha_i$. The locus of points in $\mathbb{S}^2$ at distance α_{i+1} from u_i is a circle; the polar angles of the points in the circle form an interval $C_{i+1}(\varphi)$. Specifically (see Fig. 6 (right)), we have

$$C_{i+1}(\varphi) = [\min\{\varphi \ominus \alpha_{i+1}, \varphi \oplus \alpha_{i+1}\}, \max\{\varphi \ominus \alpha_{i+1}, \varphi \oplus \alpha_{i+1}\}].$$

By rotational symmetry, $U_{i+1} = [0, 2\pi) \times I_{i+1}$, where $I_{i+1} = \bigcup_{\varphi \in I_i} C_{i+1}(\varphi)$. Consequently, $I_{i+1} \subset [0, \pi]$ is connected, and hence, I_{i+1} is an interval. Therefore U_{i+1} is a spherical zone. As $\varphi \oplus \alpha_{i+1}$ and $\varphi \ominus \alpha_{i+1}$ are piecewise linear functions of φ, we can compute I_{i+1} using $O(1)$ arithmetic operations.

We can construct the intervals $I_0, \ldots, I_{n-1} \subset [0, \pi]$ as described above. If $0 \notin I_{n-1}$, then $\mathbf{n} \notin U_{n-1}$ and A is not realizable. Otherwise, we can compute the vertices of a spherical realization $P' \subset \mathbb{S}^2$ by backtracking. Put $\mathbf{u}_{n-1} = \mathbf{n} = (0, 0, 1)$. Given $\mathbf{u}_i = \mathbf{v}(\psi, \varphi)$, we choose $\mathbf{u}_{i-1}$ as follows. Let $\mathbf{u}_{i-1}$ be $\mathbf{v}(\psi, \varphi \oplus \alpha_i)$ or $\mathbf{v}(\psi, \varphi \ominus \alpha_i)$ if either of them is in U_{i-1} (break ties arbitrarily). Else the spherical circle of radius α_i centered at $\mathbf{u}_i$ intersects the boundary of U_{i-1}, and then we choose $\mathbf{u}_{i-1}$ to be an arbitrary such intersection point. The decision algorithm (whether $0 \in I_{n-1}$) and the backtracking both use $O(n)$ arithmetic operations.   □

Enclosing the Origin. Theorem 3 provides an efficient algorithm to test whether an angle sequence can be realized by a spherical polygon, however, Lemma 4 requires a spherical polygon P' whose convex hull contains the origin in its relative interior. We show that this is always possible if a realization exists and $\sum_{i=0}^{n-1} \alpha_i \geq 2\pi$. The general strategy in the inductive proof of this claim is to incrementally modify P' by changing the turning angle at one of its vertices to 0 or π. This allows us to reduce the number of vertices of P' and apply induction. (The proof of the following lemma is deferred to the appendix.)

Lemma 5. *Given a spherical polygon P' that realizes an angle sequence $A = (\alpha_0, \ldots, \alpha_{n-1})$, $n \geq 3$, with $\sum_{i=0}^{n-1} \alpha \geq 2\pi$, we can compute in polynomial time a spherical polygon P'' realizing A such that $\mathbf{0} \in \mathrm{relint}(\mathrm{conv}(P''))$.*

The combination of Theorem 3 with Lemmas 4–5 yields Theorems 2 and 4. The proof of Lemma 5 can be turned into an algorithm with running time polynomial in n if we assume that every arithmetic operation can be carried out in $O(1)$ time. Nevertheless, we get only a weakly polynomial running time, since we are unable to guarantee a polynomial size encoding of the numerical values that are computed in the process of constructing a spherical polygon realizing A that contains $\mathbf{0}$ in its convex hull in the proof of Lemma 5.

4   Conclusion

We devised efficient algorithms to realize a consistent angle cycle with the minimum number of crossings in 2D. In 3D, we can test efficiently whether a given angle sequence is realizable, and find a realization if one exists. However, it remains an open problem to find an efficient algorithms that computes the minimum number of crossings in generic realizations. There exist angle sequences that are realizable in 3D, but every generic realization has crossings. It is not difficult to see that crossings are unavoidable only if every 3D realization of an angle sequence A is contained in a plane, which is the case, for example, when $A = (\pi - \varepsilon, \ldots, \pi - \varepsilon, (n-1)\varepsilon)$, for odd $n \geq 5$ which is the length of A. Thus, an efficient algorithm for this problem would follow by Theorem 1, once one can test efficiently whether A admits a fully 3D realization. The evidence that we have points to the following conjecture, whose "only if" part we can prove.

Conjecture 1. An angle sequence $A = (\alpha_0, \ldots, \alpha_{n-1})$, where $\alpha_i \in (0, \pi)$ and $n \geq 3$, that can be realized by a polygon in $\mathbb{R}^3$, has a realization by a self-intersection free polygon in $\mathbb{R}^3$ if and only if n is odd or $\sum_{i=0}^{n-1}(\pi - \alpha_i) \neq \pi$.

It can be seen that Conjecture 1 is equivalent to the claim that every realization A in $\mathbb{R}^3$ has a self-intersection if and only if A can be realized in $\mathbb{R}^2$ as a *thrackle*, that is, a polygon where every pair of nonadjacent edges cross each other. Here, we keep all the angles in A positive.

Can our results in $\mathbb{R}^2$ or $\mathbb{R}^3$ be extended to broader interesting classes of graphs? A natural analog of our problem in $\mathbb{R}^3$ would be a construction of triangulated spheres with prescribed dihedral angles, discussed in a recent paper by Amenta and Rojas [1]. For convex polyhedra, Mazzeo and Montcouquiol [18] proved, settling Stoker's conjecture, that dihedral angles determine face angles.

References

1. Amenta, N., Rojas, C.: Dihedral deformation and rigidity. Comput. Geom. Theor. Appl. **90**, 101657 (2020)
2. Bekos, M.A., Förster, H., Kaufmann, M.: On smooth orthogonal and octilinear drawings: relations, complexity and Kandinsky drawings. Algorithmica **81**(5), 2046–2071 (2019)
3. Richard, E., Buckman, R., Schmitt, N.: Spherical polygons and unitarization, Preprint (2002). http://www.gang.umass.edu/reu/2002/polygon.html
4. Clinch, K., Jackson, B., Keevash, P.: Global rigidity of direction-length frameworks. J. Combi. Theor. Ser. B **145**, 145–168 (2020)
5. Di Battista, G., Kim, E., Liotta, G., Lubiw, A., Whitesides, S.: The shape of orthogonal cycles in three dimensions. Discrete Comput. Geom. **47**(3), 461–491 (2012)
6. Di Battista, G., Vismara, L.: Angles of planar triangular graphs. SIAM J. Discrete Math. **9**(3), 349–359 (1996)
7. Disser, Y., Mihalák, M., Widmayer, P.: A polygon is determined by its angles. Comput. Geom. Theor. Appl. **44**, 418–426 (2011)

8. Driscoll, T.A., Vavasis, S.A.: Numerical conformal mapping using cross-ratios and Delaunay triangulation. SIAM J. Sci. Comput. **19**(6), 1783–1803 (1998)
9. Fenchel, W.: On the differential geometry of closed space curves. Bull. Am. Math. Soc. **57**(1), 44–54 (1951)
10. Garg, A.: New results on drawing angle graphs. Comput. Geom. Theor. Appl. **9**(1–2), 43–82 (1998)
11. Grünbaum, B., Shephard, G.C.: Rotation and winding numbers for planar polygons and curves. Trans. Am. Math. Soc. **322**(1), 169–187 (1990)
12. Guibas, L., Hershberger, J., Suri, S.: Morphing simple polygons. Discrete Comput. Geom. **24**(1), 1–34 (2000)
13. Jackson, B., Jordán, T.: Globally rigid circuits of the direction-length rigidity matroid. J. Comb. Theor. Ser. B **100**(1), 1–22 (2010)
14. Jackson, B., Keevash, P.: Necessary conditions for the global rigidity of direction-length frameworks. Discrete Comput. Geom. **46**(1), 72–85 (2011)
15. Lee-St. John, S., Streinu, I.: Angular rigidity in 3D: Combinatorial characterizations and algorithms. In: Proceedings of 21st Canadian Conference on Computational Geometry (CCCG), pp. 67–70 (2009)
16. Kapovich, M., Millson, J.J.: On the moduli space of a spherical polygonal linkage. Can. Math. Bull. **42**, 307–320 (1999)
17. Matoušek, J.: Lectures on Discrete Geometry. Graduate Texts in Mathematics, vol. 212, Springer-Verlag, New York (2002)
18. Mazzeo, R., Montcouquiol, G.: Infinitesimal rigidity of cone-manifolds and the Stoker problem for hyperbolic and Euclidean polyhedra. J. Differ. Geom. **87**(3), 525–576 (2011)
19. Panina, G., Streinu, I.: Flattening single-vertex origami: the non-expansive case. Comput. Geom. Theor. Appl. **43**(8), 678–687 (2010)
20. Patrignani, M.: Complexity results for three-dimensional orthogonal graph drawing. J. Discrete Algorithms **6**(1), 140–161 (2008)
21. Saliola, F., Whiteley, W.: Constraining plane configurations in CAD: Circles, lines, and angles in the plane. SIAM J. Discrete Math. **18**(2), 246–271 (2004)
22. Snoeyink, J.: Cross-ratios and angles determine a polygon. Discrete Comput. Geom. **22**(4), 619–631 (1999)
23. Streinu, I., Whiteley, W.: Single-vertex origami and spherical expansive motions. In: Akiyama, J., Kano, M., Tan, X. (eds.) JCDCG 2004. LNCS, vol. 3742, pp. 161–173. Springer, Heidelberg (2005). https://doi.org/10.1007/11589440_17
24. Sullivan, J.M.: Curves of finite total curvature. In: Bobenko, A.I., Sullivan, J.M., Schröder, P., Ziegler, G.M. (eds.) Discrete Differential Geometry, pp. 137–161. Birkhäuser, Basel (2008)
25. Thomassen, C.: Planarity and duality of finite and infinite graphs. J. Comb. Theor. Ser. B **29**, 244–271 (1980)
26. Tutte, W.T.: How to draw a graph. Proc. London Math. Soc. **3**(1), 743–767 (1963)
27. Vijayan, G.: Geometry of planar graphs with angles. In: Proceedings of 2nd ACM Symposium on Computational Geometry, pp. 116–124 (1986)
28. Wiener, C.: Über Vielecke und Vielflache. Teubner, Leipzig (1864)
29. Zayer, R., Rössler, C., Seidel, H.P.: Variations on angle based flattening. In: Dodgson, N.A., Floater, M.S., Sabin, M.A. (eds.) Advances in Multiresolution for Geometric Modelling, Mathematics and Visualization, pp. 187–199. Springer, Heidelberg (2005). https://doi.org/10.1007/3-540-26808-1_10

Restricted Drawings of Special Graph Classes

On Mixed Linear Layouts
of Series-Parallel Graphs

Patrizio Angelini[1] ⓘ, Michael A. Bekos[2,5(✉)] ⓘ, Philipp Kindermann[3,5] ⓘ,
and Tamara Mchedlidze[4] ⓘ

[1] John Cabot University, Rome, Italy
`pangelini@johncabot.edu`
[2] Universität Tübingen, Tübingen, Germany
`bekos@informatik.uni-tuebingen.de`
[3] Universität Würzburg, Würzburg, Germany
`philipp.kindermann@uni-wuerzburg.de`
[4] Karlsruhe Institute of Technology (KIT), Karlsruhe, Germany
`mched@iti.uka.de`
[5] Universität Passau, Passau, Germany

Abstract. A mixed s-stack q-queue layout of a graph consists of a linear order of its vertices and of a partition of its edges into s stacks and q queues, such that no two edges in the same stack cross and no two edges in the same queue nest. In 1992, Heath and Rosenberg conjectured that every planar graph admits a mixed 1-stack 1-queue layout. Recently, Pupyrev disproved this conjectured by demonstrating a planar partial 3-tree that does not admit a 1-stack 1-queue layout. In this note, we strengthen Pupyrev's result by showing that the conjecture does not hold even for 2-trees, also known as series-parallel graphs.

Keywords: Mixed linear layouts · Queue layouts · Book embeddings · Series-parallel graphs

1 Introduction

Over the years, linear layouts of graphs have been a fruitful subject of intense research, which has resulted in several remarkable results both of combinatorial and of algorithmic nature; see, e.g., [7,14,19,21,27,29]. A linear layout of graph is defined by a total order of its vertex-set and by a partition of its edge-set into a number of subsets, called *pages*. By imposing different constraints on the edges that may reside in the same page, one obtains different types of linear layouts; see [1,8,21,25,29]. The most notable ones are arguably the stack and the queue layouts (the former are commonly referred to as *book embeddings* in the literature), as is evident from the numerous papers that have been published over the years; see [15] for a short introduction.

In a *stack* (*queue*) *layout* of a graph, no two indepedent edges of the same page, called *stack* (*queue*) in this context, are allowed to cross (nest, resp.)

D. Auber and P. Valtr (Eds.): GD 2020, LNCS 12590, pp. 151–159, 2020.
https://doi.org/10.1007/978-3-030-68766-3_12

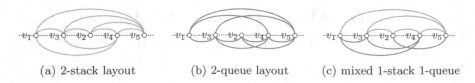

(a) 2-stack layout   (b) 2-queue layout   (c) mixed 1-stack 1-queue

Fig. 1. Illustration of different linear layouts of the complete graph on five vertices $v_1, \ldots, v_5$ minus the edge (v_1, v_2).

with respect to the underlying linear order; see [7] and [21]. In other words, the endpoints of the edges assigned to the same stack follow the last-in-first-out model in the underlying linear order, while the endpoints of the edges assigned to the same queue follow the first-in-first-out model; see Fig. 1. The minimum number of stacks (queues) required by any of the stack (queue) layouts of a graph is commonly referred to as its *stack-number* (*queue-number*, resp.). Accordingly, the stack-number (queue-number) of a class of graphs is the maximum stack-number (queue-number, resp.) over all its members.

Known Results. A large body of the literature is devoted to the study of bounds on the stack- and the queue-number of different classes of graphs.

For stack layouts, the most remarkable result is due to Yannakakis, who back in 1986 showed that every planar graph admits a 4-stack layout [28,29]. Recently, Bekos et al. [6] and Yannakakis [30] independently established that the stack-number of the class of planar graphs is 4, by demonstrating planar graphs that do not admit 3-stack layouts. Certain subclasses of planar graphs, however, allow for layouts with fewer than four stacks, e.g., 4-connected planar graphs [24], series-parallel graphs [26], planar 3-trees [19], and others [5,9,16–18,22,23].

For queue layouts, Dujmović et al. [14] recently showed that every planar graph admits a 49-queue layout, improving over previously known logarithmic bounds [4,11–13]. However, the exact queue-number of the class of planar graphs is not yet known, as the currently best-known lower bound is 4 [2]. Again, several subclasses of planar graphs allow for layouts with significantly fewer than 49 queues, e.g., outerplanar graphs [20], series-parallel graphs [26] and planar 3-trees [2].

Motivation. Back in 1992, Heath and Rosenberg [21] proposed a natural generalization of stack and queue layouts, called *mixed s-stack q-queue layout*, that supports s stack-pages and q queue-pages. In their seminal paper [21], they conjectured that every planar graph admits a mixed 1-stack 1-queue layout. However, Pupyrev [25] recently showed that the conjecture does not hold even for partial planar 3-trees. This negative result naturally raises the question whether the conjecture holds for other subclasses of planar graphs. To this end, Pupyrev conjectured that bipartite planar graphs admit mixed 1-stack 1-queue layouts.

Our Contribution. We make a step forward in understanding which subclasses of planar graphs admit mixed 1-stack 1-queue layouts by providing a negative certificate for the class of 2-trees (also known as maximal series-parallel graphs).

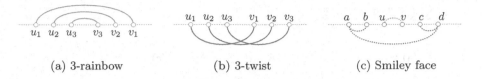

(a) 3-rainbow   (b) 3-twist   (c) Smiley face

Fig. 2. Illustration of: (a) a 3-rainbow, (b) a 3-twist, and (c) a smiley face.

This improves upon the partial planar 3-tree negative example by Pupyrev [25]. Note that 2-trees admit both 2-stack layouts and 3-queue layouts [26].

Preliminaries. A *linear order* $\prec$ of a graph G is a total order of its vertices. Let $F = \{(u_i, v_i); \ i = 1, \ldots, k\}$ be a set of $k \geq 2$ independent edges such that $u_i \prec v_i$, for all $1 \leq i \leq k$. If the order is $[u_1, \ldots, u_k, v_k, \ldots, v_1]$, then we say that the edges of F form a k-*rainbow*, while if the order is $[u_1, \ldots, u_k, v_1, \ldots, v_k]$, then the edges of F form a k-*twist*. Two edges that form a 2-twist (2-rainbow) are referred to as *crossing* (*nested*, resp.). A *stack* (*queue*) is a set of pairwise non-crossing (non-nested, resp.) edges. A *mixed s-stack q-queue layout* $\mathcal{L}$ of G consists of a linear order $\prec$ of G and a partition of the edges of G into s stacks and q queues; for short, we refer to $\mathcal{L}$ as *mixed layout* when $s = q = 1$. An edge in a stack (queue) in $\mathcal{L}$ is called a *stack-edge* (*queue-edge*, resp.).

The operation of *attaching* a vertex u to an edge (v, w) of a graph G consists of adding to G vertex u and edges (u, v) and (u, w). Vertex u is said to be *attached* or being an *attachment* of (v, w). A 2-*tree* is a graph obtained from an edge by repeatedly attaching a vertex to an edge. Consider a mixed s-stack q-queue layout $\mathcal{L}$ of a 2-tree. We say that a vertex u attached to an edge (v, w) is a *stack-attachment* (*queue-attachment*) of (v, w) if both (u, v) and (u, w) are stack-edges (queue-edges, resp.) in $\mathcal{L}$. Vertex u is a *mixed-attachment* of (v, w) if one of (u, v) and (u, w) is a queue-edge and the other is a stack-edge in $\mathcal{L}$.

2   The Main Result

In this section, we define a family $\{G(k, \ell); \ k, \ell \in \mathbb{N}^+\}$ of 2-trees, and we prove that infinitely many members of it do not admit mixed layouts. For $\ell \geq 1$, $G(1, \ell)$ is an edge; for $k > 1$, $G(k, \ell)$ is obtained from $G(k - 1, \ell)$ by attaching ℓ vertices to each edge of it. For convenience, we let $\overline{G}(k, \ell)$ be the graph $G(k, \ell) \setminus G(k - 1, \ell)$, that is, the graph induced by the edges that belong to $G(k, \ell)$ but not to $G(k - 1, \ell)$. In the following lemmas, we study properties of a mixed layout of graph $G(k, \ell)$.

Lemma 1. *Let $\mathcal{L}$ be a mixed layout of $G(k, \ell)$ with $k > 1, \ell > 2$. Then, every edge of $G(k - 1, \ell)$ has at most two stack-attachments in $\mathcal{L}$.*

Proof. Let (a, b) be an edge of $G(k - 1, \ell)$ and assume to the contrary that there exist three stack-attachments u, v and w of $\overline{G}(k, \ell)$ attached to (a, b) in $\mathcal{L}$. Neglecting edge (a, b), vertices a, b, u, v and w induce a $K_{2,3}$ in $G(k, \ell)$, whose

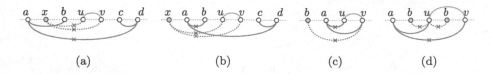

Fig. 3. Illustrations for the proofs (a–b) of Lemma 2, and (c–d) of Lemma 3.

edges are all stack-edges in $\mathcal{L}$. This is a contradiction, since the subgraph induced by the stack-edges of $G(k, \ell)$ must be outerplanar [7], while $K_{2,3}$ is not.   □

A *smiley face* $\langle a, b, u, v, c, d \rangle$ in a mixed layout consists of six vertices $a \prec b \prec u \prec v \prec c \prec d$ and four edges (a, b), (c, d), (a, d), and (u, v), such that (a, b), (c, d), and (a, d) are queue-edges, and thus (u, v) is a stack-edge; see Fig. 2c.

Lemma 2. *Let $\mathcal{L}$ be a mixed layout of $G(k, \ell)$ with $k > 1, \ell > 2$. Then, a smiley face cannot be formed by the vertices of $G(k - 1, \ell)$ in $\mathcal{L}$.*

Proof. Assume to the contrary that a smiley face $\langle a, b, u, v, c, d \rangle$ is formed in $\mathcal{L}$ by vertices of $G(k-1, \ell)$. Consider any vertex x of $\overline{G}(k, \ell)$ attached to the stack-edge (u, v). If $a \prec x \prec d$, then the queue-edge (a, d) forms a 2-rainbow both with (u, x) and with (v, x); see Fig. 3a. If $x \prec a$, then the queue-edge (a, b) forms a 2-rainbow both with (u, x) and with (v, x); see Fig. 3b. If $d \prec x$, then the queue-edge (c, d) forms a 2-rainbow both with (u, x) and with (v, x). Hence, neither (u, x) nor (v, x) is a queue-edge, so x is a stack-attachment. Since $\ell > 2$, (u, v) has more than two stack-attachments in $\mathcal{L}$, contradicting Lemma 1.   □

Lemma 3. *Let $\mathcal{L}$ be a mixed layout of $G(k, \ell)$ with $k > 1, \ell > 2$. Let a, b, c be queue-attachments of an edge (u, v) of $G(k - 1, \ell)$ with $u \prec v$. Then $u \prec a, b, c \prec v$.*

Proof. Assume to the contrary that $a \prec u$ (the case $v \prec a$ is symmetric). We first prove that $a \prec u$ implies $v \prec b, c$. Indeed, if $b \prec a$, then the queue-edges (b, v) and (a, u) form a 2-rainbow; see Fig. 3c. If $a \prec b \prec v$, then the queue-edges (a, v) and (b, u) form a 2-rainbow; see Fig. 3d. Thus, $v \prec b$ and analogously $v \prec c$. Symmetrically, $v \prec c$ implies $b \prec u$. Hence, $b \prec u \prec v \prec b$; a contradiction.   □

Lemma 4. *Let $\mathcal{L}$ be a mixed layout of $G(k, \ell)$ with $k > 4, \ell > 6$. Then, every queue-edge of $G(k - 3, \ell)$ has at most six queue-attachments in $\mathcal{L}$.*

Proof. Assume for a contradiction that there is a queue-edge (u, v) in $G(k-3, \ell)$ with seven queue-attachments $x_1, \ldots, x_7$ in $\overline{G}(k - 2, \ell)$. By Lemma 3, all seven vertices have to lie between u and v; w.l.o.g. assume that $u \prec x_1 \prec \ldots \prec x_7 \prec v$.

For any edge (u, x_i) or (v, x_i) with $2 \leq i \leq 6$ belonging to $\overline{G}(k - 1, \ell)$, consider an attachment w of this edge. By Lemma 1, we can assume that w is not a stack attachment. Further, if (w, x_i) is a queue-edge, then it forms a

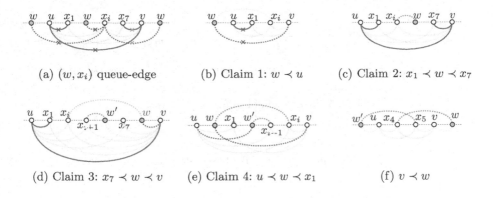

(a) (w, x_i) queue-edge   (b) Claim 1: $w \prec u$ (c) Claim 2: $x_1 \prec w \prec x_7$

(d) Claim 3: $x_7 \prec w \prec v$ (e) Claim 4: $u \prec w \prec x_1$ (f) $v \prec w$

Fig. 4. Illustrations for the proof of Lemma 4.

2-rainbow with either (u, v) , (u, x_1) , or (v, x_7) ; see Fig. 4a. Hence, we assume that every selected attachment w of (u, x_i) or (v, x_i) with $2 \leq i \leq 6$ in $\overline{G}(k-1, \ell)$ is a mixed-attachment with stack-edge (w, x_i). We prove Claims 1–4 for edges (v, x_i); for (u, x_i) symmetric arguments work; see Fig. 4.

Claim 1. There is no mixed-attachment w of (v, x_i) with $2 \leq i \leq 6$ and $w \prec u$ and there is no mixed-attachment w of (u, x_i) with $2 \leq i \leq 6$ and $v \prec w$.

Proof. Otherwise, the queue-edges (v, w) and (u, x_1) form a 2-rainbow.   □

Claim 2. There is no mixed-attachment w of (v, x_i) or (u, x_i) with $2 \leq i \leq 6$ and $x_1 \prec w \prec x_7$.

Proof. Otherwise, there is a smiley face $\langle u, x_1, x_i, w, x_7, v \rangle$ or $\langle u, x_1, w, x_i, x_7, v \rangle$ in $G(k-1, \ell)$, based on whether $x_i \prec w$ or $w \prec x_i$, contradicting Lemma 2.   □

Claim 3. There is no mixed-attachment w of (v, x_i) with $2 \leq i \leq 6$ and $x_7 \prec w \prec v$ and no mixed-attachment w of (u, x_i) with $2 \leq i \leq 6$ and $u \prec w \prec x_1$.

Proof. Let to the contrary w' be a mixed-attachment of (v, x_{i+1}). We have $x_i \prec w' \prec w$, as otherwise the stack-edges (w', x_{i+1}) and (x_i, w) would cross. Then a smiley face $\langle u, x_1, x_{i+1}, w', w, v \rangle$ exists in $G(k-1, \ell)$, contradicting Lemma 2. □

Claim 4. There is no mixed-attachment w of (v, x_i) with $3 \leq i \leq 5$ and $u \prec w \prec x_1$ and no mixed-attachment w of (u, x_i) with $3 \leq i \leq 5$ and $x_7 \prec w \prec v$.

Proof. Let to the contrary w' be a mixed-attachment of (u, x_{i-1}). We have $u \prec w \prec w' \prec x_i$, as otherwise the stack-edges (w', x_{i-1}) and (x_i, w) would cross. However, by Claims 2 and 3, this leads to a contradiction.   □

Now consider a mixed-attachment w of (v, x_4) and a mixed-attachment w' of (u, x_5). By Claims 1–4, we must have $v \prec w$ and $w' \prec u$; see Fig. 4f. However, then the stack-edges (x_4, w) and (x_5, w') cross. This concludes the proof.

Lemmas 1 and 4 imply the following

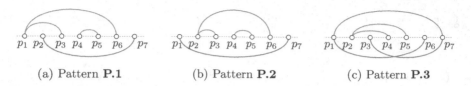

| (a) Pattern **P.1** | (b) Pattern **P.2** | (c) Pattern **P.3** |

Fig. 5. Illustration of different patterns.

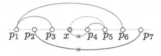

Fig. 6. Illustration for the proof of Pattern **P.1** in Lemma 5.

Corollary 1. *Let $\mathcal{L}$ be a mixed layout of $G(k, \ell)$ with $k > 4, \ell > 8$. Then, every queue-edge of $G(k - 4, \ell)$ has at least $\ell - 8$ mixed-attachments in $\mathcal{L}$.*

Next we define three patterns **P.1**–**P.3** and prove that they are forbidden in a mixed layout. Each pattern is denoted by $\langle p_1, \ldots, p_7 \rangle$, as it is defined on a set of seven vertices for which either $p_1 \prec \ldots \prec p_7$ or $p_7 \prec \ldots \prec p_1$ holds in $\mathcal{L}$; see Fig. 5. The involved edges in each pattern and their types are as follows.

P.1 Stack-edges (p_1, p_3), (p_1, p_6) and (p_4, p_5), and a queue-edge (p_2, p_7).
P.2 Stack-edges (p_2, p_3), (p_2, p_6) and (p_4, p_5), and a queue-edge (p_1, p_7).
P.3 Stack-edges (p_1, p_7), (p_2, p_4) and (p_2, p_5), and queue-edges (p_1, p_6) and (p_3, p_7).

Lemma 5. *Let $\mathcal{L}$ be a mixed layout of $G(k, \ell)$ with $k > 1, \ell > 4$. Then, $G(k-1, \ell)$ does not contain Patterns **P.1**–**P.3** in $\mathcal{L}$.*

Proof sketch. For a contradiction, let $\langle p_1, \ldots, p_7 \rangle$ be Pattern **P.1** contained in $G(k - 1, \ell)$; see Fig. 6. We first argue that at least one of the $\ell > 4$ vertices attached to (p_4, p_5) in $\overline{G}(k, \ell)$ has to be a mixed-attachment. By Lemma 1, at most two of them can be stack-attachments. If more than two of these vertices are queue-attachments, then by Lemma 3, they all appear between p_4 and p_5 in $\mathcal{L}$, and thus any queue-edge incident to them creates a 2-rainbow with the queue-edge (p_2, p_7). Hence, there is at least one mixed-attachment x of (p_4, p_5). Let e and e' be the stack- and queue-edge incident to x, respectively. Then, $p_3 \prec x \prec p_6$, as otherwise e would cross one of the stack-edges (p_1, p_3) and (p_1, p_6). However, then e' forms a 2-rainbow with the queue-edge (p_2, p_7); a contradiction. Similarly we argue for Pattern **P.2**. For Pattern **P.3** see [3]. □

We are now ready to prove the main result of this paper.

Theorem 1. *$G(k, \ell)$ does not admit a mixed layout if $k \geq 5, \ell \geq 33$.*

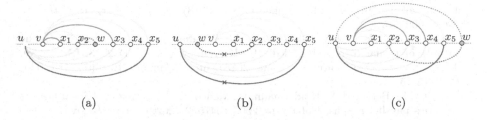

Fig. 7. Illustration for the first case of Theorem 1.

Proof sketch. Assume to the contrary that $G(5, 33)$ admits a mixed layout $\mathcal{L}$. By Lemma 1, there is at least one queue-edge (u, v) in $G(2, 33)$. W.l.o.g., let $u \prec v$ in $\mathcal{L}$. By Corollary 1, $G(3, 33)$ contains at least 25 mixed-attachments, say $x_1, \ldots, x_{25}$, of (u, v). For every $i = 1, \ldots, 25$, one of the following applies: $x_i \prec u$, or $u \prec x_i \prec v$, or $v \prec x_i$. For each of the cases, we further distinguish whether the edge (u, x_i) is a stack-edge or a queue-edge. This defines six configurations for x_i. Thus, at least five vertices, say w.l.o.g., $x_1, \ldots, x_5$, are attached with the same configuration to (u, v); we assume w.l.o.g. that $x_1 \prec \ldots \prec x_5$. We show a contradiction in the case when $v \prec x_i$ and (u, x_i) is a queue-edge for all $i = 1, \ldots, 5$; for the remaining cases refer to [3].

By Corollary 1, $G(4, 33)$ contains at least one mixed-attachment w of (u, x_2). Thus, either (x_2, w) or (u, w) is a stack-edge. In the former case, the stack-edges (v, x_1) and (v, x_3) enforce $x_1 \prec w \prec x_3$; see Fig. 7a. Hence, $\langle u, v, x_1, x_2, w, x_3, x_5 \rangle$ or $\langle u, v, x_1, w, x_2, x_3, x_5 \rangle$ of $G(4, 33)$ form Pattern **P.2** in $\mathcal{L}$. This contradicts Lemma 5. In the latter case, the stack-edge (v, x_5) enforces either $w \prec v$ or $x_5 \prec w$. We consider three subcases. If $w \prec u$, then the queue-edges (w, x_2) and (u, x_1) form a 2-rainbow. If $u \prec w \prec v$, then the queue-edges (w, x_2) and (u, x_5) form a 2-rainbow; see Fig. 7b. Otherwise, $x_5 \prec w$ holds. It follows that $\langle u, v, x_2, x_3, x_4, x_5, w \rangle$ of $G(4, 33)$ form Pattern **P.3** in $\mathcal{L}$; see Fig. 7c. $\qquad\square$

3 Open Problems

In this paper, we proved that 2-trees do not admit mixed 1-stack 1-queue layouts. Since 2-trees admit 2-stack layouts and 3-queue layouts [26], it is natural to ask whether they admit mixed 1-stack 2-queue layouts. We conclude with an algorithmic question, namely, what is the complexity of recognizing graphs that admit mixed 1-stack 1-queue layouts, even for 2-trees? Note that recently de Col et al. [10] showed that testing whether a (not necessarily planar) graph admits a mixed 2-stack 1-queue layout is NP-complete.

References

1. Alam, J.M., Bekos, M.A., Gronemann, M., Kaufmann, M., Pupyrev, S.: On dispersable book embeddings. In: Brandstädt, A., Köhler, E., Meer, K. (eds.) WG.

LNCS, vol. 11159, pp. 1–14. Springer, Cham (2018). https://doi.org/10.1007/978-3-030-00256-5_1

2. Alam, J.M., Bekos, M.A., Gronemann, M., Kaufmann, M., Pupyrev, S.: Queue layouts of planar 3-trees. In: Biedl, T., Kerren, A. (eds.) GD 2018. LNCS, vol. 11282, pp. 213–226. Springer, Cham (2018). https://doi.org/10.1007/978-3-030-04414-5_15

3. Angelini, P., Bekos, M.A., Kindermann, P., Mchedlidze, T.: On mixed linear layouts of series-parallel graphs. CoRR abs/2008.10475v2 (2020). http://arxiv.org/abs/2008.10475v2

4. Bannister, M.J., Devanny, W.E., Dujmović, V., Eppstein, D., Wood, D.R.: Track layouts, layered path decompositions, and leveled planarity. Algorithmica (2018). https://doi.org/10.1007/s00453-018-0487-5

5. Bekos, M.A., Gronemann, M., Raftopoulou, C.N.: Two-page book embeddings of 4-planar graphs. Algorithmica **75**(1), 158–185 (2016). https://doi.org/10.1007/s00453-015-0016-8

6. Bekos, M.A., Kaufmann, M., Klute, F., Pupyrev, S., Raftopoulou, C.N., Ueckerdt, T.: Four pages are indeed necessary for planar graphs. J. Comput. Geom. **11**(1), 332–353 (2020). https://journals.carleton.ca/jocg/index.php/jocg/article/view/504

7. Bernhart, F., Kainen, P.C.: The book thickness of a graph. J. Comb. Theory Ser. B **27**(3), 320–331 (1979). https://doi.org/10.1016/0095-8956(79)90021-2

8. Binucci, C., Di Giacomo, E., Hossain, M.I., Liotta, G.: 1-page and 2-page drawings with bounded number of crossings per edge. Eur. J. Comb. **68**, 24–37 (2018). https://doi.org/10.1016/j.ejc.2017.07.009

9. Cornuéjols, G., Naddef, D., Pulleyblank, W.R.: Halin graphs and the travelling salesman problem. Math. Program. **26**(3), 287–294 (1983). https://doi.org/10.1007/BF02591867

10. de Col, P., Klute, F., Nöllenburg, M.: Mixed linear layouts: complexity, heuristics, and experiments. In: Archambault, D., Tóth, C.D. (eds.) GD 2019. LNCS, vol. 11904, pp. 460–467. Springer, Cham (2019). https://doi.org/10.1007/978-3-030-35802-0_35

11. Di Battista, G., Frati, F., Pach, J.: On the queue number of planar graphs. SIAM J. Comput. **42**(6), 2243–2285 (2013). https://doi.org/10.1137/130908051

12. Dujmović, V.: Graph layouts via layered separators. J. Comb. Theory Ser. B **110**, 79–89 (2015). https://doi.org/10.1016/j.jctb.2014.07.005

13. Dujmović, V., Frati, F.: Stack and queue layouts via layered separators. J. Graph Algorithms Appl. **22**(1), 89–99 (2018). https://doi.org/10.7155/jgaa.00454

14. Dujmović, V., Joret, G., Micek, P., Morin, P., Ueckerdt, T., Wood, D.R.: Planar graphs have bounded queue-number. In: Zuckerman, D. (ed.) FOCS. pp. 862–875. IEEE Computer Society (2019). https://doi.org/10.1109/FOCS.2019.00056

15. Dujmović, V., Wood, D.R.: On linear layouts of graphs. Discrete Math. Theoret. Comput. Sci. **6**(2), 339–358 (2004). http://dmtcs.episciences.org/317

16. Ewald, G.: Hamiltonian circuits in simplicial complexes. Geometriae Dedicata **2**(1), 115–125 (1973). https://doi.org/10.1007/BF00149287

17. de Fraysseix, H., de Mendez, P.O., Pach, J.: A left-first search algorithm for planar graphs. Discrete Comput. Geometry **13**, 459–468 (1995). https://doi.org/10.1007/BF02574056

18. Guan, X., Yang, W.: Embedding planar 5-graphs in three pages. Discret. Appl. Math. (2019). https://doi.org/10.1016/j.dam.2019.11.020

19. Heath, L.S.: Embedding planar graphs in seven pages. In: FOCS, pp. 74–83. IEEE Computer Society (1984). https://doi.org/10.1109/SFCS.1984.715903

20. Heath, L.S., Leighton, F.T., Rosenberg, A.L.: Comparing queues and stacks as mechanisms for laying out graphs. SIAM J. Discrete Math. **5**(3), 398–412 (1992). https://doi.org/10.1137/0405031
21. Heath, L.S., Rosenberg, A.L.: Laying out graphs using queues. SIAM J. Comput. **21**(5), 927–958 (1992). https://doi.org/10.1137/0221055
22. Hoffmann, M., Klemz, B.: Triconnected planar graphs of maximum degree five are subhamiltonian. In: Bender, M.A., Svensson, O., Herman, G. (eds.) ESA. LIPIcs, vol. 144, pp. 58:1–58:14. Schloss Dagstuhl - Leibniz-Zentrum für Informatik (2019). https://doi.org/10.4230/LIPIcs.ESA.2019.58
23. Kainen, P.C., Overbay, S.: Extension of a theorem of Whitney. Appl. Math. Lett. **20**(7), 835–837 (2007). https://doi.org/10.1016/j.aml.2006.08.019
24. Nishizeki, T., Chiba, N.: Hamiltonian cycles. In: Planar Graphs: Theory and Algorithms, chap. 10, pp. 171–184. Dover Books on Mathematics, Courier Dover Publications (2008)
25. Pupyrev, S.: Mixed linear layouts of planar graphs. In: Frati, F., Ma, K.-L. (eds.) GD 2017. LNCS, vol. 10692, pp. 197–209. Springer, Cham (2018). https://doi.org/10.1007/978-3-319-73915-1_17
26. Rengarajan, S., Veni Madhavan, C.E.: Stack and queue number of 2-trees. In: Du, D.-Z., Li, M. (eds.) COCOON 1995. LNCS, vol. 959, pp. 203–212. Springer, Heidelberg (1995). https://doi.org/10.1007/BFb0030834
27. Wiechert, V.: On the queue-number of graphs with bounded tree-width. Electr. J. Comb. **24**(1), P1.65 (2017). http://www.combinatorics.org/ojs/index.php/eljc/article/view/v24i1p65
28. Yannakakis, M.: Four pages are necessary and sufficient for planar graphs (extended abstract). In: Hartmanis, J. (ed.) ACM Symposium on Theory of Computing, pp. 104–108. ACM (1986). https://doi.org/10.1145/12130.12141
29. Yannakakis, M.: Embedding planar graphs in four pages. J. Comput. Syst. Sci. **38**(1), 36–67 (1989). https://doi.org/10.1016/0022-0000(89)90032-9d
30. Yannakakis, M.: Planar graphs that need four pages. CoRR abs/2005.14111 (2020). https://arxiv.org/abs/2005.14111

Schematic Representation of Biconnected Graphs

Giuseppe Di Battista(ID), Fabrizio Frati(✉)(ID), Maurizio Patrignani(ID),
and Marco Tais

Roma Tre University, Rome, Italy
{gdb,frati,patrigna,tais}@dia.uniroma3.it

Abstract. Suppose that a biconnected graph is given, consisting of a
large component plus several other smaller components, each separated
from the main component by a separation pair. We investigate the existence and the computation time of schematic representations of the structure of such a graph where the main component is drawn as a disk, the
vertices that take part in separation pairs are points on the boundary of
the disk, and the small components are placed outside the disk and are
represented as non-intersecting lunes connecting their separation pairs.
We consider several drawing conventions for such schematic representations, according to different ways to account for the size of the small
components. We map the problem of testing for the existence of such
representations to the one of testing for the existence of suitably constrained 1-page book-embeddings and propose several polynomial-time
and pseudo-polynomial-time algorithms.

1 Introduction

Many of today's applications are based on large-scale networks, having billions of
vertices and edges. This spurred an intense research activity devoted to finding
methods for the visualization of very large graphs. Several recent contributions
focus on algorithms that produce drawings where either the graph is only partially represented or it is schematically visualized. Examples of the first type
are proxy drawings [7,13], where a graph that is too large to be fully visualized is represented by the drawing of a much smaller proxy graph that preserves
the main features of the original graph. Examples of the second type are graph
thumbnails [16], where each connected component of a graph is represented by
a disk and biconnected components are represented by disks contained into the
disk of the connected component they belong to.

Among the characteristics that are emphasized by the above mentioned drawings, a crucial role is played by connectivity. Following this line of thought, we

This research was supported in part by MIUR Project "AHeAD" under PRIN
20174LF3T8, by H2020-MSCA-RISE Proj. "CONNECT" n° 734922, and by Roma
Tre University Azione 4 Project "GeoView".

D. Auber and P. Valtr (Eds.): GD 2020, LNCS 12590, pp. 160–172, 2020.
https://doi.org/10.1007/978-3-030-68766-3_13

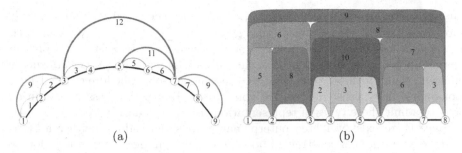

Fig. 1. Schematic representations of biconnected graphs. (a) A MAX-constrained book-embedding. (b) A two-dimensional book-embedding; for simplicity, the vertices lie on a straight line rather than on the boundary of a disk.

study schematic representations of graphs that emphasize their connectivity features. We start from the following observation: quite often, real-life very large graphs have one large connected component and several much smaller other components (see, e.g., [5,12]). This happens to biconnected and triconnected components too (see, e.g., [2] for an analysis of the graphs in [9]).

Hence, we concentrate on a single biconnected graph (that can be a biconnected component of a larger graph) consisting of a large component plus several other smaller components, each separated from the large component by a separation pair. We propose to represent the large component as a disk, to draw the vertices of such a component that take part in separation pairs as points on the boundary of the disk, and to represent the small components as non-intersecting lunes connecting their separation pairs placed outside the disk. See Fig. 1. This representation is designed to emphasize the arrangement of the components with respect to the separation pairs. For simplicity, we assume that each separation pair separates just one small component from the large one.

More formally, our input is a weighted graph $G = (V, E, \omega)$, where each vertex in V participates in at least one separation pair, each edge (u, v) of E represents a small component separated from the large one by the separation pair $\{u, v\}$, and ω assigns a positive weight to each edge. The weight of an edge represents a feature that should be emphasized in the schematic representation. As an example, it might represent the number of vertices or edges of the corresponding small component. We study one-dimensional and two-dimensional representations. In both cases, the vertices of G are linearly ordered points that are placed along the boundary of a disk. In the one-dimensional representations, we draw each edge as an arc and impose that arcs do not cross. Also, consider two edges (u, v) and (x, y) and suppose that the weight of (u, v) is larger than that of (x, y). Then we impose that (u, v) is drawn outside (x, y), so to represent the weight by means of the edge length. We call MAX-*constrained book-embedding* this type of representation (see Fig. 1(a)). In Sect. 3 we present a polynomial-time algorithm that tests whether a graph admits such a representation. We also study a more constrained type of representation. Namely, let (u, v) be an edge and consider the sequence of edges $(u_1, v_1), \ldots, (u_k, v_k)$ that are drawn immediately below (u, v); then, we may want that $\omega(u, v) > \sum_{i=1}^{k} \omega(u_i, v_i)$. We call SUM-*constrained book-embedding* this type of representation. In Sect. 4 we present a pseudo-polynomial-time algorithm that tests whether a graph admits such a representation. Both MAX- and SUM-constrained book-embeddings are 1-page book-embeddings satisfying specific constraints. Hence, a necessary condition for G to admit these types of representations is outerplanarity [1].

Since there exist weighted outerplanar graphs that admit neither a MAX-constrained nor a SUM-constrained book-embedding, in Sect. 5 we study how to represent planarly a weighted outerplanar graph with edges that have, in addition to a length, also a thickness: each edge is represented with a lune with area proportional to its weight. We call these representations two-dimensional book-embeddings. See Fig. 1(b). First, we show that all weighted outerplanar graphs admit a two-dimensional book-embedding and discuss the area requirements of such representations. Second, we show that, if a finite resolution rule is imposed, then there are graphs that do not admit any two-dimensional book-embedding and we present a polynomial-time algorithm to test whether a graph admits such a representation. Conclusions are presented in Sect. 6. Because of space limitations, complete proofs are deferred to the full version of the paper [4].

2   Preliminaries

Block-Cut-Vertex Tree. A *cut-vertex* in a connected graph G is a vertex whose removal disconnects G. A graph with no cut-vertex is *biconnected*. A *block* of G is a maximal subgraph of G which is biconnected. The *block-cut-vertex tree* T of G [6,8] has a *B-node* for each block of G and a *C-node* for each cut-vertex of G; a B-node b and a C-node c are adjacent if c is a vertex of the block of G represented by b. We denote by $G(b)$ the block of G represented by a B-node b. We often identify a C-node of T and the corresponding cut-vertex of G. Suppose that T is rooted at some B-node; then, for any node x of T (either a B-node or a C-node), we denote by $G^+(x)$ the subgraph of G consisting of all the blocks $G(b)$ such that b is a B-node in the subtree of T rooted at x.

Planar Drawings. A *drawing* of a graph maps each vertex to a point in the plane and each edge to a Jordan arc between its end-vertices. A drawing is *planar* if no two edges intersect, except at common end-vertices. A planar drawing partitions the plane into connected regions, called *faces*. The unbounded face is the *outer face*, while all the other faces are *internal*.

Outerplanar Graphs. An *outerplanar drawing* is a planar drawing such that all the vertices are incident to the outer face. An *outerplanar graph* is a graph that admits an outerplanar drawing. Two outerplanar drawings are *equivalent* if the clockwise order of the edges incident to each vertex is the same in both drawings. An *outerplane embedding* is an equivalence class of outerplanar drawings. A biconnected outerplanar graph has a unique outerplane embedding [11,14]. Given the outerplane embedding Γ of an n-vertex biconnected outerplanar graph G, the *extended dual tree* $\mathcal{T}$ of Γ is obtained from the dual graph $\mathcal{D}$ of Γ by splitting the vertex of $\mathcal{D}$ corresponding to the outer face of Γ into n degree-1 vertices. Note that $\mathcal{T}$ is an ordered tree and can be constructed in $O(n)$ time. Further, each edge of $\mathcal{T}$ is dual to an edge of G; moreover, the edges incident to leaves of $\mathcal{T}$ are dual to edges incident to the outer face of Γ.

Book-Embeddings. Given a graph G and a linear order $\mathcal{L}$ of its vertices, we write $u \prec_{\mathcal{L}} v$ to represent that u precedes v in $\mathcal{L}$. Two edges (u, v) and (w, z) of G *cross* if $u \prec_{\mathcal{L}} w \prec_{\mathcal{L}} v \prec_{\mathcal{L}} z$; then $\mathcal{L}$ is a *1-page book-embedding* of G if no two edges cross. The *flip* of $\mathcal{L}$ is a linear order $\mathcal{L}'$ such that, for any pair of vertices u and v, we have $u \prec_{\mathcal{L}'} v$ if and only if $v \prec_{\mathcal{L}} u$. By $u \preceq_{\mathcal{L}} v$ we mean that $u \prec_{\mathcal{L}} v$ or $u = v$. For a pair of distinct edges $e_1 = (u_1, v_1)$ and $e_2 = (u_2, v_2)$ of G such that $u_1 \preceq_{\mathcal{L}} u_2 \prec_{\mathcal{L}} v_2 \preceq_{\mathcal{L}} v_1$, we say that e_2 is *nested into* e_1 and e_1 *wraps around* e_2. Further, a subgraph G' of G *lies under* (resp. *lies strictly under*) an edge (u, v) of G, where $u \prec_{\mathcal{L}} v$, if for every vertex w of G', we have $u \preceq_{\mathcal{L}} w \preceq_{\mathcal{L}} v$ (resp. $u \prec_{\mathcal{L}} w \prec_{\mathcal{L}} v$). The *lowest-right edge* incident to a vertex v is the edge (u, v) such that $v \prec_{\mathcal{L}} u$ and there is no neighbor w of v such that $v \prec_{\mathcal{L}} w \prec_{\mathcal{L}} u$.

A *weighted graph* $G = (V, E, \omega)$ is a graph equipped with a function ω that assigns a positive weight to each edge in E.

3   MAX-**Constrained Book-Embeddings**

In this section, we study a first type of one-dimensional representations. We are given a weighted graph $G = (V, E, \omega)$. We draw the vertices in V as points linearly ordered on the boundary of a disk and the edges in E as non-intersecting arcs positioned outside the disk, placing edges with larger weight outside edges of smaller weight. Formally, a MAX-*constrained book-embedding* of a weighted graph $G = (V, E, \omega)$ is a 1-page book-embedding $\mathcal{L}$ such that, for any two distinct edges $e_1 = (u, v)$ and $e_2 = (x, y)$ in E with $u \preceq_{\mathcal{L}} x \prec_{\mathcal{L}} y \preceq_{\mathcal{L}} v$, we have that $\omega(e_1) > \omega(e_2)$. That is, if e_1 wraps around e_2, then $\omega(e_1) > \omega(e_2)$. We do not specify the actual drawing of the edges since, if G has a MAX-constrained book-embedding, then they can be easily represented with non-crossing Jordan arcs. An example is in Fig. 1(a). In this section we prove the following theorem.

Theorem 1. *Let $G = (V, E, \omega)$ be an n-vertex weighted outerplanar graph. There exists an $O(n \log n)$-time algorithm that tests if G admits a MAX-constrained book-embedding and, in the positive case, constructs such an embedding.*

We call MAX-BE-DRAWER the algorithm in the statement of Theorem 1. We can assume that G is connected; otherwise, it admits a MAX-constrained book-embedding if and only if every connected component of it admits one. We start by computing in $O(n)$ time the block-cut-vertex tree T of G [6,8]. We root T at any B-node b^* such that $G(b^*)$ contains an edge with maximum weight. For each B-node b of T, we compute in overall $O(n)$ time the value $W^+(b)$ of the maximum weight of an edge of $G^+(b)$.

We now visit T in arbitrary order. For each B-node b, we do what follows.

First, we check if $G(b)$ admits a MAX-constrained book-embedding. In the negative case, we conclude that G admits no MAX-constrained book-embedding, while in the positive case we compute such an embedding and call it $\mathcal{L}(b)$. This check is done in time linear in the number of vertices of $G(b)$, as follows. First, we check whether there exists a single edge $(u,v) \in E$ of maximum weight. If not, we conclude that G admits no MAX-constrained book-embedding, otherwise we compute in linear time [3,10,15] the unique 1-page book-embedding $\mathcal{L}(b)$ of $G(b)$ such that u and v are the first and the last vertex of $\mathcal{L}(b)$, respectively; note that, in any MAX-constrained book-embedding of $G(b)$, the edge (u,v) does not nest into any other edge of $G(b)$, given that it has maximum weight, hence it has to wrap around every other edge of $G(b)$. We construct in linear time the extended dual tree T of the outerplane embedding of $G(b)$ and we root T at the leaf whose incident edge is dual to (u,v). We visit T and, for every edge (α, β) of T where α is the parent of β and β has children $\gamma_1, \ldots, \gamma_k$, we check whether the dual edge e_β of (α, β) has weight larger than that of the edge e_{γ_i} that is dual to the edge (β, γ_i), for $i = 1, \ldots, k$. If one of these checks fails, we conclude that $G(b)$ admits no MAX-constrained book-embedding, since e_β wraps around e_{γ_i} in $\mathcal{L}(b)$, otherwise $\mathcal{L}(b)$ is a MAX-constrained book-embedding of $G(b)$.

Second, if $b \neq b^*$, consider the C-node c that is the parent of b in T. We check in constant time whether c is the first or the last vertex of $\mathcal{L}(b)$. If not, we conclude that G admits no MAX-constrained book-embedding, given that T is rooted at a node b^* such that $G(b^*)$ contains an edge with maximum weight, hence $G(b^*)$ does not lie under any edge incident to c. Otherwise, we possibly flip in constant time $\mathcal{L}(b)$ so that c is the first vertex of $\mathcal{L}(b)$.

Third, for each C-node c of T that is adjacent to b, we store two values $\ell_b(c)$ and $r_b(c)$. These are the weights of the edges (u,c) and (c,w) such that u and w are the vertices immediately preceding and following c in $\mathcal{L}(b)$, respectively; if a vertex preceding or following c in $\mathcal{L}(b)$ does not exist, then we set $\ell_b(c)$ or $r_b(c)$ to ∞, respectively. This is done in constant time for each C-node.

We now perform a bottom-up visit of T. After visiting a B-node b, we either conclude that G admits no MAX-constrained book-embedding or we determine a MAX-constrained book-embedding $\mathcal{L}^+(b)$ of $G^+(b)$ such that, if $b \neq b^*$, the parent of b in T is the first vertex of $\mathcal{L}^+(b)$. In more detail, we act as follows.

If b is a leaf of T, then we set in constant time $\mathcal{L}^+(b) = \mathcal{L}(b)$.

If b is an internal node of T, then we proceed as follows. We initialize $\mathcal{L}^+(b)$ to $\mathcal{L}(b)$; recall that the parent of b in T, if $b \neq b^*$, is the first vertex of $\mathcal{L}(b)$.

Let $c_1, \ldots, c_k$ be the C-nodes that are children of b in T. For $i = 1, \ldots, k$, let $b_{i,1}, \ldots, b_{i,m_i}$ be the B-nodes that are children of c_i. Since we already visited each node $b_{i,j}$, we have a MAX-constrained book-embedding $\mathcal{L}^+(b_{i,j})$ of $G^+(b_{i,j})$ whose first vertex is c_i. We now process each C-node c_i independently.

We order (and possibly relabel) the B-nodes $b_{i,1}, \ldots, b_{i,m_i}$ that are children of c_i in decreasing order of value $W^+(b_{i,j})$; that is, $W^+(b_{i,1}) \geq W^+(b_{i,2}) \geq \cdots \geq W^+(b_{i,m_i})$. This can be done in $O(m_i \log m_i)$ time. We now process the B-nodes $b_{i,1}, \ldots, b_{i,m_i}$ in this order. We use two variables, $L(c_i)$ and $R(c_i)$, and initialize them to $\ell_b(c_i)$ and $r_b(c_i)$, respectively. When processing a node $b_{i,j}$, for $j = 1, \ldots, m_i$, we insert the vertices of $G^+(b_{i,j})$ into the ordering $\mathcal{L}^+(b)$, by replacing c_i either with $\mathcal{L}^+(b_{i,j})$ (that is, $\mathcal{L}^+(b_{i,j})$ is inserted *to the right* of c_i) or with the flip of $\mathcal{L}^+(b_{i,j})$ (that is, $\mathcal{L}^+(b_{i,j})$ is inserted *to the left* of c_i). This operation can be performed in constant time. Further, the choice of whether we insert $\mathcal{L}^+(b_{i,j})$ to the left or to the right of c_i is performed as follows.

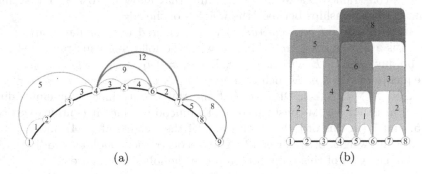

(a)   (b)

Fig. 2. Schematic representations of biconnected graphs. (a) A 1-page SUM-constrained book-embedding. (b) A MINRES-constrained two-dimensional book-embedding; for simplicity, the vertices are aligned on a straight-line.

- If $W^+(b_{i,j}) \geq L(c_i)$ and $W^+(b_{i,j}) \geq R(c_i)$, then we conclude that G admits no MAX-constrained book-embedding.
- If $W^+(b_{i,j}) < R(c_i)$, then we insert the vertices of $G^+(b_{i,j})$ into $\mathcal{L}^+(b)$, by replacing c_i with $\mathcal{L}^+(b_{i,j})$; we update $R(c_i)$ to the value of $r_{b_{i,j}}(c_i)$.
- If $W^+(b_{i,j}) \geq R(c_i)$ and $W^+(b_{i,j}) < L(c_i)$, then we insert the vertices of $G^+(b_{i,j})$ into $\mathcal{L}^+(b)$, by replacing c_i with the flip of $\mathcal{L}^+(b_{i,j})$; we update $L(c_i)$ to the value of $r_{b_{i,j}}(c_i)$.

After visiting the root b^* of T, if MAX-BE-DRAWER did not conclude that G admits no MAX-constrained book-embedding, it computed a MAX-constrained book-embedding $\mathcal{L} := \mathcal{L}^+(b^*)$ of G. The running time of MAX-BE-DRAWER is dominated by the $O(m_i \log m_i)$-time sorting performed on the m_i children of each C-node c_i, hence it is in $O(n \log n)$.

The upper bound in Theorem 1 is essentially tight, as computing a MAX-constrained book-embedding has a time complexity that is lower-bounded by that of a sorting algorithm. Indeed, given a set S of n distinct real numbers, one can construct a star T with a center c whose n edges have the weights in S. Any MAX-constrained book-embedding of T partitions the edges into two ordered sequences, one to the left of c and one to the right of c; a total ordering of S can be constructed by merging these sequences in $O(n)$ time.

4   SUM-Constrained Book-Embeddings

Even if in a MAX-constrained book-embedding no edge can wrap around an edge with larger weight, an edge e might still wrap around a sequence of edges $e_1, \ldots, e_k$ with $\omega(e) < \sum_{i=1}^{k} \omega(e_i)$. This might cause the resulting visualization to not effectively convey the information related to the edge weights. Hence, we study a second type of one-dimensional representations that are more restrictive than MAX-constrained book-embeddings and that allow us to better take into account the relationships between the weights of the edges.

A SUM-*constrained book-embedding* of a weighted outerplanar graph $G = (V, E, \omega)$ is a 1-page book-embedding $\mathcal{L}$ with the following constraint. Let $e = (u, v)$ be any edge in E with $u \prec_{\mathcal{L}} v$. Let $e_1 = (u_1, v_1), \ldots, e_k = (u_k, v_k)$ be any sequence of edges in E such that $u \preceq_{\mathcal{L}} u_1 \prec v_1 \preceq_{\mathcal{L}} \cdots \preceq_{\mathcal{L}} u_k \prec v_k \preceq_{\mathcal{L}} v$. Then, $\omega(e) > \sum_{i=1}^{k} \omega(e_i)$. Observe that the MAX-constrained book-embedding in Fig. 1(a) is not a SUM-constrained book-embedding since it contains vertices 3, 4, 5, and 7 (in this order) and the sum of the weights of $(3,4)$ and $(5,7)$ is 14, while the weight of $(3,7)$ is 12. A SUM-constrained book-embedding is in Fig. 2(a). The goal of this section is to prove the following theorem.

Theorem 2. *Let $G = (V, E, \omega)$ be an n-vertex weighted outerplanar graph and let Φ be the maximum weight of any edge in E. There exists an $O(\Phi^2 n^3 \log(\Phi n))$-time algorithm that tests whether G admits a SUM-constrained book-embedding and, in the positive case, constructs such an embedding.*

We call SUM-BE-DRAWER the algorithm in the statement of Theorem 2. As for MAX-constrained book-embeddings, we can assume that G is connected.

First, we compute in $O(n)$ time the block-cut-vertex tree T of G [6,8]. We root T at any B-node b^* containing an edge with maximum weight. Further, we equip each B-node b with the maximum weight $W(b)$ of any edge of $G(b)$.

Second, we visit (in arbitrary order) T. For each B-node b, the algorithm SUM-BE-DRAWER performs the following checks and computations.

1. We check whether $G(b)$ admits a SUM-constrained book-embedding. This can be done in time linear in the number of vertices of $G(b)$, and hence in $O(n)$ time for all the blocks of G, similarly as in Algorithm MAX-BE-DRAWER. If the check fails, then we conclude that G admits no SUM-constrained book-embedding. Otherwise, we compute such an embedding and call it $\mathcal{L}(b)$.
2. If $b \neq b^*$, we consider the C-node c of T that is the parent of b. We check in constant time whether c is the first or the last vertex of $\mathcal{L}(b)$. If not, then we conclude that G admits no SUM-constrained book-embedding. If yes, we possibly flip in constant time $\mathcal{L}(b)$ so that c is the first vertex of $\mathcal{L}(b)$.

We introduce some definitions (refer to Fig. 3). Let $\mathcal{L}$ be a 1-page book-embedding of G. A vertex c is *visible* in $\mathcal{L}$ if there exists no edge e of G such that c lies strictly under e in $\mathcal{L}$; for example, the vertices 1, 4, and 9 in Fig. 3(a) are visible. The *total extension* $\tau_{\mathcal{L}}$ of $\mathcal{L}$ is the sum of the weights of all the edges e such that there is no edge e' that wraps around e in $\mathcal{L}$. Let c be a visible vertex

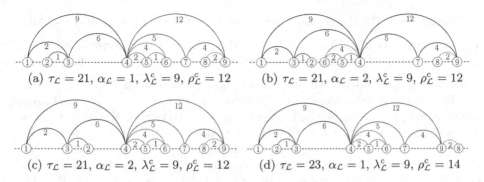

(a) $\tau_{\mathcal{L}} = 21$, $\alpha_{\mathcal{L}} = 1$, $\lambda_{\mathcal{L}}^c = 9$, $\rho_{\mathcal{L}}^c = 12$ (b) $\tau_{\mathcal{L}} = 21$, $\alpha_{\mathcal{L}} = 2$, $\lambda_{\mathcal{L}}^c = 9$, $\rho_{\mathcal{L}}^c = 12$

(c) $\tau_{\mathcal{L}} = 21$, $\alpha_{\mathcal{L}} = 2$, $\lambda_{\mathcal{L}}^c = 9$, $\rho_{\mathcal{L}}^c = 12$ (d) $\tau_{\mathcal{L}} = 23$, $\alpha_{\mathcal{L}} = 1$, $\lambda_{\mathcal{L}}^c = 9$, $\rho_{\mathcal{L}}^c = 14$

Fig. 3. (a) and (b) are left-right equivalent w.r.t. 4; (c) left-right dominates (d) w.r.t. 4; (b) and (c) are up-down equivalent; (b) up-down dominates (a).

of $\mathcal{L}$. The *extension* $\lambda_{\mathcal{L}}^c$ *of* $\mathcal{L}$ *to the left of* c is the sum of the weights of all the edges e such that: (i) there is no edge e' that wraps around e in $\mathcal{L}$; and (ii) for each end-vertex v of e, we have $v \preceq_{\mathcal{L}} c$. The *extension* $\rho_{\mathcal{L}}^c$ *of* $\mathcal{L}$ *to the right of* c is defined analogously. Let u be the first vertex of $\mathcal{L}$. The *free space* $\alpha_{\mathcal{L}}$ *of* $\mathcal{L}$ is the weight of the lowest-right edge (u, v) of u in $\mathcal{L}$ minus the total extension of the subgraph of G induced by v and by the vertices that are strictly under (u, v).

Now, let $\mathcal{L}$ and $\mathcal{L}'$ be two 1-page book-embeddings of G and let c be a vertex of G that is visible both in $\mathcal{L}$ and in $\mathcal{L}'$. We say that $\mathcal{L}$ and $\mathcal{L}'$ are *left-right equivalent with respect to* c if $\lambda_{\mathcal{L}}^c = \lambda_{\mathcal{L}'}^c$ and $\rho_{\mathcal{L}}^c = \rho_{\mathcal{L}'}^c$. We also say that $\mathcal{L}$ *left-right dominates* $\mathcal{L}'$ *with respect to* c if $\lambda_{\mathcal{L}}^c \leq \lambda_{\mathcal{L}'}^c$, $\rho_{\mathcal{L}}^c \leq \rho_{\mathcal{L}'}^c$, and at least one of the two inequalities is strict. Finally, let $\mathcal{L}$ and $\mathcal{L}'$ be two 1-page book-embeddings of G whose first vertex is the same. We say that $\mathcal{L}$ is *up-down equivalent* to $\mathcal{L}'$ if $\tau_{\mathcal{L}} = \tau_{\mathcal{L}'}$ and $\alpha_{\mathcal{L}} = \alpha_{\mathcal{L}'}$. We also say that $\mathcal{L}$ *up-down dominates* $\mathcal{L}'$ if $\tau_{\mathcal{L}} \leq \tau_{\mathcal{L}'}$, $\alpha_{\mathcal{L}} \geq \alpha_{\mathcal{L}'}$, and at least one of the two inequalities is strict.

The algorithm SUM-BE-DRAWER now performs a bottom-up visit of T.

After visiting a C-node c, SUM-BE-DRAWER either concludes that G admits no SUM-constrained book-embedding or determines a sequence of SUM-constrained book-embeddings $\mathcal{L}_1^+(c), \ldots, \mathcal{L}_k^+(c)$ of $G^+(c)$ such that:

(C1) for any $i = 1, \ldots, k$, we have that c is visible in $\mathcal{L}_i^+(c)$;
(C2) $\lambda_{\mathcal{L}_1^+(c)}^c < \cdots < \lambda_{\mathcal{L}_k^+(c)}^c$ and $\rho_{\mathcal{L}_1^+(c)}^c > \cdots > \rho_{\mathcal{L}_k^+(c)}^c$; and
(C3) $G^+(c)$ admits no SUM-constrained book-embedding that respects (C1) and that left-right dominates $\mathcal{L}_i^+(c)$ with respect to c, for some $i \in \{1, \ldots, k\}$.

After visiting a B-node $b \neq b^*$, SUM-BE-DRAWER either concludes that G admits no SUM-constrained book-embedding or determines a sequence of SUM-constrained book-embeddings $\mathcal{L}_1^+(b), \ldots, \mathcal{L}_k^+(b)$ of $G^+(b)$ such that:

(B1) the parent c of b in T is the first vertex of $\mathcal{L}_i^+(b)$, for $i = 1, \ldots, k$;
(B2) $\alpha_{\mathcal{L}_1^+(b)} < \cdots < \alpha_{\mathcal{L}_k^+(b)}$ and $\tau_{\mathcal{L}_1^+(b)} < \cdots < \tau_{\mathcal{L}_k^+(b)}$; and
(B3) $G^+(b)$ admits no SUM-constrained book-embedding that respects (B1) and that up-down dominates $\mathcal{L}_i^+(b)$, for some $i \in \{1, \ldots, k\}$.

We now describe the bottom-up visit of T performed by SUM-BE-DRAWER.

Processing a Leaf. If b is a leaf of T, then the only SUM-constrained book-embedding of $G^+(b)$ constructed by SUM-BE-DRAWER is $\mathcal{L}_1^+(b) = \mathcal{L}(b)$.

Processing a C-node. We process a C-node c as follows. Let $b_1, \ldots, b_h$ be the B-nodes that are children of c. For each b_i with $i = 1, \ldots, h$, we have a sequence $\mathcal{L}_1^+(b_i), \ldots, \mathcal{L}_{k_i}^+(b_i)$ satisfying Properties (B1)–(B3). We relabel the B-nodes $b_1, \ldots, b_h$ in such a way that $W(b_i) \leq W(b_{i+1})$, for $i = 1, \ldots, h-1$; this takes $O(n \log n)$ time. We now process the B-nodes $b_1, \ldots, b_h$ in this order. While processing these nodes, we construct h sequences $\mathcal{S}_1, \ldots, \mathcal{S}_h$, where $\mathcal{S}_i$ contains $O(\Phi n)$ SUM-constrained book-embeddings of $G^+(b_1) \cup \cdots \cup G^+(b_i)$. Roughly speaking, $\mathcal{S}_i$ is a sequence of "optimal" SUM-constrained book-embeddings of $G^+(b_1) \cup \cdots \cup G^+(b_i)$ with respect to left-right dominance.

When processing b_1, we let $\mathcal{S}_1$ consist of $\mathcal{L}_{k_i}^+(b_1)$ and its flip.

Suppose that, for some $i \in \{2, \ldots, h\}$, the B-node b_{i-1} has been processed. We process b_i as follows. We initialize $\mathcal{S}_i = \emptyset$. We individually consider each of the $O(\Phi n)$ embeddings in $\mathcal{S}_{i-1}$; let $\mathcal{L}$ be one of these embedding. We consider each embedding $\mathcal{L}_j^+(b_i)$, with $j = 1, \ldots, k_i$ and try to place the vertices of $\mathcal{L}_j^+(b_i)$ different from c to the right and/or to the left of $\mathcal{L}$. More precisely (1) if $\alpha_{\mathcal{L}_j^+(b_i)} > \rho_{\mathcal{L}}^c$ (that is, if the part of $\mathcal{L}$ to the right of c "fits" immediately to the right of c in $\mathcal{L}_j^+(b_i)$), then we construct a SUM-constrained book-embedding $\mathcal{L}'$ of $G^+(b_1) \cup \cdots \cup G^+(b_i)$ by placing the vertices of $\mathcal{L}_j^+(b_i) \setminus \{c\}$ to the right of $\mathcal{L}$, in the same order as they appear in $\mathcal{L}_j^+(b_i)$, and we insert $\mathcal{L}'$ into $\mathcal{S}_i$; (2) if $\alpha_{\mathcal{L}_j^+(b_i)} > \lambda_{\mathcal{L}}^c$, we construct a SUM-constrained book-embedding $\mathcal{L}'$ of $G^+(b_1) \cup \cdots \cup G^+(b_i)$ by placing the vertices of $\mathcal{L}_j^+(b_i) \setminus \{c\}$ to the left of $\mathcal{L}$, in the opposite order as they appear in $\mathcal{L}_j^+(b_i)$, and we insert $\mathcal{L}'$ into $\mathcal{S}_i$.

Since $k_i \in O(\Phi n)$, after we considered each of the $O(\Phi n)$ embeddings in $\mathcal{S}_{i-1}$, we have that $\mathcal{S}_i$ contains $O(\Phi^2 n^2)$ embeddings. If $\mathcal{S}_i$ is actually empty, then we conclude that G admits no SUM-constrained book-embedding. Otherwise, we order and polish $\mathcal{S}_i$ by removing left-right dominated embeddings and by leaving only one copy of left-right equivalent embeddings (this brings the number of embeddings in $\mathcal{S}_i$ down to $O(\Phi n)$). The complexity of this step is dominated by the ordering of the elements in $\mathcal{S}_i$, which takes $O(\Phi^2 n^2 \log(\Phi n))$ time.

After processing b_h, we have that $\mathcal{S} := \mathcal{S}_h$ contains the required sequence of SUM-constrained book-embeddings of $G^+(c)$ satisfying Properties (C1)–(C3).

Processing an Internal B-node Different from the Root. Let $c_1, \ldots, c_k$ be the C-nodes that are children of a B-node b, labeled in the order as they appear in the SUM-constrained book-embedding $\mathcal{L}(b)$ of $G(b)$. For each c_i with $i = 1, \ldots, h$, we have a sequence $\mathcal{L}_1^+(c_i), \ldots, \mathcal{L}_{k_i}^+(c_i)$ of SUM-constrained book-embeddings of $G^+(c_i)$ satisfying Properties (C1)–(C3). We consider each of the $O(\Phi n)$ embeddings $\mathcal{L}_1^+(c_1), \ldots, \mathcal{L}_{k_1}^+(c_1)$ of $G^+(c_1)$ and plug it into $\mathcal{L}(b)$, if possible. For each of these choices, we process the C-nodes $c_2, \ldots, c_h$ in this order. When processing c_i, we choose a SUM-constrained book-embedding $\mathcal{L}_j^+(c_i)$ for $G^+(c_i)$ so that the extension of $\mathcal{L}_j^+(c_i)$ to the right of c_i is minimum, subject to the constraint that $\mathcal{L}_j^+(c_i)$ "fits" on the left. This results in the construction of at most one embedding of $G^+(b)$ for each embedding of $G^+(c_1)$. The set of embedding of $G^+(b)$ is then simplified by removing up-down dominated embeddings and by leaving a single copy of up-down equivalent embeddings.

Processing the Root. The way we deal with the root b^* of T is similar, and actually simpler, than the way we deal with a B-node $b \neq b^*$. We choose for $G^+(c_1)$ the embedding that fits into the embedding $\mathcal{L}(b)$ of $G(b)$ and whose extension to the right is minimum. Then, for every C-node c_i with $i = 2, \ldots, k$, we select a single embedding as in the case of a B-node different from the root. We produce at most one book-embedding for $G^+(b^*) = G$.

Running Time. Algorithm SUM-BE-DRAWER processes a B-node in $O(\Phi^2 n^2)$ time and a C-node in $O(h\Phi^2 n^2 \log(\Phi n))$ time, where h is the number of children of the C-node. Since the BC-tree T has $O(n)$ nodes and edges, the running time of the algorithm SUM-BE-DRAWER is in $O(\Phi^2 n^3 \log(\Phi n))$.

5   Two-Dimensional Book-Embeddings

In order to deal with weighted outerplanar graphs that admit no MAX-constrained and no SUM-constrained 1-page book-embedding (a cycle with three edges that all have the same weight is an example of such a graph), a possibility is to give to each edge not only a length but also a thickness, so that the area of the lune representing an edge is proportional to its weight.

Given a weighted outerplanar graph $G = (V, E, \omega)$, a *two-dimensional book-embedding* Γ of G consists of a 1-page book-embedding $\mathcal{L}$ (which is said to *support* Γ) and of a representation of G with the following features:

1. Each vertex $v \in V$ is assigned an x-coordinate $x(v)$ such that $u \prec_{\mathcal{L}} v$ if and only if $x(u) < x(v)$;
2. Each edge $e = (u, v) \in E$ such that $u \prec_{\mathcal{L}} v$ is represented by an axis-parallel rectangle $R(e) := [x_{\min}(e), x_{\max}(e)] \times [y_{\min}(e), y_{\max}(e)]$ whose area is equal to $\omega(e)$, where $x_{\min}(e) = x(u)$ and $x_{\max}(e) = x(v)$. Further, let $e_1, \ldots, e_k$ be the edges in E that are nested into e and let $Y_{\max} = \max_{i=1,\ldots,k}\{y_{\max}(e_i)\}$. Then we have $y_{\min}(e) = Y_{\max}$.

The *area* of Γ is the area of the smallest axis-parallel rectangle enclosing it.

In Sect. 1, we proposed to represent vertices as points on the boundary of a disk and edges as lunes with area proportional to the edge weights. In the above definition instead, to simplify the geometric constructions, vertices are placed along a straight-line and edges are represented as rectangles. However, it is easy to connect the rectangle representing an edge (u, v) with the points representing u and v, without intersecting the internal points of any other rectangle, implying the topological equivalence of the two representations. See Fig. 1(b).

The following theorems state that all weighted outerplanar graphs admit a two-dimensional book-embedding. The algorithms in their proofs exploit a suitable visit of the extended dual tree of G.

Theorem 3. *Let $G = (V, E, \omega)$ be an n-vertex weighted biconnected outerplanar graph and let $L > 0$ be a prescribed width. There exists an $O(n)$-time algorithm that constructs a two-dimensional book-embedding in area $L \times H = \sum_{e \in E} \omega(e)$.*

Theorem 4. *For any constant $\varepsilon > 0$, every n-vertex weighted outerplanar graph $G = (V, E, \omega)$ admits a two-dimensional book-embedding with area less than or equal to $\sum_{e \in E} \omega(e) + \varepsilon$. Such an embedding can be computed in $O(n)$ time.*

Theorems 3 and 4 do not give any guarantee in terms of minimum height and width of the rectangles in the constructed two-dimensional book-embeddings. We now study two-dimensional book-embedding with finite resolution.

A MINRES-*constrained two-dimensional book-embedding* of a weighted outerplanar graph $G = (V, E, \omega)$ is a two-dimensional book-embedding such that: (1) For each edge e in E, we have that $x_{\max}(e) - x_{\min}(e) \geq 1$ and $y_{\max}(e) - y_{\min}(e) \geq 1$. (2) For each pair u, v of vertices, we have that $|x(v) - x(u)| \geq 1$.

Let $\mathcal{L}$ be a 1-page book-embedding of a graph G and let e be an edge of G. We call the number of vertices that lie strictly under e the *burden* of e in $\mathcal{L}$, and denote it by $\beta(e)$. We have the following characterization.

Theorem 5. *A n-vertex weighted outerplanar graph $G = (V, E, \omega)$ admits a* MINRES-*constrained two-dimensional book-embedding if and only if it admits a 1-page book-embedding $\mathcal{L}$ such that $\omega(e) \geq \beta(e) + 1$, for each edge $e \in E$.*

Figure 2(b) shows a MINRES-constrained two-dimensional book-embedding produced by the algorithm used to prove Theorem 5. We can prove the following theorem by means of a variation of Algorithm SUM-BE-DRAWER from Sect. 4.

Theorem 6. *Let $G = (V, E, \omega)$ be an n-vertex weighted outerplanar graph. There exists an $O(n^5 \log n)$-time algorithm that tests whether G admits a* MINRES-*constrained two-dimensional book-embedding and, in the positive case, constructs such an embedding.*

6   Conclusions and Open Problems

With the aim of constructing schematic representations of biconnected graphs consisting of a large component plus several smaller components, we studied several types of constrained 1-page book-embeddings and presented polynomial-time or pseudo-polynomial-time algorithms for testing if a graph admits such book-embeddings. All the algorithms presented in this paper have been implemented; Figs. 1 and 2 have been generated by means of such implementations.

Our paper opens several problems.

First, our algorithms allow us to represent only an outerplanar arrangement of small components around a large component. How to generalize the approach to the non-outerplanar case? One could study the problem of minimizing the crossings between components and/or minimizing the violations to the constraints on the weights of the nesting components.

Second, we propose to linearly arrange the vertices of the separation pairs of the large component on the boundary of a disk. What happens if such an arrangement is instead circular? It is probably feasible to generalize our techniques in this direction, but extra effort is required.

Third, we concentrate on a "flat" decomposition of a graph with one large component plus many small components. What happens if the small components have their own separation pairs? In other words, how to represent the decomposition of a biconnected graph in its triconnected components?

Finally, our algorithms for constructing two-dimensional book-embeddings with finite resolution may output drawings whose area is not minimum. Can we minimize the area of such drawings in polynomial time?

Acknowledgments. Thanks to an anonymous reviewer for observing that computing a MAX-constrained book-embedding has a time complexity that is lower-bounded by the one of sorting.

References

1. Bernhart, F., Kainen, P.C.: The book thickness of a graph. J. Comb. Theory Ser. B **27**(3), 320–331 (1979)
2. Ceccarelli, S.: Tecniche per la visualizzazione di grafi di grandi dimensioni basate sulla connettività. Università degli Studi Roma Tre, Thesis for the Master Degree in Computer Science, October 2018). (in Italian)
3. Deng, T.: On the implementation and refinement of outerplanar graph algorithms. Master's thesis, University of Windsor, Ontario, Canada (2007)
4. Di Battista, G., Frati, F., Patrignani, M., Tais, M.: Schematic representation of large biconnected graphs. Technical report, Cornell University arXiv:2008.09414 (2020)
5. Fornito, A., Zalesky, A., Bullmore, E.T.: Fundamentals of Brain Network Analysis. Academic Press, Cambridge (2016)
6. Harary, F.: Graph Theory. Addison-Wesley Pub. Co., Reading (1969)
7. Hong, S., Nguyen, Q.H., Meidiana, A., Li, J., Eades, P.: BC tree-based proxy graphs for visualization of big graphs. In: IEEE Pacific Visualization Symposium (PacificVis 2018), pp. 11–20. IEEE Computer Society (2018)
8. Hopcroft, J.E., Tarjan, R.E.: Algorithm 447: efficient algorithms for graph manipulation. Commun. ACM **16**(6), 372–378 (1973)
9. Leskovec, J., Krevl, A.: SNAP Datasets: Stanford large network dataset collection, June 2014. http://snap.stanford.edu/data
10. Mitchell, S.L.: Linear algorithms to recognize outerplanar and maximal outerplanar graphs. Inf. Process. Lett. **9**(5), 229–232 (1979)
11. Moran, S., Wolfstahl, Y.: One-page book embedding under vertex-neighborhood constraints. SIAM J. Discrete Math. **3**(3), 376–390 (1990)
12. Newman, M.: Networks. Oxford University Press, Oxford (2018)
13. Nguyen, Q.H., Hong, S., Eades, P., Meidiana, A.: Proxy graph: visual quality metrics of big graph sampling. IEEE Trans. Vis. Comput. Graph. **23**(6), 1600–1611 (2017)
14. Syslo, M.M.: Characterizations of outerplanar graphs. Discrete Math. **26**(1), 47–53 (1979)

15. Wiegers, M.: Recognizing outerplanar graphs in linear time. In: Tinhofer, G., Schmidt, G. (eds.) WG 1986. LNCS, vol. 246, pp. 165–176. Springer, Heidelberg (1987). https://doi.org/10.1007/3-540-17218-1_57
16. Yoghourdjian, V., Dwyer, T., Klein, K., Marriott, K., Wybrow, M.: Graph thumbnails: identifying and comparing multiple graphs at a glance. IEEE Trans. Vis. Comput. Graph. 24(12), 3081–3095 (2018)

Drawing Tree-Based Phylogenetic Networks with Minimum Number of Crossings

Jonathan Klawitter[1]([✉])(iD) and Peter Stumpf[2](iD)

[1] University of Würzburg, Würzburg, Germany
jo.klawitter@gmail.com
[2] University of Passau, Passau, Germany

Abstract. In phylogenetics, tree-based networks are used to model and visualize the evolutionary history of species where reticulate events such as horizontal gene transfer have occurred. Formally, a tree-based network N consists of a phylogenetic tree T (a rooted, binary, leaf-labeled tree) and so-called reticulation edges that span between edges of T. The network N is typically visualized by drawing T downward and planar and reticulation edges with one of several different styles. One aesthetic criteria is to minimize the number of crossings between tree edges and reticulation edges. This optimization problem has not yet been researched. We show that, if reticulation edges are drawn x-monotone, the problem is NP-complete, but fixed-parameter tractable in the number of reticulation edges. If, on the other hand, reticulation edges are drawn like "ears", the crossing minimization problem can be solved in quadratic time.

Keywords: Phylogenetic network · Tree-based · Crossing minimization

1 Introduction

The evolution of a set of species is usually depicted by a *phylogenetic tree* [12]. More precisely, a *phylogenetic tree* T is a rooted, binary tree where the leaves are labeled bijectively by the set of species. The internal vertices of T, each having two children, represent bifurcation events in the evolution of the taxa. The heights assigned to vertices indicate the flow of time from the root, lying furthest in the past, to the present-day species.

Evolutionary histories can however not always be fully represented by a tree [3]. Indeed, reticulate events such as hybridization, horizontal gene transfer, recombination, and reassortment require the use of vertices with higher indegree [8,13]. A *phylogenetic network* N generalizes a phylogenetic tree in exactly this sense, that is, besides the root, leaves and vertices with indegree one and outdegree two, N may contain vertices with indegree two and outdegree one.

© Springer Nature Switzerland AG 2020
D. Auber and P. Valtr (Eds.): GD 2020, LNCS 12590, pp. 173–180, 2020.
https://doi.org/10.1007/978-3-030-68766-3_14

Tree-Based Networks. Motivated by the question of whether the evolutionary history of the taxa is fundamentally tree-like, Francis and Steel [4] introduced a class of phylogenetic networks called *tree-based networks*, which are "merely phylogenetic trees with additional edges". Formally, a *tree-based network* N is a phylogenetic network that has a subdivision T' of a phylogenetic tree T as spanning tree. Then T is called the *base tree* of N and T' the *support tree* of N. Lately, tree-based networks have received a lot of attention in combinatorial phylogenetics [1,4,9,11] and while drawings of several other types of phylogenetics networks have been investigated in the past [2,7,8,14], this has, to the best of our knowledge, not been done for tree-based networks. In this paper, we look at drawings of tree-based networks with different drawing styles inspired by drawings in the literature.

For a tree-based network N, we assume that both the base tree T and the support tree T' as spanning tree of N are fixed. We call an edge not contained in the embedding of T' into N a *reticulation edge*. Therefore, we can perceive a drawing of N as a drawing of T (or T') and the reticulation edges. A vertex of N that is also in T is called a *tree vertex*.

Drawing Styles. Our drawing conventions are that N is drawn downwards with vertices at their fixed associated height and T is drawn planar in the style of a dendrogram, that is, each tree edge (u, v) consists of a horizontal line segment starting at u and a vertical line segment ending at v. For reticulation edges, we have different drawing styles; see Fig. 1. In the *horizontal style* – the only style where the two endpoints of a reticulation edge must have the same height – reticulation edges are drawn as horizontal line segments. This style has for example been used by Kumar et al. [10, Figure 4]. We assume that all horizontal edges come with slightly different heights. The next two styles are inspired by Figures 3 and 6 by Vaughan et al. [15]. There, a reticulation edge (u, v) is drawn with two horizontal and one vertical line segment and thus with two bends. The styles differ in where the vertical line segment is placed. We define vertex $\ell(u, v)$ as follows. If the lowest common ancestor (lca) w in T' of u and v is a tree vertex, set $\ell(u, v) = w$. Otherwise, set $\ell(u, v)$ to be the first tree vertex below w. In the *ear style*, the vertical line segment is placed to the right of the subtree rooted at $\ell(u, v)$. In the *snake style*, the vertical line segment lies between u and v and, in particular, its x-coordinate lies between the x-coordinates of the left and right subtree of $\ell(u, v)$.

The aesthetic criteria to optimize for when constructing a drawing of N, with either of the styles, is the number of crossings. Our focus is on crossings between reticulation edges and tree edges. Crossings between pairs of reticulation edges may be minimized in a post-processing step.

We make the following important observation. The number of crossings in a drawing of N is fully determined by the order of the leaves or, equivalently, by the rotation of each tree vertex. Formally, we use a map $c \colon V(T) \to V(T)$ that assigns to each non-leaf vertex v of T one of its children. In a drawing of N, we then consider v to be *rotated left*, if $c(v)$ is its left child, and *rotated right*, if $c(v)$ is its right child. Two vertices are *rotated the same way* if they are both

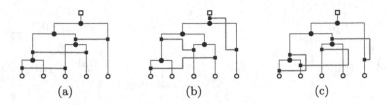

Fig. 1. Drawings of tree-based networks with the (a) horizontal, (b) snake, and (c) ear style for the red reticulation edges.

rotated left or if they are both rotated right. Let $\bar{c}(v)$ denote the child of v that is not $c(v)$.

Contribution and Outline. First, we show that the number of crossings can be minimized in quadratic time for ear-style drawings. Second, we prove that the problem is NP-hard for the horizontal style. On the positive side, we devise fixed-parameter tractable (fpt) algorithms for the horizontal and the snake style.

2   Ear-Style Drawings: Polynomial-Time Algorithm

Consider an ear-style drawing of a tree-based network N. Let $e = (u, v)$ be a reticulation edge of N and $f = (x, y)$ a tree edge of N. First, note that the vertical line segment of e is placed such that it does not cross any tree edge. Next, note that if the subtree $T(\ell(u, v))$ rooted at $\ell(u, v)$ does not contain f, then e and f cannot cross. Let l be the horizontal line segment of e starting at v. Assume $T(\ell(u, v))$ contains f and the y-coordinate range of f contains the y-coordinate of v. Observe that l and f cross if and only if f is in the right subtree of $\ell(v, y)$; see Fig. 2(a). (An analogous condition holds for the horizontal line segment starting at u.) Rotating $\ell(u, v)$ thus changes whether f and l cross. Furthermore, in general, the existence of each possible crossing depends on the rotation of a single tree vertex. We can thus minimize the number of crossings in an ear-style drawing of N by deciding for each tree vertex which orientation results in less crossings. We show that this can be done efficiently.

Theorem 1. *Let N be a tree-based network with n leaves and k reticulation edges. Then an ear-style drawing of N with minimum number of crossings can be computed in $\mathcal{O}(nk)$ time.*

Proof. The idea of the algorithm is to sweep upwards through N and, whenever an endpoint v of a reticulation edge is met, to tell v's ancestor tree vertices how many crossings it costs to have v in the left subtree. Each tree vertex is thus equipped with two counters that inform about which rotation is less favorable; see Fig. 2(a).

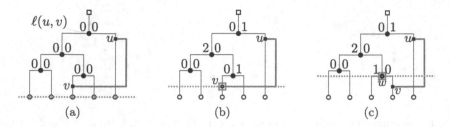

Fig. 2. (a) Start of sweep line algorithm with counters at 0; (b) adding potential crossings to counters; (c) rotating v based on counters.

Let $e = (u, v)$ be a reticulation edge. Above we observed that a horizontal segment of e can only have crossings with tree edges below $\ell(u, v)$. Therefore, we first compute and store the lca for each pair of endpoints of each reticulation edge in $\mathcal{O}(n + k)$ time with an algorithm by Gabow and Tarjan [5, Section 4.6]. We then start the sweep from the leaves towards the root of N. At every endpoint v of a reticulation edge (u, v) (or (v, u)), determine in $\mathcal{O}(n)$ time for every vertex u of T the width of its left and right subtree at the height of v; for example with a post-order traversal of T. Then from v up to $\ell(u, v)$, add for each tree vertex w the width of the subtree not containing v to the respective counter; see Fig. 2(b). This way, we count potential crossings of the horizontal segment at v with the vertical segments of all edges at the height of v in this subtree at once. When the sweep reaches a tree vertex w, as in Fig. 2(c), pick the best rotation for w based on its counters. In total we have $2k$ steps for endpoints of reticulation edges taking $\mathcal{O}(n)$ time and $\mathcal{O}(n)$ steps for tree vertices taking $\mathcal{O}(1)$ time. Hence, the algorithm runs in $\mathcal{O}(nk)$ time.   □

To minimize crossings between pairs of reticulation edges in a post-processing step, we only have to consider pairs of reticulation edges that have the vertical segment to the right of the same subtree and that are nested, that is, two reticulation edges (u, v) and (x, y) with u above x and y above v. The vertical segment of (u, v) should then be to the right of the vertical segment of (x, y).

3   Horizontal-Style Drawings: NP-Completeness

In this section, we show that the crossing minimization decision problem for horizontal-style drawings is NP-complete. We prove the NP-hardness with a reduction from MAX-CUT, which is known to be NP-complete [6]. Recall that in an instance of MAX-CUT we are given a graph $G = (V, E)$ and a parameter $p \in \mathbb{N}$, and have to decide whether there exists a bipartition (A, B) of V with at least p edges with one end in A and one end in B.

Theorem 2. *The crossing minimization problem for horizontal-style drawings of a tree-based network is NP-complete.*

Proof. Firstly, since we can non-deterministically generate all the drawings of N and count the number of crossings of a drawing in polynomial time, the problem is in NP. Concerning the hardness, we polynomial-time reduce a MAX-CUT instance with a graph $G = (V, E)$ to crossing minimization on a tree-based network N. In the following construction of N, assume that leaves are always (re)assigned the height 0.

The main idea is to have one *edge gadget* N_e for each $e \in E$ that induces a crossing if and only if e is not in our cut; see Fig. 3. Let $h \colon V \to \mathbb{N}$ be an arbitrary vertex ordering. Let $e = \{u, v\} \in E$ and suppose $h(u) < h(v)$. The construction of N_e then works as follows. We have a tree vertex u_e with two leaves as children and a tree vertex v_e with u_e and a leaf as children. We set $c(v_e) = u_e$ and the heights of u_e and v_e to $h(u)$ and $h(v)$ respectively. We add a reticulation edge f_e between $u_e c(u_e)$ and $v_e \bar{c}(v_e)$. Note that f_e and $u_e \bar{c}(u_e)$ cross if and only if u_e and v_e are rotated the same way. To connect all edge gadgets, we replace the leaves of an arbitrary rooted, binary tree with $|E|$ leaves and a downward planar embedding with the edge gadgets; see Fig. 4.

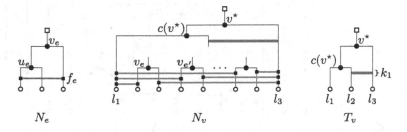

Fig. 3. An edge gadget N_e; a vertex gadget N_v based on the tree T_v.

We want to ensure that the tree vertices $v_1, \ldots, v_{\deg(v)}$ corresponding to the same node $v \in V$ are all rotated the same way. If this is enforced, we can consider all nodes in V where the corresponding tree vertices are rotated left as one partition set and all nodes in V where the corresponding tree vertices are rotated right as the other partition set. If on the other hand a cut is given, we simply choose for each vertex the rotation of the corresponding tree vertices accordingly. Now, to ensure the same rotation for all corresponding tree vertices, we construct a *vertex gadget* N_v for each node $v \in V$ (in some order); see Fig. 3. We start with a rooted, binary tree T_v on three leaves l_1, l_2, l_3 such that l_1 and l_2 have a common parent. Let v^* denote the child of the root of T_v and let $c(c(v^*)) = l_1$. Add a bundle of $k_1 = 2(|V| + 1) \cdot |E|$ reticulation edges between l_2 and l_3. We will see that k_1 is large enough such that this bundle does not induce crossings in a crossing minimum drawing. It thus enforces that l_2 lies between l_1 and l_3. We substitute l_2 by our current construction; see Fig. 3.

Lastly, for $1 \le i \le \deg(v)$, we add a reticulation edge between $v_i c(v_i)$ and the incoming edge of l_1, and a reticulation edge between $v_i \bar{c}(v_i)$ and l_3. Note that if v^* and v_i are rotated the same way, we get two crossings less than otherwise.

However, different rotations can save at most one crossing in the edge gadget containing v_i. Hence, in a crossing minimum drawing, $v^\star$ and v_i are rotated the same way. In fact, $v_1, \ldots, v_{\deg(v)}, v^\star$ are rotated the same way. This completes the construction of N. Note that N has a size polynomial in the size of G.

Note that the order of the edge gadgets does not influence the number of crossings with the two reticulation edges added for v_i; this number is fixed for crossing minimum drawings. Therefore, we can compute the total number k_2 of crossings induced by vertex gadgets. Furthermore, since $k_2 \leq 2|V|\,|E|$ and thus $k_1 \geq k_2 + |E| + 1$, we get that crossing one edge bundle would induce more crossings than we obtain from the vertex gadgets and from the edge gadgets. Hence, no bundle induces crossings in a crossing minimum drawing.

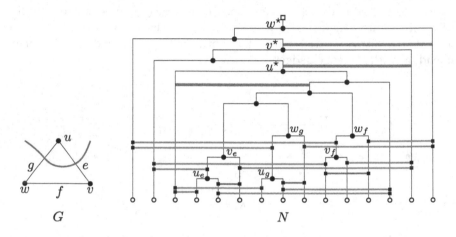

Fig. 4. A crossing-minimum drawing of N inducing a max-cut on G.

We conclude that minimizing crossings boils down to minimizing crossings in edge gadgets. Finally, by the construction of N and our observations, we get that N admits a horizontal-style drawing with $k \leq k_2 + |E| - p$ crossings if and only if G admits a cut of size at least p. The statement follows.   □

A snake-style drawing where endpoints of reticulation edges have the same height is a horizontal-style drawing; the reduction thus also works for this style.

Corollary 1. *The crossing minimization problem for snake-style drawings of a tree-based network is NP-complete.*

4   Snake-Style Drawings: FPT Algorithm

For the ear style, we have seen that whether a reticulation edge and a tree edge cross, depends on the rotation of at most one tree vertex, since horizontal line segments always go to the right. This is not the case for horizontal-style and snake-style drawings. However, fixing the rotation of $\ell(u, v)$ for each reticulation edge (u, v), also fixes for the horizontal line segments of (u, v) whether they go to the left or right. Further, while the vertical line segment may have a single crossing, this crossing occurs if and only if one endpoint of the reticulation edge is the lca of both endpoints. We can again conclude that the existence of each crossing of a horizontal line segment with a tree edge depends on the rotation of a single tree vertex – with two differences to the ear style: (i) A horizontal line segment can now also go towards the left. (ii) A horizontal line segment of a reticulation edge (u, v) ends between the two subtrees of $\ell(u, v)$, i.e., one of the two subtrees can have crossings with only one of the horizontal line segments of (u, v). With these observations we can now devise a fixed-parameter tractable algorithm.

Theorem 3. *Let N be a tree-based network with n leaves and k reticulation edges. Then a snake-style drawing of N with minimum number of crossings can be computed in $\mathcal{O}(2^k \cdot nk)$ time. The computation is thus fixed-parameter tractable when parametrized by k.*

Proof. Let $L = \{\ell(u, v) \mid (u, v)$ is a reticulation edge$\}$. Suppose the rotation for all $v \in L$ is fixed. With the observation above, we can slightly adapt our algorithm from Theorem 1 to compute for every $v \notin L$ the rotation that induces less crossings. Namely, the algorithm has to differentiate whether line segments go to the left or right, and pick a rotation only for $v \notin L$.

We try this for all possible combinations of rotations of vertices in L and then pick the drawing with the least crossings. Since there are $\mathcal{O}(2^k)$ such combinations, the statement on the running-time follows.   □

Note that this implies the same statement for the horizontal style.

References

1. Anaya, M., et al.: On determining if tree-based networks contain fixed trees. Bull. Math. Biol. **78**(5), 961–969 (2016). https://doi.org/10.1007/s11538-016-0169-x
2. Calamoneri, T., Di Donato, V., Mariottini, D., Patrignani, M.: Visualizing co-phylogenetic reconciliations. Theoret. Comput. Sci. **815**, 228–245 (2020). https://doi.org/10.1016/j.tcs.2019.12.024
3. Doolittle, W.F.: Phylogenetic classification and the universal tree. Science **284**(5423), 2124–2128 (1999). https://doi.org/10.1126/science.284.5423.2124
4. Francis, A.R., Steel, M.: Which phylogenetic networks are merely trees with additional arcs? Syst. Biol. **64**(5), 768–777 (2015). https://doi.org/10.1093/sysbio/syv037
5. Gabow, H.N., Tarjan, R.E.: A linear-time algorithm for a special case of disjoint set union. J. Comput. Syst. Sci. **30**(2), 209–221 (1985). https://doi.org/10.1016/0022-0000(85)90014-5
6. Garey, M.R., Johnson, D.S.: Computers and intractability, vol. 174. Freeman San Francisco (1979)
7. Huson, D.H.: Drawing rooted phylogenetic networks. IEEE/ACM Trans. Comput. Biol. Bioinform. **6**(1), 103–109 (2009)

8. Huson, D.H., Rupp, R., Scornavacca, C.: Phylogenetic Networks: Concepts, Algorithms and Applications. Cambridge University Press (2010). https://doi.org/10.1093/sysbio/syr055

9. Jetten, L., van Iersel, L.: Nonbinary tree-based phylogenetic networks. IEEE/ACM Trans. Comput. Biol. Bioinform. **15**(1), 205–217 (2018). https://doi.org/10.1109/TCBB.2016.2615918

10. Kumar, V., et al.: The evolutionary history of bears is characterized by gene flow across species. Sci. Rep. **7**(1) (2017). https://doi.org/10.1038/srep46487

11. Pons, J.C., Semple, C., Steel, M.: Tree-based networks: characterisations, metrics, and support trees. J. Math. Biol. **78**(4), 899–918 (2018). https://doi.org/10.1007/s00285-018-1296-9

12. Semple, C., Steel, M.A.: Phylogenetics, vol. 24. Oxford University Press, Oxford (2003)

13. Steel, M.: Phylogeny: Discrete and Random Processes in Evolution. Society for Industrial and Applied Mathematics (2016)

14. Tollis, I.G., Kakoulis, K.G.: Algorithms for visualizing phylogenetic networks. In: Hu, Y., Nöllenburg, M. (eds.) GD 2016. LNCS, vol. 9801, pp. 183–195. Springer, Cham (2016). https://doi.org/10.1007/978-3-319-50106-2_15

15. Vaughan, T.G., Welch, D., Drummond, A.J., Biggs, P.J., George, T., French, N.P.: Inferring ancestral recombination graphs from bacterial genomic data. Genetics **205**(2), 857–870 (2017). https://doi.org/10.1534/genetics.116.193425

A Tipping Point for the Planarity of Small and Medium Sized Graphs

Emanuele Balloni, Giuseppe Di Battista⬝, and Maurizio Patrignani(✉)⬝

Roma Tre University, Rome, Italy
ema-bal93@hotmail.it,
{giuseppe.dibattista,maurizio.patrignani}@uniroma3.it

Abstract. This paper presents an empirical study of the relationship between the density of small-medium sized random graphs and their planarity. It is well known that, when the number of vertices tends to infinite, there is a sharp transition between planarity and non-planarity for edge density $d = 0.5$. However, this asymptotic property does not clarify what happens for graphs of reduced size. We show that an unexpectedly sharp transition is also exhibited by small and medium sized graphs. Also, we show that the same "tipping point" behavior can be observed for some restrictions or relaxations of planarity (we considered outerplanarity and near-planarity, respectively).

Keywords: Planarity · Random graphs · Outerplanarity · Near-planarity

1 Introduction

Several popular Graph Drawing algorithms devised to draw graphs of small-medium size assume that the graph to be drawn is planar both in the static setting [11,16,17] and in the dynamic one [3,5,8]. Hence, to assess the practical applicability of such algorithms it is crucial to study the probability that a small-medium sized graph (say of about 100–200 vertices) is planar. In particular, it is interesting to consider how this probability varies as a function of the density of the graph. We might have that the probability of planarity changes smoothly or that it changes abruptly, exhibiting a tipping-point behaviour.

A *tipping point* is a threshold that, when exceeded, leads to a sharp change in the state of a system. In sociology, for example, a tipping point is a time when most of the members of a group suddenly change their behavior by adopting a practice that before was considered rare. In climate study, a tipping point is a quick and irreversible change in the climate, triggered by some specific cause, like the growth of the global mean surface temperature. Even in graph theory, tipping

This research was supported in part by MIUR Project "MODE" under PRIN 20157 EFM5C, by MIUR Project "AHeAD" under PRIN 20174LF3T8, and by Roma Tre University Azione 4 Project "GeoView".

D. Auber and P. Valtr (Eds.): GD 2020, LNCS 12590, pp. 181–188, 2020.
https://doi.org/10.1007/978-3-030-68766-3_15

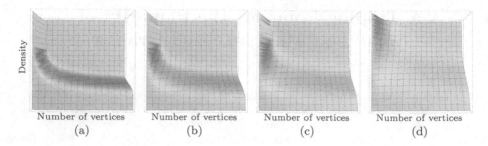

Fig. 1. Function $\zeta(n,d)$ for $n \in [1,400]$ and $d \in [0,3]$ in four cases: (a) $c_1 = 5$, $c_2 = 0.5$, $c_3 = 20$, $c_4 = 0.5$; (b) $c_1 = 5$, $c_2 = 0.5$, $c_3 = 8$, $c_4 = 0.5$; (c) $c_1 = 5$, $c_2 = 0.5$, $c_3 = 4$, $c_4 = 0.5$; and (d) $c_1 = 10$, $c_2 = 0.5$, $c_3 = 1$, $c_4 = 0.5$.

points have been found. As an example, in 1960 Erdös and Rènyi established that a random graph $G(n,m)$ with n vertices and m edges undergoes an abrupt change when the average vertex degree is equal to one, that is when $m \approx n/2$ [10]. Namely, when $m = cn/2$ and $c < 1$, asymptotically almost surely the connected components are all of size $O(\log n)$, and are either trees or unicyclic graphs. Conversely, when $c > 1$, almost surely there is a unique giant component of size $\Theta(n)$. The density $d = m/n = 1/2$ is sometimes referred to as the *critical density* or *phase transition density*. See [4,12] for a discussion of these concepts.

In this paper we investigate whether the density plays a similar role for the planarity of small-medium sized graphs. Namely, when the density of such graphs increases, does the probability of planarity change smoothly or abruptly?

To answer this question one could think of using the result of Łuczak *et al.* [14] who show that a random graph is almost surely non-planar if and only if the number of edges is $n/2 + O(n^{2/3})$. From the point of view of the density this means that a graph is almost surely non-planar if the density is $1/2 + O(n^{-1/3})$. However, the result shows only an asymptotic bound and does not clarify what happens for small-medium sized graphs. Essentially, this means that, for $n \to \infty$ graphs with density greater than $1/2$ are almost surely non-planar and that the "transition range" of density within which the probability of planarity falls from 1 to 0 is $\Theta(n^{-1/3})$. This result has been confirmed in [15], where it is proved that a graph with infinitely many vertices and density $1/2$ has probability ≈ 0.998 to be planar. Again, this gives no hint about how large is in practice this transition range for small values of n. For example, Fig. 1 shows four plots for different values of the constants $c_1, \ldots, c_4$ of the function $\zeta(n,d)$ which has both the asymptotic behaviors described in [14] (see [1]).

$$\zeta(n,d) = \frac{1}{2^{(d-(0.5+c_1/n^{c_2}))\cdot(c_3+c_4 n^{1/3})} + 1}$$

Depending on the values of $c_1, \ldots, c_4$ the function shows quite different behaviours in the range $n \in [1,400]$.

In this paper we adopt a pragmatic point of view. Namely, we are interested into investigating what are the properties of a random graph of small-medium size n when its density increases. In particular, we experimentally measured that, for each graph size $n \leq 400$, there is a value of density that marks a sharp transition from planar graphs to non-planar ones. This behavior is shared also by restrictions or relaxations of planarity, such as outerplanarity, and near-planarity.

The paper is structured as follows. Section 2 describes the methodology used for all experiments. Section 3 describes each experiment in detail. Our conclusions are given in Sect. 4.

2   Experimental Setting

All the experiments described in Sect. 3 are composed of three phases: generation of graphs; measurement; and analysis. In this section we describe the characteristics of the three phases common to all experiments.

Generation of Graphs. In all experiments (but for near-planarity) we used graphs with a number n of vertices that varies from 1 to 400, increasing at each step by one. The density $d = \frac{m}{n}$, where m is the number of edges, varies in a range that depends on the type of property that we are investigating. In fact, given a specific property, there always exists an interval of densities, that we call the *significant interval*, such that for a graph outside the significant interval either the property is granted or the property is ruled out, while inside the significant interval there are both graphs that have the property and graphs that do not. This is the interval of densities that we aim to experimentally explore[1].

For each combination of size n and density d we determined the number of edges $m = \text{Round}(n \cdot d)$ of the graphs to be generated, and generated 10,000 random graphs with n vertices and m edges[2]. In particular, we used function `randomSimpleGraph` of the OGDF library [7] for uniformly-at-random generating labeled graphs with a given number of vertices and edges. All graphs were simple (no loops or multiple edges allowed).

Measurement. For each combination of size and density we counted how many graphs have the desired property.

Analysis. We used Wolfram Mathematica 12.0.0.0 for producing the plots that are in this paper. In particular, we used function `ListPlot3D` that joins points with flat polygons. For the property of acyclicity it is also possible to compute the exact percentage of random graphs that are acyclic. This allowed us to compare the measured frequency distribution with its probability counterpart (see [1]). We used Mathematica also for sampling contour lines of surfaces and for computing fitting functions of sets of value pairs.

---

[1] For the smallest graphs we may not have all densities. For example, there is no graph with 5 vertices and density greater than 2.

[2] Function Round() rounds a value to the nearest integer, where Round(0.5) = 1.0.

3   Experimental Results

In this section we report the results of the experiments to determine how density and size impact graph-theoretic properties of random graphs of small-medium size. Since the purpose of the experiments is to show that planarity exhibits a tipping point behavior when the density increases, we start our experiments with acyclicity, a property that notoriously does not have tipping points [4, p. 118]. Then, we consider planarity, outerplanarity, and near-planarity, the main targets of our investigation.

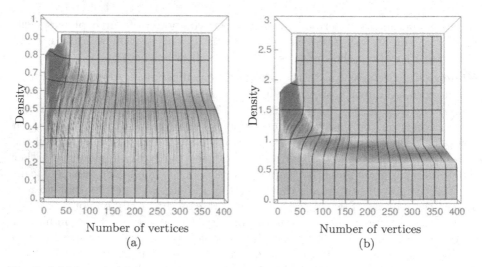

Fig. 2. (a) Measured fraction of random graphs that are acyclic. (b) Measured fraction of random graphs that are planar.

Acyclicity in Random Graphs. Simple graphs with less than three edges are acyclic. Conversely, since a tree has $m = n - 1$ edges, when $m = n$ a graph has at least one cycle. Hence, the significant interval of densities for acyclicity is $[\frac{3}{n}, 1 - \frac{1}{n}]$. We used densities ranging from 0.0 to 1.0, with a step of 0.05 performing a total of 84×10^6 tests. The plot in Fig. 2(a) shows the measured frequency of acyclic graphs as a function of density and size. Is it apparent that the density is the main cause of the loss of acyclicity, while the size of the graph seems to have weaker effects. In particular, bigger graphs tend to loose acyclicity earlier than smaller graphs.

Overall, the percentage of acyclic graphs seems to decrease smoothly through the significant interval of densities, without any quick transition or drop. Acyclic graphs allow us to compare a case where the tipping point is absent with the cases discussed in the next sections where a tipping point is present. Also, for acyclicity we were able to compute the actual probability of a graph of having this property and we used the comparison between experimental and theoretical values to validate the experimental pipeline (see [1]).

Planarity in Random Graphs. We now consider the property of the graph of being planar. All graphs with less than 9 edges are planar and there is no planar graph with more than $3n-6$ edges. Hence, the significant interval of densities for planarity is $[\frac{9}{n}, \frac{3n-6}{n}]$. For our experiments we used densities from 0.0 to 3.0, with a step of 0.1, performing a total of 124×10^6 planarity tests. In order to test the generated graphs for planarity we first used the OGDF function `makeConnected` that adds the minimum number of edges to make the graph connected and then called a single planarity test on the obtained graph: it can be easily seen that the minimality of the added edges implies that the connected graph is planar if and only if the connected components of the original graph were all planar.

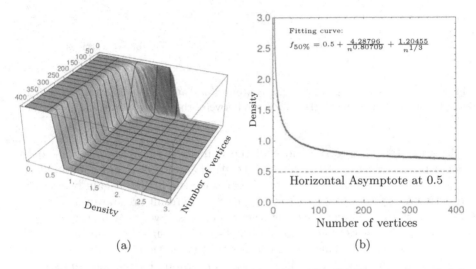

(a)   (b)

Fig. 3. (a) View from the side of the same graph of Fig. 2(b). (b) The samples at height 50% (red dots) and a possible fitting curve (solid blue line). (Color figure online)

Figures 2(b) and 3(a) show a plot of the frequency of planar graphs in random simple graphs as a function of density and size. It is apparent that the percentage of planar graphs drops from 100% to 0% in a short range of density values. As an example, for $n = 200$ we have that the fraction of planar graphs drops from 99% to 1% in the interval of densities $[0.915, 0.598]$, that corresponds to the 10.6% of the significant interval. In contrast, for the same value of n, the fraction of acyclic graphs depicted in Fig. 2(a) drops from 99% to 1% in the 53% of the significant interval. The tipping point is strongly related with density and appears earlier in larger graphs. Figure 7 in [1] shows a plot of 9 equally spaced contour lines at height 10%, 20%, ..., 90%.

In order to quantitatively study the behavior of the plot we determined the sample points of the contour line at height 50% and computed a fitting of such points. For the fitting, because of the results in [14], we selected a function of type $d = 1/2 + c_1/n^{c_2} + c_3/n^{1/3}$. The result of the fitting is shown in Fig. 3(b). Observe that the value of c_2 is consistent with the theory.

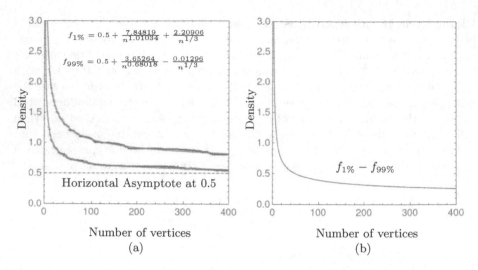

Fig. 4. (a) The sample points of the contour lines at height 1% and 99% and the corresponding fitting curves. (b) Difference between the fitting curves in (a).

In order to evaluate the width of the transition range we determined the sample points of the contour lines at height 1% and 99% and computed two fittings, one for each set of such points. For both the fittings, again, we selected a function of type $d = 1/2 + c_1/n^{c_2} + c_3/n^{1/3}$. The result are shown in Fig. 4(a). Observe how the difference between the two curves is very small (Fig. 4(b)).

Surprisingly, for random graphs of small-medium size the drop value for the measured fraction of planar graph is much smaller than it would have been hoped for: if you grow the density of a random graph of small-medium size you very likely loose planarity way before you have any chance to get connectivity ($d = 1$). Practically speaking, if you were interested into graphs with density one, planarity is almost granted for number of vertices in the range [1, 40] but is almost absent above 100 vertices. For density 1.5, instead, a random graph with more than 25 vertices is very likely non-planar.

Outerplanarity in Random Graphs. An *outerplanar* graph is a graph that admits a planar drawing where all vertices are on the external face. All graphs with less than 6 edges are outerplanar—the smallest non-outerplanar graphs being K_4 and $K_{2,3}$—and there is no outerplanar graph with more than $2n - 3$ edges. Hence, the significant interval of densities for outerplanarity is $[\frac{6}{n}, \frac{2n-3}{n}]$. For our experiments we used densities from 0.0 to 2.0, with a step of 0.1.

Figure 5(a) shows the fraction of outerplanar graphs as a function of the number of vertices and density.

Near-Planarity in Random Graphs. A *near-planar* graph is a graph that can be made planar by removing (at most) one edge [6]. Near-planar graphs are also called *skewness*-1 or *almost planar* graphs [9]. The smallest not near-planar graph is $K_{3,4}$, with 12 edges. From the definition of near-planar graphs it follows

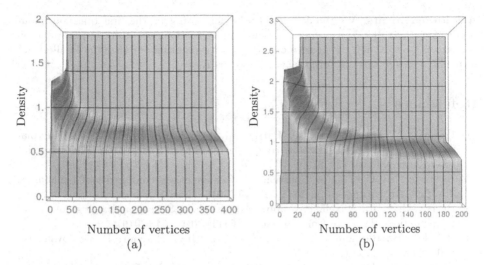

Fig. 5. (a) Measured fraction of random graphs that are outerplanar. (b) Measured fraction of random graphs that are near-planar.

that such graphs have a maximum of $3n - 6 + 1$ vertices. Hence, the significant interval of densities for near-planarity is $[\frac{14}{n}, \frac{3n-5}{n}]$. In our experiments we used densities ranging from 0.0 to 3.0 increasing by 0.1. The recognition of near-planar graphs can be made in quadratic-time: it suffices to test for planarity any graph obtained by removing one edge.

Figure 5(b) shows the measured fraction of random graphs that are near-planar as a function of the number of vertices (from 1 to 200) and the density. Observe that the transition from near-planar graphs to non-near-planar ones is sharper than what we measured for planarity or quasi-planarity, although it occurs for higher values of densities.

4   Conclusion and Future Work

We reported empirical evidence of the existence of a tipping point for planarity in random graphs of small-medium size. The same phenomenon appears to be present for restrictions and relaxations of planarity as outerplanarity and near-planarity. It would be interesting to measure whether other popular families of 'beyond planar' graphs, as 1-planar or quasiplanar graphs, also feature the same abrupt transition in their distribution in random graphs. Unfortunately, testing 1-planarity is NP-complete [13] even for near-planar graphs [6] and, to our knowledge, no implementation of the FPT algorithm in [2] for testing 1-planarity is available. Also, no testing algorithm has been proposed for quasiplanarity. Finally, we could consider other types of graphs, as random bipartite, biconnected, or triconnected graph, as well as other graph models like small-world graphs or scale-free graphs.

Acknowledgments. We thank Carlo Batini for posing us the first question about rapid transitions of graph properties. Sometimes questions are more important than answers. We also thank the anonymous reviewer for pointing out that the smallest not near-planar graph in terms of number of edges is $K_{3,4}$.

References

1. Balloni, E., Di Battista, G., Patrignani, M.: A tipping point for the planarity of small and medium sized graphs. Technical report, Cornell University arXiv:2008.09405v2 (2020)
2. Bannister, M., Cabello, S., Eppstein, D.: Parameterized complexity of 1-planarity. J. Graph Algorithms Appl. **22**(1), 23–49 (2018)
3. Barrera-Cruz, F., Haxell, P., Lubiw, A.: Morphing Schnyder drawings of planar triangulations. Discrete Comput. Geom. **61**(1), 161–184 (2019)
4. Bollobás, B.: Random Graph. Academic Press Inc., Harcourt Brace Jovanovich Publishers, London (1985)
5. Borrazzo, M., Da Lozzo, G., Di Battista, G., Frati, F., Patrignani, M.: Graph stories in small area. J. Graph Algorithms Appl. **24**(3), 269–292 (2020). https://doi.org/10.7155/jgaa.00530
6. Cabello, S., Mohar, B.: Adding one edge to planar graphs makes crossing number and 1-planarity hard. SIAM J. Comput. **42**(5), 1803–1829 (2013)
7. Chimani, M., Gutwenger, C., Jünger, M., Klau, G.W., Klein, K., Mutzel, P.: Open graph drawing framework (OGDF). In: Tamassia, R. (ed.) Handbook of Graph Drawing and Visualization, chap. 17. CRC Press (2014)
8. Da Lozzo, G., Di Battista, G., Frati, F., Patrignani, M., Roselli, V.: Upward planar morphs. Algorithmica **82**(10), 2985–3017 (2020). https://doi.org/10.1007/s00453-020-00714-6
9. Didimo, W., Liotta, G., Montecchiani, F.: A survey on graph drawing beyond planarity. ACM Comput. Surv. **52**(1), 4:1–4:37 (2019). https://doi.org/10.1145/3301281
10. Erdöos, P., Réenyi, A.: On the evolution of random graphs. Magyar Tud. Akad. Mat. Kutatóo Int. Közl. **5**, 17–61 (1960)
11. de Fraysseix, H., Pach, J., Pollack, R.: How to draw a planar graph on a grid. Combinatorica **10**(1), 41–51 (1990). https://doi.org/10.1007/BF02122694
12. Janson, S., Łuczak, T., Rucinski, A.: Random Graphs. Wiley-Interscience Series in Discrete Mathematics and Optimization. Wiley-Interscience, New York (2000)
13. Korzhik, V.P., Mohar, B.: Minimal obstructions for 1-immersions and hardness of 1-planarity testing. J. Graph Theory **72**(1), 30–71 (2013). https://doi.org/10.1002/jgt.21630
14. Łuczak, T., Pittel, B., Wierman, J.C.: The structure of a random graph at the point of the phase transition. Trans. Am. Math. Soc. **341**(2), 721–748 (1994)
15. Noy, M., Ravelomanana, V., Rué, J.: The probability of planarity of a random graph near the critical point. In: International Conference on Formal Power Series and Algebraic Combinatorics (FPSAC 2013), pp. 791–802 (2013)
16. Schnyder, W.: Embedding planar graphs on the grid. In: Johnson, D.S. (ed.) Proceedings of the First Annual ACM-SIAM Symposium on Discrete Algorithms, pp. 138–148. SIAM (1990)
17. Tutte, W.T.: How to Draw a Graph. Proc. London Math. Soc. **s3–13**(1), 743–767 (1963). https://doi.org/10.1112/plms/s3-13.1.743

Orthogonality

Characterization and a 2D Visualization of B_0-VPG Cocomparability Graphs

Sreejith K. Pallathumadam$^{(\boxtimes)}$ (ID) and Deepak Rajendraprasad (ID)

Indian Institute of Technology Palakkad, Palakkad, India
111704002@smail.iitpkd.ac.in, deepak@iitpkd.ac.in

Abstract. B_0-*VPG graphs* are intersection graphs of vertical and horizontal line segments on a plane. Cohen, Golumbic, Trotter, and Wang [Order, 2016] pose the question of characterizing B_0-VPG permutation graphs. We respond here by characterizing B_0-VPG cocomparability graphs. This characterization also leads to a polynomial time recognition and B_0-VPG drawing algorithm for the class. Our B_0-VPG drawing algorithm starts by fixing any one of the many posets P whose cocomparability graph is the input graph G. The drawing we obtain not only visualizes G in that one can distinguish comparable pairs from incomparable ones, but one can also identify which among a comparable pair is larger in P from this visualization.

Keywords: Poset visualization · Permutation graph ·
Cocomparability graph · B_0-VPG · Graph drawing

1 Introduction

Representing a graph as an intersection graph of two-dimensional geometric objects like strings, line segments, rectangles and disks is a means to depict a graph on the plane. When the graph being represented is a comparability or cocomparability graph, one can also ask whether the "direction" of the comparability relation in the associated poset can also be inferred from the drawing. In this paper we characterize cocomparability graphs which can be represented as intersection graphs of vertical and horizontal line segments in a plane. For the posets whose cocomparability graphs can be represented thus, we describe a representation from which one can also infer the direction of the comparability relation. Our drawing algorithm runs in polynomial time.

B_k-VPG graphs are intersection graphs of simple paths with at most k bends on a two-dimensional grid. Here, a path is simple if it does not pass through any grid vertex twice, and two paths are said to intersect if they share a vertex of the grid. The name B_k-VPG is an abbreviation for Vertex-intersection graphs of k-Bend Paths on a Grid. In particular, B_0-VPG graphs are intersection graphs of vertical and horizontal line segments on a plane. The *bend number* of a graph G is the minimum k for which G belongs to B_k-VPG.

© Springer Nature Switzerland AG 2020
D. Auber and P. Valtr (Eds.): GD 2020, LNCS 12590, pp. 191–204, 2020.
https://doi.org/10.1007/978-3-030-68766-3_16

The *dimension* of a poset $P = (X, \prec)$ is the smallest k such that $\prec$ is the intersection of k total orders on X. The *comparability graph of P* is the undirected graph on the vertex set X with edges between the pairs of elements comparable in P. A graph G is a *comparability graph* if it is the comparability graph of a poset. If two posets have the same comparability graph, then they have the same dimension [30]. Hence we can unambiguously define the *dimension of a comparability graph G* as the dimension of any poset P whose comparability graph is G. The complement of a comparability graph is a *cocomparability* graph. A *permutation graph* is a comparability graph of dimension at most two. It is known that a graph G is a permutation graph if and only if G is both comparability and cocomparability [27].

Cohen et al. [11] illustrated, via an elegant picture-proof, that if G is a comparability graph of dimension k ($k \geq 1$), the bend number of its complement $\overline{G}$ is at most $k - 1$. In particular therefore, the bend number of a permutation graph is either 0 or 1. They posed the problem of characterizing permutation graphs with bend number 0 as an open question (Qn 4.2 in [11]). We settle this question with a stronger result. We characterize cocomparability graphs with bend number 0 as follows (Theorem 1).

The simple cycle on k vertices is denoted by C_k. A C_4 together with an additional edge e between two non-consecutive vertices of the C_4 is a *diamond* and the edge e is a *diamond diagonal*. Two vertices x and y in a graph G are *diamond related* if there exists a path from x to y in G made up of diamond diagonals alone. This is easily verified to be an equivalence relation that refines the connectivity relation in G.

Theorem 1. *A cocomparability graph G is B_0-VPG if and only if*

(i) *No two vertices of an induced C_4 in G are diamond related, and*
(ii) *G does not contain an induced subgraph isomorphic to $\overline{C_6}$, the complement of C_6.*

A poset $P = (X, \prec)$ is an *interval order* if all the elements of X can be mapped to intervals on $\mathbb{R}$ such that $\forall x, y \in X$, $x \prec y$ if and only if the interval representing x is disjoint from and to the left of the interval representing y. Complements of the comparability graphs of interval orders form the well known class of *interval graphs*. While interval graphs are trivially B_0-VPG, it is known that there exists interval orders of arbitrarily high dimension [3]. Hence the class of B_0-VPG cocomparability graphs is richer than the class of B_0-VPG permutation graphs. In fact, since permutation graphs are $\overline{C_6}$-free, the first of the two conditions in Theorem 1 characterizes B_0-VPG permutation graphs.

Corollary 1. *A permutation graph G is B_0-VPG if and only if no two vertices of an induced C_4 in G are diamond related.*

A naive check for the conditions in Theorem 1 can be done in $O(n^6)$ time. Combining this with any of the known polynomial time recognition algorithms for cocomparability graphs [12, 18] will give a polynomial time recognition algorithm for B_0-VPG cocomparability graphs. We do not try to optimize the recognition algorithm here, but only note that this is in contrast to the NP-completeness of recognizing B_0-VPG graphs.

The above algorithm starts by fixing a partial order P_G whose cocomparability graph is G and a linear extension σ of P_G. The resulting drawing D ends up being a representation of $P_G = (V(G), \prec_P)$ in the following sense. We define three binary relations $\prec_D^{v,h}$, $\prec_D^{h,v}$ and $\prec_D$ among vertices of G based on the drawing D as follows.

Definition 1. *Let x and y be two vertices in $V(G)$ and let I_x and I_y be the line segments representing them in D, respectively.*

- *$x \prec_D^{v,h} y$ if I_x is either vertically below I_y or if both are intersected by a horizontal line, I_x is to the left of I_y.*
- *$x \prec_D^{h,v} y$ if I_x is either to the left of I_y or if both are intersected by a vertical line, I_x is vertically below I_y.*
- *$x \prec_D y$ if and only if $x \prec_D^{v,h} y$ when I_x is horizontal and $x \prec_D^{h,v} y$ when I_x is vertical.*

While the above relations are not even partial orders in general, in our drawing D, the relation $\prec_D$ faithfully captures $\prec_P$. Theorem 2 states this formally and Fig. 1 illustrates an example.

Theorem 2. *Any poset $P_G = (V_G, \prec_P)$ whose cocomparability graph G is B_0-VPG has a two dimensional visualization D such that $x \prec_P y$ if and only if $x \prec_D y$.*

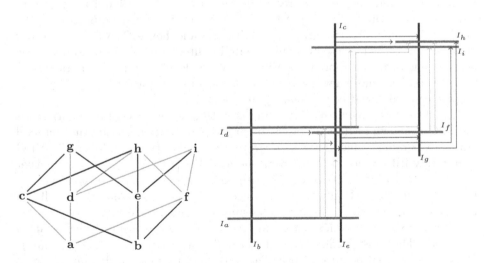

Fig. 1. Hasse diagram of a poset corresponding to a B_0-VPG cocomparability graph and a B_0-VPG representation D in which the covering relation is indicated by thin directed paths for clarity. The directed paths with blue color and green color respectively depict the relations $\prec_D^{h,v}$ and $\prec_D^{v,h}$.

1.1   Literature

B_k-VPG graphs were introduced by Asinowski et al. in 2012 [2] as a parameter-ized generalization for string graphs (intersection graphs of curves in a plane) and grid intersection graphs (bipartite graphs which are intersection graphs of vertical and horizontal segments in the plane in which all the vertices in one part are represented by vertical segments and all the vertices in the other part by horizontal line segments). Grid Intersection graphs (GIGs) are equivalent to bipartite B_0-VPG graphs. Similarly one can show that B_k-VPG graphs with an unrestricted k, which are called VPG graphs simpliciter, are equivalent to string graphs. B_0-VPG graphs are also equivalent to 2-DIR graphs, where a k-DIR graph is an intersection graph of line segments lying in at most k directions in the plane. All these equivalences were formally established in [2].

The NP-completeness of the recognition problem for VPG graphs follows from that of string graphs [23,28]. For B_0-VPG graphs, it follows from that of 2-DIR graphs [24]. Chaplick et al. showed that, $\forall k \geq 0$, it is NP-complete to recognize whether a given graph G is in B_k-VPG even when G is guaranteed to be in B_{k+1}-VPG and represented as such [7]. This also shows that $\forall k \geq 0$ the classes B_k-VPG and B_{k+1}-VPG are separated. Cohen et al. showed that, $\forall k \geq 0$, there exists a cocomparability graph with bend number k (Theorem 3.1 in [11]). This shows that, $\forall k \geq 0$, the classes B_k-VPG and B_{k+1}-VPG are separated within cocomparability graphs. The question of a similar separation within chordal graphs was left open in [7] and a partial answer was given in [4].

Since the B_k-VPG representation is a kind of planar representation, the bend number of planar graphs have received special attention. Chaplick and Ueck-erdt showed, disproving a conjecture in [2], that every planar graph is B_2-VPG [8]. Every planar bipartite graph is a GIG [21] and hence B_0-VPG. The order dimension of GIGs has also been investigated in literature [6]. A polynomial time decision algorithm for chordal B_0-VPG graphs is developed in [5]. Characteriza-tions for B_0-VPG are known within the classes of split graphs, chordal bull-free graphs, chordal claw-free graphs [20] and block graphs [1].

Subclasses of cocomparability graphs within which a characterization for B_0-VPG is known include cographs, bipartite permutation graphs, and interval graphs. Cographs, which form a subgraph of permutation graphs are B_0-VPG if and only if they do not contain an induced W_4 [10]. A W_4 is a C_4 together with a universal fifth vertex. All bipartite permutation graphs are B_0-VPG [11]. Interval graphs are trivially B_0-VPG, since the interval representation itself is a B_0-VPG representation. Theorem 1 subsumes these three results.

Towards the end of this paper, we discuss a two dimensional visualization of posets. The most common way to visualize a poset $P = (X, \prec)$ so far is *Hasse Diagram* (also called Order Diagram). The problem of drawing a Hasse diagram algorithmically was addressed by many algorithms *e.g. upward planar drawing* [13], *dominance drawing* [14], *confluent drawing* [15], *weak dominance drawing* [22]. The key concern here is to get a crossing-free drawing in which no two upward edges cross at a non vertex point. The first three algorithms can only handle posets of dimension at most two and a few other cases. Though our drawing handles only those posets whose cocomparability graph is B_0-VPG irrespective of its dimension, crossing-freeness is not a concern in our drawing since comparability is inferred from the relative position of lines.

1.2 Terminology and Notation

The complement of a graph G is denoted as $\overline{G}$. We denote a path and a cycle on n vertices, respectively, by P_n and C_n. A graph G is said to be *H-free* if G contains no induced subgraph isomorphic to the graph H. A poset P is said to be *T-free* if P contains no induced subposet isomorphic to the poset T. In this case, P is also said to *exclude* T.

The *closed neighborhood* of a vertex v is the set of neighbors of v together with v. An *Asteroidal Triple* (*AT*) is a set of three independent vertices such that there exists a path between each two of them not passing through any vertex from the closed neighborhood of the third. We have defined interval orders, interval, comparability, cocomparability, and permutation graphs in the introduction. We will make use of the facts that C_4-free cocomparability graphs are interval graphs [17] and all cocomparability graphs are AT-free [19].

2 Proof of Theorem 1

The necessity of the two conditions in Theorem 1 is relatively easier to establish, and hence we do that first. A C_4 has a unique B_0-VPG representation as shown in Fig. 2(a) [2]. Notice that no two vertices of an induced C_4 can be represented by collinear paths in a B_0-VPG representation. In contrast, one can see that in any B_0-VPG representation of a C_3, at least two of its three vertices have to be represented by collinear paths. Moreover, in any B_0-VPG representation of a diamond, the endpoints of the diamond diagonal have to be represented by collinear paths as shown in Fig. 2(b) [20]. Since collinearity is transitive, any two vertices which are diamond related have to be represented by collinear paths. This shows the necessity of the first condition in Theorem 1. Notice that $\overline{C_6}$ has a C_3 in which each pair of vertices is part of an induced C_4. Being part of C_3 forces two of the corresponding three paths to be collinear which prevents a B_0-VPG representation of the corresponding induced C_4. Hence the necessity of the second condition.

The three step algorithm (Algorithm 1) and the proof of its correctness (in the full version[1]) are devoted to showing that these two necessary conditions are sufficient to construct a B_0-VPG representation of a cocomparability graph. The construction is completed in three steps. We start with a cocomparability graph G satisfying the two conditions of Theorem 1. In the first step, we contract a subset of edges of G to obtain a bipartite minor R_G of G with a couple of additional properties. A set of vertices in G which gets represented by a single vertex in R_G after all the edge contractions is referred to as a *branch set* of

---

[1] Details of the proof can be found in the Appendix section of the full version [26].

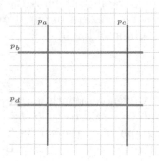
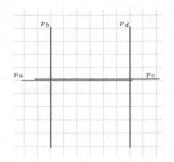

(a) A C_4 induced by $\{a,b,c,d\}$.

(b) A diamond induced by $\{a,b,c,d\}$ with ac as its diagonal.

Fig. 2. The unique B_0-VPG representation of C_4 and a diamond.

R_G. We will denote the vertices in R_G by the corresponding branch sets. In the second step, for each of the subgraphs of G induced by each branch set of R_G, we find an interval representation, again with a few additional properties. In the third and final step, we fit all the above interval representations together to get a B_0-VPG representation of G.

Before proceeding to the algorithm, we state in the next section some of the known results which ease our construction.

2.1 Preliminaries

An ordering σ of V of a graph $G(V,E)$ is called a *cocomparability ordering* or an *umbrella-free ordering* if for all three vertices in $x <_\sigma y <_\sigma z$, adjacency of x and z implies that at least one of the other pairs are adjacent. If not, (x,y,z) is called an *umbrella* in σ.

Lemma 1 ([25]). *A graph G is a cocomparability graph if and only if there is a cocomparability ordering σ of the vertices of G.*

Definition 2. *Given a graph G and a total ordering σ of $V(G)$, a triple (u,v,w) of vertices of G where $u \prec_\sigma v \prec_\sigma w$ is called a* forbidden triple *if there exists a path from u to w without containing a vertex from the closed neighborhood of v.*

Lemma 2. *Any umbrella free ordering is forbidden triple free.*

Proof. Let σ be an umbrella free ordering. Assume a forbidden triple $u \prec_\sigma v \prec_\sigma w$ exists. Thus there exists a path from u to w without containing a vertex from the closed neighborhood of v. If we arrange the vertices of the path together with v in an order respecting σ, there exists two adjacent vertices u_1 and w_1 among them such that $u_1 \prec_\sigma v \prec_\sigma w_1$. Thus (u_1, v, w_1) forms an umbrella in σ which is a contradiction.

Lemma 3 ([9]). *Cocomparability is preserved under edge contraction.*

A $2 + 2$ is a poset containing four elements where every element is comparable with exactly one element. A $2 + 2$ poset corresponds to an induced C_4 in the complement of its comparability graph. Thus an interval order cannot contain a $2 + 2$. Similarly, if a poset does not have a $2 + 2$, then the complement of all of its comparability graphs will be C_4-free cocomparability graphs.

Theorem 3 (Fishburn-1970) [[16], Theorem 6.29 in [31]]. *A poset is an interval order if and only if it excludes* $2 + 2$.

Consider a bipartite graph $G(A \cup B, E)$. An ordering σ of A is said to have *adjacency property* if the neighborhood of every vertex of B is consecutive in σ. Here G is called *convex* if there exists an ordering σ of A with the adjacency property and *biconvex* if it is convex and there exists an ordering τ of B with the adjacency property. Bipartite permutation graphs are biconvex graphs [29].

Theorem 4 ([11]). *Bipartite permutation graphs are* B_0-*VPG*.

2.2 B₀-VPG Algorithm

We see a three-step algorithm to construct a B_0-VPG representation for any arbitrary cocomparability graph satisfying the conditions of Theorem 1. Figure 3 helps to understand the algorithm easily. The first step is depicted in Fig. 3b, 3c, 3e and the final drawing in the third step is shown in Fig. 3f.

In the following definition, we assume that any self loop produced by an edge contraction is removed and any parallel edges formed by an edge contraction is represented by a single edge in the minor.

Definition 3 (dd-minor). *A dd-minor of graph G is the graph obtained by contracting every diamond diagonal in G.*

Definition 4 (Reduced dd-minor). *A reduced dd-minor of graph G is a minimal graph R_G that can be obtained by edge contractions of the dd-minor of G such that no branch-set of R_G contains more than one vertex of an induced C_4 in G.*

Remark 1. Though the dd-minor exists for every graph, a reduced dd-minor does not exist for every graph. A necessary and sufficient condition for the existence of a reduced dd-minor for a graph G is that no two vertices of an induced C_4 in G should be diamond related.

If an edge xy ($x \prec_\sigma y$) is contracted to a new vertex, then the new vertex is placed at the position of x in σ and labeled as x itself. This results in a new order σ' which is a subsequence of σ. By Lemma 3, σ' is an umbrella-free ordering. Thus after all the edge contractions to get the minimal graph R_G, we get an umbrella-free ordering σ_{R_G} of $V(R_G)$ which is a subsequence of σ. This is sufficient to say that R_G is also a cocomparability graph by Lemma 1. In fact, every vertex B of R_G is represented in σ_{R_G} by the leftmost (under σ) vertex b in the branch set B.

For any two branch sets B_i and B_j of R_G, $B_{j,i}$ denotes the vertices in B_j which have a neighbor in B_i.

Lemma 4 (Proof in the full version [26]). *The following claims on the cocomparability graph R_G are true.*

1. *For any two adjacent branch sets B_1 and B_2, the set $B_{1,2} \cup B_{2,1}$ induces a clique in G.*
2. *If B_0, B_1, B_2 form consecutive vertices of a C_3 or an induced C_4 in R_G then $B_{1,0} \cap B_{1,2} \neq \emptyset$. Moreover, if $B_0, \ldots, B_{k-1}$ is a C_3 or an induced C_4 in R_G, then there exists an induced cycle $b_0, \ldots, b_{k-1}$ in G where each $b_i \in B_i$.*
3. *R_G is a bipartite permutation graph.*

Relabeling of $V(R_G)$. It is clear from Lemma 4.3 that the reduced dd-minor R_G of the cocomparability graph G is a bipartite permutation graph. Given the umbrella-free ordering σ of G, we inherited the umbrella-free ordering σ_{R_G} for R_G which respects σ. In the algorithm, we label each branch set of the left part of R_G with $B_1, B_3, \ldots$ (odd indices) such that $i < j$ implies that in the order σ, the leftmost vertex in B_i is to the left of the leftmost vertex in B_j. Similarly we label each branch set of the right part of R_G with $B_0, B_2, \ldots$ (even indices) such that $i < j$ implies that in the order σ, the leftmost vertex in B_i is to the left of the leftmost vertex in B_j.

Henceforth, we slightly abuse the notation $\prec_\sigma$ for the branch sets of the reduced dd-minor of G in the following way. For any two such branch sets B_i and B_j, $B_i \prec_\sigma B_j$ if $\forall x \in B_i, \forall y \in B_j$, $x \prec_\sigma y$. Thus B_i and B_j are said to be *separated in σ* if either $B_i \prec_\sigma B_j$ or $B_j \prec_\sigma B_i$.

Lemma 5 (Proof in the full version [26]). *For any two branch sets B_i, B_j of the same parity if $i < j$, then $B_i \prec_\sigma B_j$.*

Lemma 6 (Proof in the full version [26]). *For each branch set B_i of R_G, $G[B_i^*]$ is an interval graph. Moreover, $G[B_i^*]$ has an interval representation $\mathcal{I}_i^*$ satisfying the following properties.*

(i) *For all $x, y \in B_i^*$, we have the interval for x to the left of interval for y if and only if $x \prec_P y$.*
(ii) *For each neighbor B_j of B_i, all the intervals corresponding to the vertices in $B_{j,i}$ are point intervals at a point $p_{j,i}$.*
(iii) *If B_j and B_k are two neighbors of B_i such that $j < k$, then the point $p_{j,i}$ is to the left of the point $p_{k,i}$ in $\mathcal{I}_i^*$.*

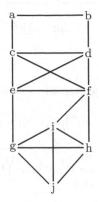

(a) The *cocomparability* graph G.

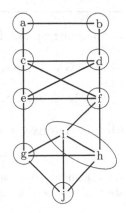

(b) The *dd-minor* of G.

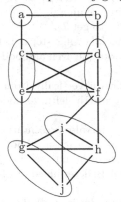

(c) A *reduced dd-minor* R_G of G (also bipartite *permutation* graph).

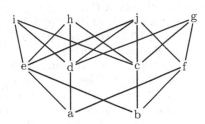

(d) Hasse diagram of an arbitrary poset $P_G(V, \prec_P)$ of G. Choose and fix a *linear extension* $\sigma = (b, a, c, d, f, e, g, i, h, j)$.

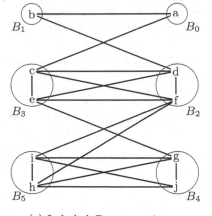

(e) Labeled R_G respecting σ.

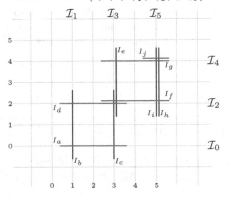

(f) The drawing D which is also a 2-D visualization of P_G by Theorem 2.

Fig. 3. Drawing a B₀-VPG representation D of a cocomparability graph G satisfying the conditions of Theorem 1. Note that the collinear intersecting line segments in D are drawn a little apart in order to distinguish them easily.

Algorithm 1

Input. G is an arbitrary but fixed cocomparability graph satisfying the two conditions of Theorem 1.

Output. A B_0-VPG representation D of G .

Assumptions.

1) $P_G(V(G), \prec_P)$ is an arbitrary but fixed partial order whose comparability graph is $\overline{G}$.

2) σ is an arbitrary but fixed linear extension of P_G and hence an umbrella-free ordering for G.

---

1. Step 1: Choose an arbitrary but fixed reduced dd-minor R_G of G and label $V(R_G)$ as described in the above mentioned relabeling procedure.
2. Step 2:
   (i) For each branch set B_i of R_G, let $B_i^* = B_i \cup \{B_{j,i} : B_i B_j \in E(R_G)\}$ and we obtain an interval representation $\mathcal{I}_i^*$ of B_i^* using Lemma 6.
   (ii) Remove intervals of vertices in $B_i^* \setminus B_i$ from $\mathcal{I}_i^*$ to get $\mathcal{I}_i$.
3. Step 3: Construction of the drawing D using the following steps. Let e and o respectively (with further subscripts if needed) denote the even and odd indices of the branch sets of R_G.
   (i) For each odd-indexed branch set B_o, $\mathcal{I}_o$ is drawn vertically from the point $(o, (e_1 - 0.5))$ to $(o, (e_2 + 0.5))$ where B_{e_1} and B_{e_2} are the leftmost and rightmost neighbors of B_o in σ_{R_G}.
   (ii) Stretch or shrink the intervals in each vertical interval representation $\mathcal{I}_o$ without changing their intersection pattern, so that for each neighbor B_e of B_o, the point $p_{e,o}$ is at (o, e). This can be done since the intersection pattern of an interval representation with n intervals is solely determined by the order of the corresponding $2n$ endpoints.
   (iii) For each even-indexed branch set B_e, $\mathcal{I}_e$ is drawn horizontally from the point $((o_1 - 0.5), e)$ to $((o_2 + 0.5), e)$ where B_{o_1} and B_{o_2} are the leftmost and rightmost neighbors of B_e in σ_{R_G}.
   (iv) Stretch or shrink the intervals in each horizontal interval representation $\mathcal{I}_e$ without changing their intersection pattern, so that for each neighbor B_o of B_e, the point $p_{o,e}$ is at (o, e).

Proposition 1. *The B_0-VPG representation D is precisely a B_0-VPG representation of the cocomparability graph G.*

The proof of the above proposition is written in the full version [26]. One can easily verify that Algorithm 1 runs in polynomial time.

3   Proof of Theorem 2

In this section, we fix $P(V, \prec_P)$ as the given input poset and G as its cocomparability graph. Let D be the B_0-VPG representation of G obtained by the construction employed in the proof of Theorem 1, where the partial order P_G

assumed in Algorithm 1 is P. We argue below that for any two vertices x and y in $V(G)$, $x \prec_D y$ (cf. Definition 1) if and only if $x \prec_P y$ and thus establish Theorem 2. First, we show that if $x \prec_P y$, then $x \prec_D y$. But a simple exchange of variables is not enough to prove the converse because $\prec_D$ is not antisymmetric in the set of all vertical and horizontal line segments. Hence in order to complete the proof we show that the relation $\prec_D$ is antisymmetric when restricted to the line segments in D.

3.1 If $x \prec_P y$ Then $x \prec_D y$

Recall that σ is a linear extension of P_G. Thus if $x \prec_P y$, then $x \prec_\sigma y$ and x is non-adjacent to y in G.

If x and y are in the same branch set B_i of R_G then clearly $I_x \prec_{\mathcal{I}_i} I_y$ by Lemma 6.(i). Here if I_x is horizontal, clearly $x \prec_D^{v,h} y$. Otherwise, $x \prec_D^{h,v} y$. If x and y are in different branch sets, B_i and B_j respectively, of the same parity, then $i < j$ (Lemma 5). Thus in D, if both are of odd parity, B_i is drawn to the left of B_j and if both are of even parity, B_i is drawn to the bottom of B_j. Thus if I_x is vertical, then $x \prec_D^{h,v} y$ and if I_x is horizontal, then $x \prec_D^{v,h} y$.

If the parity is opposite, we have two sub-cases; that is B_i and B_j are either non-adjacent or adjacent. If non-adjacent, the branch set B_j cannot have a neighbor B_h where $h < i$. Suppose there exists such a neighbor B_h for B_j such that $h < i$. Clearly since B_h is adjacent to B_j, h and i are of the same parity and hence $B_h \prec_\sigma B_i$ (Lemma 5). Since B_h is adjacent to B_j, there exists a path from a vertex $z \in B_h$ to $y \in B_j$ in $G[B_h \cup B_j]$. Moreover since B_i is disjoint from B_h and B_j, x has no neighbor in $B_h \cup B_j$, and hence the triple (z, x, y) is a forbidden triple in σ which is a contradiction as per Lemma 2. Since B_j has no opposite parity neighbor B_h for any $h \leq i$, the following property can easily be verified from our drawing. Thus if B_i is of even parity, then B_i is to the bottom of B_j in D. Otherwise, B_i to the left of B_j in D. Hence clearly if I_x is horizontal, then I_x is to the bottom of I_y, that is $x \prec_D^{v,h} y$. Similarly, if I_x is vertical, then I_x is to the left of I_y, that is $x \prec_D^{h,v} y$.

Now the remaining sub-case is that B_i and B_j are adjacent. The following observation is frequently used in the remaining part of the proof.

Observation 1. *If the interval representations $\mathcal{I}_i$ and $\mathcal{I}_j$ intersects, there exist at least two intersecting intervals $I_{u_1} \in \mathcal{I}_i$ and $I_{v_1} \in \mathcal{I}_j$. For any interval $I_u \in \mathcal{I}_i$, if $I_u \prec_{\mathcal{I}_i} p_{j,i}$, then $u \prec_\sigma v_1$ and if $p_{j,i} \prec_{\mathcal{I}_i} I_u$ then $v_1 \prec_\sigma u$. This is easily inferred from $\mathcal{I}_i^*$. We can symmetrically argue the same for I_v.*

The intervals I_x and I_y do not intersect since x and y are non-adjacent in G. If I_x contains $p_{j,i}$, then $p_{i,j}$ (geometrically coinciding with $p_{j,i}$) has to precede I_y in $\mathcal{I}_j$. Otherwise due to Observation 1, we get $y \prec_\sigma x$ which is a contradiction. Similarly if I_y contains $p_{i,j}$, then I_x has to precede $p_{j,i}$ in $\mathcal{I}_i$. In both these case, if I_x is horizontal, then $x \prec_D^{v,h} y$ and if I_x is vertical, then $x \prec_D^{h,v} y$. Henceforth we assume that neither I_x nor I_y contains the intersection point of $\mathcal{I}_i$ and $\mathcal{I}_j$.

In this case, it is easy to see that x has no neighbors in B_j and y has no neighbors in B_i.

In order to rule out the following scenarios, we show the existence of a forbidden triple in σ which leads to a contradiction as per the Lemma 2.

If $I_y \prec_{\mathcal{I}_j} p_{i,j}$, then $x \prec_\sigma y \prec_\sigma u_1$ as per Observation 1. Thus (x, y, u_1) is a forbidden triple.

If $p_{j,i} \prec_{\mathcal{I}_i} I_x$, then $v_1 \prec_\sigma x \prec_\sigma y$ as per Observation 1. Thus (v_1, x, y) is a forbidden triple.

Hence $p_{i,j} \prec_{\mathcal{I}_j} I_y$ and $I_x \prec_{\mathcal{I}_i} p_{j,i}$. In this case, if I_x is horizontal, $x \prec_D^{v,h} y$ and if I_x is vertical, we get $x \prec_D^{h,v} y$. Moreover, in both these cases, I_x is to the left and to the bottom of I_y.

Thus we have proved that when $x \prec_P y$, we get that $x \prec_D^{v,h} y$ when I_x is horizontal or $x \prec_D^{h,v} y$ when I_x is vertical in the B_0-VPG representation D of G. That is $x \prec_D y$. If x and y are incomparable in P_G, then they are adjacent in G and the corresponding intervals intersect in D.

3.2   Antisymmetry of $\prec_D$

Observation 2. *If two opposite parity branch sets are non-intersecting, then one of them is entirely to the bottom left of the other in D. Hence for all I_x in the bottom left branch set and for all I_y in the top right branch set, $x \prec_D y$.*

Justification. When two branch sets are non-intersecting, they are *separated in σ* since otherwise, there will exist a forbidden triple $u \prec_\sigma v \prec_\sigma w$ such that u and w are in one branch set and v in the other branch set. Without loss of generality, let $B_i \prec_\sigma B_j$. We claim that B_i is entirely to the bottom left of B_j. Assume not. That is either B_i has an opposite parity neighbor B_k for some $k > j$ or B_j has an opposite parity neighbor B_h for some $h < i$ or both. In the first case, there exists a forbidden triple (x, y, z) for any $x \in B_i$, $y \in B_j$ and $z \in B_k$ which is a contradiction by Lemma 2. Similarly in the second case, there exists a forbidden triple (w, x, y) for any $w \in B_h$, $x \in B_i$ and $y \in B_j$ which is again a contradiction by Lemma 2.

Observation 3. *In D, there is no line segment I_b which is to the bottom right of a line segment I_t of an opposite parity branch set.*

Justification. Assume I_t is in $\mathcal{I}_i$ and I_b is in $\mathcal{I}_j$. The branch sets B_i and B_j are of the opposite parity. If they are non-adjacent, then by Observation 2, either I_t has to be bottom left of I_b or I_b has to be bottom left of I_t. In both cases, I_b can not be bottom right of I_t. Now we consider the case when the branch sets are adjacent. Since I_t and I_b are non-intersecting, there exists intersecting intervals $I_{t_1} \in \mathcal{I}_i$ and $I_{b_1} \in \mathcal{I}_j$ as per Observation 1. In σ, either t precedes b or b precedes t. These result in the forbidden triple either (t, b, t_1) or (b, t, b_1) respectively leading to a contradiction by Lemma 2.

Lemma 7. *Any two non-intersecting line segments I_x and I_y in D satisfy either $x \prec_D y$ or $y \prec_D x$, but not both. In particular, the relation $\prec_D$ is antisymmetric.*

Proof. Since D is a B$_0$-VPG representation of G, x and y are nonadjacent in G and hence comparable in P. That is, either $x \prec_P y$ or $y \prec_P x$. Therefore, $x \prec_D y$ or $y \prec_D x$. Hence it is enough to verify that $x \prec_D y$ and $y \prec_D x$ cannot both be true. This is easily verified when I_x and I_y are both horizontal or both vertical. Hence we can assume without loss of generality that I_x is horizontal and I_y is vertical. If both $x \prec_D y$ and $y \prec_D x$, then I_x has to be to the bottom right of I_y. This contradicts Observation 3.

This concludes the proof of Theorem 2. For every two incomparable elements in $\prec_P$, the corresponding line segments intersect in D. For every two comparable elements $x, y \in V(G)$, if $x \prec_P y$, then $x \prec_D y$. By exchange of variables, if $y \prec_P x$, then $y \prec_D x$. Lemma 7 asserts that exactly one of the above is true for any two non-intersecting line segments. Hence $x \prec_D y$ only when $x \prec_P y$. Thus the relation $\prec_D$ is isomorphic to the relation $\prec_P$.

References

1. Alcón, L., Bonomo, F., Mazzoleni, M.P.: Vertex intersection graphs of paths on a grid: characterization within block graphs. Graphs Comb. **33**(4), 653–664 (2017)
2. Asinowski, A., Cohen, E., Golumbic, M.C., Limouzy, V., Lipshteyn, M., Stern, M.: Vertex intersection graphs of paths on a grid. J. Graph Algorithms Appl. **16**(2), 129–150 (2012)
3. Bogart, K.P., Rabinovich, I., Trotter Jr., W.T.: A bound on the dimension of interval orders. J. Comb. Theory Ser. A **21**(3), 319–328 (1976)
4. Chakraborty, D., Das, S., Mukherjee, J., Sahoo, U.K.: Bounds on the bend number of split and cocomparability graphs. Theory Comput. Syst. **63**(6), 1336–1357 (2019)
5. Chaplick, S., Cohen, E., Stacho, J.: Recognizing some subclasses of vertex intersection graphs of 0-bend paths in a grid. In: Kolman, P., Kratochvíl, J. (eds.) WG 2011. LNCS, vol. 6986, pp. 319–330. Springer, Heidelberg (2011). https://doi.org/10.1007/978-3-642-25870-1_29
6. Chaplick, S., Felsner, S., Hoffmann, U., Wiechert, V.: Grid intersection graphs and order dimension. Order **35**(2), 363–391 (2018)
7. Chaplick, S., Jelínek, V., Kratochvíl, J., Vyskočil, T.: Bend-bounded path intersection graphs: sausages, noodles, and waffles on a grill. In: Golumbic, M.C., Stern, M., Levy, A., Morgenstern, G. (eds.) WG 2012. LNCS, vol. 7551, pp. 274–285. Springer, Heidelberg (2012). https://doi.org/10.1007/978-3-642-34611-8_28
8. Chaplick, S., Ueckerdt, T.: Planar graphs as VPG-graphs. In: Didimo, W., Patrignani, M. (eds.) GD 2012. LNCS, vol. 7704, pp. 174–186. Springer, Heidelberg (2013). https://doi.org/10.1007/978-3-642-36763-2_16
9. Chejnovská, A.: Optimisation using graph searching on special graph classes. Bachelor Thesis, Univerzita Karlova, Matematicko-fyzikální fakulta (2015)
10. Cohen, E., Golumbic, M.C., Ries, B.: Characterizations of cographs as intersection graphs of paths on a grid. Discrete Appl. Math. **178**, 46–57 (2014)
11. Cohen, E., Golumbic, M.C., Trotter, W.T., Wang, R.: Posets and VPG graphs. Order **33**(1), 39–49 (2016)
12. Corneil, D.G., Olariu, S., Stewart, L.: LBFS orderings and cocomparability graphs. In: Proceedings of the Tenth Annual ACM-SIAM Symposium on Discrete Algorithms, pp. 883–884 (1999)

13. Di Battista, G., Tamassia, R.: Algorithms for plane representations of acyclic digraphs. Theoret. Comput. Sci. **61**(2–3), 175–198 (1988)
14. Battista, G.D., Tamassia, R., Tollis, I.G.: Area requirement and symmetry display of planar upward drawings. Discrete Comput. Geometry **7**(4), 381–401 (1992). https://doi.org/10.1007/BF02187850
15. Dickerson, M., Eppstein, D., Goodrich, M.T., Meng, J.Y.: Confluent drawings: visualizing non-planar diagrams in a planar way. J. Graph Algorithms Appl. **9**(1), 31–52 (2005)
16. Fishburn, P.C.: Intransitive indifference with unequal indifference intervals. J. Math. Psychol. **7**(1), 144–149 (1970)
17. Gilmore, P.C., Hoffman, A.J.: A characterization of comparability graphs and of interval graphs. Can. J. Math. **16**, 539–548 (1964)
18. Golumbic, M.C.: The complexity of comparability graph recognition and coloring. Computing **18**(3), 199–208 (1977)
19. Golumbic, M.C., Monma, C.L., Trotter Jr., W.T.: Tolerance graphs. Discrete Appl. Math. **9**(2), 157–170 (1984)
20. Golumbic, M.C., Ries, B.: On the intersection graphs of orthogonal line segments in the plane: characterizations of some subclasses of chordal graphs. Graphs Comb. **29**(3), 499–517 (2013)
21. Hartman, I.B.A., Newman, I., Ziv, R.: On grid intersection graphs. Discrete Math. **87**(1), 41–52 (1991)
22. Kornaropoulos, E.M., Tollis, I.G.: Weak dominance drawings for directed acyclic graphs. In: Didimo, W., Patrignani, M. (eds.) GD 2012. LNCS, vol. 7704, pp. 559–560. Springer, Heidelberg (2013). https://doi.org/10.1007/978-3-642-36763-2_52
23. Kratochvíl, J.: String graphs. II. recognizing string graphs is NP-hard. J. Comb. Theory Ser. B **52**(1), 67–78 (1991)
24. Kratochvíl, J., Matoušek, J.: Intersection graphs of segments. J. Comb. Theory Ser. B **62**(2), 289–315 (1994)
25. Kratsch, D., Stewart, L.: Domination on cocomparability graphs. SIAM J. Discrete Math. **6**(3), 400–417 (1993)
26. Pallathumadam, S.K., Rajendraprasad, D.: Characterization and a 2D visualization of B_0-VPG cocomparability graphs. arXiv preprint arXiv:2008.02173 (2020)
27. Pnueli, A., Lempel, A., Even, S.: Transitive orientation of graphs and identification of permutation graphs. Can. J. Math. **23**(1), 160–175 (1971)
28. Schaefer, M., Sedgwick, E., Štefankovič, D.: Recognizing string graphs in NP. J. Comput. Syst. Sci. **67**(2), 365–380 (2003)
29. Spinrad, J., Brandstädt, A., Stewart, L.: Bipartite permutation graphs. Discrete Appl. Math. **18**(3), 279–292 (1987)
30. Trotter, W.T., Moore, J.I., Sumner, D.P.: The dimension of a comparability graph. Proc. Am. Math. Soc. **60**(1), 35–38 (1976)
31. Trotter, W., Keller, M.: Applied Combinatorics. CreateSpace Independent Publishing Platform (2016)

Planar L-Drawings of Bimodal Graphs

Patrizio Angelini[1], Steven Chaplick[2], Sabine Cornelsen[3(✉)],
and Giordano Da Lozzo[4]

[1] John Cabot University, Rome, Italy
pangelini@johncabot.edu
[2] Maastricht University, Maastricht, The Netherlands
s.chaplick@maastrichtuniversity.nl
[3] University of Konstanz, Konstanz, Germany
sabine.cornelsen@uni-konstanz.de
[4] Roma Tre University, Rome, Italy
giordano.dalozzo@uniroma3.it

Abstract. In a *planar L-drawing* of a directed graph (digraph) each edge
e is represented as a polyline composed of a vertical segment starting at
the tail of e and a horizontal segment ending at the head of e. Distinct
edges may overlap, but not cross. Our main focus is on *bimodal graphs*,
i.e., digraphs admitting a planar embedding in which the incoming and
outgoing edges around each vertex are contiguous. We show that every
plane bimodal graph without 2-cycles admits a planar L-drawing. This
includes the class of upward-plane graphs. Finally, outerplanar digraphs
admit a planar L-drawing – although they do not always have a bimodal
embedding – but not necessarily with an outerplanar embedding.

Keywords: Planar L-drawings · Directed graphs · Bimodality

1 Introduction

In an *L-drawing* of a directed graph (digraph), vertices are represented by points
with distinct x- and y-coordinates, and each directed edge (u, v) is a polyline
consisting of a vertical segment incident to the tail u and of a horizontal segment
incident to the head v. Two edges may overlap in a subsegment with end point
at a common tail or head. An L-drawing is *planar* if no two edges cross (Fig.
1(c)). Non-planar L-drawings were first defined by Angelini et al. [2]. Chaplick
et al. [11] showed that it is NP-complete to decide whether a directed graph has
a planar L-drawing if the embedding is not fixed. However it can be decided in
linear time whether a planar st-graph has an *upward-planar L-drawing*, i.e. an
L-drawing in which the vertical segment of each edge leaves its tail from the top.

Sabine Cornelsen—The work of Sabine Cornelsen was funded by the German Research
Foundation DFG – Project-ID 50974019 – TRR 161 (B06).
Giordano Da Lozzo—The work of Giordano Da Lozzo was partially supported by MIUR
grants 20157EFM5C *"MODE: MOrphing graph Drawings Efficiently"* and 20174LF3T8
"AHeAD: efficient Algorithms for HArnessing networked Data".

© Springer Nature Switzerland AG 2020
D. Auber and P. Valtr (Eds.): GD 2020, LNCS 12590, pp. 205–219, 2020.
https://doi.org/10.1007/978-3-030-68766-3_17

A vertex v of a plane digraph G is k-*modal* ($\mathrm{mod}(v) = k$) if in the cyclic sequence of edges around v there are exactly k pairs of consecutive edges that are neither both incoming nor both outgoing. A digraph G is k-*modal* if $\mathrm{mod}(v) \leq k$ for every vertex v of G. The 2-modal graphs are often referred to as *bimodal*, see Fig. 1(a). Any plane digraph admitting a planar L-drawing is clearly 4-modal. Upward-planar and level-planar drawings induce bimodal embeddings. While testing whether a graph has a bimodal embedding is possible in linear time, testing whether a graph has a 4-modal embedding [4] and testing whether a partial orientation of a plane graph can be extended to be bimodal [8] are NP-complete.

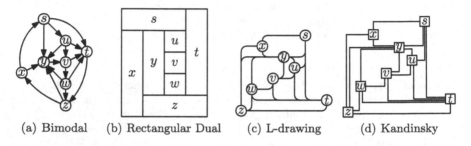

(a) Bimodal   (b) Rectangular Dual   (c) L-drawing   (d) Kandinsky

Fig. 1. Various representations of a bimodal irreducible triangulation.

A *plane digraph* is a planar digraph with a fixed rotation system of the edges around each vertex and a fixed outer face. In an L-drawing of a plane digraph G the clockwise cyclic order of the edges incident to each vertex and the outer face is the one prescribed for G. In a planar L-drawing the edges attached to the same port of a vertex v are ordered as follows: There are first the edges bending to the left with increasing length of the segment incident to v and then those bending to the right with decreasing length of the segment incident to v.

This is analogous to the Kandinsky model [14] where vertices are drawn as squares of equal size on a grid and edges as orthogonal polylines on a finer grid (Fig. 1(d)). Bend-minimization in the Kandinsky model is NP-complete [9] and can be approximated within a factor of two [3]. Each undirected simple graph admits a Kandinsky drawing with one bend per edge [10]. The relationship between Kandinsky drawings and planar L-drawings was established in [11].

L-drawings of directed graphs can be considered as bend-optimal drawings, since one bend per edge is necessary in order to guarantee the property that edges must leave a vertex from the top or the bottom and enter it from the right or the left. Planar L-drawings can be also seen as a directed version of +-contact representations, where each vertex is drawn as a + and two vertices are adjacent if the respective +es touch. If the graph is bimodal then the +es are Ts (including ⊢, ⊥, and ⊣). Undirected planar graphs always allow a T-contact representation, which can be computed utilizing Schnyder woods [12].

Biedl and Mondal [7] showed that a +-contact representation can also be constructed from a rectangular dual (Fig. 1(b)). A plane graph with four vertices on the outer face has a rectangular dual if and only if it is an inner triangulation without separating triangles [18]. Bhasker and Sahni [5] gave the first linear time algorithm for computing rectangular duals. He [15] showed how to compute a rectangular dual from a regular edge labeling and Kant and He [17] gave two linear time algorithms for computing regular edge labelings. Biedl and Derka [6] computed rectangular duals via (3,1)-canonical orderings.

Contribution: We show that every bimodal graph without 2-cycles admits a planar L-drawing respecting a given bimodal embedding. This implies that every upward-planar graph admits a planar L-drawing respecting a given upward-planar embedding. We thus solve an open problem posed in [11]. The construction is based on rectangular duals. Finally, we show that every outerplanar graph admits a planar L-drawing but not necessarily one where all vertices are incident to the outer face. We conclude with open problems.

Proofs for statements marked with ($\star$) can be found inthe full version [1], where we also provide an iterative algorithm showing that any bimodal graph with 2-cycles admits a planar L-drawing if the underlying undirected graph without 2-cycles is a planar 3-tree.

2   Preliminaries

L-Drawings. For each vertex we consider four *ports*, North, South, East, and West. An L-drawing implies a *port assignment*, i.e. an assignment of the edges to the ports of the end vertices such that the outgoing edges are assigned to the North and South port and the incoming edges are assigned to the East and West port. A port assignment for each edge e of a digraph G defines a pair $(\text{out}(e), \text{in}(e)) \in \{\text{North}, \text{South}\} \times \{\text{East}, \text{West}\}$. An L-drawing *realizes* a port assignment if each edge $e = (v, w)$ is incident to the $\text{out}(e)$-port of v and to the $\text{in}(e)$-port of w. A port assignment *admits* a planar L-drawing if there is a planar L-drawing that realizes it. Given a port assignment it can be tested in linear time whether it admits a planar L-drawing [11].

In this paper, we will distinguish between given L-drawings of a triangle.

Lemma 1 ($\star$). *Figure 5 shows all planar L-drawings of a triangle up to symmetry.*

Coordinates for the Vertices. Given a port assignment that admits a planar L-drawing, a planar L-drawing realizing it can be computed in linear time by the general compaction approach for orthogonal or Kandinsky drawings [13]. However, in this approach, the graph has to be first augmented such that each face has a rectangular shape. For L-drawings of plane triangulations it suffices to make sure that each edge has the right shape given by the port assignment, which can be achieved using topological orderings only.

Theorem 1 ($\star$). *Let $G = (V, E)$ be a plane triangulated graph with a port assignment that admits a planar L-drawing and let X and Y be the digraphs with vertex set V and the following edges. For each edge $e = (v, w) \in E$*

- *there is (v, w) in X if $in(e) = West$ and (w, v) in X if $in(e) = East$.*
- *there is (v, w) in Y if $out(e) = North$ and (w, v) in Y if $out(e) = South$.*

Let x and y be a topological ordering of X and Y, respectively. Drawing each vertex v at $(x(v), y(v))$ yields a planar L-drawing realizing the given port assignment.

Observe that we can modify the edge lengths in a planar L-drawing independently in x- and y-directions in an arbitrary way, as long as we maintain the ordering of the vertices in x- and y-direction, respectively. This will still yield a planar L-drawing. This fact implies the following remark.

Remark 1. Let G be a plane digraph with a triangular outer face, let Γ be a planar L-drawing of G, and let Γ_0 be a planar L-drawing of the outer face of G such that the edges on the outer face have the same port assignment in Γ and Γ_0. Then there exists a planar L-drawing of G with the same port assignment as in Γ in which the drawing of the outer face is Γ_0.

Generalized Planar L-Drawings. An *orthogonal polyline* $P = \langle p_1, \ldots, p_n \rangle$ is a sequence of points s.t. $\overline{p_i p_{i+1}}$ is vertical or horizontal. For $1 \leq i \leq n-1$ and a point $p \in \overline{p_i p_{i+1}}$, the polyline $\langle p_1, \ldots, p_i, p \rangle$ is a *prefix* of P and the polyline $\langle p, p_{i+1}, \ldots, p_n \rangle$ is a *suffix* of P. Walking from p_1 to p_n, consider a *bend* p_i, $i = 2, \ldots, n-1$. The rotation $rot(p_i)$ is 1 if P has a left turn at p_i, -1 for a right turn, and 0 otherwise (when $\overline{p_{i-1} p_i}$ and $\overline{p_i p_{i+1}}$ are both vertical or horizontal). The *rotation* of P is $\mathrm{rot}(P) = \sum_{i=2}^{n-1} \mathrm{rot}(p_i)$.

In a *generalized planar L-drawing* of a digraph, vertices are still represented by points with distinct x- and y-coordinates and the edges by orthogonal polylines with the following three properties. (1) Each directed edge $e = (u, v)$ starts with a vertical segment incident to the tail u and ends with a horizontal segment incident to the head v. (2) The polylines representing two edges overlap in at most a common straight-line prefix or suffix, and they do not cross.

In order to define the third property, let $\mathrm{init}(e)$ be the prefix of e overlapping with at least one other edge, let $\mathrm{final}(e)$ be the suffix of e overlapping with at least one other edge, and let $\mathrm{mid}(e)$ be the remaining individual part of e. Observe that the first and the last vertex of $\mathrm{init}(e)$, $\mathrm{final}(e)$, and $\mathrm{mid}(e)$ are end vertices of e, bends of e, or bends of some other edges. Now we define the third property: (3) For an edge e one of the following is true: (i) neither of the two end points of $\mathrm{mid}(e)$ is a bend of e and $\mathrm{rot}(e) = \pm 1$ or (ii) one of the two end points of $\mathrm{mid}(e)$, but not both, is a bend of e and $\mathrm{rot}(\mathrm{mid}(e)) = 0$. See Fig. 2. As a consequence of the flow model of Tamassia [19], we obtain the following lemma.

Lemma 2 ($\star$). *A plane digraph admits a planar L-drawing if and only if it admits a generalized planar L-drawing with the same port assignment.*

(a) not a generalized planar L-drawing   (b) underlying orthogonal drawing

Fig. 2. Cond. 3 of generalized planar L-drawings is fulfilled for all edges but for e_1 and e_2. The rotation of each edge is ± 1. However, $\mathrm{rot}(\mathrm{mid}(e_1)) = 2$ and both end vertices of $\mathrm{mid}(e_2)$ are bends of e_2.

Rectangular Dual. An *irreducible triangulation* is an internally triangulated graph without separating triangles, where the outer face has degree four (Fig. 1). A *rectangular tiling* of a rectangle R is a partition of R into a set of non-overlapping rectangles such that no four rectangles meet at the same point. A *rectangular dual* of a planar graph is a rectangular tiling such that there is a one-to-one correspondence between the inner rectangles and the vertices and there is an edge between two vertices if and only if the respective rectangles touch. We denote by R_v the rectangle representing the vertex v. Note that an irreducible triangulation always admits a rectangular dual, which can be computed in linear time [5,6,15,17].

Perturbed Generalized Planar L-Drawing. Consider a rectangular dual for a directed irreducible triangulation G. We construct a drawing of G as follows. We place each vertex of G on the center of its rectangle. Each edge is routed as a *perturbed orthogonal polyline*, i.e., a polyline within the two rectangles corresponding to its two end vertices, such that each edge segment is parallel to one of the two diagonals of the rectangle containing it. See Fig. 3(a). This drawing is called a *perturbed generalized planar L-drawing* if and only if (1) each directed edge $e = (u, v)$ starts with a segment on the diagonal $\backslash_u$ of R_u from the upper left to the lower right corner and ends with a segment on the diagonal $/_v$ of R_v from the lower left to the upper right corner. Observe that a change of directions at the intersection of R_v and R_u is not considered a bend if the two incident segments in R_v and R_u are both parallel to $\backslash$ or to $/$. The definition of rotation and Conditions (2) and (3) are analogous to generalized planar L-drawings.

In a perturbed generalized planar L-drawing, the North port of a vertex is at the segment between the center and the upper left corner of the rectangle. The other ports are defined analogously. Since we can always approximate a segment with an orthogonal polyline (Figs. 3(b) to 3(d)), we obtain the following.

Lemma 3 ($\star$). *If a directed irreducible triangulation has a perturbed generalized planar L-drawing, then it has a planar L-drawing with the same port assignment.*

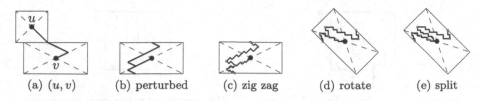

(a) (u, v) (b) perturbed   (c) zig zag   (d) rotate   (e) split

Fig. 3. (a) An edge in a perturbed generalized planar L-drawing. (b-e) From a perturbed generalized planar L-drawing to a generalized planar L-drawing.

3   Planar L-drawings of Bimodal Graphs

We study planar L-drawings of plane bimodal graphs. Our main contribution is to show that if the graph does not contain any 2-cycles, then it admits a planar L-drawing (Theorem 2). In the full version of this paper [1], we also show that if there are 2-cycles, then there is a planar L-drawing if the underlying undirected graph after removing parallel edges created by the 2-cycles is a planar 3-tree.

3.1   Bimodal Graphs Without 2-Cycles

Our approach is inspired by the work of Biedl and Mondal [7] that constructs a +-contact representation for undirected graphs from a rectangular dual. We extend their technique in order to respect the given orientations of the edges.

The idea is to triangulate and decompose a given bimodal graph G. Proceeding from the outermost to the innermost 4-connected components, we construct planar L-drawings of each component that respects a given shape of the outer face. We call a pair of edges e_1, e_2 a *pincer* if e_1 and e_2 are on a triangle T, both are incoming or both outgoing edges of its common end vertex v (i.e. v is a *sink-* or a *source switch* of T), and there is another edge e of G incident to v in the interior of T but with the opposite direction. See Fig. 4. If the outer face of a 4-connected component contains a pincer, we have to make sure that e_1 and e_2 are not assigned to the same port of v in an ancestral component. In a partial perturbed generalized planar L-drawing of G, we call a pincer *bad* if e_1 and e_2 are assigned to the same port. Observe that in a bimodal graph, a pincer must be a source or a sink in an ancestral component. Moreover, in a 4-connected component at most one pair of incident edges of a vertex can be a pincer.

Theorem 2. *Every plane bimodal graph without 2-cycles admits a planar L-drawing. Moreover, such a drawing can be constructed in linear time.*

Proof. Triangulate the graph as follows: Add a new directed triangle in the outer face. Augment the graph by adding edges to obtain a plane bimodal graph in which each face has degree at most four as shown in the full version [1]. More precisely, now each non-triangular face is bounded by a 4-cycle consisting of alternating source and sink switches of the face. We finally insert a 4-modal vertex of degree 4 into each non-triangular face maintaining the 2-modality of

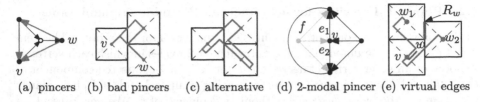

(a) pincers   (b) bad pincers   (c) alternative   (d) 2-modal pincer   (e) virtual edges

Fig. 4. (a) The blue edges incident to v and w, respectively, are pincers that are bad in the drawing of the blue triangle in (b) and not bad in (c). (d) shows the only case (up to reversing directions) of a graph H in Sect. 3.2 with a pincer that is incident to a 2-modal vertex (the orientation of the undirected outer edge is irrelevant). (e) Avoiding bad pincers with virtual edges. (Color figure online)

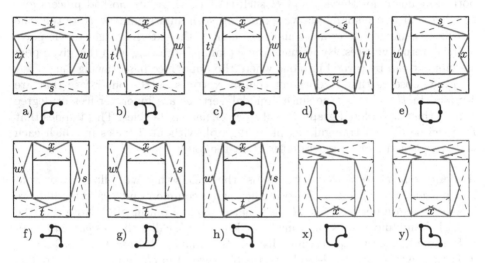

Fig. 5. Realization in the rectangular dual for any kind of drawings of the outer face up to symmetries.

the neighbors. Let G be the obtained triangulated graph. We construct a port assignment that admits a planar L-drawing of G as follows. Decompose G at separating triangles into 4-connected components. Proceeding from the outermost to the innermost components, we compute a port assignment for each 4-connected component H, avoiding bad pincers and such that the ports of the outer face of H are determined by the corresponding inner face of the parent component of H. See Sect. 3.2. By Theorem 1, we compute a planar L-drawing realizing the given port assignment. Finally, we remove the added vertices and edges from Γ. Since the augmentation of G and its decomposition into 4-connected components [16] can be performed in linear time, the total running time is linear.

Theorem 2 yields the following implication, solving an open problem in [11].

Corollary 1. *Every upward-plane graph admits a planar L-drawing.*

3.2   Planar L-Drawings for 4-Connected Bimodal Triangulations

In this subsection, we present the main algorithmic tool for the proof of Theorem 2. Let G be a triangulated plane digraph without 2-cycles in which each vertex is 2-modal or an inner vertex of degree four. Let H be a 4-connected component of G (obtained by decomposing G at its separating triangles) and let Γ_0 be a planar L-drawing of the outer face of H without bad pincers of G. We now present an algorithm that constructs a planar L-drawing of H in which the drawing of the outer face is Γ_0 and no face contains bad pincers of G.

Port Assignment Algorithm. The aim of the algorithm is to compute a port assignment for the edges of H such that (i) there are no bad pincers and (ii) there exists a planar L-drawing realizing such an assignment. Note that the drawing Γ_0 already determines an assignment of the external edges to the ports of the external vertices. By Remark 1 any planar L-drawing with this given port assignment can be turned into one where the outer face has drawing Γ_0.

First, observe that H does not contain vertices on the outer face that are 4-modal in H: This is true since 4-modal vertices are inner vertices of degree four in the triangulated graph G and since G has no 2-cycles. This implies that H, likewise G, is a triangulated plane digraph without 2-cycles in which each vertex is 2-modal or an inner vertex of degree four.

Avoiding Bad Pincers. Next, we discuss the means that will allow us to avoid bad pincers. Let e_1 and e_2 be two edges with common end vertex v that are incident to an inner face f of H such that e_1, e_2 is a pincer of G. Note that the triangle bounding f is a separating triangle of G. We call f the *designated face* of v. In the following we can assume that v is 0-modal in H: In fact, if v is 2-modal in H then v was an inner 4-modal vertex of degree 4 in G, and e_1 and e_2 are two non-consecutive edges incident to v. It follows that H is a K_4 where the outer face is not a directed cycle. See Fig. 4(d). For any given drawing Γ_0 of the outer face (see Fig. 5 and Lemma 1 for the possible drawings of a triangle), the inner vertex can always be added such that no bad pincer is created. Finally, observe that v cannot be 4-modal in H otherwise it would be at least 6-modal in G.

Hence, in the following, we only have to take care of pincers where the common end vertex is 0-modal in H. Since each 0-modal vertex was 2-modal in G, it has at most one designated face. In the following, we assume that all 0-modal vertices are assigned a designated incident inner face where no 0° angle is allowed.

Constructing the Rectangular Dual. As an intermediate step towards a perturbed generalized planar L-drawing, we have to construct a rectangular dual of H, more precisely of an irreducible triangulation obtained from H as follows. Let s, t, and w be the vertices on the outer face of H. Depending on the given drawing of the outer face, subdivide one of the edges of the outer face by a new vertex x according to the cases given in Fig. 5 – up to symmetries. Let f be the inner face incident to x. Then f is a quadrangle. Triangulate f by adding an edge e incident to x: Let y be the other end vertex of e. If y was 2-modal, we can

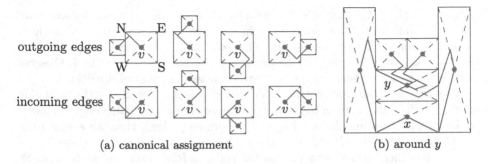

outgoing edges

incoming edges

(a) canonical assignment   (b) around y

Fig. 6. Port assignment: (b) around neighbor y of outer subdivision vertex x.

orient e such that y is still 2-modal. If y was 0-modal and f was its designated face, then orient e such that y is now 2-modal. Otherwise, orient e such that y remains 0-modal. Observe that if y had degree 4 in the beginning it has now degree 5.

The resulting graph H_x is triangulated, has no separating triangles and the outer face is bounded by a quadrangle, hence it is an irreducible triangulation. Thus, we can compute a rectangular dual R for H_x. Up to a possible rotation of a multiple of 90°, we can replace the four rectangles on the outer face with the configuration depicted in Fig. 5 that corresponds to the given drawing of the outer face. Let R_v be the rectangle of a vertex v.

Port Assignment. We now assign edges to the ports of the incident vertices. For the edges on the outer face the port assignment is given by Γ_0. Figure 5 shows the assignments for the outer face.

Let v be a vertex of H_x. We define the *canonical assignment* of an edge incident to v to a port around v as follows (see Fig. 6(a)). An outgoing edge (v, u) is assigned to the North port, if R_u is to the left or the top of R_v. Otherwise it is assigned to the South port. An incoming edge (u, v) is assigned to the West port, if R_u is to the left or the bottom of R_v. Otherwise it is assigned to the East port.

In the following we will assign the edges to the ports of their end vertices such that each edge is assigned in a canonical way to at least one of its end points and such that crossings between edges incident to the same vertex can be avoided within the rectangle of the common end vertex. We exploit this property alongside with the absence of 2-cycles to prove that such an assignment determines a perturbed generalized planar L-drawing of the plane graph H_x.

0-Modal Vertices. We consider each 0-modal vertex v to be 2-modal by adding a *virtual edge* inside its designated face f. Namely, suppose v is a source and let $e_1 = (v, w_1)$ and $e_2 = (v, w_2)$ be incident to f. We add a virtual edge (w, v) between e_1 and e_2 from a new *virtual vertex* w. Of course, there is not literally a rectangle R_w representing w, but for the assignments of the edges to the ports of v, we assume that R_w is the degenerate rectangle corresponding to the segment on the intersection of R_{w_1} and R_{w_2}. See Fig. 4(e).

2-Modal Vertices. Let now v be a 2-modal vertex. We discuss the cases where we have to deviate from the canonical assignment. We call a side s of a rectangle in the rectangular dual to be *mono-directed*, *bi-directed*, or *3-directed*, respectively, if there are 0, 1, or 2 changes of directions of the edges across s. See Fig. 7. Observe that by 2-modality there cannot be more than two changes of directions.

Consider first the case that R_v has a side that is 3-directed, say the right side of R_v. See Fig. 7(a). If from top to bottom there are first outgoing edges followed by incoming edges and followed again by outgoing edges, then we assign from top to bottom first the North port, then the East port, and then the South port to the edges incident to rectangles on the right of R_v (*counterclockwise switch*). Otherwise, we assign from top to bottom first the East port, then the South port, and then the West port (*clockwise switch*). All other edges are assigned in a canonical way to the ports of v; observe that there is no other change of directions.

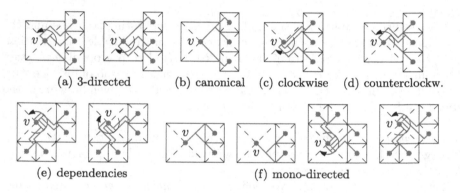

(a) 3-directed   (b) canonical   (c) clockwise   (d) counterclockw.

(e) dependencies   (f) mono-directed

Fig. 7. Assignment of ports when the direction of edges incident to one side of a rectangle changes a) twice b-e) once, or f) never.

Consider now the case that R_v has one side that is bi-directed, say again the right side of R_v. If the order from top to bottom is first incoming then outgoing then assign the edges incident to the right side of R_v in a canonical way (*canonical switch*, Fig. 7(b)). Otherwise (*unpleasant switch*), we have two options, we either assign the outgoing edges to the North port and the incoming edges to the East port (*counter-clockwise switch*, Fig. 7(d)) or we assign the outgoing edges to the South port and the incoming edges to the West port (*clockwise switch*, Fig. 7(c)).

Observe that if there is an unpleasant switch on one side of R_v then there cannot be a canonical switch on an adjacent side. Assume now that there are two adjacent sides s_1 and s_2 of R_v in this clockwise order around R_v with unpleasant switches. Then we consider both switches as counterclockwise or both as clockwise. See Fig. 7(e). Observe that due to 2-modality two opposite sides of R_v are neither both involved in unpleasant switches nor both in canonical switches.

Consider now the case that one side s of R_v is mono-directed, say again the right side of R_v. See Fig. 7(f). In most cases, we assign the edges incident to s in a canonical way. There would be – up to symmetry – the following exceptions: The top side of R_v was involved in a clockwise switch and the edges at the right side are incoming edges. In that case we have a *clockwise switch* at s, i.e., the edges at the right side are assigned to the West port of v. The bottom side of R_v was involved in a counter-clockwise switch and the edges at the right side are outgoing edges. In that case we have a *counter-clockwise switch* at s, i.e., the edges at the right side are assigned to the North port of v.

In order to avoid switches at mono-directed sides, we do the following: Let s_1, s, s_2 be three consecutive sides in this clockwise order around the rectangle R_v such that there is an unpleasant switch on side s; say s is the right side of R_v, s_1 is the top and s_2 is the bottom, and the edges on the right side are from top to bottom first outgoing and then incoming. By 2-modality, there cannot be a switch of directions on both, s_1 and s_2, i.e., s_1 contains no incoming edges, or s_2 contains no outgoing edges. In the first case, we opt for a counterclockwise switch for s, otherwise, we opt for a clockwise switch.

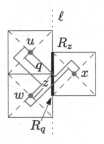

Fig. 8. EXTRA RULE

There is one exception to the rule in the previous paragraph (which we refer to as EXTRA RULE): Let u and w be two adjacent 0-modal vertices with the same designated face f such that the two virtual end vertices are on one line ℓ. See Fig. 8. Let s_u be the side of R_u intersecting R_w and let s_w be the side of R_w intersecting R_u. Assume that u has an unpleasant switch at s_u and, consequently, w has an unpleasant switch at s_w. Let x be the third vertex on f and let s_x be the side of R_x intersecting R_u and R_w. Observe that $s_x \subset \ell$. Do the switch at s_u and s_w in clockwise direction if and only if the switch at s_x is in clockwise direction, otherwise in counterclockwise direction.

Property 1. There is neither a clockwise nor a counter-clockwise switch at a mono-directed side of a rectangle R_v except if v is one of the 0-modal vertices to which the EXTRA RULE was applied.

4-Modal Vertices. If v is an inner 4-modal vertex of degree 4, then each side of R_v is incident to exactly one rectangle, and we always use the canonical assignment.

If v is a 4-modal vertex of degree 5, then v is the inner vertex y adjacent to the subdivision vertex x. Note that we do not have to draw the edge between x and y. However, this case is still different from the previous one, since there are two rectangles incident to the same side s of R_y. If the switch at s is canonical then there is no problem. Otherwise we do the assignment as in Fig. 6(b).

Observe that we get one edge between y and a vertex on the outer face that is not assigned in a canonical way at y. But this edge is assigned in a canonical way at the vertex in the outer face. This completes the port assignments.

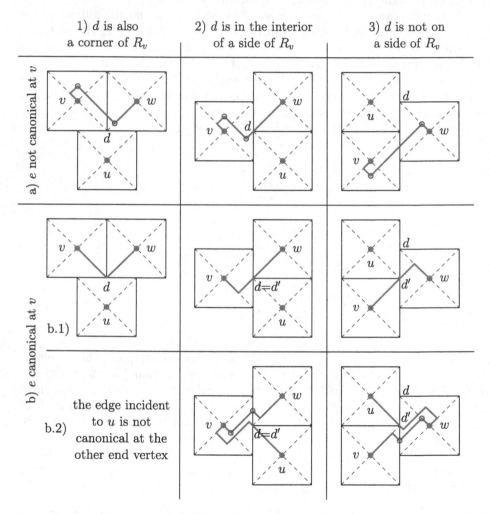

Fig. 9. How to route the edge e between v and w. Point d is the corner at the end of the diagonal of R_w to which e is assigned.

Correctness. In the full version [1], we give a detailed proof that the constructed port assignment admits a perturbed generalized planar L-drawing and, thus, a planar L-drawing of H. The proof starts with the observation that each edge is assigned in a canonical way at one end vertex at least. Then we route the edges as indicated in Fig. 9 where each part of a segment that is not on a diagonal of a rectangle represents a perturbed orthogonal polyline of rotation 0. Finally, we show that the encircled bends are not contained in any other edge.

Lemma 4. *A planar L-drawing of H in which the drawing of the outer face is Γ_0 and no face contains bad pincers of G can be constructed in linear time.*

Proof. The construction guarantees a planar L-drawing of H. The port assignment is such that there are no bad pincers and for the outer face it is the same as in Γ_0. A rectangular dual can be constructed in linear time [5,17]. The port assignment can also be done in linear time. Finally, the coordinates can be computed in linear time using topological ordering – see Theorem 1.

4 Outerplanar Digraphs

Since there exist outerplanar digraphs that do not admit any bimodal embedding [4], we cannot exploit Theorem 2 to construct planar L-drawings for the graphs in this class. However, we are able to prove the following.

Theorem 3. *Every outerplanar graph admits a planar L-drawing.*

Proof. Put all vertices on a diagonal in the order in which they appear on the outer face – starting from an arbitrary vertex. The drawing of the edges is determined by the direction of the edges. This implies that some edges are drawn above and some below the diagonal. By outerplanarity, there are no crossings.

We remark that Theorem 3 provides an alternative proof to the one in [4] that any outerplanar digraph admits a 4-modal embedding. Observe that the planar L-drawings constructed in the proof of Theorem 3 are not necessarily outerplanar. In the following, we prove that this may be unavoidable.

Theorem 4. *Not every outerplanar graph admits an outerplanar L-drawing.*

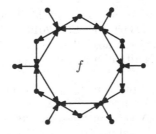

Proof. Consider the graph depicted above. It has a unique outerplanar embedding. Let f be the inner face of degree 6. Each vertex incident to f is 4-modal and is a source switch or a sink switch of f. Thus, the angle at each vertex is $0°$. The angle at each bend is at most $3/2\pi$. Thus, the angular sum around f would imply $(2 \cdot \deg f - 2) \cdot \pi \leq 3/2 \cdot \deg f \cdot \pi$, which is not possible for $\deg f = 6$.

There are even 4-modal biconnected internally triangulated outerplane digraphs that do not admit an outerplanar L-drawing. See the full version [1].

5 Open Problems

- Are there bimodal graphs with 2-cycles that do not admit a planar L-drawing (with or without the given embedding)?
- What is the complexity of testing whether a 4-modal graph admits a planar L-drawing with a fixed embedding?
- In the directed Kandinsky model where edges leave a vertex to the top or the bottom and enter a vertex from the left or the right, for which k is there always a drawing with at most $1 + 2k$ bends per edge for any 4-modal graph? $k = 0$ does not suffice. What about $k = 1$?
- Can it be tested efficiently whether an outerplanar graph with a given 4-modal outerplanar embedding admits an outerplanar L-drawing?

References

1. Angelini, P., Chaplick, S., Cornelsen, S., Da Lozzo, G.: Planar L-drawings of bimodal graphs. Technical report arXiv:2008.07834v1, Cornell University Library (2020)
2. Angelini, P., et al.: Algorithms and bounds for L-drawings of directed graphs. Int. J. Found. Comput. Sci. **29**(4), 461–480 (2018). https://doi.org/10.1142/S0129054118410010
3. Barth, W., Mutzel, P., Yıldız, C.: A new approximation algorithm for bend minimization in the Kandinsky model. In: Kaufmann, M., Wagner, D. (eds.) GD 2006. LNCS, vol. 4372, pp. 343–354. Springer, Heidelberg (2007). https://doi.org/10.1007/978-3-540-70904-6_33
4. Besa Vial, J.J., Da Lozzo, G., Goodrich, M.T.: Computing k-modal embeddings of planar digraphs. In: Bender, M.A., Svensson, O., Herman, G. (eds.) ESA 2019. LIPIcs, vol. 144, pp. 19:1–19:16. Schloss Dagstuhl - Leibniz-Zentrum für Informatik (2019). https://doi.org/10.4230/LIPIcs.ESA.2019.19
5. Bhasker, J., Sahni, S.: A linear algorithm to find a rectangular dual of a planar triangulated graph. Algorithmica **3**, 247–278 (1988). https://doi.org/10.1007/BF01762117
6. Biedl, T.C., Derka, M.: The (3,1)-ordering for 4-connected planar triangulations. JGAA **20**(2), 347–362 (2016). https://doi.org/10.7155/jgaa.00396
7. Biedl, T.C., Mondal, D.: A note on plus-contacts, rectangular duals, and box-orthogonal drawings. Technical report arXiv:1708.09560v1, Cornell University Library (2017)

8. Binucci, C., Didimo, W., Patrignani, M.: Upward and quasi-upward planarity testing of embedded mixed graphs. Theoret. Comput. Sci. **526**, 75–89 (2014). https://doi.org/10.1016/j.tcs.2014.01.015

9. Bläsius, T., Brückner, G., Rutter, I.: Complexity of higher-degree orthogonal graph embedding in the Kandinsky model. In: Schulz, A.S., Wagner, D. (eds.) ESA 2014. LNCS, vol. 8737, pp. 161–172. Springer, Heidelberg (2014). https://doi.org/10.1007/978-3-662-44777-2_14

10. Brückner, G.: Higher-degree orthogonal graph drawing with flexibility constraints. Bachelor thesis, Department of Informatics, Karlsruhe Institute of Technology (2013). https://i11www.iti.kit.edu/_media/teaching/theses/ba-brueckner-13.pdf

11. Chaplick, S., et al.: Planar L-drawings of directed graphs. In: Frati, F., Ma, K.-L. (eds.) GD 2017. LNCS, vol. 10692, pp. 465–478. Springer, Cham (2018). https://doi.org/10.1007/978-3-319-73915-1_36

12. de Fraysseix, H., Ossona de Mendez, P., Rosenstiehl, P.: On triangle contact graphs. Comb. Probab. Comput. **3**(2), 233–246 (1994). https://doi.org/10.1016/j.comgeo.2017.11.001

13. Eiglsperger, M., Kaufmann, M.: Fast compaction for orthogonal drawings with vertices of prescribed size. In: Mutzel, P., Jünger, M., Leipert, S. (eds.) GD 2001. LNCS, vol. 2265, pp. 124–138. Springer, Heidelberg (2002). https://doi.org/10.1007/3-540-45848-4_11

14. Fößmeier, U., Kaufmann, M.: Drawing high degree graphs with low bend numbers. In: Brandenburg, F.J. (ed.) GD 1995. LNCS, vol. 1027, pp. 254–266. Springer, Heidelberg (1996). https://doi.org/10.1007/BFb0021809

15. He, X.: On finding the rectangular duals of planar triangular graphs. SIAM J. Comput. **22**(6), 1218–1226 (1993). https://doi.org/10.1137/0222072

16. Kant, G.: A more compact visibility representation. Int. J. Comput. Geometry Appl. **7**(3), 197–210 (1997). https://doi.org/10.1142/S0218195997000132

17. Kant, G., He, X.: Two algorithms for finding rectangular duals of planar graphs. In: van Leeuwen, J. (ed.) WG 1993. LNCS, vol. 790, pp. 396–410. Springer, Heidelberg (1994). https://doi.org/10.1007/3-540-57899-4_69

18. Koźmiński, K., Kinnen, E.: Rectangular duals of planar graphs. Networks **15**(2), 145–157 (1985). https://doi.org/10.1002/net.3230150202

19. Tamassia, R.: On embedding a graph in the grid with the minimum number of bends. SIAM J. Comput. **16**, 421–444 (1987). https://doi.org/10.1137/0216030

Layered Drawing of Undirected Graphs with Generalized Port Constraints

Julian Walter[1], Johannes Zink[1](✉) [ID], Joachim Baumeister[1,2] [ID],
and Alexander Wolff[1] [ID]

[1] Institut für Informatik, Universität Würzburg, Würzburg, Germany
zink@informatik.uni-wuerzburg.de
[2] denkbares GmbH, Würzburg, Germany

Abstract. The aim of this research is a practical method to draw cable plans of complex machines. Such plans consist of electronic components and cables connecting specific ports of the components. Since the machines are configured for each client individually, cable plans need to be drawn automatically. The drawings must be well readable so that technicians can use them to debug the machines. In order to model plug sockets, we introduce *port groups*; within a group, ports can change their position (which we use to improve the aesthetics of the layout), but together the ports of a group must form a contiguous block.

We approach the problem of drawing such cable plans by extending the well-known Sugiyama framework such that it incorporates ports and port groups. Since the framework assumes directed graphs, we propose several ways to orient the edges of the given undirected graph. We compare these methods experimentally, both on real-world data and synthetic data that carefully simulates real-world data. We measure the aesthetics of the resulting drawings by counting bends and crossings. Using these metrics, we compare our approach to *Kieler* [JVLC 2014], a library for drawing graphs in the presence of port constraints.

Keywords: Sugiyama framework · Port constraints · Experimental evaluation

1 Introduction

Today, the development of industrial machinery implies a high interdependency of mechanical, electrical, hydraulic, and software-based components. The continuous improvement of these machines yielded an increased complexity in all these domains, but also in their interrelations. In the case of a malfunction, a human technician needs to understand the particular interdependencies. Only then, (s)he will be able to find, understand, and resolve errors. Different types of schematics play a key role in this diagnosis task for depicting dependencies between the involved components, e.g., electric or functional schematics. The intuitive understanding and comprehensibility of these schematics is critical for finding errors efficiently.

J.Z. acknowledges support by BMWi (ZIM project iPRALINE – grant ZF4117505).

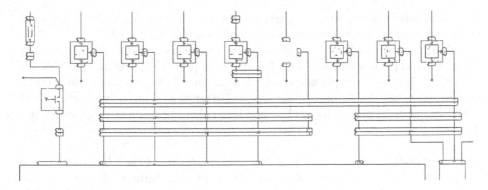

Fig. 1. Extract of a hand-drawn plan. The labels have been intentionally obfuscated or removed.

Due to the increased complexity of machinery, such schematics cannot be drawn manually anymore: The high variance of machine configurations nowadays requires the ad-hoc computation and visualization of schematics appropriate for the requested diagnosis case. To support technicians, algorithms for drawing schematics should adhere to the visual "laws" of the manual drawings that the technicians are familiar with; see Fig. 1 for an example. Such drawings route connections between components in an orthogonal manner. Manual drawings often use few layers and seem to avoid crossings and bends as much as possible.

In many applications (such as UML diagrams or data flow diagrams), connections are directed from left to right or from top to bottom. This setting is supported by the framework introduced by Sugiyama et al. [12]. Given a directed graph, their approach arranges the edges mainly in the same direction by organizing the nodes in subsequent layers (or levels). The layer-based approach solves the graph-layout problem by dividing it into five phases: cycle elimination, layer assignment, crossing minimization, node placement, and edge routing.

There are also algorithms for practical applications purely based on the orthogonal drawing paradigm, where all vertices are rectangles on a regular grid and the edges are routed along the horizontal and vertical lines of the grid. There, a classic three-phase method dates back to Biedl et al. [1].

In many technical drawings (such as cable plans, UML diagrams, or data flow diagrams), components are drawn as axes-aligned rectangles, connections between the components are drawn as axes-aligned polygonal chains that are attached to a component using a *port*, that is, a geometric icon that is small relative to a component and whose shape has a specific meaning for the domain expert. Using so-called *port constraints*, a user can insist that a connection enters a component on a specific side—a natural requirement in many applications.

The well-established Kieler library [11] implements the Sugiyama framework. Kieler is particularly interesting for our application as Kieler allows the user to specify several types of port constraints; namely, on which side of a vertex rectangle should a port be placed, and, for each side, the exact order in which the ports should be arranged. Alternatively, the order is variable and can be exploited to improve the layouts in terms of crossings and bends.

We have chosen to build our algorithm for undirected graphs on the (directed) layer-based approach instead of an (undirected) purely orthogonal one because the typical hand-drawn plans use only few distinct layers to place the vertices on, the layer-based approach seems to be better investigated in practice, and Kieler has already proven to yield by and large pleasing results in the considered domain.

Our Contribution. First, we propose two methods to direct the edges of the given undirected graph so that we can apply the Sugiyama framework (see Sect. 3); one is based on breadth-first search, the other on a force-directed layout. We compare the two methods experimentally with a simple baseline method that places the nodes of the given graph randomly and directs all edges upward (see Sect. 4.3), both on real-world and synthetic cable plans (see Sect. 4.2). We claim that our approach to generate realistic test graphs is of independent interest. We "perturb" real-world instances such that, statistically, they have similar features as the original instances.

Second, we extend the set of port constraints that the aforementioned Kieler library allows the user to specify. In order to model plug sockets, we introduce *port groups*; within a group, the position of the ports is either fixed or variable. In either case, the ports of a group must form a contiguous block. Port groups can be nested. If the order of a port group is variable, our algorithm exploits this to improve the aesthetics of the layout.

Apart from such hierarchical constraints, we also give the user the possibility to specify pairings between ports that belong to opposite sides of a vertex rectangle (top and bottom). Such a pairing constraint enforces that the two corresponding ports are placed at the same x-coordinates on opposite sides of the vertex rectangle. Pairing constraints model pairs of sockets of equal width that are plugged into each other.

After formally defining the problem (Sect. 2), we describe our algorithm (Sect. 3). Finally, we present our experimental evaluation (Sect. 4).

2   Preliminaries

We define the problem LAYERED GRAPH DRAWING WITH GENERALIZED PORT CONSTRAINTS as follows. For an illustration refer to Fig. 2b.

Given: An undirected *port graph* $G = (V, P, PG, PP, E)$, where

-   V is the set of vertices—each vertex v is associated with two positive numbers $w(v)$ and $h(v)$; v will be represented by a rectangle of width at least $w(v)$ and height at least $h(v)$ (to ensure a given vertex label can be accommodated),
-   P is the set of ports—each port belongs either directly to a vertex or indirectly through a port group (or a nested sequence of port groups),

- PG is the set of port groups—each port group belongs to a side (BOTTOM, TOP, FREE)[1] of exactly one vertex and contains a set of ports and port groups (not contained in another port group) whose order is fixed or variable,
- PP is the set of port pairings—each port pairing consists of two unique ports from P that belong to the same vertex (directly or via port groups), and
- E is the set of edges—each edge connects two unique ports from P that are contained in different vertices, and
- the graph where all ports are contracted into their vertices is connected.

Find: A drawing of G such that

- no drawing elements overlap each other except that edges may cross each other in one point,
- each vertex $v \in V$ is drawn as an axis-aligned rectangle of width at least $w(v)$ and height at least $h(v)$ on a horizontal layer,
- each port $p \in P$ is drawn as a (small, fixed-size) rectangle attached to the boundary of its vertex rectangle (on the specified side unless set to FREE),
- when walking along the boundary of a vertex, the ports of a port group (or subgroup) form a contiguous block; and for a port group with fixed order, its ports and port groups appear in that order,
- for each port pair $\{p, p'\} \in PP$, ports p and p' are drawn on the same vertical line on opposite sides of their vertex,
- each edge $\{p, p'\} \in E$ is drawn as a polygonal chain of axis-aligned line segments (*orthogonal polyline*) that connects the drawings of p and p', and
- the total number of layers, the width of the drawing, the lengths of the edges, and the number of bend points of the edges are kept close to a minimum.

We have chosen this problem definition to be both, simple and extendable to more complex settings by using the described elements as building blocks. For instance, if there are multiple edges per port, then in a preprocessing we can assign each edge its own port and keep them together using a port group. In a post-processing, we draw just one of these ports and we re-draw the ends of the edges incident to the other ports of this group. Or if there are bundles of edges (e.g. a cable with twisted wires), we can keep their ports together by introducing port groups.

Note that our problem definition generalizes the LAYERED GRAPH DRAW-ING problem that is formalized and solved heuristically by the Sugiyama frame-work [12]. Several subtasks of the framework correspond to NP-hard optimization problems such as ONE-SIDED CROSSING MINIMIZATION [4]. Hence, we have to make do with a heuristic for our problem, too. This heuristic is coming up next.

3 Algorithm

We assume that we are given a graph as described in Sect. 2. (Otherwise we can preprocess accordingly.) Similarly to the algorithm of Sugiyama et al. [12], our algorithm proceeds in phases, which we treat in the following subsections.

---

[1] We can also handle sides LEFT and RIGHT, which we describe in the full version [13]. We do not have constraints for ports on the left or the right side in our experiments.

3.1 Orienting Undirected Edges

Classical algorithms for layered graph drawing expect as input a directed acyclic graph, whose vertices are placed onto layers such that all edges point upwards. For directed cyclic graphs, some edges may be reversed or removed to make the graph acyclic. In our case of undirected graphs, we suggest the following procedures to orient the undirected edges, making the graph simultaneously directed and acyclic. (Hence, we don't need the cycle elimination phase of the Sugiyama framework.) We ignore the ports in this step.

BFS: We execute a breadth-first search from a random start vertex. Edges are oriented from vertices discovered earlier to vertices discovered later.

FD: We run a force-directed graph drawing algorithm. In the resulting drawing, edges are oriented upwards.

RAND: We place the vertices randomly into the drawing area, uniformly distributed. In the resulting drawing, we orient the edges as in FD.

The runtime of this phase is dominated by the force-directed algorithm. We also suggest to execute the force-directed algorithm more than once, say k times, with different random start positions and then to use the drawing admitting the fewest crossings. This is less time consuming than re-iterating the whole algorithm. In our experiments, we used a classical spring embedder [6] with the speed-up technique as described by Lipp et al. [8]. The resulting runtime is in $O(k \cdot I \cdot |V| \log |V|)$, where I is the number of iterations per execution of the force-directed algorithm.

3.2 Assigning Vertices to Layers

In this step we seek for an assignment of vertices to layers, such that all directed edges point upwards. We use a network simplex algorithm as described by Gansner et al. [7]. The algorithm is optimal in the sense that the sum of layers the edges span is minimized. With respect to the runtime of their algorithm, the authors state: "Although its time complexity has not been proven polynomial, in practice it takes few iterations and runs quickly."

3.3 Orienting Ports and Inserting Dummy Vertices

Consider the ports of a vertex. If a port group is of a type different than FREE, we assign all ports of this port group or a port group containing this port group to the specified vertex side, e.g., the bottom side.[2] If this leads to contradicting assignments of the same port, we reject the instance. We treat port pairings analogously. We assign ports that are in no port group to the top or the bottom side depending on whether they have an outgoing or incoming edge. If ports of a port group of type FREE remain unassigned, we make a majority decision for the top-level port group—if there are more outgoing than incoming edges, we set its ports to the top side; otherwise to the bottom side.

---

[2] See the full version [13] for handling port groups of type LEFT and RIGHT.

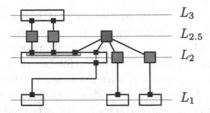 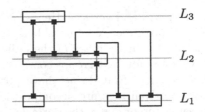

(a) We insert an extra layer $L_{2.5}$ to host a dummy vertex (red) as turning point. All edges traversing a layer are subdivided by dummy vertices (green).

(b) Each port of the vertex on L_2 is in a port group or in a port pairing. Thus, the two rightmost ports are placed on the top side, although they have incoming edges from below.

Fig. 2. Example for the insertion of dummy vertices. (Color figure online)

In any case, we may end up with ports being on the "wrong" side in terms of incident edges, e.g., a port on the top side has an incoming edge. To make such edges reach their other endpoints without running through the vertex rectangle, we introduce an extra layer directly above the layer at hand. On the extra layer, we then place a dummy vertex that will serve as a "turning point" for these edges; see Fig. 2. We will refer to them as *turning dummy vertices*.

In contrast, KIELER [11] appends effectively, for each port that lies on the "wrong" side, a dummy port on the opposite side of the vertex rectangle, to the very right or left of the ports there. The edges will later be routed around the vertex to this dummy port. Our new approach therefore provides a somewhat greater flexibility in routing edges around vertices.

As in the classical algorithms for layered graph drawing, we subdivide edges traversing a layer (which may also be an extra layer) by a new dummy vertex on each such layer. Hence, we have only edges connecting neighboring layers. As for all algorithms that rely on decomposing the edges, this phase runs in time $O(\lambda \cdot |E| + |P|)$, where λ is the number of layers. Note that $\lambda \in O(|V|)$.

3.4 Reducing Crossings by Swapping Vertices

We employ the layer sweep algorithm using the well-known barycenter heuristic proposed by Sugiyama et al. [12]. However, we also have to take the ports and the port constraints into account. We suggest three ways to incorporate them.

VERTICES: We first ignore ports. We arrange the vertices as follows. Since there may be many edges between the same pair of vertices, we compute the vertex barycenters weighted by edge multiplicities. After having arranged all vertices, we arrange the ports at each vertex to minimize edge crossings. Finally, we rearrange the ports according to port pairings and port groups by computing barycenters of the ports of each port group.

PORTS: We use indices for the ports instead of the vertices and apply the barycenter heuristic to the ports. This may yield an invalid ordering with respect to port groups and vertices. Hence, we sort the vertices by the arithmetic mean of the port indices computed before. Within a vertex, we sort the port groups by the arithmetic mean of the indices of their ports. We recursively proceed in this way for port groups contained in port groups and finally for the ports.

MIXED: Vertices that do not have port pairings are kept as a whole, vertices with port pairings are decomposed into their ports. The idea is that, when sweeping up or down, the ports do not influence the ordering on the other side and can be handled in the end—unless they are paired. After each iteration, we force the ports from decomposed vertices to be neighbors by computing their barycenters, and we arrange the paired ports above each other. Finally, we arrange all ports that are not included in the ordering as in VERTICES.

In all cases, if a port group has fixed order, we cannot re-permute its elements, but we take the order as described from left to right. We use random start permutations for vertices and ports. We execute this step r times for some constant r (in our experiments $r = 10$) and take the solution that causes the fewest crossings. KIELER [11] also computes barycenters depending on the order of ports of the previous layer. Similar to PORTS they describe a *layer-total* approach and similar to MIXED they describe a *node-relative* approach. However, they compute barycenters only for vertices as a whole. We use barycenters of ports to recursively determine also an ordering of port groups.

This phase runs in time $O(r \cdot J \cdot \lambda \cdot |E|)$, where J is the number of (top-down or bottom-up) sweeps within one execution of the layer sweep algorithm.

3.5 Determining Vertex Coordinates

To position both vertices and ports, we decompose the vertices into ports and edges. An example is given in Fig. 3. We duplicate each layer L_i (except for the extra layers introduced in Sect. 3.3) to an upper layer L_{i+} and a lower layer L_{i-}. For a vertex on layer L_i, we place all ports of the TOP side in the previously computed order onto L_{i+} and all ports of the BOTTOM side in the previously computed order onto L_{i-}. To separate the vertices from each other and to assign them a rectangular drawing area, we insert a path of length one with the one port on L_{i-} and the other port on L_{i+} at the beginning and the end of each layer and between every two consecutive vertices (gray in Fig. 3(b)). Moreover, we may insert dummy ports without edges within the designated area of a vertex, to increase the width of a vertex. This can be seen as "padding" the width of a vertex v via ports to obtain the desired minimum width $w(v)$. For each port pairing $\{p, p'\}$, where p is on L_{i-} and p' is on L_{i+}, insert a dummy edge connecting p and p'. Observe that we do not have edge crossings between L_{i-} and L_{i+}. Therefore, using the algorithm of Brandes and Köpf [2] (see below), these edges will end up as vertical line segments. This fulfills our requirement for vertices being rectangular and for ports of port pairings being vertically aligned.

Now we have a new graph G' with ports being assigned to layers, but without vertices and without port constraints. So, in the following we consider the ports as vertices. This is precisely the situation as in the classical algorithms

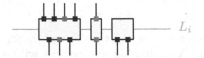
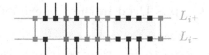

(a) three vertices with two port pairings on one layer before transforming
them to ports only

(b) only ports on two layers; port pairings are
connected by a dummy edge, the rightmost vertex is "padded" to be wider using dummy ports

Fig. 3. Example of the transformation of vertices with ports on one layer to ports and edges on two layers; port pairings are indicated by color. (Color figure online)

for layered graph drawing when determining coordinates of vertices. After the current coordinate assignment step, we will re-transform the drawing into our setting with vertices, ports, and edges.

The y-coordinate of a vertex is given by its layer. For assigning x-coordinates, we use the well-established linear-time algorithm of Brandes and Köpf [2]. It heuristically tries to straighten long edges vertically and balancing the position of a vertex with respect to its upper and lower neighbors. It guarantees to pre-serve the given vertex order on each layer and a minimum distance δ between consecutive vertices. Moreover, it guarantees that uncrossed edges are drawn as vertical line segments, which is crucial for our application.

We note that the original algorithm of Brandes and Köpf [2] contained two flaws that came up in our experiments. Subsequently, they were fixed [3].

This phase runs in time linear in the number of ports and edges.

3.6   Constructing the Drawing and Orthogonal Edge Routing

First, we obtain vertices drawn as rectangles from (dummy) ports and edges by reversing the transformation described in Sect. 3.5. Then, we transform the dummy vertices inserted in Sect. 3.3 into bend points of their edges. Finally, we draw the edges orthogonally. In the full version [13], we describe how to route the edge pieces between two consecutive layers including edges going through turning dummy vertices. The total runtime of this phase is $O(\lambda \cdot |E|^2)$ in the worst case (while in practice we would rather expect a linear runtime behavior).

4   Experimental Evaluation

For our experiments we got access to 380 real cable plans of a large German machine manufacturing company. To obfuscate these plans and to have more data for our experiments, we generated 1139 pseudo cable plans from the real cable plans—almost always three from each real cable plan. For replicability, we have made all of our algorithms, data structures, and data described here publicly available on github [9,10]—except for the original (company-owned) plans.

4.1   Graphs Used in the Experiments

First, we discuss the structure of these cable plans and how we transformed them to the format that is expected by our algorithm. A cable plan has vertices with ports and vertex groups that comprise multiple vertices. Moreover, there can be edges connecting two or more ports (that is, hyperedges) and a port can be incident to an arbitrary number of edges. In a vertex group, there are port pairings between two vertices and these vertices should be drawn as touching rectangles. In our model, we do not have vertex groups and port pairings between different vertices. Instead, we model a vertex group as a single vertex with (internal) port pairings and a port group for the ports of each vertex. Moreover, we split ports of degree d into d separate ports and enforce that they are drawn next to each other and on the same side of the vertex by an (unordered) port group. We replace hyperedges by a dummy vertex having an edge to each of the ports of the hyperedge. We don't have ports on the left or the right side of a vertex.

4.2   Generating a Large Pseudo Data Set from Original Data

Now, we describe briefly how we generated the pseudo cable plans. This can be seen as a method to extend and disguise a set of real-world graphs. A drawing of an original cable plan and derived pseudo cable plan is depicted in the appendix of the full version [13]. There, we also show larger examples of drawings generated by our algorithm. First, we preprocess the real-world input data by extracting only the largest connected component of each graph as we draw each connected component independently anyways. Then, we generate a pseudo plan by removing and inserting elements from/to an original plan. Elements of the plans are the vertex groups, vertices, ports, port pairings, and edges. As a requirement we had to replace or remove at least a q-fraction of the original elements (in our case $q = 0.1$). We proceed in three phases.

1. We determine target values for most elements of the graph (number of vertex groups, vertices, ports, port pairings) and more specific parameters (distribution of edge–port incidences, arithmetic mean of parallel edges per edge, number of self loops, distribution of ports per edge, distribution of edges per port). We pick each target value randomly using a normal distribution, where the mean is this value in the original plan and the standard deviation is the standard deviation of this value across all graphs of the original data set divided by the number of plans in the original data set times a constant.
2. We remove a q-fraction of the original elements uniformly at random in the following order: vertex groups (incl. contained vertices and incident edges), vertices (incl. ports and incident edges), port pairings (incl. ports and incident edges), ports (incl. incident edges), and edges.
3. In the same order, we add as many new elements as needed to reach the respective target values. For the insertion of edges we are a bit more careful. In case the graph became disconnected during the deletion phase, we first reconnect the graph by connecting different components. Then, we insert the

remaining edges according to the distributions of edge–port incidences while trying to reduce the gaps between the target value and the current value for parallel edges per edge and for the number of self loops. Parallel edges have the same terminal vertices but not necessarily the same terminal ports. We mostly use ports that do not have edges (they are new or their edges were removed or they had no edges initially) and assign for each one the number of edges it should get in the end. This gives us a set of candidate ports. Next, we iteratively add a (hyper)edge e connecting d ports. In each iteration, we pick c sets of d ports from our set of candidate ports uniformly at random—each set is a candidate for the end points of the new edge. We choose the set where we approach the aforementioned target values the best if we would add the corresponding edge to the current graph. We used $c = 1000$, which means we took one out of 1000 randomly generated edge candidates.

Our generated pseudo cable plans are good if they are similar to and have similar characteristics as the original cable plans, and if the corresponding original cable plans cannot easily be reconstructed from the pseudo cable plans.

For our purposes, we can compare the results of the experiments using the original data set and the generated data set or we can compute explicit graph characterization parameters. The numbers of vertices, ports, edges, . . . are similar by using the target values. For example, the arithmetic mean (median) of the number of vertices in the original data set is 104.16 (105), while it is 103.98 (105) in the generated data set. The arithmetic mean (median) across the arithmetic means of parallel edges per edge in the original data set is 1.592 (1.429), while it is 1.493 (1.402) in the generated data set. Some characteristic parameters where we did not have target values exhibit at least some similarities, which indicates a similar structure of the graphs of both sets. For example, the arithmetic mean (median) of the diameters across all graphs in the original data set is 9.508 (10), while it is 8.128 (8) in the generated data set.

4.3   Experiments

Our experiments were run in Java on an Intel Core i7 notebook with 8 cores and 24 GB RAM under Linux and took about 14 h.

Orienting Undirected Edges. For each graph and each of FD, BFS, RAND, we oriented the edges and executed the algorithm 5 times. For crossing minimization, we used the variant PORTS with 10 repetitions. For FD, we used only one execution of the force-directed algorithm (so $k = 1$) to make it better comparable to the other methods. We recorded

- the number n_{cr} of crossings in the final drawing,
- the number n_{bp} of bends created when executing the algorithm,
- the number n_{dv} of dummy vertices created when executing the algorithm,
- the total area and the ratio of the bounding box of the drawing, and
- the time to orient the edges and run the algorithm.

Table 1. Comparison of the methods for orienting the edges. The mean μ is relative to RAND (standard deviation in the range $[.2, .6]$); β measures (in %) how often a method provides the best result ($\sum \beta > 100$ possible due to ties).

	Original cable plans						Generated artificial cable plans					
	FD		BFS		RAND		FD		BFS		RAND	
	μ	β	μ	β	μ	β	μ	β	μ	β	μ	β
n_{cr}	.55	**89**	.67	25	1	8	.68	**89**	.80	21	1	11
n_{bp}	.80	**85**	.86	20	1	10	1.01	**60**	1.03	29	1	21
n_{dv}	1.03	9	.81	**91**	1	9	1.13	6	.93	**89**	1	11
area	1.14	20	1.05	42	1	**42**	1.30	10	1.13	37	1	**55**
w:h	.51	**85**	.73	16	1	3	.65	**85**	.86	14	1	3
time	1.47	8	.88	**74**	1	26	1.66	4	1.03	**51**	1	48

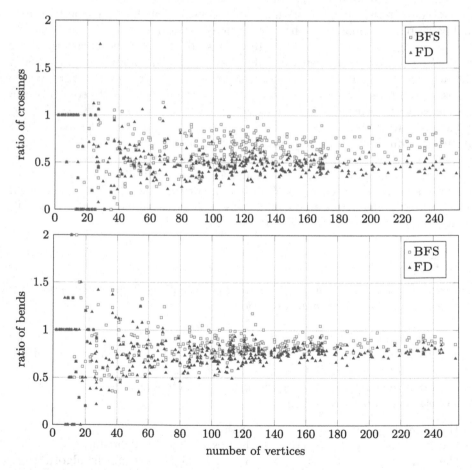

Fig. 4. Comparison of the edge-orientation methods FD and BFS relative to RAND. In each color, each dot represents one of the 380 original plans. (Color figure online)

For each graph and each criterion, we took for each method the best of the five results and normalized by the best value of RAND. The means (μ) of these values are listed in Table 1. The winner percentage β measures how often a specific method achieved the best objective value (usually the smallest, but for the aspect ratio (w:h) the one closest to 1). Ties are not broken, so over the three methods, the β-values add up to more than 100. We have a plot relating the normalized values of n_{cr} and n_{bp} to the number of vertices in Fig. 4 for the original plans and in the appendix of the full version [13] for the generated plans.

Crossing Minimization. We used the same settings as when we compared the methods for orienting the edges, but here we exclusively used FD for orienting the edges. We compared the methods VERTICES, MIXED, and PORTS, each with 10 repetitions in the crossing reduction phase. KIELER joined the comparison as the base line method to which we relate our results.

The variant KIELER uses instead of our algorithm the algorithm ElkLayered in eclipse.elk (formerly known as: KLayered in KIELER) [5]. As our algorithm, ElkLayered does Sugiyama-based layered drawing using ports at vertices. ElkLayered, however, expects a directed graph as input and its port constraints are less powerful. ElkLayered offers free placement of the ports around a vertex, fixed side at a vertex, fixed order around a vertex, and fixed position at a vertex. After orienting the given undirected graph, we used this algorithm as a black box (and hence, we did not record the number of dummy vertices for KIELER) when we set the port constraints to the most flexible value for each vertex. So, for vertices having multiple port groups or port pairings, we set the order of ports to be fixed, while we allow free port placement for all other vertices. As both algorithms expect different input, use different subroutines and ElkLayered uses more additional steps for producing aesthetic drawings, this comparison should be treated with caution. For our results, see Table 2, Fig. 5, and the appendix of the full version [13].

Table 2. Comparison of the methods for crossing reduction. The mean μ is relative to KIELER (except for n_{dv}, where it is relative to the best); the standard deviation is in the range [.2, 1.1] (except for time, where it is higher); β is as in Table 1.

	Original cable plans								Generated artificial cable plans							
	VERTICES		MIXED		PORTS		KIEL		VERTICES		MIXED		PORTS		KIEL	
	μ	β	μ	β	μ	β	μ	β	μ	β	μ	β	μ	β	μ	β
n_{cr}	.83	19	.83	16	.65	**84**	1	12	.87	39	.96	15	.82	**62**	1	14
n_{bp}	.46	13	.44	29	.42	**72**	1	1	.56	40	.56	34	.56	**41**	1	0
n_{dv}	1.11	38	1.10	**40**	1.11	37	–	–	1.10	34	1.08	**40**	1.08	39	–	–
area	3.20	3	3.40	2	3.44	3	1	**97**	3.70	1	4.03	1	4.06	1	1	**99**
w:h	1.05	**31**	1.11	21	1.23	15	1	37	1.20	18	1.25	14	1.32	10	1	**62**
time	14.31	2	39.66	1	51.71	1	1	**100**	18.18	1	45.82	1	68.52	1	1	**100**

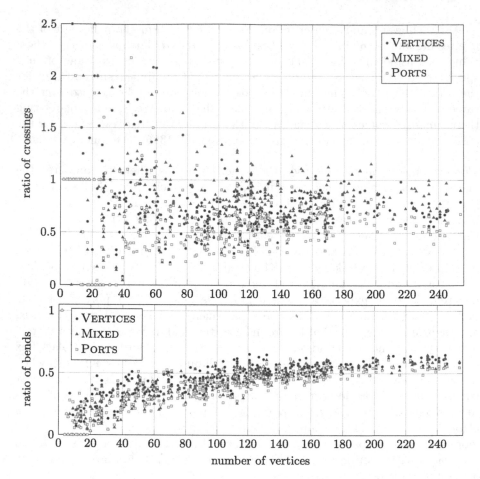

Fig. 5. Comparison of the three crossing-reduction methods relative to KIELER. In each color, each dot represents one of the 380 original cable plans. (Color figure online)

5   Discussion and Conclusion

FD almost always yields orientations of the undirected graphs that lead to drawings with fewer crossings than the orientations obtained from BFS and RAND. Surprisingly, the oriented graphs obtained from BFS mostly use fewer dummy vertices than FD. Although RAND performs rather poorly for most criteria, it often uses the smallest drawing area. The savings in the total area by RAND can be attributed almost exclusively to a lower height, which corresponds to fewer layers. We discuss the direction assignment phase in more detail in the full version [13]. Since we consider the numbers of crossings and bends, and a balanced aspect ratio the most relevant parameters for obtaining visually pleasant drawings, we recommend FD for orienting edges.

For the crossing reduction phase, PORTS performs clearly better than VER-TICES and MIXED in terms of n_{cr} and n_{bp}. This is in line with our expectation that incorporating distinct port orderings during the whole crossing reduction procedure helps to avoid edge crossings, which crucially depend on the precise order of ports. Rather surprisingly, VERTICES performs slightly better than MIXED. We discuss the crossing reduction phase and the comparison of our algorithm to KIELER in more detail in the full version [13].

We concede that the artificial plans that we generated are not perfect as they behave somewhat differently from the original plans for certain criteria. For instance, for the artificial plans the relative advantage of PORTS in terms of n_{cr} and n_{bp} is smaller than for the original plans. Nevertheless, the obfuscation allowed us to make somewhat realistic cable plans publicly available, so that others can validate our experiments in the future. Our generation procedure may also serve as an entry point for more research in generating pseudo data from original data. As suggested by a reviewer, we intend to integrate our algorithm into the software of our industrial partner to see whether this statistical improvement yields advantages in practice. Last but not least we refer to the appendix (full version [13]) for a cute combinatorial problem that we have not solved exactly.

References

1. Biedl, T.C., Madden, B., Tollis, I.G.: The three-phase method: a unified approach to orthogonal graph drawing. Int. J. Comput. Geom. Appl. **10**(6), 553–580 (2000). https://doi.org/10.1142/S0218195900000310
2. Brandes, U., Köpf, B.: Fast and simple horizontal coordinate assignment. In: Mutzel, P., Jünger, M., Leipert, S. (eds.) GD 2001. LNCS, vol. 2265, pp. 31–44. Springer, Heidelberg (2002). https://doi.org/10.1007/3-540-45848-4_3
3. Brandes, U., Walter, J., Zink, J.: Erratum: fast and simple horizontal coordinate assignment. CoRR abs/2008.01252 (2020). http://arxiv.org/abs/2008.01252
4. Eades, P., Whitesides, S.: Drawing graphs in two layers. Theor. Comput. Sci. **131**(2), 361–374 (1994). https://doi.org/10.1016/0304-3975(94)90179-1
5. Eclipse layout kernel (ELK) (2020). https://www.eclipse.org/elk/
6. Fruchterman, T.M.J., Reingold, E.M.: Graph drawing by force-directed placement. Softw. Pract. Exp. **21**(11), 1129–1164 (1991). https://doi.org/10.1002/spe.4380211102
7. Gansner, E.R., Koutsofios, E., North, S.C., Vo, K.: A technique for drawing directed graphs. IEEE Trans. Softw. Eng. **19**(3), 214–230 (1993). https://doi.org/10.1109/32.221135
8. Lipp, F., Wolff, A., Zink, J.: Faster force-directed graph drawing with the well-separated pair decomposition. Algorithms **9**(3), 53 (2016). https://doi.org/10.3390/a9030053
9. PRALINE data structure and layouting algorithm (2020). https://github.com/j-zink-wuerzburg/praline
10. PRALINE pseudo plans - algorithm and data sets (2020). https://github.com/j-zink-wuerzburg/pseudo-praline-plan-generation

11. Schulze, C.D., Spönemann, M., von Hanxleden, R.: Drawing layered graphs with port constraints. J. Vis. Lang. Comput. **25**(2), 89–106 (2014). https://doi.org/10.1016/j.jvlc.2013.11.005
12. Sugiyama, K., Tagawa, S., Toda, M.: Methods for visual understanding of hierarchical system structures. IEEE Trans. Syst. Man Cybern. **11**(2), 109–125 (1981). https://doi.org/10.1109/TSMC.1981.4308636
13. Walter, J., Zink, J., Baumeister, J., Wolff, A.: Layered drawing of undirected graphs with generalized port constraints. CoRR abs/2008.10583 (2020). http://arxiv.org/abs/2008.10583

An Integer-Linear Program
for Bend-Minimization in Ortho-Radial
Drawings

Benjamin Niedermann[1]([✉]) [ID] and Ignaz Rutter[2] [ID]

[1] Universität Bonn, 53115 Bonn, Germany
niedermann@uni-bonn.de
[2] Universität Passau, 94032 Passau, Germany
rutter@fim.uni-passau.de

Abstract. An ortho-radial grid is described by concentric circles and straight-line spokes emanating from the circles' center. An ortho-radial drawing is the analog of an orthogonal drawing on an ortho-radial grid. Such a drawing has an unbounded outer face and a central face that contains the origin. Building on the notion of an ortho-radial representation [1], we describe an integer-linear program (ILP) for computing bend-free ortho-radial representations with a given embedding and fixed outer and central face. Using the ILP as a building block, we introduce a pruning technique to compute bend-optimal ortho-radial drawings with a given embedding and a fixed outer face, but freely choosable central face. Our experiments show that, in comparison with orthogonal drawings using the same embedding and the same outer face, the use of ortho-radial drawings reduces the number of bends by 43.8% on average. Further, our approach allows us to compute ortho-radial drawings of embedded graphs such as the metro system of Beijing or London within seconds.

Keywords: Ortho-radial drawing · Integer-linear program

1 Introduction

Planar orthogonal drawings are arguably one of the most popular drawing styles. Their aesthetic appeal derives from their good angular resolution and the restriction to only the horizontal and the vertical slope, which makes it easy to trace the edges. They naturally correspond to embeddings into the standard grid, where edges are mapped to paths between their endpoints. The most important aesthetic criterion for orthogonal drawings is the number of bends. Consequently, a large body of literature deals with optimizing the number of bends [3–5,7,8,14]. *Ortho-radial drawings* are a natural analog of orthogonal drawings but on an *ortho-radial grid*, which is formed by concentric circles and straight-line spokes emanating from the circles' center. Besides their aesthetic appeal and the fact

© Springer Nature Switzerland AG 2020
D. Auber and P. Valtr (Eds.): GD 2020, LNCS 12590, pp. 235–249, 2020.
https://doi.org/10.1007/978-3-030-68766-3_19

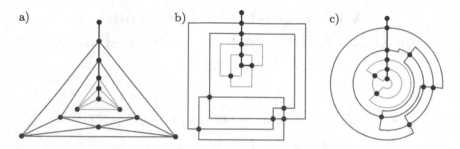

Fig. 1. Graph with (a) straight-line, (b) orthogonal and (c) ortho-radial layout. The ortho-radial layout has been created by our approach and has 14 bends. The orthogonal layout has been proposed by Biedl and Kant [2] and has 23 bends.

that they inherit favorable properties of orthogonal drawings like a good angular resolution, they have the potential to save on the number of bends; see Fig. 1.

The corner-stone of the whole theory of bend minimization is the notion of an *orthogonal representation*, which for a *plane graph* (i.e., a graph with a fixed embedding) describes for each vertex the angles between consecutive incident edges and for each edge the order and directions of its bends. It is a seminal result of Tamassia [14] that characterizes the orthogonal representations in terms of local conditions and shows that every orthogonal representation admits a drawing. The usefulness of this result hinges on the fact that it turns the seemingly geometric problem of computing a bend-optimal drawing into a purely combinatorial one. Geometric aspects of the drawing, such as choosing edge lengths, can then be dealt with separately, and after deciding the bends on the edges.

Today, this is usually described as a pipeline consisting of three steps, the topology-shape-metrics framework (TSM for short). The topology step chooses a planar embedding of the input graph. The shape step computes an orthogonal representation for this embedding (e.g., using flow-based methods), and the metrics step computes edge lengths so that a crossing-free drawing is obtained.

Recently, this framework has been adapted to ortho-radial drawings. There is a natural analog of ortho-radial representation that satisfies analogous local conditions to orthogonal representations. However, unlike the orthogonal case, there exist ortho-radial representations that satisfy all the local conditions, but do not correspond to an ortho-radial drawing; see Fig. 2a,b for an example. After initial results on the characterization of ortho-radial representations of cycles [11] and ortho-radial representations of maxdeg-3 graphs, where all faces are rectangles [10], Barth et al. [1] gave a characterization of the drawable ortho-radial representations in terms of a third, more global condition. Niedermann et al. [13] further showed that, given an ortho-radial representation that satisfies the third condition, its ortho-radial drawing can be computed in quadratic time.

Up to now, however, there are no algorithms for computing ortho-radial representations, even if the graph comes with a fixed planar embedding, including

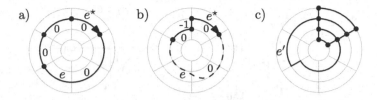

Fig. 2. An ortho-radial drawing of a 4-cycle (a) and an ortho-radial representation of it that is not drawable (b), though the sum of the angles around each vertex and around each face is the same as in (a). A bend-optimal drawing of a graph (c), where the edge e' has bends in different directions.

the central and the outer face. It is an open question whether a bend-optimal valid ortho-radial representation can be computed efficiently in this setting. The example from Fig. 2a, b already shows that such an ortho-radial representation cannot be characterized in terms of purely local conditions. Figure 2c is an example of a bend-optimal drawing where an edge bends in two different directions. This shows that a straightforward adaption of existing techniques that are based on min-cost flows, is unlikely to succeed. In this paper, we develop a method for computing ortho-radial representations with few bends based on an integer-linear program (ILP). This yields the first practical algorithm that takes an arbitrary plane maxdeg-4 graph as input and computes an ortho-radial drawing. We use it to evaluate the usefulness of ortho-radial drawings, in particular with respect to the potential of saving bends in comparison to orthogonal drawings.

Contribution and Outline. We start with preliminary results in Sect. 2 introducing notions and facts on orthogonal and ortho-radial representations. In Sect. 3, we present an ILP for computing bend-free ortho-radial representations for graphs with a fixed embedding. In Sect. 4 we extend that ILP to optimize the number of bends. To that end, we provide theoretical insights into the number of bends required for ortho-radial drawings. Moreover, we describe a pruning strategy that allows us to quickly compute a bend-optimal drawing. In Sect. 5 we evaluate our algorithms and compare them to standard approaches for computing orthogonal drawings.

2   Preliminaries

A graph of maximum degree 4 is a *4-graph*. Unless stated otherwise, all graphs occurring in this paper are 4-graphs. Let $G = (V, E)$ be a connected planar 4-graph with a fixed combinatorial embedding $\mathcal{E}$ and let $v \in V$ be a vertex. We call the counterclockwise order of edges around v in the embedding the *rotation of v*, and we denote it by $\mathcal{E}(v)$. An *angle at v* is a pair of edges (e_1, e_2) that are both incident to v and such that e_1 immediately precedes e_2 in $\mathcal{E}(v)$.

Let Δ be an orthogonal (or ortho-radial) drawing of G with embedding $\mathcal{E}$. By turning bends into vertices, we can assume that the drawing is bend-free.

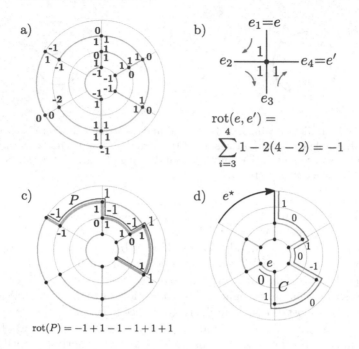

a)

b) $e_1=e$

$$\mathrm{rot}(e,e') = \sum_{i=3}^{4} 1 - 2(4-2) = -1$$

c) $\mathrm{rot}(P) = -1+1-1-1+1+1$

d)

Fig. 3. (a) Rotations between angles. (b) Rotations between two edges e, e'. (c) Rotation of path P. (d) Label of edge e with respect to essential cycle C.

We now derive a labeling of the angles of v with labels in $\{-2, -1, 0, 1\}$ with so-called rotation values. For an angle (e_1, e_2) at v, we set $\mathrm{rot}(e_1, e_2) = 2 - 2\alpha/\pi$, where α is the counterclockwise geometric angle between e_1 and e_2 in Δ; see Fig. 3a. Intuitively, this counts the number of right turns one takes when traversing e_1 towards v and afterwards e_2 away from v, where negative numbers correspond to left turns. Note that if $e_1 = e_2$, then $\mathrm{rot}(e_1, e_2) = -2$, i.e., v contributes two left turns.

For a face f of G, we denote by $\mathrm{rot}(f)$ the sum of the rotations of all angles incident to f. Formally, if $v_0, \ldots, v_{n-1}$ is the facial walk around f (oriented such that f lies to the right of the facial walk), we define $\mathrm{rot}(f) = \sum_{i=0}^{n-1} \mathrm{rot}(v_{i-1}v_i, v_iv_{i+1})$ where indices are taken modulo n. Intuitively, this counts the number of right turns minus the number of left turns one takes when traversing the face boundary such that the face f lies to the right. Since Δ is an orthogonal drawing with some outer face f_o, it satisfies the following conditions [14].

(I) For each vertex, the sum of the rotations around v is $2(\deg(v) - 2)$.
(II) For each face $f \neq f_o$ it is $\mathrm{rot}(f) = 4$ and it is $\mathrm{rot}(f_o) = -4$.

We call an assignment Γ of rotation values to the angles that satisfy these two rules an *orthogonal representation*. Every orthogonal drawing Δ induces an orthogonal representation. An orthogonal representation Γ is *drawable* if there exits a drawing Δ that induces it.

For ortho-radial drawings a similar situation occurs. Here, we have two special faces; an unbounded face, called the *outer face* f_o, and a *central face* f_c, which contains the origin. If the central and the outer face are identical, then the drawing does not enclose the origin, and the ortho-radial drawing does in fact lie on some patch of the ortho-radial grid that can be conformally mapped to an orthogonal grid (i.e., without changing any angles). An ortho-radial drawing Δ similarly defines rotation values that satisfy the following conditions.

(I)' For each vertex, the sum of the rotations around v is $2(\deg(v) - 2)$.
(II)' For each face $f \neq f_o, f_c$ it is $\mathrm{rot}(f) = 4$, if $f_o \neq f_c$, then $\mathrm{rot}(f_o) = \mathrm{rot}(f_c) = 0$ and $\mathrm{rot}(f_o) = -4$ if $f_o = f_c$.

Similar to the orthogonal case, an ortho-radial drawing Δ therefore induces an *ortho-radial representation* Γ that defines rotation values satisfying these conditions. An ortho-radial representation is *drawable* if there exists an ortho-radial drawing that induces it.

Tamassia [14] proved that every orthogonal representation is drawable. In contrast, there exist ortho-radial representations that are not drawable; see e.g., Fig. 2b, which illustrates a so-called strictly monotone cycle. Its ortho-radial representation satisfies conditions (I)' and (II)', yet it is not drawable.

To characterize the drawable ortho-radial representations, Barth et al. [1] introduce a labeling concept. Since the horizontal and vertical directions on an ortho-radial grid are not interchangeable (one is circular, the other is not), additional information is required. The information which is the horizontal direction is given by a *reference edge* e^* which is assumed to lie on the outer face and that is directed such that it points in the clockwise direction. To present the characterization of Barth et al. [1], we extend the notion of rotation. For two edges e, e' incident to a vertex v, let $e = e_1, \ldots, e_k = e'$ be the edges between them in $\mathcal{E}(v)$ so that (e_i, e_{i+1}) is an angle for $i = 1, \ldots, k - 1$. To measure the rotation between e and e', we convert the rotation values between them into geometric angles, sum them up, and convert them back to a rotation, which gives $\mathrm{rot}(e, e') = \sum_{i=1}^{k-1} \mathrm{rot}(e_i, e_{i+1}) - 2(k - 2)$; see Fig. 3b. Note that for an angle (e, e'), it is $k = 2$, and therefore the two definitions of $\mathrm{rot}(e, e')$ coincide. For a path $P = v_0, \ldots, v_{n-1}$ in G, we define its rotation $\mathrm{rot}(P) = \sum_{i=1}^{n-1} \mathrm{rot}(v_{i-1}v_i, v_iv_{i+1})$, and for a cycle C in G, we define its rotation $\mathrm{rot}(C) = \sum_{i=1}^{n} \mathrm{rot}(v_{i-1}v_i, v_iv_{i+1})$, where indices are taken modulo n; see Fig. 3c. A cycle C of G is called *essential* if it separates the central and the outer face. A cycle is essential if and only if $\mathrm{rot}(C) = 0$ [1]. We assume that C is directed such that the central face lies to its right. Let e be an edge on C. A *reference path* for e on C is a (not necessarily simple) walk P that starts with the edge e^*, ends with the edge e and does not contain an edge or a vertex that lies to the right of C. We define $\ell_C(e) = \mathrm{rot}(P)$ as the label of e on C; see Fig. 3d. Barth et al. [1] show that the label does not depend on the actual path P (however the same edge may have different labels for different cycles). With this, Barth et al.

formulate a third condition. An ortho-radial representation is called *valid* if, for each essential cycle C, either $\ell_C(e) = 0$ for all edges $e \in E(C)$, or there exist edges e^-, e^+ in $E(C)$ with $\ell_C(e^-) < 0$ and $\ell_C(e^+) > 0$. A cycle C that does not satisfy this condition is called *strictly monotone*. Thus, an ortho-radial representation is valid if and only if it has no strictly monotone cycle. In Fig. 2a,b the edges of the 4-cycle are labeled with their labels with respect to the reference edge $e^\star$; Fig. 2b is a strictly monotone cycle. The following two results form the combinatorial and algorithmic basis for our work.

Theorem 1 (Barth et al. [1]). *An ortho-radial representation is drawable if and only if it is valid.*

Theorem 2 (Niedermann et al. [13]). *There is an $O(n^2)$-time algorithm that, given an ortho-radial representation Γ of an n-vertex graph G, either outputs a drawing of Γ, or a strictly monotone cycle C in Γ.*

3   ILP for Bend-Free Ortho-Radial Drawings

In this section we are given a planar 4-graph $G = (V, E)$ with a combinatorial embedding $\mathcal{E}$, an outer face f_o, a central face f_c and a reference edge $e^\star$ on f_o; we denote that instance by $\mathcal{G} = (G, \mathcal{E}, f_o, f_c, e^\star)$. We present an algorithm based on an ILP that yields a valid ortho-radial representation of $\mathcal{G}$, if it exists.

Basic Formulation. For each vertex u and each of its angles (e, e') we introduce an integer variable $r_{e,e'} \in \{-2, -1, 0, 1\}$, which describes the rotation $\mathrm{rot}(e, e')$ between e and e'. Condition I' is enforced by the following constraint for u.

$$\sum_{i=1}^{k} r_{e_i, e_{i+1}} = 2(\deg(v) - 2), \tag{1}$$

where $e_1, \ldots, e_k$ are the incident edges of u in counter-clockwise order; we define $e_{k+1} = e_1$. For each face f of $\mathcal{G}$ Condition II' is enforced by the next constraint.

$$\sum_{i=1}^{k} r_{e_i, e_{i+1}} = \begin{cases} 4 & \text{if } f \text{ is a regular face,} \\ 0 & \text{if } f \text{ is the outer or central face but not both,} \\ -4 & \text{if } f \text{ is both the central and outer face,} \end{cases} \tag{2}$$

where $e_1, \ldots, e_k$ are the edges of the facial walk around f such that f lies to the right; we define $e_{k+1} = e_1$. We denote that formulation by $\mathcal{F}_{\mathrm{or}}$. By construction a valid assignment of the variables in $\mathcal{F}_{\mathrm{or}}$ *induces* an ortho-radial representation Γ. In particular, assuming that $e^\star$ is directed such that it points clockwise, we can derive from the variable assignment the directions of the other edges in G. The next theorem summarizes this result.

Theorem 3. *An ortho-radial representation exists for $\mathcal{G}$ if and only if $\mathcal{F}_{\mathrm{or}}$ induces an ortho-radial representation.*

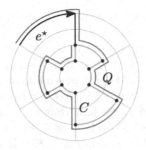

Fig. 4. Illustration of path Q used for the labeling of C.

However, the induced ortho-radial representation Γ is not necessarily valid, but may contain strictly monotone cycles. We therefore extend $\mathcal{F}_{or}$ by constraints for each essential cycle C of $\mathcal{G}$. To that end, let P be a path that starts at e^* and ends at C such that it does not use any vertex or edge that lies to the right of C. Further, let Q be the path $e^* + P + C$ that follows C in clockwise order from the endpoint of P and ends at the end point of P; see Fig. 4. For each edge e of Q we introduce an integer variable l_e with $-m \leq l_e \leq m$, which models a label with respect to C. Here m denotes the number of edges of G. We require that the label of the reference edge is 0, i.e., $l_{e^*} = 0$. Moreover, for an edge $e = vw$ of $Q \setminus \{e^*\}$ and its predecessor $e' = uv$ on Q let $e' = e_1, \ldots, e_k = e$ be the edges between them in $\mathcal{E}(v)$ so that (e_i, e_{i+1}) is an angle for $i = 1, \ldots, k-1$. We introduce the constraint

$$l_e = l_{e'} + \sum_{i=1}^{k-1} r_{e_i, e_{i+1}} - 2(k-2) \tag{3}$$

Hence, the values l_e for $e \in E(C)$ describe a labeling of C, where $E(C)$ denotes the edges of C. To exclude strictly monotone cycles, we ensure that either $l_e = 0$ for all edges $e \in E(C)$, or there exist edges e^-, e^+ in $E(C)$ with $l_{e^-} > 0$ and $l_{e^+} < 0$. We first observe that C can only be strictly monotone if $\sum_{e \in E(C)} l_e \neq 0$. We introduce a single binary variable z that is 1 if and only if $\sum_{e \in E(C)} l_e = 0$. Additionally, for each edge e of C we introduce two binary variables x_e and y_e, which are used to enforce that l_e is negative or positive, respec-

tively. $$\sum_{e \in E(C)} l_e \leq M \cdot (1 - z) \quad (4) \qquad \sum_{e \in E(C)} l_e \geq -M \cdot (1 - z) \quad (5)$$

$$\sum_{e \in E(C)} x_e + z \geq 1 \quad (6) \qquad \sum_{e \in E(C)} y_e + z \geq 1 \quad (7)$$

$$l_e \leq -1 + M \cdot (1 - x_e) \; \forall e \in E(C) \qquad l_e \geq 1 - M \cdot (1 - y_e) \; \forall e \in E(C)$$
$$(8) \qquad\qquad\qquad\qquad\qquad (9)$$

We define M as a constant with $M > m$ so that the corresponding constraints are trivially satisfied for $z = 0$, $x_e = 0$ and $y_e = 0$, respectively. If $z = 1$, we obtain by Constraint 4 and Constraint 5 that $\sum_{e \in E(C)} l_e = 0$. Hence, C is not strictly monotone. Otherwise, if $z = 0$, by Constraint 6 there is an edge $e^- \in E(C)$ with $x_{e^-} = 1$. By Constraint 8 we obtain $l_{e^-} < 0$. Similarly, by Constraint 7 there is an edge $e^+ \in E(C)$ with $y_{e^+} = 1$. By Constraint 9 we obtain $l_{e^+} > 0$. Altogether, we find that C is not strictly monotone. We emphasize that for each essential cycle C of $\mathcal{G}$ we introduce a fresh set of variables and constraints; which we denote by $\mathcal{F}_C$. Hence, we consider the ILP $\mathcal{F}(\mathcal{G}) = \mathcal{F}_{\mathrm{or}} \cup \bigcup_{C \in \mathcal{C}} \mathcal{F}_C$, where $\mathcal{C}$ is the set of all essential cycles in $\mathcal{G}$. The next theorem summarizes this.

Theorem 4. *If $\mathcal{G}$ has an ortho-radial representation, then the formulation $\mathcal{F}(\mathcal{G})$ induces a valid ortho-radial representation.*

Separation of Constraints. Adding $\mathcal{F}_C$ for each essential cycle C of $\mathcal{G}$ is not feasible in practice, as there can be exponentially many of these in $\mathcal{G}$. Hence, instead, we propose an algorithm that adds $\mathcal{F}_C$ on demand. The algorithm first checks whether $\mathcal{G}$ has an ortho-radial representation Γ_1 using the formulation $\mathcal{F}_1 := \mathcal{F}_{\mathrm{or}}$ (Theorem 3). If this is not the case, the algorithm stops and returns that there is no ortho-radial representation for $\mathcal{G}$. Otherwise, starting with $\mathcal{F}_1$ and Γ_1 it applies the following iterative procedure. In the i-th iteration (with $2 \le i$) it checks whether Γ_{i-1} is valid (Theorem 2). If it is, the algorithm stops and returns Γ_{i-1}. Otherwise, the validity test yields a strictly monotone cycle C as a certificate proving that Γ_{i-1} is not valid. The algorithm creates then the formulation $\mathcal{F}_i = \mathcal{F}_{i-1} \cup \mathcal{F}_C$ and induces the ortho-radial representation Γ_i, in which it is enforced that C is not strictly monotone. The algorithm stops at the latest when the formulation $\mathcal{F}_C$, which prohibits that C is a strictly monotone cycle, has been added for each essential cycle $C \in \mathcal{C}$. Hence, in theory an exponential number of iterations may be necessary. However, in our experiments the procedure stopped after few iterations for all test instances; see Sect. 5.

Bend Optimization. We also can use the ILP to optimize the ortho-radial representation. In Sect. 4 we consider bend minimization by modeling bends as degree-2 vertices. We therefore extend $\mathcal{F}_{\mathrm{or}}$ such that it allows us to optimize the change of direction at such nodes. For each degree-2 vertex we introduce a binary variable c_u, which is 1 if and only if one of the two incident edges of u lies on a concentric circle and the other lies on a ray of the grid. The two incident edges e_1 and e_2 of u form the two angles (e_1, e_2) and (e_2, e_1). For these we introduce the constraints $c_u \ge r_{e_1,e_2}$ and $c_u \ge r_{e_2,e_1}$. Subject to these constraints we minimize $\sum_{u \in V_2} c_u$, where $V_2 \subseteq V$ denotes the degree-2 vertices of G. We can easily restrict the optimization to a subset of V_2 distinguishing between degree-2 vertices that originally belong to G and those that we use for modeling bends.

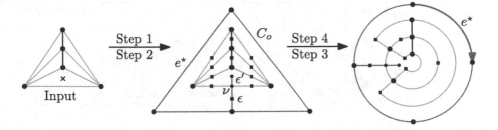

Fig. 5. Illustration of the layout algorithm. Subdivision vertices are squares.

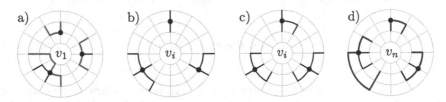

Fig. 6. Constructions for the proof of Theorem 5

4   Optimizing Bends and the Choice of the Central Face

In this section we are given a graph G with embedding $\mathcal{E}$ and designated outer face f_o. We describe an algorithm that returns a bend-optimal ortho-radial drawing for $\mathcal{G} = (G, \mathcal{E}, f_o)$, i.e., there is no other ortho-radial drawing $\mathcal{G}$ that has fewer bends for any choice of the central face f_c and the reference edge $e^\star$ on f_o. The algorithm uses the ILP from Sect. 3 as a building block. The ILP does not directly allow to express bends; rather, we subdivide the edges with degree-2 vertices, which can then be used as bends. See Fig. 5 for an illustration.

1. Insert a cycle C_o in $\mathcal{E}$ that encloses G and connect C_o via an edge ϵ to a newly inserted vertex ν on the original outer face of $\mathcal{E}$. Insert edge ϵ' on the opposite side enforcing that ν has degree 4. Hence, C_o is the new boundary of the outer face f_o. Choose an arbitrary edge of C_o as reference edge $e^\star$.
2. Subdivide each edge of G with degree-2 vertices such that each maximally long chain of degree-2 vertices consists of at least K vertices.
3. Create a valid ortho-radial representation Γ_f for face f as the central face. To that end, apply the ILP formulation of Sect. 3 with separated constraints and bend optimization on $\mathcal{G}_f = (G, \mathcal{E}, f_o, f, e^\star)$ charging the newly inserted degree-2 vertices with costs; the subdivision vertices on ϵ are not charged.
4. For the representation Γ_f with fewest bends compute a drawing (Theorem 2).

For bend-optimal drawings an appropriately large K is decisive. For biconnected graphs $K = 2n + 4$ is sufficient even for a fixed central and outer face.

Theorem 5. *Every biconnected plane 4-graph on n vertices with designated central and outer faces has a planar ortho-radial drawing with at most $2n + 4$ bends*

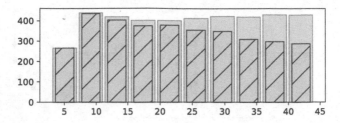

Fig. 7. The node distribution (gray bars) of the graphs in $\mathcal{I}_{\text{Rome}}$ distributed on 10 equally sized bins. The number of vertices ranges between 3 and 44. The blue, tiled bars indicate the number of optimally solved instances. (Color figure online)

and at most two bends per edge with the exception of up to two edges that may have three bends.

The proof, which is deferred to the full version [12], uses similar constructions as in the orthogonal case; see also Fig. 6. It seems plausible that the bound can be transferred to non-biconnected graphs as in the work by Biedl and Kant [3]; as we use a different bound, we refrain from the rather technical proof. Moreover, we insert C_o to make the layout independent from the choice of the reference edge $e^\star$. This does not impact the number of bends needed for the original part of G, because we subdivide ϵ with sufficiently many 2-degree vertices that can be bent for free. Further, as ν has degree 4, the drawing cannot be bent at ν.

Replacing each edge by $2n+4$ degree-2 vertices increases the size of the graph drastically. However, the ILP can be solved much faster if fewer subdivision vertices are used. Next, we describe a pruning strategy that uses upper and lower bounds on the optimal drawing to exclude central faces and to limit the number of subdivision vertices and the number of times we solve the ILP.

We first compute the minimum number U of bends that is necessary for a bend-optimal orthogonal drawing of $\mathcal{G}$. This also bounds the number of bends in a bend-optimal ortho-radial drawing of $\mathcal{G}$. Hence, it is sufficient to subdivide each edge with U vertices in Step 2. Initially, we run $\mathcal{F}_{\text{or}}$ on each face f of $\mathcal{G}$ as central face. This gives us a lower bound l_f for the bends in the case that f is the central face. In Step 3 we then consider the faces in increasing order of their lower bounds. If the lower bound l_f of the current face f exceeds the upper bound U we prune f and continue with the next face. Otherwise, we iteratively compute a valid ortho-radial representation Γ_f for f as described in Sect. 3 and update U if it is improved by the current solution. Further, when we update the ILP due to strictly monotone cycles, we skip f if its number of bends exceeds U and continue with the next face.

5   Experimental Evaluation

In this section we present our experimental evaluation which we have conducted to show the potential of our approach as a general graph drawing tool.

Fig. 8. Examples of bend-optimal orthogonal and ortho-radial drawings for the Rome graphs. The outer face was fixed, but the central face was optimized.

5.1   Feasibility of Approach

We first pursue the issue of whether our approach is feasible. It is far from clear whether prohibiting strictly monotone cycles on demand is practical, as we may need to insert an exponential number of constraints into the ILP formulation. To answer this question we have conducted the first experiments on a subset of the *Rome graphs*[1], which is a widely accepted benchmark set. We have replaced each vertex v with degree $k > 4$ with a cycle of k vertices, which we connected to the neighbors of v correspondingly. Further, we applied a heuristic from OGDF [6] to embed the remaining graphs such that the size of the outer face is maximized. We replaced all edge crossings with degree-4 vertices. A preliminary analysis showed that the graphs contain many degree-2 vertices. To ensure for the purpose of the evaluation that our approach is forced to introduce bends with costs, we normalized each instance by removing all degree-2 vertices. We only considered instances up to 44 nodes. In total we obtained a set $\mathcal{I}_{\text{Rome}}$ of 4048 instances. Figure 7 shows the size distribution of the resulting instances. We implemented our approaches in Python and solved the ILP formulations using Gurobi 9.0.2 [9] using a timeout of 2 min in each iteration. We ran the experiments on an Intel(R) Xeon(R) W-2125 CPU clocked 4.00 GHz with 128 GiB RAM.

For each of the instances in $\mathcal{I}_{\text{Rome}}$ we applied the algorithm described in Sect. 4; see Fig. 8 for four examples. For 3462 instances we obtained bend-optimal ortho-radial drawings. For 586 instances the solver returned a not necessarily optimal result due to timeouts. The number of not optimally solved instances

---

[1] http://www.graphdrawing.org/data.

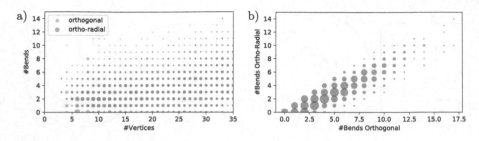

Fig. 9. Overview of the considered Rome graphs. A disk with radius r and position (x, y) corresponds to r instances (a) with x vertices and y bends in the ortho-radial (blue) and the orthogonal (red) drawing, (b) with x bends in the orthogonal drawing and y bends in the ortho-radial drawing. (Color figure online)

increases with the number of nodes; see Fig. 7 for more details. For 1081 instances the algorithm took less than half a second. Only for 861 instance it took more than 10 s; 628 of them took more than one minute. Further, when searching for the best choice of the central face about 76.5% of the faces are pruned in advance on average due to exceeding upper bounds. Hence, for more than three quarters of the faces we do not need to solve the ILP formulation, still guaranteeing that we obtain a drawing with minimum number of bends. Moreover, when the algorithm runs for a fixed central face, it needs less than 3.8 iterations on average until it finds a valid ortho-radial representation. Put differently, we insert the formulation $\mathcal{F}_C$ prohibiting a strictly monotone cycle C into the ILP formulation 3.8 times on average. Altogether, the evaluation shows the practical feasibility of the approach. It supports the rather strong hypothesis that prohibiting strictly monotone cycles on demand is sufficient, but considering all essential cycles is not necessary.

5.2   Ortho-Radial Drawings vs. Orthogonal Drawings

In this part we compare ortho-radial drawings with orthogonal drawings with respect to the necessary number of bends. We expect a reduction of the number of bends in an ortho-radial drawing compared with its orthogonal drawing.

Figure 9a shows that independent of the size of the graphs the ortho-radial drawings often have fewer bends than the orthogonal drawings. Further, Fig. 9b shows that for many of the instances we achieve a reduction between 1 to 3 bends in the ortho-radial drawings. To investigate this in greater detail we consider for each instance $I \in \mathcal{I}_{\mathrm{Rome}}$ the *bend reduction* $r_I = \frac{b_{\mathrm{og}} - b_{\mathrm{or}}}{b_{\mathrm{og}}} \cdot 100\%$, where b_{og} is the minimum number of bends of an orthogonal drawing of I and b_{or} is the number of bends of the ortho-radial drawing created with our approach; note that for both drawings we assume the same embedding and the same outer face. From this comparison we have excluded any instance with zero bends. The bend reduction is 43.8% on average and the median is at 40.0%. We emphasize that for 550 instances there are bend-free ortho-radial drawings, whereas only 129 admit

a)   b)

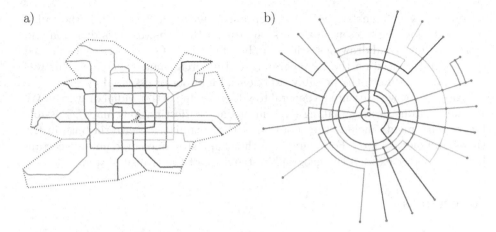

Fig. 10. The metro system in Beijing, China. (a) The input graph derived from vectorizing a metro map of Beijing. The outer and central faces are dashed. (b) The ortho-radial layout induced by our approach within 7 s.

bend-free orthogonal drawings. Thus, our experiments support our hypothesis that ortho-radial drawings lead to a substantial bend reduction.

5.3 Case Study on Metro Maps

Ortho-radial drawings are particularly used to represent metro systems [15]. We tested our algorithm on the metro system of Beijing, which is a comparably large and complex transit system; see Fig 10. We have vectorized a metro map of the city that shows 21 lines; for details see the full version [12]. The created graph has 224 vertices, 289 edges and 67 faces. We fixed the central face by hand to intentionally determine the appearance of the final layout. We subdivided the edges such that each chain consisting of degree-2 vertices has at least three intermediate vertices. Our algorithm created the layout shown in Fig. 10b within seven seconds. It has 21 bends. We emphasize that the outer loop line is represented as a circle and the inner loop line has only two bends. Altogether, the layout reflects the main geometric features of the system well, although we have only optimized the number of bends, e.g., outgoing metro lines are mainly drawn as straight-lines emanating from the center. In a second run, which took three minutes, we proved that 21 bends is optimal. Further metro systems are found in the full version [12].

6 Conclusion

Barth et al. [1] and Niedermann et al. [13] carried over the metrics step of the TSM framework from orthogonal to ortho-radial drawings explaining how to obtain such a drawing from a valid ortho-radial representation. However, they let open how to transfer the shape step constructing such a valid ortho-radial representation. We presented the first algorithm that answers this question and creates ortho-radial drawings, which are bend-optimal. Our experiments showed its feasibility based on the Rome graphs and different metro systems. This was far from clear due to the possibly exponential number of essential cycles.

Altogether, we presented a general tool for creating ortho-radial drawings. We see applications in map making (e.g., metro maps, destinations maps). Possible future refinements include the adaption of the optimization criteria both in the shape and metrics step. For example in the shape step one could enforce certain bends to better express the geographic structure of the transit system.

References

1. Barth, L., Niedermann, B., Rutter, I., Wolf, M.: Towards a topology-shape-metrics framework for ortho-radial drawings. In: Aronov, B., Katz, M.J. (eds.) Computational Geometry (SoCG 2017). Leibniz International Proceedings in Informatics (LIPIcs), vol. 77, pp. 14:1–14:16. Schloss Dagstuhl-Leibniz-Zentrum fuer Informatik (2017)
2. Biedl, T., Kant, G.: A better heuristic for orthogonal graph drawings. In: van Leeuwen, J. (ed.) ESA 1994. LNCS, vol. 855, pp. 24–35. Springer, Heidelberg (1994). https://doi.org/10.1007/BFb0049394
3. Biedl, T., Kant, G.: A better heuristic for orthogonal graph drawings. Comput. Geom. **9**(3), 159–180 (1998)
4. Bläsius, T., Rutter, I., Wagner, D.: Optimal orthogonal graph drawing with convex bend costs. ACM Trans. Algorithms **12**(3), 33 (2016)
5. Bläsius, T., Lehmann, S., Rutter, I.: Orthogonal graph drawing with inflexible edges. Comput. Geom. **55**, 26–40 (2016)
6. Chimani, M., Gutwenger, C., Jünger, M., Klau, G.W., Klein, K., Mutzel, P.: The open graph drawing framework (OGDF). In: Tamassia, R. (ed.) Handbook of Graph Drawing and Visualization, pp. 543–569. CRC Press (2013). Chap. 17
7. Duncan, C.A., Goodrich, M.T.: Planar Orthogonal and Polyline Drawing Algorithms. In: Handbook of Graph Drawing and Visualization, pp. 223–246. CRC Press (2013)
8. Felsner, S., Kaufmann, M., Valtr, P.: Bend-optimal orthogonal graph drawing in the general position model. Comput. Geom. **47**(3, Part B), 460–468 (2014). Special Issue on the 28th European Workshop on Computational Geometry (EuroCG 2012)
9. Gurobi Optimization, L.: Gurobi optimizer reference manual (2020). http://www.gurobi.com
10. Hasheminezhad, M., Hashemi, S.M., McKay, B.D., Tahmasbi, M.: Rectangular-radial drawings of cubic plane graphs. Comput. Geom. Theory Appl. **43**, 767–780 (2010)
11. Hasheminezhad, M., Hashemi, S.M., Tahmasbi, M.: Ortho-radial drawings of graphs. Australas. J. Comb. **44**, 171–182 (2009)
12. Niedermann, B., Rutter, I.: An integer-linear program for bend-minimization in ortho-radial drawings. CoRR abs/2008.10373v2 (2013)

13. Niedermann, B., Rutter, I., Wolf, M.: Efficient algorithms for ortho-radial graph drawing. In: Barequet, G., Wang, Y. (eds.) Proceedings of the 35th International Symposium on Computational Geometry (SoCG 2019). Leibniz International Proceedings in Informatics (LIPIcs), vol. 129, pp. 53:1–53:14. Schloss Dagstuhl-Leibniz-Zentrum fuer Informatik (2019)

14. Tamassia, R.: On embedding a graph in the grid with the minimum number of bends. J. Comput. **16**(3), 421–444 (1987)

15. Wu, H.Y., Niedermann, B., Takahashi, S., Roberts, M.J., Nöllenburg, M.: A survey on transit map layout - from design, machine, and human perspectives. Comput. Graph. Forum **39**(3), 619–646 (2020)

On Turn-Regular Orthogonal Representations

Michael A. Bekos[1] , Carla Binucci[2]([✉]) , Giuseppe Di Battista[3] ,
Walter Didimo[2] , Martin Gronemann[4] , Karsten Klein[5] ,
Maurizio Patrignani[3] , and Ignaz Rutter[6]

[1] Department of Computer Science, University of Tübingen, Tübingen, Germany
`bekos@informatik.uni-tuebingen.de`
[2] Department of Engineering, University of Perugia, Perugia, Italy
`{carla.binucci,walter.didimo}@unipg.it`
[3] Department of Engineering, Roma Tre University, Rome, Italy
`gdb@dia.uniroma3.it, maurizio.patrignani@uniroma3.it`
[4] Theoretical Computer Science, Osnabrück University, Osnabrück, Germany
`martin.gronemann@uni-osnabrueck.de`
[5] Department of Computer and Information Science, University of Konstanz,
Konstanz, Germany
`karsten.klein@uni-konstanz.de`
[6] Department of Computer Science and Mathematics, University of Passau,
Passau, Germany
`rutter@fim.uni-passau.de`

Abstract. An interesting class of orthogonal representations consists of
the so-called *turn-regular* ones, i.e., those that do not contain any pair
of reflex corners that "point to each other" inside a face. For such a representation H it is possible to compute in linear time a minimum-area
drawing, i.e., a drawing of minimum area over all possible assignments of
vertex and bend coordinates of H. In contrast, finding a minimum-area
drawing of H is NP-hard if H is non-turn-regular. This scenario naturally motivates the study of which graphs admit turn-regular orthogonal
representations. In this paper we identify notable classes of biconnected
planar graphs that always admit such representations, which can be computed in linear time. We also describe a linear-time testing algorithm
for trees and provide a polynomial-time algorithm that tests whether a
biconnected plane graph with "small" faces has a turn-regular orthogonal
representation without bends.

Keywords: Orthogonal drawings · Turn-regularity · Compaction

Work partially supported by DFG grants Ka 812/17-1 and Ru 1903/3-1; by MIUR
Project "MODE" under PRIN 20157EFM5C; by MIUR Project "AHeAD" under PRIN
20174LF3T8; and by Roma Tre University Azione 4 Project "GeoView". This work
started at the Bertinoro Workshop on Graph Drawing BWGD 2019.

1   Introduction

Computing *orthogonal drawings* of graphs is among the most studied problems in graph drawing [9,13,20,24], because of its direct application to several domains, such as software engineering, information systems, and circuit design (e.g., [2, 11,14,19,22]). In an orthogonal drawing, the vertices of the graph are mapped to distinct points of the plane and each edge is represented as an alternating sequence of horizontal and vertical segments between its end-vertices. A point in which two segments of an edge meet is called a *bend*. An orthogonal drawing is a *grid* drawing if its vertices and bends have integer coordinates.

One of the most popular and effective strategies to compute a readable orthogonal grid drawing of a graph G is the so-called *topology-shape-metrics* (or *TSM*, for short) approach [26], which consists of three steps: (i) compute a planar embedding of G by possibly adding dummy vertices to replace edge crossings if G is not planar; (ii) obtain an *orthogonal representation* H of G from the previously determined planar embedding; H describes the "shape" of the final drawing in terms of angles around the vertices and sequences of left/right bends along the edges; (iii) assign integer coordinates to vertices and bends of H to obtain the final non-crossing orthogonal grid drawing Γ of G.

If G is planar, the TSM approach computes a planar orthogonal grid drawing Γ of G. Such a planar drawing exists if and only if G is a *4-graph*, i.e., of maximum vertex-degree at most four. To increase the readability of Γ, a typical optimization goal of Step (ii) is the minimization of the number of bends. In Step (iii) the goal is to minimize the area or the total edge length of Γ; a problem referred to as *orthogonal compaction*. Unfortunately, while the computation of an embedding-preserving bend-minimum orthogonal representation H of a plane 4-graph is polynomial-time solvable [8,26], the orthogonal compaction problem for a planar orthogonal representation H is NP-complete in the general case [25]. Nevertheless, Bridgeman et al. [6] showed that the compaction problem for the area requirement can be solved optimally in linear time for a subclass of orthogonal representations called *turn-regular*. A similar polynomial-time result for the minimization of the total edge length in this case is proved by Klau and Mutzel [21]. Esser showed that these two approaches are equivalent [15].

Informally speaking, a face of a planar orthogonal representation H is turn-regular if it does not contain a pair of reflex corners (i.e., turns of 270°) that point to each other (see Sect. 2 for the formal definition); H is turn-regular if all its faces are turn-regular. For a turn-regular representation H, every pair of vertices or bends has a unique orthogonal relation (left/right or above/below) in any planar drawing of H. Conversely, different orthogonal relations are possible for a pair of opposing reflex corners, which makes it computationally hard to optimally compact non-turn-regular representations. For example, Figs. 1(a) and 1(b) show two different drawings of a non-turn-regular orthogonal representation; the drawing in Fig. 1(b) has minimum area. Figure 1(c) depicts a minimum-area drawing of a turn-regular orthogonal representation of the same graph.

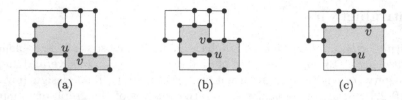

Fig. 1. (a) Drawing of a non-turn-regular orthogonal representation H; vertices u and v point to each other in the gray face. (b) Another drawing of H with smaller area. (c) Drawing of a turn-regular orthogonal representation of the same graph.

The aforementioned scenario naturally motivates the problem of computing orthogonal representations that are turn-regular, so to support their subsequent compaction. To the best of our knowledge, this problem has not been studied so far (a related problem is studied for upward planar drawings only [5,10,12]). Heuristics have been described to make any given orthogonal representation H turn-regular, by adding a minimal set of dummy edges [6,18]; however, such edges impose constraints that may yield a drawing of sub-optimal area for H.

Our contribution is as follows:

(i) We identify notable classes of planar graphs that always admit turn-regular orthogonal representations. We prove that biconnected planar 3-graphs and planar Hamiltonian 4-graphs (which include planar 4-connected 4-graphs [23]) admit turn-regular representations with at most two bends per edge and at most three bends per edge, respectively. For these graphs, a turn-regular representation can be constructed in linear time. We also prove that every biconnected planar graph admits an orthogonal representation that is *internally* turn-regular, i.e., its internal faces are turn-regular (Sect. 3). We leave open the question whether every biconnected planar 4-graph admits a turn-regular representation.

(ii) For 1-connected planar graphs, including trees, there exist infinitely many instances for which a turn-regular representation does not exist. Motivated by this scenario, and since the orthogonal compaction problem remains NP-hard even for orthogonal representations of paths [16], we study and characterize the class of trees that admit turn-regular representations. We then describe a corresponding linear-time testing algorithm, which in the positive case computes a turn-regular drawing without bends (Sect. 4). Finally, we prove that such drawings are "convex" (i.e., all edges incident to leaves can be extended to infinite crossing-free rays). We remark that a linear-time algorithm to compute planar straight-line convex drawings of trees is described by Carlson and Eppstein [7]. However, in general, the drawings they compute are not orthogonal.

(iii) We address the problem of testing whether a given biconnected plane graph admits a turn-regular *rectilinear* representation, i.e., a representation without bends. For this problem we give a polynomial-time algorithm for plane graphs with "small" faces, namely faces of degree at most eight (Sect. 5).

2   Preliminary Definitions and Basic Results

We consider connected graphs and assume familiarity with basic concepts of orthogonal graph drawing and planarity [9] (see also [3]). Let G be a plane 4-graph and H be an orthogonal representation of G. If H has no bends, then it is called *rectilinear*. W.l.o.g., we assume that H comes with a given orientation, i.e., for each edge segment $\overline{pq}$ of H (where p and q are vertices or bends), it is fixed if p is to the left, to the right, above, or below q in every (orthogonal) drawing of H. Let f be a face of H. We assume that the boundary of f is traversed counterclockwise (clockwise) if f is internal (external). The *rectilinear image* of H is the orthogonal representation $\overline{H}$ obtained from H by replacing each bend with a degree-2 vertex. For any face f of H, let $\overline{f}$ denote the corresponding face of $\overline{H}$. For each occurrence of a vertex v of $\overline{H}$ on the boundary of $\overline{f}$, let $\text{prec}(v)$ and $\text{succ}(v)$ be the edges preceding and following v, respectively, on the boundary of $\overline{f}$ ($\text{prec}(v) = \text{succ}(v)$ if $\deg(v) = 1$). Let α be the value of the angle internal to $\overline{f}$ between $\text{prec}(v)$ and $\text{succ}(v)$. We associate with v one or two *corners* based on the following cases: If $\alpha = 90°$, associate with v one *convex* corner; if $\alpha = 180°$, associate with v one *flat* corner; if $\alpha = 270°$, associate with v one *reflex* corner; if $\alpha = 360°$, associate with v an ordered pair of *reflex* corners. For example, in the (internal) face of Fig. 2(a), a convex corner is associated with v_1, a flat corner with v_2, a reflex corner with v_3, and an ordered pair of reflex corners with v_4.

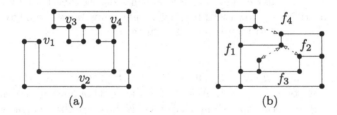

Fig. 2. Illustration of (a) convex, flat and reflex corners, and (b) kitty corners.

Based on the definition above, there is a circular sequence of corners associated with (the boundary) of $\overline{f}$. For a corner c of $\overline{f}$, we define: $\text{turn}(c) = 1$ if c is convex; $\text{turn}(c) = 0$ if c is flat; $\text{turn}(c) = -1$ if c is reflex. For any ordered pair (c_i, c_j) of corners of $\overline{f}$, we define the following function: $\text{rot}(c_i, c_j) = \sum_c \text{turn}(c)$ for all corners c along the boundary of $\overline{f}$ from c_i (included) to c_j (excluded). For example, in Fig. 2(a) let c_1, c_2, and c_3 be the corners associated with v_1, v_2, and v_3, respectively, and let (c_4, c_4') be the ordered pair of reflex corners associated with v_4. We have $\text{rot}(c_1, c_2) = 3$, $\text{rot}(c_3, c_4) = 1$, $\text{rot}(c_3, c_4') = 0$, and $\text{rot}(c_3, c_1) = -3$. The properties below are consequences of results in [26,27].

Property 1. For each face $\overline{f}$ of $\overline{H}$ and for each corner c_i of $\overline{f}$, we have $\text{rot}(c_i, c_i) = 4$ if $\overline{f}$ is internal and $\text{rot}(c_i, c_i) = -4$ if $\overline{f}$ is external.

Property 2. For each ordered triplet of corners (c_i, c_j, c_k) of a face of $\overline{H}$, we have $\mathrm{rot}(c_i, c_k) = \mathrm{rot}(c_i, c_j) + \mathrm{rot}(c_j, c_k)$.

Property 3. Let c_i and c_j be two corners of $\overline{f}$. If $\overline{f}$ is internal then $\mathrm{rot}(c_i, c_j) = 2 \iff \mathrm{rot}(c_j, c_i) = 2$. If $\overline{f}$ is external then $\mathrm{rot}(c_i, c_j) = 2 \iff \mathrm{rot}(c_j, c_i) = -6$.

Let c be a reflex corner of $\overline{H}$ associated with either a degree-2 vertex or a bend of H. Let s_h and s_v be the horizontal and vertical segments incident to c and let ℓ_h and ℓ_v be the lines containing s_h and s_v, respectively. We say that c (or equivalently its associated vertex/bend of H) points *up-left*, if s_h is to the right of ℓ_v and s_v is below ℓ_h. The definitions of c that points *up-right*, *down-left*, and *down-right* are symmetric (see Figs. 3(a)–3(d)). If v is a degree-1 vertex in H, then it has two associated reflex corners in $\overline{H}$. In this case, v points *upward* (*downward*) if its incident segment is vertical and below (above) the horizontal line passing through v. The definitions of a degree-1 vertex that points *leftward* or *rightward* are symmetric (see Figs. 3(e)–3(h)).

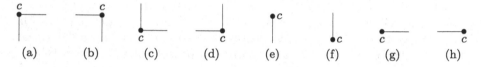

Fig. 3. Directions of a reflex corner associated with a degree-2 vertex or with a bend: (a) up-left; (b) up-right; (c) down-left; (d) down-right. Directions of a degree-1 vertex: (e) upward; (f) downward; (g) leftward; (h) rightward.

Two reflex corners c_i and c_j of a face of $\overline{H}$ are called *kitty corners* if $\mathrm{rot}(c_i, c_j) = 2$ or $\mathrm{rot}(c_j, c_i) = 2$. A face f of an orthogonal representation H is *turn-regular*, if the corresponding face $\overline{f}$ of $\overline{H}$ has no kitty corners. If every face of H is turn-regular, then H is *turn-regular*. For example, the orthogonal representation in Fig. 2(b) is not turn-regular as the faces f_1 and f_3 are turn-regular, while the internal face f_2 and the external face f_4 are not turn-regular (the pairs of kitty corners in each face are highlighted with dotted arrows). A graph G is *turn-regular*, if it admits a turn-regular orthogonal representation. If G admits a turn-regular rectilinear representation, then G is *rectilinear turn-regular*. The next lemma (whose proof can be found in [3]), provides a sufficient condition for the existence of a kitty-corner pair in the external face.

Lemma 1. *Let $\overline{H}$ be the rectilinear image of an orthogonal representation H of a plane graph G. Let (c_1, c_2) be two corners of the external face such that $\mathrm{rot}(c_1, c_2) \geq 3$ or c_1 is a reflex corner and $\mathrm{rot}(c_1, c_2) \geq 2$. Then, the external face contains a pair of kitty corners.*

Corollary 1. *Let H be an orthogonal representation of a plane graph G. If the external face of $\overline{H}$ has three consecutive convex corners, H is not turn-regular.*

3   Turn-Regular Graphs

The theorems in this section can be proven by modifying a well-known linear-time algorithm by Biedl and Kant [4] that produces an orthogonal drawing Γ with at most two bends per edge of a biconnected planar 4-graph G with a fixed embedding $\mathcal{E}$. Such an algorithm exploits an st-ordering $s = v_1, v_2, \ldots, v_n = t$ of the vertices of G, where s and t are two distinct vertices on the external face of $\mathcal{E}$. We recall that an st-ordering $s = v_1, v_2, \ldots, v_n = t$ is a linear ordering of the vertices of G such that any vertex v_i distinct from s and t has at least two neighbors v_j and v_k in G with $j < i < k$ [17]. The orthogonal drawing Γ is constructed incrementally by adding vertex v_k, for $k = 1, \ldots, n$, into the drawing Γ_{k-1} of $\{v_1, \ldots, v_{k-1}\}$, while preserving the embedding $\mathcal{E}$. Some invariants are maintained when vertex v_k is placed above Γ_{k-1}: (i) vertex v_k is attached to Γ_{k-1} with at least one edge incident to v_k from the bottom; (ii) after v_k is added to Γ_{k-1}, some extra columns are introduced into Γ_k to ensure that each edge (v_i, v_j), such that $i \leq k < j$ has a dedicated column in Γ_k that is reachable from v_i with at most one bend and without introducing crossings.

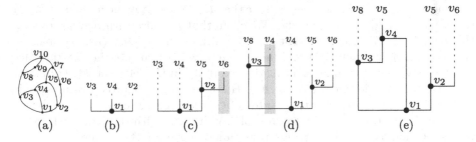

Fig. 4. The first four steps of the algorithm in the proof of Theorem 1 for the construction of a turn-regular orthogonal drawing of the biconnected planar 3-graph shown in (a).

Theorem 1. *Every biconnected planar 3-graph admits a turn-regular representation with at most two bends per edge, which can be computed in linear time.*

Proof. Let G be a biconnected planar 3-graph and let $\mathcal{E}$ be any planar embedding of G. Let s and t be two distinct vertices on the external face of $\mathcal{E}$. As in [4], based on an st-ordering of G, we incrementally construct an orthogonal drawing Γ of G by adding v_k into the drawing Γ_{k-1} of graph G_{k-1}, for $k = 1, \ldots, n$. Besides the invariants (i) and (ii) described above, we additionally maintain the invariant (iii): each reflex corner introduced in the drawing points either down-right or up-right with the possible exception of the reflex corners of the edges on the external face that are incident to s or t. Drawing Γ_1 consists of the single vertex v_1. Since $\deg(v_1) \leq 3$, the columns assigned to its three incident edges are the column where v_1 lies and the two columns immediately on its left and on its right (see Fig. 4(b)). These columns are assigned to the edges incident to v_1 in the order they appear in $\mathcal{E}$. Now, suppose you have to add vertex v_k to Γ_{k-1}. Observe that, since G has maximum degree three, v_k has a maximum of three edges (v_k, v_h), (v_k, v_i), and (v_k, v_j), where we may assume, without loss of generality, that $h < i < j$. To complete the proof, we consider three cases:

Case 1 ($h < k < i$): We place v_k on the first empty row above Γ_{k-1} and on the column assigned to (v_k, v_h). Also, to preserve the invariant (ii), we introduce an extra column immediately to the right of v_k and we assign the column of v_k and the newly added extra column to (v_k, v_i) and (v_k, v_j) in the order that is given by $\mathcal{E}$. For example, Fig. 4(c) shows the placement of v_2 directly above v_1, with one extra column inserted to the right of v_2 and assigned to the edge (v_2, v_6).

Case 2 ($i < k < j$): We place v_k on the first empty row above Γ_{k-1} and on the leftmost column between the columns assigned to (v_k, v_h) and (v_k, v_i). Also, we assign the column of v_k to (v_k, v_j), e.g., Fig. 4(e) shows the placement of v_4 on the leftmost column assigned to its incoming edges (v_3, v_4) and (v_1, v_4).

Case 3 ($j < k$): Here, v_k is t. We place v_k on the first empty row above Γ_{k-1} and on the middle column among those assigned to (v_k, v_h), (v_k, v_i), and (v_k, v_j).

The discussion in [4] suffices to prove that Γ is a planar orthogonal drawing of G with at most two bends per edge. We claim that Γ is also turn-regular. In fact, the invariant (iii) guarantees that all internal faces have reflex corners pointing either down-right or up-right and, hence, are turn-regular. On the external face we may have reflex corners pointing down-left (from the leftmost edge of v_1) or up-left (from the leftmost edge of v_n). However, since there is a y-monotonic path leading from s to any other vertex of G, such corners correspond to bends lying on the bottom or on the top row of any drawing with the same orthogonal representation as Γ and, therefore, they cannot form a kitty corner.   □

The proofs of the next two theorems exploit a similar technique as in Theorem 1. The full proof of Theorem 2 can be found in [3].

Theorem 2. *Every planar Hamiltonian 4-graph G admits a turn-regular representation H with at most 3 bends per edge, and such that only one edge of H gets 3 bends and only if G is 4-regular. Given the Hamiltonian cycle, H can be computed in linear time.*

Sketch of proof. We use as st-ordering for the Biedl and Kant approach [4] the ordering given by the Hamiltonian cycle. We choose a suitable vertex v_1 from which we start the construction. The construction rules are given in Fig. 5. If G has a vertex of degree less than four, then we choose such a vertex as v_1. Otherwise, G is 4-regular and we prove that G has at least one vertex such that the configuration of Fig. 5(c) is ruled out by the embedding $\mathcal{E}$. We maintain the invariant that all the reflex corners introduced in the drawing point (i) down-left or up-left, if they are contained in a face that is on the left side of the portion of the Hamiltonian cycle traversing Γ_k, and (ii) down-right or up-right, if they are contained in a face that is on the right side. Possible exceptions are the reflex corners on the external face and that occur on edges incident to v_1 or to v_n.   □

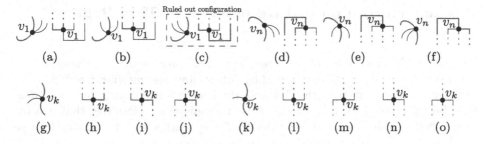

Fig. 5. Drawing rules for the algorithm in the proof of Theorem 2. The Hamiltonian path is drawn red and thick. Figures (g)–(j) are to be intended up to a horizontal flip.

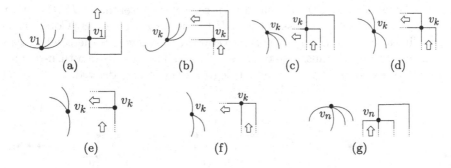

Fig. 6. The construction rules for the algorithm in the proof of Theorem 3.

Theorem 3. *Every biconnected planar 4-graph has a representation with $O(n)$ bends per edge that is internally turn-regular and that is computed in $O(n)$ time.*

Proof. We modify the algorithm of Biedl and Kant [4] again, where instead of the standard bottom-up construction, we adopt a spiraling one. The vertices are inserted in the drawing according to an st-ordering, based on the rules depicted in Fig. 6. For an internal face f let $s(f)$ ($d(f)$, resp.) be the index of the first (last, resp.) inserted vertex incident to f. By construction, f is bounded by two paths P_ℓ and P_r that go from $v_{s(f)}$ to $v_{d(f)}$, where P_ℓ precedes P_r in the left-to-right list of outgoing edges of $v_{s(f)}$. The construction rules imply that P_r has only convex corners. On the other hand, each convex corner of P_ℓ is always immediately preceded or immediately followed by a reflex corner. This rules out kitty corners in f. Indeed, consider to reflex corners c_i and c_j of P_ℓ and the counter-clockwise path from c_i to c_j all contained into P_ℓ. When computing $\mathrm{rot}(c_i, c_j)$ a positive amount $+1$ is always followed by a negative amount -1, and the sum is never equal to 2. Since f is an internal face, $\mathrm{rot}(c_i, c_j) + \mathrm{rot}(c_j, c_i) = 4$, and $\mathrm{rot}(c_i, c_j) \neq 2$ implies $\mathrm{rot}(c_j, c_i) \neq 2$. Note that an edge (v_i, v_j) contains $O(j - i)$ bends, which yields $O(n)$ bends per edge in the worst case. $\square$

4   Characterization and Recognition of Turn-Regular Trees

We give a characterization of the trees that admit turn-regular representations, which we use to derive a corresponding linear-time testing and drawing algorithm. For a tree T, let smooth(T) denote the tree obtained from T by smoothing all subdivision vertices, i.e., smooth(T) is the unique smallest tree that can be subdivided to obtain a tree isomorphic to T. We start with an auxiliary lemma which is central in our approach (see [3] for details).

Lemma 2. *T is turn-regular if and only if* smooth(T) *is rectilinear turn-regular.*

Sketch of proof. Suppose that T has a turn-regular representation H (the other direction is obvious). We can assume that H has no zig-zag edges. By Corollary 1, the rectilinear image of H has at most two consecutive convex corners, which can be removed with local transformations as in Fig. 7. This results in a turn-regular representation with only flat corners at degree-2 vertices as desired.   □

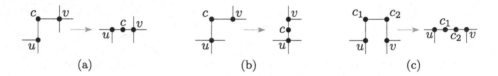

(a)   (b)   (c)

Fig. 7. Illustrations for the proof of Lemma 2.

Unless otherwise specified, from now on we will assume by Lemma 2 that T is a tree without degree-2 vertices. We will further refer to a tree as turn-regular if and only if it is rectilinear turn-regular. This implies that the class of turn-regular trees coincides with the class of trees admitting planar straight-line *convex drawings*, i.e., all edges incident to leaves can be extended to infinite crossing-free rays, whose edges are horizontal or vertical segments. The next property directly follows from Lemma 2 and the absence of degree-2 vertices.

Property 4. Let H be a turn-regular rectilinear representation of a tree T. Then, the reflex corners of H are formed by the leaves of T.

While turn-regularity is not a hereditary property in general graphs, the next lemma, whose proof can be found in [3], shows that it is in fact hereditary for trees.

Lemma 3. *If a tree T is turn-regular, then any subtree of T is turn-regular.*

A *trivial tree* is a single edge; otherwise, it is *non-trivial*. For $k \in \{2, 3\}$, a k-*fork* in a tree T consists of a vertex v whose degree is $k + 1$ and at least k leaves adjacent to it in T.

For $k \in \{2, 3\}$, a *k-fork* at a vertex v in a tree T consists of vertex v whose degree is $k + 1$ and at least k leaves adjacent to it in T. Due to the degree restriction, a 2-fork is not a 3-fork, and vice versa. Also, notice that by definition $K_{1,4}$ is a 3-fork. The next lemma follows from [7, Lem. 7]; a simplified proof is given in [3].

Lemma 4. *A turn-regular tree has (i) at most four 2-forks and no 3-fork, or (ii) two 3-forks and no 2-fork, or (iii) one 3-fork and at most two 2-forks.*

Lemma 5. *A non-trivial tree T contains at least one 2- or 3-fork.*

Proof. Since T is non-trivial and contains no vertices of degree two, there exists a non-leaf vertex v with degree either three or four, such that v is adjacent to exactly two or three leaves, respectively. Thus, the claim follows.   □

Corollary 2. *A turn-regular tree has at most four non-trivial disjoint subtrees.*

A vertex v of a tree T is a *splitter* if v is adjacent to at least three non-leaf vertices.

Lemma 6. *A turn-regular tree T contains at most two splitters.*

Proof. Assume to the contrary that T contains at least three splitters v_1, v_2 and v_3. We first claim that it is not a loss of generality to assume that v_1, v_2 and v_3 appear on a path in T. If this is not the case, then there is a vertex, say u, such that v_1, v_2 and v_3 lie in three distinct subtrees rooted at u. Hence, u is a splitter that lies on the path from v_1 to v_3. If we choose v_2 to be u, the claim follows.

Let P be the path containing v_1, v_2 and v_3 in T, and assume w.l.o.g. that v_1 and v_3 are the two end-vertices of P. Since v_1 is a splitter, it is adjacent to at least three vertices that are not leaves and two of them do not belong to P. Let T_1 and T_2 be the subtrees of T rooted at these two vertices, which by definition are non-trivial and do not contain v_2 and v_3. By a symmetric argument on v_3, we obtain two non-trivial subtrees T_3 and T_4 of T that do not contain v_1 and v_2. The third splitter v_2 may have only one neighbor that is not a leaf and does not belong to P. The (non-trivial) subtree T_5 rooted at this vertex contains neither v_1 nor v_3. Hence, $T_1, \ldots, T_5$ contradict Corollary 2.   □

By Lemma 6, a turn-regular tree contains either zero or one or two splitters (see Lemmas 7–9). A *caterpillar* is a tree, whose leaves are within unit distance from a path, called *spine*. For $k \in \{3, 4\}$, a *k-caterpillar* is a non-trivial caterpillar (i.e., not a single edge), whose spine vertices have degree at least 3 and at most k.

Lemma 7. *A tree T without splitters is a 4-caterpillar and turn-regular.*

Proof. In the absence of splitters in T, all inner vertices of T form a path. Hence, T is a 4-caterpillar and thus turn-regular; see Fig. 8(a).   □

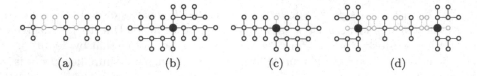

Fig. 8. Illustration of (a) a 4-caterpillar, (b) a 4-windmill, (c) a 3-windmill, and (d) a double-windmill. Possible extensions are highlighted in gray.

A tree with one splitter v is (i) a 4-*windmill*, if v is the root of four 3-caterpillars (Fig. 8(b)), (ii) a 3-*windmill*, if v is the root of two 3-caterpillars and one 4-caterpillar (Fig. 8(c)). Note that in the latter case, v can be adjacent to a leaf if it has degree four. The operation of *pruning* a rooted tree T *at* a degree-k vertex v with $k \in \{3, 4\}$ that is not the root of T, removes the $k - 1$ subtrees of T rooted at the children of v without removing these children, and yields a new subtree T' of T, in which v and its children form a $(k - 1)$-fork in T'.

Lemma 8. *A tree T with one splitter is turn-regular if and only if it is a 3- or 4-windmill.*

Sketch of proof. Every 3- or 4-windmill is turn-regular; see Figs. 8(b)–8(c). Now, let u be the splitter of T. If u has four non-leaf neighbors, then u is the root of four non-trivial subtrees $T_1, \ldots, T_4$, which by Lemma 7 are 4-caterpillars. We claim that none of them has a degree-4 vertex. Assume to the contrary that T_1 contains such a vertex $v \neq u$. We root T at u and prune at v, resulting in a (turn-regular, by Lemma 3) subtree T' of T that contains a 3-fork at v. By Lemma 5, each of the non-trivial trees $T_2, \ldots, T_4$ contains a fork. By Lemma 4, these three forks together with the 3-fork formed at v contradict the turn-regularity of T'. The case in which u has three non-leaf neighbors can be found in [3]. □

A tree T with exactly two splitters u and v is a *double-windmill* if (i) the path from u to v in T forms the spine of a 4-caterpillar in T, (ii) each of u and v is the root of exactly three non-trivial subtrees, and (iii) the two non-trivial subtrees rooted at u (v) that do not contain v (u) are 3-caterpillars; see Fig. 8(d). The proof of the next lemma is similar to the one of Lemma 8; see [3].

Lemma 9. *A tree T with two splitters is turn-regular if and only if it is a double-windmill.*

Lemmas 7–9 imply the next theorem. Note that for the recognition, one can test if a (sub-)tree is a 3- or a 4-caterpillar in linear time (for details, see [3]).

Theorem 4. *A tree T is turn-regular if and only if* smooth(T) *is (i) a 4-caterpillar, or (ii) a 3- or a 4-windmill, or (iii) a double-windmill. Moreover, recognition and drawing can be done in linear time.*

5 Turn-Regular Rectilinear Representations

Here we focus on rectilinear planar representations and prove the following.

Theorem 5. *Let G be an n-vertex biconnected plane graph with faces of degree at most eight. There exists an $O(n^{1.5})$-time algorithm that decides whether G admits an embedding-preserving turn-regular rectilinear representation and that computes such a representation in the positive case.*

Proof. We describe a testing algorithm based on a constrained version of Tamassia's flow network $N(G)$, which models the space of orthogonal representations of G within its given planar embedding [26]. Let V, E, and F be the set of vertices, edges, and faces of G, respectively. Tamassia's flow network $N(G)$ is a directed multigraph having a *vertex-node* ν_v for each vertex $v \in V$ and a *face-node* ν_f for each face $f \in F$. $N(G)$ has two types of edges: (i) for each vertex v of a face f, there is a directed edge (ν_v, ν_f) with capacity 3; (ii) for each edge $e \in E$, denoted by f and g the two faces incident to e, there is a directed edge (ν_f, ν_g) and a directed edge (ν_g, ν_f), both with infinite capacity.

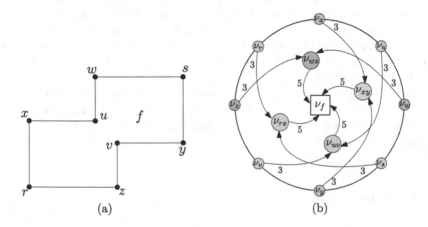

Fig. 9. (a) A pair of kitty corners in a face of degree eight. (b) The modification of the flow network around a face-node corresponding to an internal face. The labels on the directed edge represent capacities.

A feasible flow on $N(G)$ corresponds to an orthogonal representation of G: a flow value $k \in \{1, 2, 3\}$ on an edge (ν_v, ν_f) represents an angle of $90 \cdot k$ degrees at v in f (since G is biconnected, there is no angle larger than $270°$ at a vertex); a flow value $k \geq 0$ on an edge (ν_f, ν_g) represents k bends on the edge of G associated with (ν_f, ν_g), and all these bends form an angle of $90°$ inside f. Hence, each vertex-node ν_v supplies 4 units of flow in $N(G)$, and each face-node ν_f in $N(G)$ demands an amount of flow equal to $c_f = (2 \deg(f) - 4)$ if f is internal and to $c_f = (2 \deg(f) + 4)$ if f is external. The value c_f represents the *capacity of* f.

It is proved in [26] that the total flow supplied by the vertex-nodes equals the total flow demanded by the face-nodes; if a face-node ν_f cannot consume all the flow supplied by its adjacent vertex-nodes (because its capacity c_f is smaller), it can send the exceeding flow to an adjacent face-node ν_g, through an edge (ν_f, ν_g), thus originating bends.

Our algorithm has to test the existence of an orthogonal representation H such that: (a) H has no bend; (b) H is turn-regular. To this aim, we suitably modify $N(G)$ so that the possible feasible flows only model the set of orthogonal representations that verify Properties (a) and (b). To enforce Property (a), we just remove from $N(G)$ the edges between face-nodes. To enforce Property (b), we enhance $N(G)$ with additional nodes and edges. Consider first an internal face f of G. By hypothesis $\deg(f) \leq 8$. It is immediate to see that if $\deg(f) \leq 7$ then f cannot have a pair of kitty corners. If $\deg(f) = 8$, a pair $\{u, v\}$ of kitty corners necessarily requires three vertices along the boundary of f going from u to v (and hence also from v to u); see Fig. 9(a). Therefore, for such a face f, we locally modify $N(G)$ around ν_f as shown in Fig. 9(b). Namely, for each potential pair $\{u, v\}$ of kitty corners, we introduce an intermediate node ν_{uv}; the original edges (ν_u, ν_f) and (ν_v, ν_f) are replaced by the edges (ν_u, ν_{uv}) and (ν_v, ν_{uv}), respectively (each still having capacity 3); finally, an edge (ν_{uv}, ν_f) with capacity 5 is inserted, which avoids that u and v form a reflex corner inside f at the same time. For the external face f, it can be easily seen that a pair of kitty corners is possible only if the face has degree at least 10. Since we are assuming that $\deg(f) \leq 8$, we do not need to apply any local modification to $N(G)$ for the external face.

Hence, a rectilinear turn-regular representation of G corresponds to a feasible flow in the modified version of $N(G)$. Since $N(G)$ can be easily transformed into a sparse unit capacity network, this problem can be solved in $O(n^{1.5})$ time by applying a maximum flow algorithm (the value of the maximum flow must be equal to $4|V|$) [1].   □

6   Open Problems

Our work raises several open problems. (i) A natural question is if all biconnected planar 4-graphs are turn-regular (not only internally). (ii) While we suspect the existence of non-turn regular biconnected planar 4-graphs, we conjecture that triconnected planar 4-graphs are turn-regular. (iii) It would be interesting to extend the result of Theorem 5 to more general classes of plane graphs.

References

1. Ahuja, R.K., Magnanti, T.L., Orlin, J.B.: Network Flows - Theory, Algorithms and Applications. Prentice Hall, Upper Saddle River (1993)
2. Batini, C., Talamo, M., Tamassia, R.: Computer aided layout of entity relationship diagrams. J. Syst. Softw. 4(2–3), 163–173 (1984)

3. Bekos, M.A., et al.: On turn-regular orthogonal representations. Technical report arXiv:2008.09002, Cornell University (2020)
4. Biedl, T.C., Kant, G.: A better heuristic for orthogonal graph drawings. Comput. Geom. **9**(3), 159–180 (1998)
5. Binucci, C., Di Giacomo, E., Didimo, W., Rextin, A.: Switch-regular upward planarity testing of directed trees. J. Graph Algorithms Appl. **15**(5), 587–629 (2011)
6. Bridgeman, S.S., Di Battista, G., Didimo, W., Liotta, G., Tamassia, R., Vismara, L.: Turn-regularity and optimal area drawings of orthogonal representations. Comput. Geom. **16**(1), 53–93 (2000)
7. Carlson, J., Eppstein, D.: Trees with convex faces and optimal angles. In: Kaufmann, M., Wagner, D. (eds.) GD 2006. LNCS, vol. 4372, pp. 77–88. Springer, Heidelberg (2007). https://doi.org/10.1007/978-3-540-70904-6_9
8. Cornelsen, S., Karrenbauer, A.: Accelerated bend minimization. J. Graph Algorithms Appl. **16**(3), 635–650 (2012)
9. Di Battista, G., Eades, P., Tamassia, R., Tollis, I.G.: Graph Drawing: Algorithms for the Visualization of Graphs. Prentice-Hall, Upper Saddle River (1999)
10. Di Battista, G., Liotta, G.: Upward planarity checking: "faces are more than polygons". In: Whitesides, S.H. (ed.) GD 1998. LNCS, vol. 1547, pp. 72–86. Springer, Heidelberg (1998). https://doi.org/10.1007/3-540-37623-2_6
11. Didimo, W., Liotta, G.: Graph visualization and data mining. In: Mining Graph Data, pp. 35–64. Wiley (2007)
12. Didimo, W.: Upward planar drawings and switch-regularity heuristics. J. Graph Algorithms Appl. **10**(2), 259–285 (2006)
13. Duncan, C.A., Goodrich, M.T.: Planar orthogonal and polyline drawing algorithms. In: Handbook of Graph Drawing and Visualization, pp. 223–246. Chapman and Hall/CRC (2013)
14. Eiglsperger, M., et al.: Automatic layout of UML class diagrams in orthogonal style. Inf. Vis. **3**(3), 189–208 (2004)
15. Esser, A.M.: Orthogonal compaction: turn-regularity, complete extensions, and their common concept. In: Cláudio, A.P., et al. (eds.) VISIGRAPP 2019. CCIS, vol. 1182, pp. 179–202. Springer, Cham (2020). https://doi.org/10.1007/978-3-030-41590-7_8
16. Evans, W.S., Fleszar, K., Kindermann, P., Saeedi, N., Shin, C.S., Wolff, A.: Minimum rectilinear polygons for given angle sequences. Technical report arXiv:1606.06940, Cornell University (2020)
17. Even, S., Tarjan, R.E.: Corrigendum: Computing an st-numbering. TCS 2(1976), 339–344. Theoret. Comput. Sci. **4**(1), 123 (1977)
18. Hashemi, S.M., Tahmasbi, M.: A better heuristic for area-compaction of orthogonal representations. App. Math. Comput. **172**(2), 1054–1066 (2006)
19. Jünger, M., Mutzel, P. (eds.): Graph Drawing Software. Springer, Heidelberg (2004)
20. Kaufmann, M., Wagner, D. (eds.): Drawing Graphs. LNCS, vol. 2025. Springer, Heidelberg (2001). https://doi.org/10.1007/3-540-44969-8
21. Klau, G.W., Mutzel, P.: Optimal compaction of orthogonal grid drawings (extended abstract). In: Cornuéjols, G., Burkard, R.E., Woeginger, G.J. (eds.) IPCO 1999. LNCS, vol. 1610, pp. 304–319. Springer, Heidelberg (1999). https://doi.org/10.1007/3-540-48777-8_23
22. Lengauer, T.: Combinatorial Algorithms for Integrated Circuit Layout. B. G. Teubner/Wiley, Stuttgart (1990)
23. Nishizeki, T., Chiba, N.: Hamiltonian cycles (chap. 10). In: Planar Graphs: Theory and Algorithms, pp. 171–184. Courier Dover Publications (2008)

24. Nishizeki, T., Rahman, M.S.: Planar Graph Drawing. Lecture Notes Series on Computing, vol. 12. World Scientific (2004)
25. Patrignani, M.: On the complexity of orthogonal compaction. Comput. Geom. $19(1)$, 47–67 (2001)
26. Tamassia, R.: On embedding a graph in the grid with the minimum number of bends. SIAM J. Comput. $16(3)$, 421–444 (1987)
27. Vijayan, G., Wigderson, A.: Rectilinear graphs and their embeddings. SIAM J. Comput. $14(2)$, 355–372 (1985)

Extending Partial Orthogonal Drawings

Patrizio Angelini[1], Ignaz Rutter[2], and T. P. Sandhya[2][(✉)]

[1] John Cabot University, Rome, Italy
pangelini@johncabot.edu
[2] Universität Passau, 94032 Passau, Germany
{rutter,thekkumpad}@fim.uni-passau.de

Abstract. We study the planar orthogonal drawing style within the framework of partial representation extension. Let (G, H, Γ_H) be a partial orthogonal drawing, i.e., G is a graph, $H \subseteq G$ is a subgraph and Γ_H is a planar orthogonal drawing of H.

We show that the existence of an orthogonal drawing Γ_G of G that extends Γ_H can be tested in linear time. If such a drawing exists, then there also is one that uses $O(|V(H)|)$ bends per edge. On the other hand, we show that it is NP-complete to find an extension that minimizes the number of bends or has a fixed number of bends per edge.

Keywords: Planar orthogonal drawing · Partial representation extension · Bend minimization

1 Introduction

One of the most popular drawing styles are *orthogonal drawings*, where vertices are represented by points and edges are represented by chains of horizontal and vertical segments connecting their endpoints. Such a drawing is *planar* if no two edges share an interior point. An interior point of an edge where a horizontal and a vertical segment meet is called a *bend*. The main aesthetic criterion for planar orthogonal drawings is the number of bends on the edges.

A large body of literature is devoted to optimizing the number of bends in planar orthogonal drawings. The complexity of the problem strongly depends on the particular input. If the combinatorial embedding can be chosen freely, then it is NP-complete to decide whether there exists a drawing without bends [17]. If the input graph comes with a fixed combinatorial embedding, then a bend-optimal drawing that preserves the given embedding can be computed efficiently by a classical result of Tamassia [26]. A recent trend has been to investigate under which conditions the variable-embedding case becomes tractable. For maxdeg-3 graphs a bend-optimal drawing can be computed efficiently [10], which has recently been improved to linear time [12]. The problem is also FPT with respect to the number of degree-4 vertices [11], and if one discounts the first bend on

---

The full version of this article is available at ArXiv [3].

© Springer Nature Switzerland AG 2020
D. Auber and P. Valtr (Eds.): GD 2020, LNCS 12590, pp. 265–278, 2020.
https://doi.org/10.1007/978-3-030-68766-3_21

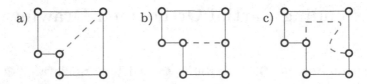

Fig. 1. An instance of the partial representation extension problem (G, H, Γ_H) is given. The graph H is solid black and the edges of $E(G) \setminus E(H)$ are dashed red. (a) (G, H, Γ_H) admits a planar extension, but not an orthogonal extension. (b) (G, H, Γ_H) admits an orthogonal extension with no bends (c) An orthogonal representation of G (the curved part of the dashed edge has no bends) that extends the description of the solid black drawing of H. There exists no drawing of G with this representation that extends the given drawing of H. (Color figure online)

each edge, an optimal solution can be computed even for individual convex cost functions on the edges [4,5]. We refer to the survey [13] for further references.

In light of this popularity and the existence of a strongly developed theory, it is surprising that the planar orthogonal drawings have not been investigated within the framework of partial representation extension. Especially so, since it has been considered in the related context of simultaneous representations [1].

In the partial representation extension problem, the input graph G comes together with a subgraph $H \subseteq G$ and a representation (drawing) Γ_H of H. One then seeks a drawing Γ_G of G that *extends* Γ_H, i.e., whose restriction to H coincides with Γ_H. The partial representation extension problem has recently been considered for a large variety of different types of representations. For planar straight-line drawings, it is NP-complete [25], whereas for topological drawings there exists a linear-time algorithm [2] as well as a characterization via forbidden substructures [18]. Moreover, it is known that, if a topological drawing extension exists, then it can be drawn with polygonal curves such that each edge has a number of bends that is linear in the complexity of Γ_H [6]. Here the complexity of Γ_H is the number of vertices and bends in Γ_H. Most recently the problem has been investigated in the context of 1-planarity [14]. Besides classical drawing styles, it has also been studied for contact representations [7] and for geometric intersection representations, e.g., for (proper/unit) interval graphs [19,21], chordal graphs [20], circle graphs [8], and trapezoid graphs [23].

In this paper, we provide an in-depth study of partial representation extension problems for the orthogonal drawing style. Since the aesthetics are of particular importance for the quality of such a drawing, we put a major emphasis on extension questions in relation to the number of bends. It is worth noting that even the seminal work of Tamassia [26] already mentions the idea of preserving the shape of a given subgraph by maintaining its orthogonal representation via modifications in his flow network. However, this approach only preserves the shape of the subgraph as described by an orthogonal representation, and not necessarily its drawing. Figure 1 shows that there are partial planar orthogonal drawings that can be extended in a planar way, but not orthogonally (Fig. 1a) and that, even if an orthogonal representation O_G of G preserves a given orthogonal representation O_H of a drawing Γ_H of H, there does not necessarily exist a drawing Γ_G of G realizing O_G that extends Γ_H (Fig. 1b).

Contribution and Outline. After presenting preliminaries in Sect. 2, we give a linear-time algorithm for deciding the existence of an orthogonal drawing extension in Sect. 3. Then, we consider the realizability problem, where we are given an orthogonal extension in the form of a suitable planar embedding, and we seek an orthogonal drawing extension that optimizes the number of bends. Along the lines of a result by Chan et al. [6], we show that there always exists an orthogonal drawing extension such that each edge has a number of bends that is linear in the complexity of Γ_H in Sect. 4. We complement these findings in Sect. 5 by showing that it is NP-hard to minimize the number of bends and NP-complete to test whether there exists an orthogonal drawing extension with a fixed number of bends per edge. For the proofs of theorems marked with a [∗], please refer to the full version of this paper [3].

2 Preliminaries

We call the circular clockwise ordering of the edges around a vertex v in an embedding the *rotation* at v. Let $G = (V, E)$ be a simple undirected graph and let $H \subseteq G$ be a subgraph. We refer to the vertices and edges of H as *H-vertices* and *H-edges*, respectively. Similarly, we refer to the vertices of $V(G) \setminus V(H)$ and to the edges of $E(G) \setminus E(H)$ as *G-vertices* and *G-edges*, respectively.

Let (G, H, Γ_H) be a triple composed of a graph G, a subgraph $H \subseteq G$, and an orthogonal drawing Γ_H of H. We denote by REPEXT(ORTHO) (REPEXT stands for representation extension) the problem of testing whether G admits an orthogonal drawing Γ_G that extends Γ_H. In Γ_H, we say that an H-edge is *attached to* one of the four *ports* of its end vertices. If there is no H-edge attached to a port of a vertex, then this port is *free*; note that the free ports are those at which the G-edges can be attached in Γ_G. For two edges e and e' that are consecutive in the rotation at a vertex v in Γ_H, we denote by $\mathcal{P}_H(e, e') = k$ the fact that there exist exactly k free ports of v when moving from e to e' in clockwise order around their common endvertex. We call $\mathcal{P}_H(e, e') = k$ a *port constraint*, and we denote by $\mathcal{P}_H$ the set of all port constraints in Γ_H. Note that, for a vertex v with rotation $e_1, \dots, e_h$ in Γ_H, with $h \leq 4$, we have $\sum_{i=1}^{h} \mathcal{P}_H(e_i, e_{i+1}) = 4 - \deg(v)$ (defining $e_{h+1} := e_1$).

We now show that to solve an instance (G, H, Γ_H) of the REPEXT(ORTHO) problem, it suffices to only consider the port constraints determined by Γ_H together with the embedding $\mathcal{E}_H$ of H in Γ_H. More specifically, we prove the following characterization, which could also be deduced from [1].

Theorem 1 (⋆). *Let (G, H, Γ_H) be an instance of* REPEXT(ORTHO). *Let $\mathcal{E}_H$ be the embedding of H in Γ_H, and let $\mathcal{P}_H$ be the port constraints induced by Γ_H. Then, (G, H, Γ_H) admits an orthogonal drawing extension if and only if G admits a planar embedding $\mathcal{E}_G$ that extends $\mathcal{E}_H$ and such that, for every port* constraint $\mathcal{P}_H(e, e') = k$, there exist at most k G-edges between e and e' in the rotation at v in $\mathcal{E}_G$, where v is the common vertex of the H-edges e and e'.

In view of Theorem 1, we define a new problem, called REPEXT(TOP+PORT), which is linear-time equivalent to REPEXT(ORTHO). An instance of this problem is a 4-tuple $(G, H, \mathcal{E}_H, \mathcal{P}_H)$ and the goal is to test whether G admits an embedding $\mathcal{E}_G$ that satisfies the conditions of Theorem 1. In order to unify the terminology, we also refer to the *Partially Embedded Planarity* problem studied in [2] as REPEXT(TOP) (TOP stands for topological drawing). Recall that an instance of this problem is a triple $\langle G, H, \mathcal{E}_H \rangle$, and the goal is to test whether G admits an embedding $\mathcal{E}_G$ that extends $\mathcal{E}_H$. As proved in [2], REPEXT(TOP) can be solved in linear time.

3   Testing Algorithm

In this section we show that REPEXT(ORTHO) can be solved in linear time. By Theorem 1, it suffices to prove that REPEXT(TOP+PORT) can be solved in linear time. The algorithm is based on constructing in linear time, starting from an instance $(G, H, \mathcal{E}_H, \mathcal{P}_H)$ of REPEXT(TOP+PORT), an instance $(G', H', \mathcal{E}_{H'})$ of REPEXT(TOP) that admits a solution if and only if $(G, H, \mathcal{E}_H, \mathcal{P})$ does.

In order to construct the instance $(G', H', \mathcal{E}_{H'})$ of REPEXT(TOP+PORT), we initialize $G' = G$, $H' = H$, and $\mathcal{E}_{H'} = \mathcal{E}_H$. Then, for each vertex v such that $1 < \deg_H(v) < \deg_G(v)$, we perform the following modifications; see Fig. 2.

Case 1: Suppose first that $\deg_H(v) = 3$ and $\deg_G(v) = 4$, and let $e = vw$ be the unique G-edge incident to v; refer to Fig. 2(a). Since $\deg_H(v) = 3$, there exist exactly two H-edges e_1 and e_2 such that e_1 immediately precedes e_2 in the rotation at v in $\mathcal{E}_H$ and $\mathcal{P}(e_1, e_2) = 1$. Note that, to respect the port constraint, we have to guarantee that e is placed between e_1 and e_2 in the rotation at v in $\mathcal{E}_G$. For this, we subdivide e with a new vertex w', that is, we remove e from G', and we add the vertex w' and the edges vw' and $w'w$ to G'. Also, we add w' and vw' to H', and insert vw' between e_1 and e_2 in the rotation at v in $\mathcal{E}_{H'}$.

Case 2: Suppose now that $\deg_H(v) = 2$ and $\deg_G(v) \geq 3$. Let e_1 and e_2 be the two H-edges incident to v, and let $e = vw$ and $e^* = vz$ be the at most two G-edges incident to v. We distinguish two cases, based on whether $\mathcal{P}(e_1, e_2) = 2$ and $\mathcal{P}(e_2, e_1) = 0$ (or vice versa), or $\mathcal{P}(e_1, e_2) = \mathcal{P}(e_2, e_1) = 1$.

Case 2.a: If $\mathcal{P}(e_1, e_2) = 2$, then we need to guarantee that both e and e^* (if it exists) are placed between e_1 and e_2 in the rotation at v in $\mathcal{E}_G$; refer to Fig. 2(b). For this, we remove e and e^* from G', and we add a new vertex w' and the edges vw', $w'w$, and $w'z$ to G'. Also, we add w' and vw' to H', and insert vw' between e_1 and e_2 in the rotation at v in $\mathcal{E}_{H'}$. Note that, if e^* does not exist, this is the same procedure as in the previous case. Case 2.b: If $\mathcal{P}(e_1, e_2) = \mathcal{P}(e_1, e_2) = 1$, then we need to guarantee that e and e^* (if it exists) appear on different sides of the path composed of the edges e_1 and e_2; refer to Fig. 2(c). Note that, if e^* does not exist, then e can be on any of the two sides of this path, and thus in this case we do not perform any modification. If e^* exists, we subdivide e, e^*, e_1, and e_2 with a new vertex each, that is, we remove these edges from G' (e_1

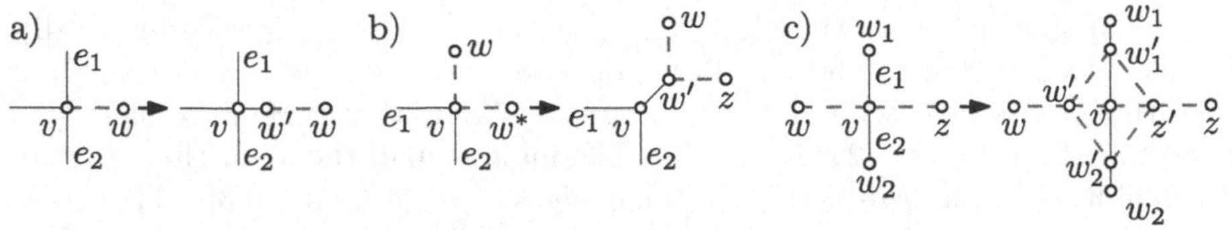

Fig. 2. Gadgets for H-vertices

and e_2 also from H'), and we add four new vertices w', z', w'_1, and w'_2. Also, we add to G' the edges vw', vz', vw'_1, and vw'_2, and the edges $w'w$, $z'z$, w'_1w_1, and w'_2w_2, where w_1 and w_2 are the endpoints of e_1 and e_2, respectively, different from v. Further, we add the edges $w'w'_1$, w'_1z', $z'w'_2$, and w'_2w' to G'. Finally, we add the edges vw'_1, w'_1w_1, vw'_2, and w'_2w_2 also to H'; in $\mathcal{E}_{H'}$, we place w'_1w_1 and w'_2w_2 in the rotations at w_1 and at w_2, respectively, in the same position as e_1 and e_2, respectively, in $\mathcal{E}_H$. The rotations at v, w', z', w'_1, and w'_2 in $\mathcal{E}_{H'}$ do not need to be set, since each of these vertices has at most two incident H'-edges. The above construction leads to the following lemma.

Lemma 1 ($\star$). *The instance $(G', H', \mathcal{E}_{H'})$ has an embedding extension if and only if $(G, H, \mathcal{E}_H, \mathcal{P})$ has an embedding extension satisfying the port constraints.*

Theorem 2. *The* REPEXT(TOP+PORT) *problem can be solved in linear time.*

Proof. Given an instance $I = (G, H, \mathcal{E}_H, \mathcal{P})$ of REPEXT(TOP+PORT), we construct the instance $I' = (G', H', \mathcal{E}_{H'})$ of REPEXT(TOP) that has linear size as described above. This takes $O(1)$ time per vertex, and hence total linear time. By Lemma 1, I has a solution if and only if I' has one. Since the existence of a solution of I' can be tested in linear time [2], the statement follows.

As a consequence of Theorems 1 and 2, we conclude the following.

Theorem 3. *The* REPEXT(ORTHO) *problem can be solved in linear time.*

4   Realizability with Bounded Number of Bends

In this section we prove that, if there exists an orthogonal drawing extension for an instance (G, H, Γ_H) of REPEXT(ORTHO), then there also exists one in which the number of bends per edge is linear in the complexity of the drawing Γ_H. By subdividing H at the bends of Γ_H, we can assume that Γ_H is a bend-free drawing of H. To achieve the desired edge complexity, it then suffices to show that $O(|V(H)|)$ bends per edge suffice. This result can be considered as the counterpart for the orthogonal setting of the one by Chan et al. [6] for the polyline setting. In their work, in fact, they show that a positive instance (G, H, Γ_H) of the REPEXT(TOP) problem can always be realized with at most $O(|V(H)|)$ bends per edge when Γ_H is a planar straight-line drawing of H.

Our approach follows the algorithm given in [6], with a main technical difference which is due to the peculiar properties of orthogonal drawings. Their algorithm first constructs a planar supergraph G' of G that is Hamiltonian using a method of Pach et al. [24, Lemma 5]. The main step of the algorithm of Chan et al. [6] involves the *contraction* of some edges of G' [6, Lemma 3]). This operation identifies the two end-vertices of the contracted edge and merges their adjacency lists. However, both the construction of the supergraph G' and the contractions may produce vertices of degree greater than 4, which implies that the resulting graph does not admit an orthogonal drawing any longer. As such, these operations are not suitable for the realization of orthogonal drawings. In order to overcome this problem, we consider instead the *Kandinsky* model [16], which extends the orthogonal drawing model to also allow for vertices of large degree. Once the drawing has been computed, we remove the previously added parts and by adding a small amount of additional bends on the G-edges, we arrive at a orthogonal drawing of the initial graph G. More specifically, we prove the following theorem:

Theorem 4 ($\star$). *Let (G, H, Γ_H) be an instance of* REPEXT(ORTHO). *Suppose that G admits an orthogonal drawing Γ_G that extends Γ_H, and let $\mathcal{E}_G$ be the embedding of G in Γ_G. Then we can construct a planar Kandinsky drawing of G in $O(n^2)$-time, where n is the number of vertices of G, that realizes $\mathcal{E}_G$, extends $\mathcal{H}$, and has at most $262|V(H)|$ bends per edge.*

An overview of the algorithm to construct the desired Kandinsky orthogonal drawing Γ_G^* of G, whose main steps follow the method in [6], is given below.

Step 1: Consider a face F of Γ_H with facial walks $W_1, W_2, \ldots, W_k$. Construct an ε-approximation of F and let W_i' be the orthogonal polygon that approximates W_i, $1 \le i \le k$. Let F' be the face bounded by the approximated boundary components of F.

Step 2: Partition F' into rectangles [15] and construct a graph K by placing a vertex at the center of each rectangle and by joining the vertices of adjacent rectangles. Let T be a spanning tree of K. For each facial walk W_i, add a new vertex near to W_i as a leaf of T (see Fig. 3).

Step 3: Construct the multigraph G_F induced by the vertices lying inside or on the boundary of F and by contracting each facial walk of F to a single vertex. Then draw G_F along T. Now, reconstruct the edges of $G \setminus H$ and the edges between G_F and other components of G inside F. Refer to Fig. 4.

We then transform Γ_G^* into an orthogonal drawing Γ_G of G with $O(|V(H)|)$ bends per edge that extends Γ_H. An illustration is given in Fig. 5.

Theorem 5 ($\star$). *Let (G, H, Γ_H) be an instance of* REPEXT(ORTHO). *Suppose that G admits an orthogonal drawing Γ_G that extends Γ_H, and let $\mathcal{E}_G$ be the embedding of G in Γ_G. Then we can construct a planar orthogonal drawing of G in $O(n^2)$-time, where n is the number of vertices of G, that realizes $\mathcal{E}_G$, extends $\mathcal{H}$, and has at most $270|V(H)|$ bends per edge.*

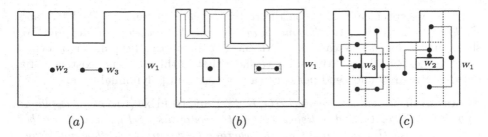

Fig. 3. (a) A face with outer walk W_1 and, inner facial walks W_2 and W_3. (b) An approximation F' of F. (c) A face and a corresponding tree T

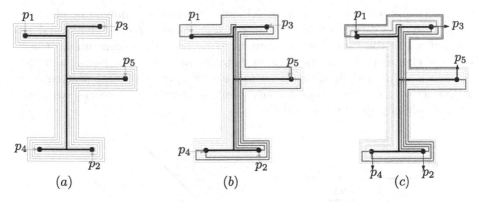

Fig. 4. (a) An orthogonal drawing of a tree T together with approximations along T (b) An orthogonal drawing of the Hamiltonian cycle C with respect to T (c) The edge p_3p_5 is drawn using approximations of T

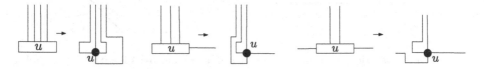

Fig. 5. Re-routing the edges incident to a vertex u in the Kandinsky drawing Γ_G^K to obtain the orthogonal drawing Γ_G.

5 Bend-Optimal Extension

In this section we study the problem of computing an orthogonal drawing extension of an instance $I = (G, H, \Gamma_H)$ of REPEXT(ORTHO) with the minimum number of bends. Observe that, if H is empty, this is equivalent to computing a bend-minimal drawing of G, which is NP-complete if the embedding of G is not fixed. We thus assume that G comes with a fixed planar embedding $\mathcal{E}_G$ that satisfies the port constraints of Γ_H, and we study the complexity of computing a bend-optimal drawing Γ_G of G with embedding $\mathcal{E}_G$ that extends Γ_H.

Here, we specifically focus on the restricted case where $V(H) = V(G)$ and $E(H) = \emptyset$, which we call *orthogonal point set embedding with fixed mapping*. We show that, even in this case, it is NP-hard to minimize the number of bends on the edges. On the positive side, we show that in this case the existence of a drawing that uses one bend per edge can be tested in polynomial time.

Theorem 6. *Given an instance (G, H, Γ_H) of* REPEXT(ORTHO), *a planar embedding $\mathcal{E}_G$ of G that satisfies the port constraints of Γ_H, and a number $k \in \mathbb{N}_0$, it is NP-complete to decide whether G admits an orthogonal drawing Γ_G with embedding $\mathcal{E}_G$ that extends H and has at most k bends. This holds even if $V(G) \setminus V(H) = \emptyset$, $E(H) = \emptyset$, and $E(G)$ is a matching.*

Proof. We give a reduction from the NP-complete problem *monotone planar 3-SAT* [9]. In this variant of 3-SAT, the variable–clause graph is planar and has a layout where the variables are represented by horizontal segments on the x-axis, the clauses by horizontal segments above and below the x-axis, and each variable is connected to each clause containing it by a vertical segment, the clauses above the x-axis contain only positive literals and the clauses below contain only negative literals; see Fig. 6a.

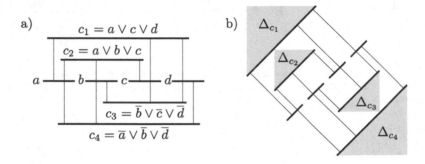

Fig. 6. A representation of an instance of monotone planar 3-SAT with four variables a, b, c, d and four clauses c_1, c_2, c_3, c_4 (a). Image of the vertically-stretched version of (a) under the mapping Φ (b).

A *box* is an axis-aligned rectangle whose bottom-left and the top-right corners contain two H-vertices, connected by a G-edge. We consider non-degenerate boxes, and thus this G-edge requires at least one bend; when this edge is drawn with one bend, there is a choice whether it contains the top-left or the bottom-right corner of the box. In these cases we say that the box is *drawn top* and *drawn bottom*, respectively. We now describe our *variable*, *pipe*, and *clause gadgets*.

A *variable gadget* consists of $h > 0$ boxes $R_1, \ldots, R_h$ that are 3×3-squares, where the bottom-left corner of R_i lies at $b + (2(i-1), 2(i-1))$, for an arbitrary base point b; see Fig. 7a-b. The crucial property is that in a one-bend drawing of the gadget, R_i is drawn bottom if and only if R_{i+1} is drawn top for $i = 1, \ldots, h-1$. Thus, in such a drawing, either all the *odd boxes* (those with odd

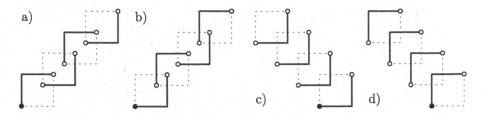

Fig. 7. Variable gadget with $h = 4$ boxes (a,b). In (a) the even boxes are drawn top and the odd boxes are drawn bottom, (b) shows the opposite. Pipe gadget (c,d). In (c) all boxes are drawn bottom, in (d) they are all drawn top. In all cases the base point is marked.

indices) are drawn top and all the *even boxes* (those with even indices) are drawn bottom, or vice versa. This will be used to encode the truth value of a variable.

A *(positive) pipe gadget* works similarly; see Fig. 7c-d. For a base point b, it consists of $h > 0$ boxes $R_1, \ldots, R_h$ that are 3×3-squares such that the bottom-left corner of R_i lies at $b + (-2(i-1), 2(i-1))$; see Fig. 7c-d. The decisive property is that in a one-bend drawing of the gadget, all the boxes are drawn the same as R_1, that is, either all bottom (see Fig. 7c) or all top (see Fig. 7d). Negative pipe gadgets are symmetric with respect to the line $y = x$ and behave symmetrically.

The last gadget we describe is the *(positive) clause gadget*; negative clause gadgets are symmetric with respect to the line $y = x$ and behave symmetrically. The positive clause gadget has three *input boxes* R_1, R_2, R_3, whose corners lie on a single line with slope 1; we assume that R_1 lies left of R_2, which in turn lies left of R_3. To simplify the description, we assume that the left lower corners of these rectangles lie at $(x, x), (y, y)$, and (z, z), respectively. Refer to Fig. 8a.

We create three *literal* boxes L_1, L_2, L_3 that are 3×3-squares. The lower left corner of L_1 is $(x - 3, y + 2)$, the lower left corner of L_2 if $(y - 2, y + 2)$, and the lower left corner of L_3 is $(y, z + 3)$. Note that the interiors of L_2 and R_2 intersect in a unit square, and therefore, if R_2 is drawn top, then L_2 must be drawn top. To obtain the same behavior for the other input and literal rectangles, we add two *transmission boxes* T_1 and T_2. The lower left corner of T_1 is $(x - 1, x + 2)$ and its upper right corner is $(x + 1, y + 4)$. The bottom-left and top-right corner of T_2 are $(y + 2, z + 2)$ and $(z + 1, z + 4)$, respectively. This guarantees that, also for $i = 1, 3$, if R_i is drawn top, then T_i and L_i are drawn top. We finally have a *blocker* box B, with corners at $(x - 1, z + 1)$ and $(x + 1, z + 4)$; and a *clause box*, whose corners are in the centers of L_1 and L_3, respectively.

Note that the G-edge connecting the two corners of the clause box, which we call the *clause edge*, requires at least two bends, as any one-bend drawing cuts horizontally through either the blocker B or the literal square L_2; see Fig. 8a. The following claim shows that the possibility of drawing it with exactly two bends depends on the drawings of the literal boxes of the clause gadget, and thus on the truth values of the literals; see Fig. 8b-c.

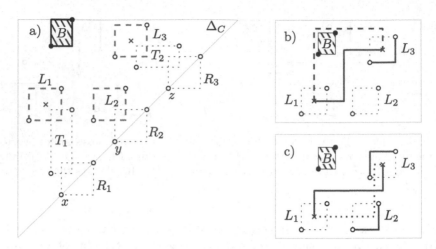

Fig. 8. Clause gadget with input rectangles R_1, R_2, R_3. The bottom-left and top-right corner of the clause box are drawn as crosses (a). The image of the triangle Δ_C under the mapping $(x, y) \mapsto (x - y, x + y)$ is drawn gray. The possibilities of routing the clause edge with two bends, if L_3 is drawn bottom (b) and if L_3 is drawn top and L_2 is drawn bottom (c).

Claim 1 ($\star$). *If the other edges are drawn with one bend, then the clause edge can be drawn with two bends if and only if not all literal boxes are drawn top.*

We are now ready to put the construction together. Consider the layout of the variable–clause graph, where each variable x is represented by a horizontal segment s_x on the x-axis, and each clause $C = (c_1, c_2, c_3)$ with only positive (only negative) literals by a horizontal segment s_C above (below) the x-axis. Further, the occurrence of a variable x in a clause C is represented by a vertical visibility segment $s_{x,C}$ that starts at an inner point of s_x and ends at an inner point of s_C; see Fig. 6a. We call these points *attachment points*. By suitably stretching the drawing horizontally, we may assume that all segments start and end at points with integer coordinates divisible by 8. We also stretch the whole construction vertically by a factor of n, which guarantees that for each clause segment s_C the right-angled triangle Δ_C, whose long side is s_C and that lies above s_C (below s_C if C consists of negative literals) does not intersect any other segments in its interior. Note that the initial drawing fits on a grid of polynomial size [22], and the transformations only increase the area polynomially. For the construction it is useful to consider this representation rotated by 45° in counterclockwise direction and scaled by a factor of $\sqrt{2}$ back to the grid. This is achieved by the affine mapping $\Phi\colon (x, y) \mapsto (x - y, x + y)$; see Fig. 6b.

For each variable segment s_x with left endpoint $(a, 0)$ and right endpoint $(b, 0)$ we create a variable gadget with $h = (b - a)/2$ boxes and base point (a, a). For each clause segment s_C above the x-axis with attachment points $(a_1, b), (a_2, b), (a_3, b)$, we create a positive clause gadget with input boxes at $(a_i - b, a_i + b)$. For each vertical segment $s_{x,C}$ above the x-axis with attachment points $(a, 0)$ and (a, b), we create a positive pipe gadget of $h = (b/2) - 2$ boxes at base point $(a - 2, a - 2)$. Note that, together with the box of the variable gadget of x at (a, a) and the input box of C at $(a - b, a + b)$, the newly placed boxes form a pipe gadget that consists of $h + 2$ boxes. Since distinct vertical segments on the same side of the x-axis have horizontal distance at least 8, the boxes of distinct pipes do not intersect, and the placement is such that only the first and last box of each pipe gadget intersect boxes that belong to the corresponding variable or clause gadget. Finally note that for each clause C, except for the input boxes, the clause gadget lies inside the image of the triangle Δ_C under the mapping Φ, since the attachment points are interior points of s_C, and the x-coordinates of its endpoints are divisible by 8. Hence, the only interaction of the clause gadget with the remainder of the construction is via the input variables The proof of the following claim is based on showing that we can draw each box with exactly one bend and each clause edge with exactly two bends, if and only if the original instance of monotone planar 3-SAT is satisfiable.

Claim 2 ($\star$). *Let φ be an instance of monotone planar 3-SAT, with γ clauses. Also, let β be the number of boxes in the instance (G, H, Γ_H) of REPEXT(ORTHO) constructed as described above. Then, the formula φ is satisfiable if and only if the instance (G, H, Γ_H) admits an extension with at most $k = \beta + \gamma$ bends.*

Since the construction has polynomially many vertices and edges on a polynomial size grid, it can be executed in polynomial time. Moreover, by construction, $V(H) = V(G)$, $E(H) = \emptyset$, and $E(G)$ is a matching. The statement of the theorem follows.

By subdividing each non-clause edge with a G-vertex, and each clause edge with two G-vertices, we get the following corollary.

Corollary 1. *It is NP-complete to decide whether a partial orthogonal drawing (G, H, Γ_H) admits an extension without bends.*

Similarly, we can ask whether an instance (G, H, Γ_H) admits an extension with at most k bends per edge for a fixed number k. The construction depicted in Fig. 9 shows how to force an edge to use k bends for any fixed number k. By making the part that enforces the first $k - 1$ bends sufficiently small, we essentially obtain the behavior of the box gadget from the proof of Theorem 6.

Corollary 2. *For any fixed $k \geq 2$, it is NP-complete to decide whether an instance (G, H, Γ_H) of REPEXT(ORTHO) admits an extension that uses at most k bends per edge, even if $V(G) = V(H)$.*

On the positive side, if all vertices are predrawn, the existence of an extension with at most k bends per edge can be tested efficiently for $k = 0$ and $k = 1$.

Theorem 7. *Let (G, H, Γ_H) be an instance of REPEXT(ORTHO) with $V(G) = V(H)$ and let $k \in \{0, 1\}$. It can be tested in polynomial time whether (G, H, Γ_H) admits an extension with at most k bends per edge.*

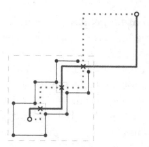

Fig. 9. Gadget for forcing an edge to use $k = 4$ bends. All vertices and the thin solid black lines are H-vertices. Up to minor geometric adjustments, the thick blue and dotted red lines show the only two ways to draw the G-edge between the two H-vertices u and v with k bends. Scaling the lower left part to make it sufficiently small results in a construction that behaves like a box. (Color figure online)

Proof. For $k = 0$ we simply draw each G-edge as the straight-line segment between its endpoints, and check whether this is a crossing-free orthogonal drawing.

For $k = 1$ we proceed as follows. While there exists a G-edge $e = uv$ whose endpoints have the same x- or the same y-coordinates, we do the following. If e must be drawn as a straight-line (if u and v have the same x- or the same y-coordinates), the instance (G, H, Γ_H) is equivalent to the instance (G, H', Γ'_H), where H' is obtained from H by adding e, and Γ'_H is obtained from inserting e as a straight-line segment. By applying this reduction rule, we eventually arrive at an instance (G'', H'', Γ''_H) such that the endpoints of each G-edge have distinct x- and distinct y-coordinates. Now for each such edge, there are precisely two ways to draw them with one bend. It is then straightforward to encode the existence of choices that lead to a planar drawing into a 2-SAT formula.

6   Conclusions

In this paper we studied the problem of extending a partial orthogonal drawing. We gave a linear-time algorithm to test the existence of such an extension, and we proved that if one exists, then there is also one whose edge complexity is linear in the size of the given drawing. On the other hand, we showed that, if we also restrict to a fixed constant the total number of bends or the number of bends per edge, then deciding the existence of an extension is NP-hard.

Concerning future work we feel that the most important questions are the following: 1) The complexity of $270|V(H)|$ bends per edge resulting from the transition to orthogonal drawings is significantly worse than the one of $72|V(H)|$ bends per edge in the case of arbitrary polygonal drawings [6]. Can this number be significantly reduced to, say, less than $100|V(H)|$? 2) As mentioned in the introduction, Tamassia [26] already observed that an orthogonal representation of H can be efficiently extended to an orthogonal representation of G.

However, drawing such an extension may require to modify the drawing Γ_H of the given subgraph. Is it possible to efficiently test whether a given orthogonal representation can be drawn such that it extends a given drawing Γ_H?

References

1. Angelini, P., et al.: Simultaneous orthogonal planarity. In: Hu, Y., Nöllenburg, M. (eds.) GD 2016. LNCS, vol. 9801, pp. 532–545. Springer, Cham (2016). https://doi.org/10.1007/978-3-319-50106-2_41
2. Angelini, P., et al.: Testing planarity of partially embedded graphs. ACM Trans. Algorithm. **11**(4), 32:1–32:42 (2015). https://doi.org/10.1145/2629341
3. Angelini, P., Rutter, I., T.P., S.: Extending partial orthogonal drawings (2020). https://arxiv.org/abs/2008.10280
4. Bläsius, T., Lehmann, S., Rutter, I.: Orthogonal graph drawing with inflexible edges. Comput. Geom. **55**, 26–40 (2016). https://doi.org/10.1016/j.comgeo.2016.03.001
5. Bläsius, T., Rutter, I., Wagner, D.: Optimal orthogonal graph drawing with convex bend costs. ACM Trans. Algorithm. **12**(3), 33:1–33:32 (2016). https://doi.org/10.1145/2838736
6. Chan, T.M., Frati, F., Gutwenger, C., Lubiw, A., Mutzel, P., Schaefer, M.: Drawing partially embedded and simultaneously planar graphs. J. Graph Algorithm. Appl. **19**(2), 681–706 (2015). https://doi.org/10.7155/jgaa.00375
7. Chaplick, S., Dorbec, P., Kratochvíl, J., Montassier, M., Stacho, J.: Contact representations of planar graphs: extending a partial representation is hard. In: Kratsch, D., Todinca, I. (eds.) WG 2014. LNCS, vol. 8747, pp. 139–151. Springer, Cham (2014). https://doi.org/10.1007/978-3-319-12340-0_12
8. Chaplick, S., Fulek, R., Klavík, P.: Extending partial representations of circle graphs. J. Graph Theory **91**(4), 365–394 (2019). https://doi.org/10.1002/jgt.22436
9. de Berg, M., Khosravi, A.: Optimal binary space partitions for segments in the place. Int. J. Comput. Geom. Appl. **22**(3), 187–205 (2012)
10. Di Battista, G., Liotta, G., Vargiu, F.: Spirality and optimal orthogonal drawings. SIAM J. Comput. **27**(6), 1764–1811 (1998). https://doi.org/10.1137/S0097539794262847
11. Didimo, W., Liotta, G.: Computing orthogonal drawings in a variable embedding setting. In: Chwa, K.-Y., Ibarra, O.H. (eds.) ISAAC 1998. LNCS, vol. 1533, pp. 80–89. Springer, Heidelberg (1998). https://doi.org/10.1007/3-540-49381-6_10
12. Didimo, W., Liotta, G., Ortali, G., Patrignani, M.: Optimal orthogonal drawings of planar 3-graphs in linear time. In: Chawla, S. (ed.) Proceedings of the 30th ACM-SIAM Symposium on Discrete Algorithms (SODA'20), pp. 806–825. SIAM (2020). https://doi.org/10.1137/1.9781611975994.49
13. Duncan, C.A., Goodrich, M.T.: Planar orthogonal and polyline drawing algorithms. In: Tamassia, R. (ed.) Handbook on Graph Drawing and Visualization, pp. 223–246. Chapman and Hall/CRC (2013)
14. Eiben, E., Ganian, R., Hamm, T., Klute, F., Nöllenburg, M.: Extending partial 1-planar drawings. In: Czumaj, A., Dawar, A., Merelli, E. (eds.) 47th International Colloquium on Automata, Languages, and Programming (ICALP 2020). Leibniz International Proceedings in Informatics (LIPIcs), vol. 168, pp. 43:1–43:19. Schloss Dagstuhl-Leibniz-Zentrum für Informatik, Dagstuhl, Germany (2020). https://doi.org/10.4230/LIPIcs.ICALP.2020.43, https://drops.dagstuhl.de/opus/volltexte/2020/12450

15. Eppstein, D.: Graph-theoretic solutions to computational geometry problems. In: Paul, C., Habib, M. (eds.) WG 2009. LNCS, vol. 5911, pp. 1–16. Springer, Heidelberg (2010). https://doi.org/10.1007/978-3-642-11409-0_1

16. Fößmeier, U., Kaufmann, M.: Drawing high degree graphs with low bend numbers. In: Brandenburg, F.J. (ed.) GD 1995. LNCS, vol. 1027, pp. 254–266. Springer, Heidelberg (1996). https://doi.org/10.1007/BFb0021809

17. Garg, A., Tamassia, R.: On the computational complexity of upward and rectilinear planarity testing. SIAM J. Comput. 31(2), 601–625 (2001). https://doi.org/10.1137/S0097539794277123

18. Jelínek, V., Kratochvíl, J., Rutter, I.: A Kuratowski-type theorem for planarity of partially embedded graphs. Comput. Geom. 46(4), 466–492 (2013). https://doi.org/10.1016/j.comgeo.2012.07.005

19. Klavík, P., et al.: Extending partial representations of proper and unit interval graphs. Algorithmica 77(4), 1071–1104 (2016). https://doi.org/10.1007/s00453-016-0133-z

20. Klavík, P., Kratochvíl, J., Otachi, Y., Saitoh, T.: Extending partial representations of subclasses of chordal graphs. Theor. Comput. Sci. 576, 85–101 (2015). https://doi.org/10.1016/j.tcs.2015.02.007

21. Klavík, P., Kratochvíl, J., Otachi, Y., Saitoh, T., Vyskočil, T.: Extending partial representations of interval graphs. Algorithmica 78(3), 945–967 (2016). https://doi.org/10.1007/s00453-016-0186-z

22. Knuth, D.E., Raghunathan, A.: The problem of compatible representatives. SIAM J. Discret. Math. 5(3), 422–427 (1992). https://doi.org/10.1137/0405033

23. Krawczyk, T., Walczak, B.: Extending partial representations of trapezoid graphs. In: Bodlaender, H.L., Woeginger, G.J. (eds.) WG 2017. LNCS, vol. 10520, pp. 358–371. Springer, Cham (2017). https://doi.org/10.1007/978-3-319-68705-6_27

24. Pach, J., Wenger, R.: Embedding planar graphs at fixed vertex locations. Graphs Comb. 17, 717–728 (2001)

25. Patrignani, M.: On extending a partial straight-line drawing. Int. J. Found. Comput. Sci. 17(5), 1061–1070 (2006). https://doi.org/10.1142/S0129054106004261

26. Tamassia, R.: On embedding a graph in the grid with the minimum number of bends. J. Comput. 16(3), 421–444 (1987)

Topological Constraints

Topological Drawings Meet Classical Theorems from Convex Geometry

Helena Bergold[1], Stefan Felsner[2], Manfred Scheucher[1,2](✉), Felix Schröder[2], and Raphael Steiner[2]

[1] Fakultät für Mathematik und Informatik, FernUniversität in Hagen, Hagen, Germany
helena.bergold@fernuni-hagen.de
[2] Institut für Mathematik, Technische Universität Berlin, Berlin, Germany
{felsner,scheucher,fschroed,steiner}@math.tu-berlin.de

Abstract. In this article we discuss classical theorems from Convex Geometry in the context of topological drawings. In a simple topological drawing of the complete graph K_n, any two edges share at most one point: either a common vertex or a point where they cross. Triangles of simple topological drawings can be viewed as convex sets. This gives a link to convex geometry.

We present a generalization of Kirchberger's Theorem, a family of simple topological drawings with arbitrarily large Helly number, and a new proof of a topological generalization of Carathéodory's Theorem in the plane. We also discuss further classical theorems from Convex Geometry in the context of simple topological drawings.

We introduce "generalized signotopes" as a generalization of simple topological drawings. As indicated by the name they are a generalization of signotopes, a structure studied in the context of encodings for arrangements of pseudolines.

Keywords: Topological drawing · Kirchberger's Theorem · Carathéodory's Theorem · Helly's Theorem · Convexity hierarchy · Generalized signotope

1 Introduction

A point set in the plane (in general position) induces a straight-line drawing of the complete graph K_n. In this article we investigate simple topological drawings of K_n and use the triangles of such drawings to generalize and study classical

Raphael Steiner was funded by DFG-GRK 2434. Stefan Felsner and Manfred Scheucher were partially supported by DFG Grant FE 340/12-1. Manfred Scheucher was partially supported by the internal research funding "Post-Doc-Funding" from Technische Universität Berlin. We thank Alan Arroyo, Emo Welzl, Heiko Harborth, and Geza Tóth for helpful discussions. A special thanks goes to Patrick Schnider for his simplification of the construction in the proof of Proposition 3.

D. Auber and P. Valtr (Eds.): GD 2020, LNCS 12590, pp. 281–294, 2020.
https://doi.org/10.1007/978-3-030-68766-3_22

Fig. 1. Forbidden patterns in topological drawings: self-crossings, double-crossings, touchings, and crossings of adjacent edges.

problems from the convex geometry of point sets. Since we only deal with simple topological drawings we omit the attribute *simple* and define a *topological drawing* D of K_n as follows:

▶ vertices are mapped to distinct points in the plane,
▶ edges are mapped to simple curves connecting the two corresponding vertices and containing no other vertices, and
▶ every pair of edges has at most one common point, which is either a common vertex or a crossing (but not a touching).

Figure 1 illustrates the forbidden patterns for topological drawings. Moreover, we assume throughout the article that no three or more edges cross in a single point. Topological drawings are also known as "good drawings" or "simple drawings".

In this article, we discuss classical theorems such as Kirchberger's, Helly's, and Carathéodory's Theorem in terms of the *convexity* hierarchy of topological drawings introduced by Arroyo, McQuillan, Richter, and Salazar [5], which we introduce in Sect. 2. In that section, we also introduce *generalized signotopes*, a combinatorial generalization of topological drawings. Our proof of a generalization of Kirchberger's Theorem in Sect. 3 makes use of this structure. Section 4 deals with a generalization of Carathéodory's Theorem. In Sect. 5, we present a family of topological drawings with arbitrarily large Helly number. We conclude this article with Sect. 6, where we discuss some open problems.

2   Preliminaries

Let D be a topological drawing and v a vertex of D. The cyclic order π_v of incident edges around v is called the *rotation* of v in D. The collection of rotations of all vertices is called the *rotation system* of D. Two topological drawings are *weakly isomorphic* if there is an isomorphism of the underlying abstract graphs which preserves the rotation system or reverses all rotations.

A triangular cell, which has no vertex on its boundary, is bounded by three edges. By moving one of these edges across the intersection of the two other edges, one obtains a weakly isomorphic drawing; see Fig. 2. This operation is called *triangle-flip*. Gioan [21], see also Arroyo et al. [6], showed that any two weakly isomorphic drawings of the complete graph can be transformed into each other with a sequence of triangle-flips and at most one reflection of the drawing.

Besides weak isomorphism, there is also the notion of strong isomorphism: two topological drawings are called *strongly isomorphic* if they induce homeomorphic cell decompositions of the sphere. Every two strongly isomorphic drawings are also weakly isomorphic.

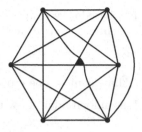 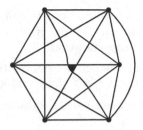

Fig. 2. Two weakly isomorphic drawings of K_6, which can be transformed into each other by a triangle-flip.

Convexity Hierarchy. Given a topological drawing D, we call the induced sub-drawing of three vertices a *triangle*. Note that the edges of a triangle in a topological drawing do not cross. The removal of a triangle separates the plane into two connected components – a bounded component and an unbounded component. We call the closure of these connected components *sides*. A side of a triangle is *convex* if every edge that has its two end-vertices in the side is completely drawn in the side. We are now ready to introduce the "convexity hierarchy" of Arroyo et al. [5]). For $1 \le i < j \le 5$, drawings with property (j) also have property (i).

(1) topological drawings;
(2) *convex* drawings: each triangle has a convex side;
(3) *hereditary-convex* drawings: if a triangle $\triangle_1$ is fully contained in the convex side of another triangle $\triangle_2$, then also its convex side is;
(4) *face-convex* drawings: there is a special face f_∞ such that, for every triangle, the side not containing f_∞ is convex;
(5) *pseudolinear* drawings: all edges of the drawing can be extended to bi-infinite curves – called *pseudolines* – such that any two cross at most once[1];
(6) *straight-line* drawings: all edges are drawn as straight-line segments connecting their endpoints.

Arroyo et al. [7] showed that the face-convex drawings where the special face f_∞ is drawn as the unbounded outer face are precisely the pseudolinear drawings (see also [4] and [2]).

Pseudolinear drawings are generalized by pseudocircular drawings. A drawing is called *pseudocircular* if the edges can be extended to pseudocircles (simple closed curves) such that any pair of non-disjoint pseudocircles has exactly two crossings. Since stereographic projections preserve (pseudo)circles, pseudocircularity is a property of drawings on the sphere.

Pseudocircular drawings were studied in a recent article by Arroyo, Richter, and Sunohara [8]. They provided an example of a topological drawing which is

---

[1] Arrangements of pseudolines obtained by such extensions are equivalent to *pseudo-configurations of points*, and can be considered as oriented matroids of rank 3 (cf. Chapter 5.3 of [17]).

not pseudocircular. Moreover, they proved that hereditary-convex drawings are precisely the *pseudospherical* drawings, i.e., pseudocircular drawings with the additional two properties that

▶ every pair of pseudocircles intersects, and
▶ for any two edges $e \neq f$ the pseudocircle γ_e has at most one crossing with f.

The relation between convex drawings and pseudocircular drawings remains open.

Convexity, hereditary-convexity, and face-convexity are properties of the weak isomorphism classes. To see this, note that the existence of a convex side is not affected by changing the outer face or by transferring the drawing to the sphere, moreover, convex sides are not affected by triangle-flips. Hence, these properties only depend on the rotation system of the drawing. For pseudolinear and straight-line drawings, however, the choice of the outer face plays an essential role.

Generalized Signotopes

Let D be a topological drawing of a complete graph in the plane. Assign an *orientation* $\chi(abc) \in \{+,-\}$ to each ordered triple abc of vertices. The sign $\chi(abc)$ indicates whether we go counterclockwise or clockwise around the triangle if we traverse the edges $(a,b),(b,c),(c,a)$ in this order.

If D is a straight-line drawing of K_n, then the underlying point set $S = \{s_1,\ldots,s_n\}$ has to be in general position (no three points lie on a line). Assuming that the points are sorted from left to right, then for every 4-tuple s_i, s_j, s_k, s_l with $i < j < k < l$ the sequence $\chi(ijk), \chi(ijl), \chi(ikl), \chi(jkl)$ (index-triples in lexicographic order) is *monotone*, i.e., there is at most one sign-change. A *signotope* is a mapping $\chi : \binom{[n]}{3} \to \{+,-\}$ with the above monotonicity property, where $[n] = \{1,2,\ldots,n\}$. Signotopes are in bijection with Euclidean pseudoline arrangements [19] and can be used to characterize pseudolinear drawings [11, Theorem 3.2].

When considering topological drawings of the complete graph we have no left to right order of the vertices, i.e., no natural labeling. Exchanging the labels of two vertices reverts the orientation of all triangles containing both vertices. This suggests to look at the *alternating* extension of χ. Formally $\chi(i_{\sigma(1)}, i_{\sigma(2)}, i_{\sigma(3)}) = \mathrm{sgn}(\sigma) \cdot \chi(i_1, i_2, i_3)$ for any distinct labels i_1, i_2, i_3 and any permutation $\sigma \in S_3$. This yields a mapping $\chi : [n]_3 \to \{+,-\}$, where $[n]_3$ denotes the set of all triples (a,b,c) with pairwise distinct $a,b,c \in [n]$. To see whether the alternating extension of χ still has a property comparable to the monotonicity of signotopes, we have to look at 4-tuples of vertices, i.e., at drawings of K_4. On the sphere there are two types of drawings of K_4: type-I has one crossing and type-II has no crossing. Type-I can be drawn in two different ways in the plane: in type-I$_a$ the crossing is only incident to bounded faces and in type-I$_b$ the crossing lies on the outer face; see Fig. 3.

A drawing of K_4 with vertices a,b,c,d can be characterized in terms of the sequence of orientations $\chi(abc), \chi(abd), \chi(acd), \chi(bcd)$. The drawing is

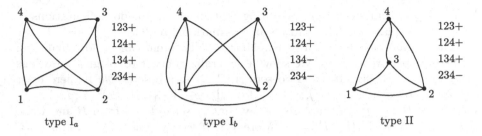

Fig. 3. The three types of topological drawings of K_4 in the plane.

▶ of type-I_a or type-I_b iff the sequence is $++++$, $++--$, $+--+$, $-++-$, $--++$, or $----$; and
▶ of type-II iff the number of $+$'s (and $-$'s respectively) in the sequence is odd.

Therefore there are at most two sign-changes in the sequence $\chi(abc), \chi(abd)$, $\chi(acd), \chi(bcd)$ and, moreover, any such sequence is in fact induced by a topological drawing of K_4. Allowing up to two sign-changes is equivalent to forbidding the two patterns $+-+-$ and $-+-+$.

If χ is alternating and avoids the two patterns $+-+-$ and $-+-+$ on sorted indices, i.e., $\chi(ijk), \chi(ijl), \chi(ikl), \chi(jkl)$ has at most two sign-changes for all $i < j < k < l$, then it avoids the two patterns in $\chi(abc), \chi(abd), \chi(acd), \chi(bcd)$ for any pairwise distinct $a, b, c, d \in [n]$. We refer to this as the *symmetry property* of the forbidden patterns.

The symmetry property allows us to define *generalized signotopes* as alternating mappings $\chi \colon [n]_3 \to \{+, -\}$ with at most two sign-changes on $\chi(abc), \chi(abd)$, $\chi(acd), \chi(bcd)$ for any pairwise different $a, b, c, d \in [n]$. We conclude:

Proposition 1. *Every topological drawing of K_n induces a generalized signotope on n elements.*

3 Kirchberger's Theorem

Two closed sets $A, B \subseteq \mathbb{R}^d$ are called *separable* if there exists a hyperplane H separating them, i.e., $A \subset H_1$ and $B \subset H_2$ with H_1, H_2 being the two closed half-spaces defined by H. It is well-known that, if two non-empty compact sets A, B are separable, then they can also be separated by a hyperplane H containing points of A and B. *Kirchberger's Theorem* (see [30] or [15]) asserts that two finite point sets $A, B \subseteq \mathbb{R}^d$ are separable if and only if for every $C \subseteq A \cup B$ with $|C| = d + 2$, $C \cap A$ and $C \cap B$ are separable.

Goodman and Pollack [23] proved duals of Kirchberger's Theorem and further theorems like Radon's, Helly's, and Carathéodory's Theorem for arrangements of pseudolines. Their results also transfer to pseudoconfigurations of points and thus to pseudolinear drawings. To be more precise, they proved a natural generalization of Kirchberger's Theorem to pseudoline-arrangements in the plane which, by duality, is equivalent to a separating statement on pseudoconfigurations of points in the plane (cf. Theorem 4.8 and Remark 5.2 in [23]).

The 2-dimensional version of Kirchberger's Theorem can be formulated in terms of triple orientations. We show a generalization for topological drawings using generalized signotopes. Two sets $A, B \subseteq [n]$ are *separable* if there exist $i, j \in A \cup B$ such that $\chi(i, j, x) = +$ for all $x \in A \setminus \{i, j\}$ and $\chi(i, j, x) = -$ for all $x \in B \setminus \{i, j\}$. In this case we say that ij *separates* A from B and write $\chi(i, j, A) = +$ and $\chi(i, j, B) = -$. Moreover, if we can find $i \in A$ and $j \in B$, we say that A and B are *strongly separable*. As an example, consider the 4-element generalized signotope of the type-I_b drawing of K_4 in Fig. 3. The sets $A = \{1, 2\}$ and $B = \{3, 4\}$ are strongly separable with $i = 2$ and $j = 3$ because $\chi(2, 3, 1) = +$ and $\chi(2, 3, 4) = -$.

Theorem 1 (Kirchberger's Theorem for Generalized Signotopes). *Let* $\chi : [n]_3 \to \{+, -\}$ *be a generalized signotope, and let* $A, B \subseteq [n]$ *be two non-empty sets. If for every* $C \subseteq A \cup B$ *with* $|C| = 4$*, the sets* $A \cap C$ *and* $B \cap C$ *are separable, then* A *and* B *are strongly separable.*

Note that, since every topological drawing yields a generalized signotope, Theorem 1 generalizes Kirchberger's Theorem to topological drawings of complete graphs. We remark that also a stronger version of the converse of the theorem is true: If A and B are separable, then for every $C \subseteq A \cup B$ with $|C| = 4$, the sets $A \cap C$ and $B \cap C$ are separable.

Proof. First, an elaborate case distinction, which we defer to [16], shows that all 4-tuples $C \subseteq A \cup B$ with $C \cap A$ and $C \cap B$ non-empty which are separable are also strongly separable. Hence in the following we assume that all such 4-tuples from $A \cup B$ are strongly separable.

By symmetry we may assume $|A| \le |B|$. First we consider the cases $|A| = 1, 2, 3$ individually and then the case $|A| \ge 4$.

Let $A = \{a\}$, let B' be a maximal subset of B such that B' is strongly separated from $\{a\}$, and let $b \in B'$ be such that $\chi(a, b, B') = -$. Suppose that $B' \ne B$, then there is a $b^* \in B \setminus B'$ with

$$\chi(a, b, b^*) = +. \tag{1}$$

By maximality of B' we cannot use the pair a, b^* for a strong separation of $\{a\}$ and $B' \cup \{b^*\}$. Hence, for some $b' \in B'$:

$$\chi(a, b^*, b') = +. \tag{2}$$

Since χ is alternating (1) and (2) together imply $b' \ne b$. Since $b' \in B'$ we have $\chi(a, b, b') = -$. From this together with (1) and (2) it follows that the four-element set $\{a, b, b', b^*\}$ has no separator. This is a contradiction, whence $B' = B$.

As a consequence we obtain:

▶ Every one-element set $\{a\}$ with $a \in A$ can be strongly separated from B. Since χ is alternating there is a unique $b(a) \in B$ such that $\chi(a, b(a), B) = -$.

Now consider the case that $A = \{a_1, a_2\}$. Let $b_i = b(a_i)$, i.e., $\chi(a_i, b_i, B) = -$ for $i = 1, 2$. If $\chi(a_1, b_1, a_2) = +$ or if $\chi(a_2, b_2, a_1) = +$, then $a_1 b_1$ or $a_2 b_2$, respectively, is a strong separator for A and B. Therefore, we may assume that $\chi(a_1, b_1, a_2) = -$, $\chi(a_2, b_2, a_1) = -$ and therefore $b_1 \neq b_2$. We get the sequence $+--+$ for the four-element set $\{a_1, a_2, b_1, b_2\}$ which has no strong separator, a contradiction.

The case $|A| = 3$ works similarly but is more technical. A proof of this case is given in [16].

For the remaining case $|A| \geq 4$ consider a counterexample (χ, A, B) minimizing the size of the smaller of the two sets. We have $4 \leq |A| \leq |B|$.

Let $a^* \in A$. By minimality $A' = A \backslash \{a^*\}$ is separable from B. Let $a \in A'$ and $b \in B$ such that $\chi(a, b, A') = +$ and $\chi(a, b, B) = -$. Hence

$$\chi(a, b, a^*) = -. \tag{3}$$

Let $b^* = b(a^*)$, i.e., $\chi(a^*, b^*, B) = -$. There is some $a' \in A'$ such that

$$\chi(a^*, b^*, a') = -. \tag{4}$$

If $a' = a$, then $b \neq b^*$ because of (3) and (4). From (3), (4), $\chi(a, b, B) = -$, and $\chi(a^*, b^*, B) = -$ it follows that the four-element set $\{a, a^*, b, b^*\}$ has the sign pattern $+--+$, hence there is no separator. This shows that $a' \neq a$.

Let $b' = b(a')$. If $b \neq b'$ we look at the four elements $\{a, b, a', b'\}$. It corresponds to $+-*-$ so that we can conclude $\chi(a, a', b') = -$. If $b = b'$, then $a' \in A'$ implies $\chi(a, b, a') = +$ which yields $\chi(a', b', a) = -$.

Hence, regardless whether $b = b'$ or $b \neq b'$ we have

$$\chi(a', b', a) = -. \tag{5}$$

Since $|A| \geq 4$, we know by the minimality of the instance (χ, A, B) that the set $\{a, b, a', b', a^*, b^*\}$, which has 3 elements of A and at least 1 element of B, is separable. It follows from $\chi(a, b, B) = \chi(a', b', B) = \chi(a^*, b^*, B) = -$ that the only possible strong separators are ab, $a'b'$, and a^*b^*. They, however, do not separate because of (3), (4) and (5) respectively. This contradiction shows that there is no counterexample. □

4   Carathéodory's Theorem

Carathéodory's Theorem asserts that, if a point x lies in the convex hull of a point set P in $\mathbb{R}^d$, then x lies in the convex hull of at most $d + 1$ points of P.

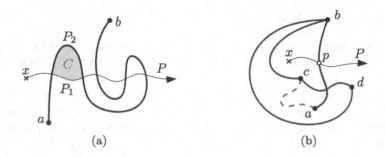

Fig. 4. (a) and (b) give an illustration of the proof of Theorem 2.

As already mentioned in Sect. 3, Goodman and Pollack [23] proved a dual of Carathéodory's Theorem, which transfers to pseudolinear drawings.

A more general version of Carathéodory's Theorem in the plane is due to Balko, Fulek, and Kynčl, who provided a generalization to topological drawings. In this section, we present a shorter proof for their theorem.

Theorem 2 (Carathéodory for Topological Drawings [11, Lemma 4.7]).
Let D be a topological drawing of K_n and let $x \in \mathbb{R}^2$ be a point contained in a bounded connected component of $\mathbb{R}^2 - D$. Then there is a triangle in D that contains x in its interior.

Proof. Suppose towards a contradiction that there is a pair (D, x) violating the claim. We choose D minimal with respect to the number of vertices n.

Let a be a vertex of the drawing. If we remove all incident edges of a from D, then, by minimality of the example, x becomes a point of the outer face. Therefore, if we remove the incident edges of a one by one, we find a last subdrawing D' such that x is still in a bounded face. Let ab be the edge such that in the drawing $D' - ab$ the point x is in the outer face.

There is a simple curve P connecting x to infinity, which does not cross any of the edges in $D' - ab$. By the choice of D', curve P has at least one crossing with ab. We choose P minimal with respect to the number of crossings with ab.

We claim that P intersects ab exactly once. Suppose that P crosses ab more than once. Then there is a *lense* C formed by P and ab, that is, two crossings of P and ab such that the simple closed curve ∂C, composed of a subcurve P_1 of P and a part P_2 of edge ab between the crossings, encloses a simply connected region C, see Fig. 4(a).

Now consider the curve P' from x to infinity which is obtained from P by replacing the subcurve P_1 by a curve P_2' which is a close copy of P_2 in the sense that it has the same crossing pattern with all edges in D and the same topological properties, but is disjoint from ab. As P was chosen minimal with respect to the number of crossings with ab, there has to be an edge of the drawing D' that intersects P_2' (and by the choice of P_2' also P_2). This edge has no crossing with P, by construction, and crosses ab at most once, so it has one of its endpoints inside the lense C and one outside C. Depending on whether

$b \in C$ or not, we choose an endpoint c_1 of that edge such that the edge bc_1 in D' intersects ∂C. But since they are adjacent, bc_1 cannot intersect ab and by the choice of P it does not intersect P. The contradiction shows that P crosses ab in a unique point p.

If a has another neighbor c_2 in the drawing D' then, since only edges incident to a have been removed there is an edge connecting b to c_2 in D'. The edges ac_2 and bc_2 do not cross P, so x is in the interior of the triangle abc_2 and we are done.

If there is no edge ac_2 in D', then $\deg(a) = 1$ in D'. As x is not in the outer face of D', there must be an edge cd in D' which intersects the partial segment of the edge ab starting in a and ending in p, in its interior. Let c be the point on the same side of ab as x; see Fig. 4(b). The edges bc and bd of D' cross neither P nor ab. Consequently, the triangle bcd (drawn blue) must contain a in its interior. We claim that the edge ac in the original drawing D (drawn red dashed) lies completely inside the triangle bcd: The bounded region defined by the edges ab, cd, and bd of D' contains a and c. Since D is a topological drawing, and ac has no crossing with ab and cd, ac has no crossing with bd. This proves the claim. Now the curve P does not intersect ac, and the only edge of the triangle abc intersected by P is ab. Therefore, x lies in the interior of the triangle abc. This contradicts the assumption that (D, x) is a counterexample. $\qquad\square$

Colorful Carathéodory Theorem

Bárány [13] generalized Carathéodory's Theorem as follows: Given finite point sets $P_0, \ldots, P_d$ from $\mathbb{R}^d$ such that there is a point $x \in \operatorname{conv}(P_0) \cap \ldots \cap \operatorname{conv}(P_d)$, then x lies in a simplex spanned by $p_0 \in P_0, \ldots, p_d \in P_d$. Such a simplex is called *colorful*. The theorem is known as the *Colorful Carathéodory Theorem*.

A strengthening, known as the *Strong Colorful Carathéodory Theorem*, was shown by Holmsen, Pach, and Tverberg [26] (cf. [27]): It is sufficient if there is a point x with $x \in \operatorname{conv}(P_i \cup P_j)$ for all $i \neq j$, to find a colorful simplex. The Strong Colorful Carathéodory Theorem was further generalized to oriented matroids by Holmsen [25]. In particular, the theorem applies to pseudolinear drawings (which are in correspondence with oriented matroids of rank 3).

There are several ways to prove Colorful Carathéodory Theorem for pseudo-linear drawings. Besides Holmsen's proof [25], which uses sophisticated methods from topology, we have also convinced ourselves that Bárány's proof [13] can be adapted to pseudoconfigurations of points in the plane. However, Bárány's proof idea does not directly generalize to higher dimensions because oriented matroids of higher ranks do not necessarily have a representation in terms of pseudoconfigurations of points in d-space (cf. [17, Chapter 1.4]).

Another way to prove the Strong Colorful Carathéodory Theorem for pseudolinear drawings is by computer assistance: Since the statement of the theorem only involves 10 points and only the relative positions play a role (not the actual coordinates), one can verify the theorem by checking all combinatorially different point configurations using the order type database (cf. [1] and [33, Section 6.1]). Alternatively, one can – similar as in [34] – formulate a SAT instance that models

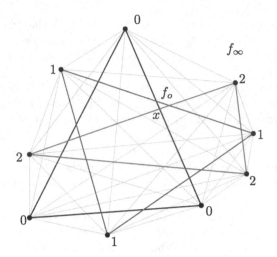

Fig. 5. A face-convex drawings of K_9. If the cell f_o is chosen as the outer face, then Colorful Carathéodory Theorem does not hold for the colored triangles and x. The special cell of the pseudolinear drawing is marked f_∞.

the statement of the Strong Colorful Carathéodory Theorem. Using modern SAT solvers one can then verify that there is no 10-point configuration that violates the theorem.

The following result shows that in the convexity hierarchy of topological drawings of K_n the Colorful Carathéodory Theorem is not valid beyond the class of pseudolinear drawings.

Proposition 2. *The Colorful Carathéodory Theorem does not hold for the face-convex drawing of Fig. 5.*

Proof. The drawing depicted in Fig. 5 is face-convex because it is obtained from a straight-line drawing by choosing f_o as outer face. The point x is contained in the three colored triangles. This point is separated from the outer face only by three colored edges. Therefore, there is no triangle containing x with a vertex of each of the three colors.   □

5   Helly's Theorem

The *Helly number* of a family of sets $\mathcal{F}$ with empty intersection is the size of the smallest subfamily of $\mathcal{F}$ with empty intersection. *Helly's Theorem* asserts that the Helly number of a family of n convex sets $S_1, \ldots, S_n$ from $\mathbb{R}^d$ is at most $d+1$, i.e., the intersection of $S_1, \ldots, S_n$ is non-empty if the intersection of every $d+1$ of these sets is non-empty.

In the following we discuss the Helly number in the context of topological drawings, where the sets S_i are triangles of the drawing.

From the results of Goodman and Pollack [23] it follows that Helly's Theorem generalizes to pseudoconfigurations of points in two dimensions, and thus for pseudolinear drawings. A more general version of Helly's Theorem was shown by Bachem and Wanka [9]. They prove Helly's and Radon's Theorem for oriented matroids with the "intersection property". Since all oriented matroids of rank 3 have the intersection property (cf. [9] and [10]) and oriented matroids of rank 3 correspond to pseudoconfigurations of points, which in turn yield pseudolinear drawings, the two theorems are valid for pseudolinear drawings.

We show that Helly's Theorem does not hold for face-convex drawings, moreover, the Helly number can be arbitrarily large in face-convex drawings. Note that the following proposition does not contradict the Topological Helly Theorem [24] (cf. [22]) because there are triangles whose intersection is disconnected.

Proposition 3. *Helly's Theorem does not generalize to face-convex drawings. Moreover, for every integer $n \geq 3$, there exists a face-convex drawing of K_{3n} with Helly number at least n, i.e., there are n triangles such that for any $n-1$ of the triangles, their bounded sides have a common interior point, but the intersection of the bounded sides of all n triangles is empty.*

Proof. Consider a straight-line drawing D of K_{3n} with n triangles T_i as shown for the case $n = 7$ in Fig. 6. With D' we denote the drawing obtained from D by making the gray cell f_o the outer face. Let O_i be the side of ∂T_i that is bounded in D'. For $1 \leq i < n$ the set O_i corresponds to the outside of ∂T_i in D while O_n corresponds to the inside of ∂T_n.

In D' we have $\bigcap_{i=1}^{n-1} O_i \neq \emptyset$, indeed any point p_n which belongs to the outer face of D is in this intersection. Since $T_n \subset \bigcup_{i=1}^{n-1} T_i$, we have $T_n \cap \bigcap_{i=1}^{n-1} O_i = \emptyset$, i.e., $\bigcap_{i=1}^{n} O_i = \emptyset$. For each $i \in \{1, \ldots, n-1\}$ there is a point $p_i \in T_i \cap T_n$ which is not contained in any other T_j. Therefore, $p_i \in \bigcap_{j=1; j \neq i}^{n} O_i$.

In summary, the intersection of any $n-1$ of the n sets $O_1, \ldots, O_n$ is non-empty but the intersection of all of them is empty. $\square$

6 Discussion

We conclude this article with three further classical theorems from Convex Geometry.

Lovász (cf. Bárány [13]) generalized Helly' Theorem as follows: Let $\mathcal{C}_0, \ldots, \mathcal{C}_d$ be families of compact convex sets from $\mathbb{R}^d$ such that for every "colorful" choice of sets $C_0 \in \mathcal{C}_0, \ldots, C_d \in \mathcal{C}_d$ the intersection $C_0 \cap \ldots \cap C_d$ is non-empty. Then, for some k, the intersection $\bigcap \mathcal{C}_k$ is non-empty. This result is known as the *Colorful Helly Theorem*. Kalai and Meshulam [28] presented a topological version of the Colorful Helly Theorem, which, in particular, carries over to pseudolinear drawings. Since Helly's Theorem does not generalize to face-convex drawings (cf. Proposition 3), neither does the Colorful Helly Theorem.

The (p, q)-*Theorem* (conjectured by Hadwiger and Debrunner, proved by Alon and Kleitman [3], cf. [29]) says that for any $p \geq q \geq d + 1$ there is a

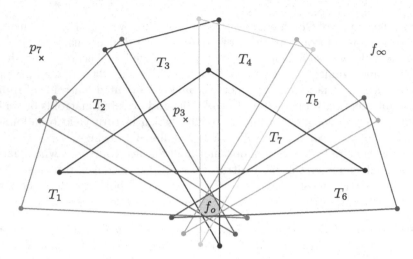

Fig. 6. A drawing D of K_{21} is obtained by adding the remaining edges as straight-line segments. Making the gray cell f_o the outer face, we obtain a face-convex drawing with Helly number 7.

finite number $c(p, q, d)$ with the following property: If C is a family of convex sets in $\mathbb{R}^d$, with the property that among any p of them, there are q that have a common point, then there are $c(p, q, d)$ points that cover all the sets in C. The case $p = q = d+1$ is Helly's Theorem, i.e., $c(d+1, d+1, d) = 1$. A (p, q)-Theorem for triangles in topological drawings can be derived from [18, Theorem 4.6]:

Theorem 3. *For $p \geq q \geq 2$, there exists a finite number $\tilde{c}(p, q)$ such that, if T is a family of triangles of a topological drawing and among any p members of T there are q that have a common point, then there are $\tilde{c}(p, q)$ points that cover all the triangles of T.*

Last but not least, we would like to mention *Tverberg's Theorem*, which asserts that every set V of at least $(d+1)(r-1)+1$ points in $\mathbb{R}^d$ can be partitioned into $V = V_1 \,\dot{\cup}\, \ldots \,\dot{\cup}\, V_r$ such that $\mathrm{conv}(V_1) \cap \ldots \cap \mathrm{conv}(V_r)$ is non-empty. A generalization of Tverberg's Theorem applies to pseudolinear drawings [32] and to drawings of K_{3r-2} if r is a prime-power [31] (cf. [12]). Also a generalization of Birch's Theorem, a weaker version of Tverberg's Theorem, was recently proven for topological drawings of complete graphs [20]. The general case, however, remains unknown. For a recent survey on generalizations of Tverberg's Theorem, we refer to [14].

In future work, we study the structure of generalized signotopes in more detail. There we show that the number of generalized signotopes on n elements is of order $2^{\Theta(n^3)}$, and deduce that most of them are not induced by a topological drawing.

References

1. Aichholzer, O., Aurenhammer, F., Krasser, H.: Enumerating order types for small point sets with applications. Order **19**(3), 265–281 (2002). https://doi.org/10. 1023/A:1021231927255
2. Aichholzer, O., Hackl, T., Pilz, A., Salazar, G., Vogtenhuber, B.: Deciding monotonicity of good drawings of the complete graph. In: Proceedings of XVI Spanish Meeting on Computational Geometry (EGC 2015), pp. 33–36 (2015). http://www. ist.tugraz.at/cpgg/downloadables/ahpsv-dmgdc-15.pdf
3. Alon, N., Kleitman, D.J.: Piercing convex sets and the Hadwiger-Debrunner (p, q)-problem. Adv. Math. **96**(1), 103–112 (1992). https://doi.org/10.1016/0001-8708(92)90052-M
4. Arroyo, A., Bensmail, J., Richter, R.B.: Extending drawings of graphs to arrangements of pseudolines. In: 36th International Symposium on Computational Geometry (SoCG 2020). LIPIcs, vol. 164, pp. 9:1–9:14. Schloss Dagstuhl (2020). https:// doi.org/10.4230/LIPIcs.SoCG.2020.9
5. Arroyo, A., McQuillan, D., Richter, R.B., Salazar, G.: Convex drawings of the complete graph: topology meets geometry (2017). arXiv:1712.06380
6. Arroyo, A., McQuillan, D., Richter, R.B., Salazar, G.: Drawings of K_n with the same rotation scheme are the same up to Reidemeister moves (Gioan's Theorem). Aust. J. Comb. **67**, 131–144 (2017). http://ajc.maths.uq.edu.au/pdf/67/ ajcv67p131.pdf
7. Arroyo, A., McQuillan, D., Richter, R.B., Salazar, G.: Levi's Lemma, pseudolinear drawings of K_n, and empty triangles. J. Graph Theory **87**(4), 443–459 (2018). https://doi.org/10.1002/jgt.22167
8. Arroyo, A., Richter, R.B., Sunohara, M.: Extending drawings of complete graphs into arrangements of pseudocircles (2020). arXiv:2001.06053
9. Bachem, A., Wanka, A.: Separation theorems for oriented matroids. Discret. Math. **70**(3), 303–310 (1988). https://doi.org/10.1016/0012-365X(88)90006-4
10. Bachem, A., Wanka, A.: Euclidean intersection properties. J. Comb. Theory, Ser. B **47**(1), 10–19 (1989). https://doi.org/10.1016/0095-8956(89)90061-0
11. Balko, M., Fulek, R., Kynčl, J.: Crossing numbers and combinatorial characterization of Monotone Drawings of K_n. Discret. Comput. Geom. **53**(1), 107–143 (2014). https://doi.org/10.1007/s00454-014-9644-z
12. Bárány, I., Shlosman, S.B., Szücs, A.: On a topological generalization of a theorem of Tverberg. J. Lond. Math. Soc. **s2–23**(1), 158–164 (1981). https://doi.org/10. 1112/jlms/s2-23.1.158
13. Bárány, I.: A generalization of Carathéodory's theorem. Discret. Math. **40**(2), 141–152 (1982). https://doi.org/10.1016/0012-365X(82)90115-7
14. Bárány, I., Soberón, P.: Tverberg's theorem is 50 years old: a survey. Bull. Amer. Math. Soc. **55**, 459–492 (2018). https://doi.org/10.1090/bull/1634
15. Barvinok, A.: A Course in Convexity, Graduate Studies in Mathematics, vol. 54. American Mathematical Society (2002). https://doi.org/10.1090/gsm/054
16. Bergold, H., Felsner, S., Scheucher, M., Schröder, F., Steiner, R.: Topological Drawings meet Classical Theorems from Convex Geometry (2020). arXiv:2005.12568
17. Björner, A., Las Vergnas, M., White, N., Sturmfels, B., Ziegler, G.M.: Oriented Matroids, Encyclopedia of Mathematics and Its Applications, vol. 46. Cambridge University Press, 2 edn. (1999). https://doi.org/10.1017/CBO9780511586507
18. Chan, T.M., Har-Peled, S.: Approximation algorithms for maximum independent set of pseudo-disks. Discret. Comput. Geom. **48**(2), 373–392 (2012). https://doi. org/10.1007/s00454-012-9417-5

19. Felsner, S., Weil, H.: Sweeps, arrangements and signotopes. Discret. Appl. Math. **109**(1), 67–94 (2001). https://doi.org/10.1016/S0166-218X(00)00232-8
20. Frick, F., Soberón, P.: The topological Tverberg problem beyond prime powers (2020). arXiv:2005.05251
21. Gioan, E.: Complete graph drawings up to triangle mutations. In: Kratsch, D. (ed.) WG 2005. LNCS, vol. 3787, pp. 139–150. Springer, Heidelberg (2005). https://doi.org/10.1007/11604686_13
22. Goaoc, X., Paták, P., Patáková, Z., Tancer, M., Wagner, U.: Bounding Helly numbers via Betti numbers. In: Loebl, M., Nešetřil, J., Thomas, R. (eds.) A Journey Through Discrete Mathematics, pp. 407–447. Springer, Cham (2017). https://doi.org/10.1007/978-3-319-44479-6_17
23. Goodman, J.E., Pollack, R.: Helly-type theorems for pseudoline arrangements in $\mathcal{P}^2$. J. Comb. Theory Ser. A **32**(1), 1–19 (1982). https://doi.org/10.1016/0097-3165(82)90061-9
24. Helly, E.: Über Systeme von abgeschlossenen Mengen mit gemeinschaftlichen Punkten. Monatshefte für Mathematik und Physik **37**(1), 281–302 (1930). https://doi.org/10.1007/BF01696777
25. Holmsen, A.F.: The intersection of a matroid and an oriented matroid. Adv. Math. **290**, 1–14 (2016). https://doi.org/10.1016/j.aim.2015.11.040
26. Holmsen, A.F., Pach, J., Tverberg, H.: Points surrounding the origin. Combinatorica **28**(6), 633–644 (2008). https://doi.org/10.1007/s00493-008-2427-5
27. Kalai, G.: Colorful Caratheodory Revisited (2009). http://gilkalai.wordpress.com/2009/03/15/colorful-caratheodory-revisited
28. Kalai, G., Meshulam, R.: A topological colorful Helly theorem. Adv. Math. **191**(2), 305–311 (2005). https://doi.org/10.1016/j.aim.2004.03.009
29. Keller, C., Smorodinsky, S., Tardos, G.: Improved bounds on the Hadwiger–Debrunner numbers. Isr. J. Math. **225**(2), 925–945 (2018). https://doi.org/10.1007/s11856-018-1685-1
30. Kirchberger, P.: Über Tschebychefsche Annäherungsmethoden. Mathematische Annalen **57**, 509–540 (1903)
31. Özaydin, M.: Equivariant maps for the symmetric group. Unpublished preprint, University of Wisconsin-Madison (1987). http://minds.wisconsin.edu/bitstream/handle/1793/63829/Ozaydin.pdf
32. Roudneff, J.P.: Tverberg-type theorems for pseudoconfigurations of points in the plane. Eur. J. Comb. **9**(2), 189–198 (1988). https://doi.org/10.1016/S0195-6698(88)80046-5
33. Scheucher, M., Schrezenmaier, H., Steiner, R.: A note on universal point sets for planar graphs. J. Graph Algorithm. Appl. **24**(3), 247–267 (2020). https://doi.org/10.7155/jgaa.00529
34. Scheucher, M.: Two disjoint 5-holes in point sets. Comput. Geom. Theory Appl. **91**(101670) (2020). https://doi.org/10.1016/j.comgeo.2020.101670

Towards a Characterization of Stretchable Aligned Graphs

Marcel Radermacher[1](✉), Ignaz Rutter[2], and Peter Stumpf[2]

[1] Department of Informatics, Karlsruhe Institute of Technology (KIT),
Karlsruhe, Germany
radermacher@kit.edu
[2] Faculty of Computer Science and Mathematics, University of Passau,
Passau, Germany
{rutter,stumpf}@fim.uni-passau.de

Abstract. We consider the problem of stretching pseudolines in a planar straight-line drawing to straight lines while preserving the straightness and the combinatorial embedding of the drawing. We answer open questions by Mchedlidze et al. [9] by showing that not all instances with two pseudolines are stretchable. On the positive side, for $k \geq 2$ pseudolines intersecting in a single point, we prove that in case that some edge-pseudoline intersection-patterns are forbidden, all instances are stretchable. For intersection-free pseudoline arrangements we show that every aligned graph has an aligned drawing. This considerably reduces the gap between stretchable and non-stretchable instances.

1 Introduction

Every planar graph $G = (V, E)$ has a straight-line drawing [8,11]. In a restricted setting one seeks a drawing of G that obeys given constraints, e.g., Biedl et al. [1,2] studied whether a bipartite planar graph has a drawing where the two sets of the partitions can be separated by a straight line; refer to Fig. 1a. Da Lozzo et al. [4] generalized this result and characterized the planar graphs with a partition $L \cup R \cup S = V$ of the vertex set that have a planar straight-line drawing such that the vertices in L and R lie left and right of a common line l, respectively, and the vertices in S lie on l. In this case S is called *collinear*. In particular, they showed that S is collinear if and only if there is a drawing of G such that there is an open simple curve $\mathcal{P}$ that starts and ends in the outer face of G, separates L from R, collects all vertices in S and that either entirely contains or intersects at most once each edge. We refer to $\mathcal{P}$ as a *pseudoline with respect to G*.

Dujmovic et al. [5] proved the following surprising result: If S is a collinear set, then for every point set P with $|S| = |P|$ there is a straight-line drawing Γ of G such that S is mapped to P. Another recent research stream considers

Work was partially supported by grant RU 1903/3-1 of the German Research Foundation(DFG).

© Springer Nature Switzerland AG 2020
D. Auber and P. Valtr (Eds.): GD 2020, LNCS 12590, pp. 295–307, 2020.
https://doi.org/10.1007/978-3-030-68766-3_23

the problem of drawing all vertices on as few lines as possible [3]. Eppstein [7] proved that for every integer l there is a cubic planar graph graph G with $O(l^3)$ vertices such that not all vertices of G can lie on l lines.

Mchedlidze et al. [9] generalized the concept of a single pseudoline with respect to an embedded graph to arrangements of pseudolines and introduced the notion of *aligned graphs*, i.e, a pair $(G, \mathcal{A})$ where G is a planar embedded graph and $\mathcal{A} = \{\mathcal{L}_1, \dots, \mathcal{L}_k\}$ is a set of pseudolines $\mathcal{L}_i$ with respect to G that intersect pairwise at most once. We cite the original definition of aligned drawings [9]. A tuple (Γ, A) is an *aligned drawing of* $(G, \mathcal{A})$ if and only if the arrangement of the union of Γ and A has same combinatorial properties as the union of G and $\mathcal{A}$. In the following, we specify these combinatorial properties. Let $A = \{L_1, L_2, \dots, L_k\}$, i.e., line L_i corresponds to pseudoline $\mathcal{L}_i$. A (pseudo)-line arrangement divides the plane into a set of *cells* $\mathcal{C}_1, \mathcal{C}_2, \dots, \mathcal{C}_\ell$. If A is homeomorphic to $\mathcal{A}$, then there is a bijection ϕ between the cells of $\mathcal{A}$ and the cells of A. If (Γ, A) is an aligned drawing of $(G, \mathcal{A})$, then it has the following properties: (i) the arrangement of A is homeomorphic to the arrangement of $\mathcal{A}$, (ii) Γ is a straight-line drawing homeomorphic to the planar embedding of G, (iii) the intersection of each vertex v and each edge e with a cell $\mathcal{C}$ of $\mathcal{A}$ is non-empty if and only if the intersection of v and e with $\phi(\mathcal{C})$ in (Γ, A), respectively, is non-empty, (iv) if an edge uv (directed from u to v) intersects a sequence of cells $\mathcal{C}_1, \mathcal{C}_2, \dots, \mathcal{C}_r$ in this order, then uv intersects in (Γ, A) the cells $\phi(\mathcal{C}_1), \phi(\mathcal{C}_2), \dots, \phi(\mathcal{C}_r)$ in this order, and (v) each line L_i intersects in Γ the same vertices and edges as $\mathcal{L}_i$ in G, and it does so in the same order.

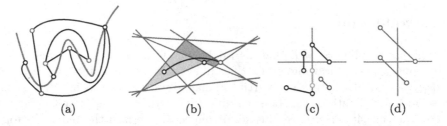

Fig. 1. (a) An aligned graph on one (blue) pseudoline. The color indicates the vertex partition $L \cup R \cup S$. (b) Aligned graph of alignment complexity $(\bot, 3, \bot)$ that does not have an aligned drawing [9]. (c) Allowed types of edges in aligned graphs of alignment complexity $(1, 0, 0)$. The green edge is aligned. The purple edge is free. (c) Aligned graph of alignment complexity $(2, 1, \bot)$. (Color figure online)

Mchedlidze et al. observed that not every aligned graph has an aligned drawing. For example, the modification of the Pappus configuration in Fig. 1b does not have an aligned drawing. Note that one endpoint of the edge is *anchored* on some pseudolines and that the edge *crosses* three pseudolines. Hence, Mchedldize et al. studied a restricted subclass of aligned graphs that only contains edges uv that are either (see Fig. 1c and Fig. 1d)

- *free*, i.e, the entire edge uv is in a single cell,
- *aligned*, i.e., the entire edge uv is on a single pseudoline,
- *one-sided anchored*, i.e., u or v is on a pseudoline but not both, and uv does not cross a pseudoline,
- *1-crossed*, i.e., u and v are in the interior of a cell and uv crosses one pseudoline.

For this restricted class Mchedlidze et al. proved that every aligned graph has an aligned drawing. For this purpose they reduced their instances to aligned graphs that do neither have free edges nor aligned edges nor separating triangles. Then the original instance has an aligned drawing if the reduced instance has an aligned drawing. Thus, the key to success is to characterize the reduced instances and to prove that every reduced instance has an aligned drawing. In the reduced setting, Mchedlidze et al. were able to show that each cell of the pseudoline arrangement contains at-most a single vertex. Since the union of two adjacent cells in the line arrangement is convex, any placement of the vertices that adheres the ordering constraints along the lines induces a valid aligned drawing of the reduced aligned graph. If we additionally allow *two-sided anchored edges*, i.e., edges where both endpoints are on pseudolines but that do not cross a pseudoline, then it is possible to construct a family of aligned graphs such that each cell can contain a number of vertices that is not bounded by the number of pseudolines.

Contribution. We show that every aligned graph on $k \geq 2$ pseudolines intersecting in a single point with free, aligned, one-sided and two-sided anchored, and 1-crossed edges has an aligned drawing. If we allow an additional edge type, we show that there is an aligned graph on two pseudolines that does not have an aligned drawing. Note that in the counterexample given in Fig. 1b, no point in the green cell is visible from the red vertex within the polygon defined by union of the (colored) cells traversed by the edge. Hence, this instance trivially does not admit an aligned drawing. In contrast, each edge in Fig. 3a can be drawn independently as a straight-line segment. We show that the entire instance does not admit a straight-line drawing. Further, we show that every aligned graph $(G, \mathcal{A})$ has an aligned drawing, if $\mathcal{A}$ does not have crossings, i.e., $\mathcal{A}$ corresponds to an arrangement A of parallel lines. This couples aligned graphs to hierarchical (level) graphs. This significantly narrows the gap in the characterization of realizable an non-realizable aligned graphs.

2   Preliminaries

We first introduce some notation used in context of aligned graph on k pseudolines intersecting in a single point. Let $\mathcal{O}$ be a point called the *origin*. Let $\mathcal{X} = \{\mathcal{X}_1, \mathcal{X}_2, \ldots, \mathcal{X}_k\}$ be a pseudoline arrangement where the pseudolines pairwise intersect in $\mathcal{O}$; refer to Fig. 2. We refer to an aligned graph $(G, \mathcal{X})$ as an k-*star aligned graph*. Correspondingly, we refer to (Γ, X), with $X = \{X_1, X_2, \ldots, X_k\}$ as an aligned drawing of $(G, \mathcal{X})$, where the lines in X

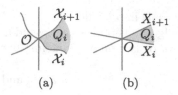

Fig. 2. (a,b) (Pesudo)-line arrangements of a 3-star aligned graph. The green region indicates a cell. (Color figure online)

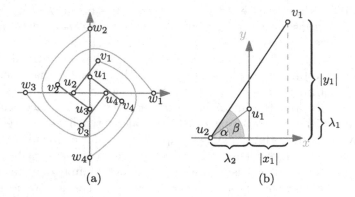

Fig. 3. (a) A 2-aligned graph that does not have an aligned drawing. (b) We have $\lambda_1/\lambda_2 = \tan(\alpha) < \tan(\beta) = |y_1|/(\lambda_2 + |x_1|)$.

pairwise intersect in the *origin* O. The curves in $\mathcal{X}$ divide the plane into a set of cells $\mathcal{Q}_1, \ldots, \mathcal{Q}_{2k}$ in counterclockwise order. These cells naturally correspond to the regions $Q_1, \ldots, Q_{2k}$ bounded by the lines in X.

We refer to an edge (vertex) as *free* if it is entirely in the interior of a cell. An *aligned edge (vertex)* is entirely on a pseudoline. For each *l-crossed* edge e there are l but not $l + 1$ pseudolines that intersect e in its interior. An edge e is *i-anchored* if i of its endpoints lie on i distinct pseudolines. Mchedlidze et al. used a triple (l_0, l_1, l_2), with $l_i \in \mathbb{N} \cup \{\bot\}$ to describe the complexity of an aligned graph $(G, \mathcal{A})$. Let E_i be the set of *i-anchored* edges; note that, the set of edges is the disjoint union $E_0 \uplus E_1 \uplus E_2$. A non-empty edge set $A \subset E$ is *l-crossed* if l is the smallest number such that every edge in A is at most *l-crossed*. An aligned graph $(G, \mathcal{A})$ has alignment complexity (l_0, l_1, l_2), if E_i is at most l_i-crossed or has to be empty, if $l_i = \bot$. In particular, Mchedlidze et al. proved that every aligned graph of alignment complexity $(1, 0, \bot)$ has an aligned drawing. Our results can be restated as that every 2-aligned graph of alignment complexity $(1, 0, 0)$ has an aligned drawing. Further, there is an aligned graph of alignment complexity $(\bot, 1, \bot)$ that does not have an aligned drawing.

3 Star Aligned Graphs

In this section, we study whether k-star aligned graphs have aligned drawings. We first prove that the 2-star aligned graph in Fig. 3a does not have an aligned drawing.

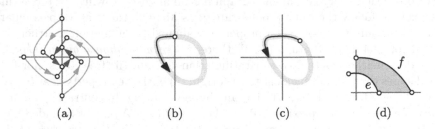

Fig. 4. (a) This 2-aligned graph does not have an aligned drawing. (b,c) The green curve indicates the Jordan curve that completes the black edge. The edge in (b) is an edge of a ccw-aligned graph. The edge depicted in (c) is forbidden in ccw-aligned graphs. (d) A comb of edges e, f. (Color figure online)

Theorem 1. *There is a 2-star aligned graph of alignment complexity $(\perp, 1, \perp)$ that does not have an aligned drawing.*

Proof. Assume that the aligned graph in Fig. 3a has an aligned drawing. For $i = 1, \ldots, 4$ let (x_i, y_i) be the point for v_i, let λ_i be the distance of u_i to the origin O and let $\lambda_5 = \lambda_1$. Since $u_2 v_1$ intersects the y-axis above u_1, edge $u_2 v_1$ has a steeper slope than the segment $u_2 u_1$; see Fig. 3b. We obtain $\lambda_1/\lambda_2 < |y_1|/(\lambda_2 + |x_1|)$ and therefore $|x_1| < \lambda_2/\lambda_1 \cdot |y_1|$. Analogously, we obtain

$$|x_i| < \frac{\lambda_{i+1}}{\lambda_i} \cdot |y_i|, i = 1, 3 \qquad |y_i| < \frac{\lambda_{i+1}}{\lambda_i} \cdot |x_i|, i = 2, 4. \tag{1}$$

Since $v_{i+1} w_i$, with $v_5 = v_1$, are embedded as straight lines, we further get estimation (2) that $|y_i| < |y_{i+1}|$ for $i = 1, 3$ and $|x_i| < |x_{i+1}|$ for $i = 2, 4$ and $x_5 = x_1$. By multiplying the left and the right sides we obtain $|x_1| \cdot |y_2| \cdot |x_3| \cdot |y_4| \overset{(1)}{<} |y_1| \cdot |x_2| \cdot |y_3| \cdot |x_4| \cdot \frac{\lambda_2 \lambda_3 \lambda_4 \lambda_1}{\lambda_1 \lambda_2 \lambda_3 \lambda_4} = |y_1| \cdot |x_2| \cdot |y_3| \cdot |x_4| \overset{(2)}{<} |y_2| \cdot |x_3| \cdot |y_4| \cdot |x_1|$. A contradiction. □

3.1 Aligned Drawings of Counterclockwise Star Aligned Graphs

We now consider aligned drawings of k-star aligned graphs $(G, \mathcal{A})$ for $k \geq 2$. Recall that the aligned graph in Fig. 4a does not have an aligned drawing. The crux is that the source of the red edges are free and the source of green edges are aligned. In the following we introduce so-called *counterclockwise aligned graphs* and show that they have aligned drawings.

We orient each non-aligned edge uv of an aligned graph $(G, \mathcal{X})$ such that it can be extended to a Jordan curve, i.e., a closed simple curve, $\mathcal{C}_{uv}$ with the property that it intersects each pseudoline exactly twice and has the origin to its left. A *counterclockwise aligned (ccw-aligned)* graph is a k-star aligned graph of alignment complexity $(1, 1, 0)$ whose orientation does not contain 1-anchored 1-crossed edges with a free source vertex.

We prove that every ccw-aligned graph has an aligned drawing. To prove this statement we follow the same proof strategy as Mchedlidze et al. In particular, we have to ensure that there is a proper ccw-aligned triangulation. Further, we use that for each aligned graph $(G, \mathcal{X})$ there is a *reduced aligned graph* $(G_R, \mathcal{X})$ (i.e., it does neither contain (i) separating triangles, nor (ii) free edges, nor (iii) aligned edges that are not incident to the origin $\mathcal{O}$) with the property that $(G, \mathcal{X})$ has an aligned drawing if $(G_R, \mathcal{X})$ has an aligned drawing. In contrast to aligned graphs of alignment complexity $(1, 0, \perp)$ the size of $(G_R, \mathcal{X})$ is not bounded by a constant. The aim of Lemma 3 and Lemma 4 is to describe the structure of the reduced instances. This helps to prove Lemma 5 that states that each reduced instance has an aligned drawing.

We first introduce further notations. A k-star aligned graph $(G, \mathcal{X})$ is a *proper k-star aligned triangulation* if each inner face is a triangle, the boundary of the outer face is a $2k$-cycle of 2-anchored edges, the outer face does not contain the origin and there is a degree-$2k$ vertex o on the origin incident to four aligned edges. We refer to a reduced proper ccw-aligned triangulation as a *reduced aligned triangulation*. We refer to 1-anchored 1-crossed and 2-anchored edges as *separating*. The region within a cell that is bounded by two separating edges e and f is an *edge region* (Fig. 4d). An inclusion-minimal edge region is a *comb*.

The following lemma is a consequence from the results by Mchedlitze et al. [9]. For further details we refer to the full version.

Lemma 2. *For every ccw-aligned graph $(G, \mathcal{X})$ there is a reduced aligned triangulation $(G_R, \mathcal{X})$ such that $(G, \mathcal{X})$ has an aligned drawing if $(G_R, \mathcal{X})$ has an aligned drawing.*

Hence, our main contribution is to characterize reduced k-star aligned triangulations and then, to prove that every such instance has an aligned drawing.

Lemma 3. *Let $(G_R, \mathcal{X})$ be a reduced aligned triangulation and let o be the vertex on the origin. Then in $(G_R - o, \mathcal{X})$ each pseudolines $\mathcal{X}_i$ alternately intersect vertices and edges, and each comb contains at most one vertex.*

Proof. Assume that there are two consecutive aligned vertices u and v. Since G is triangulated and u and v are consecutive, G contains the edge uv. This contradicts the assumption that $(G, \mathcal{X})$ does not contain aligned edges.

The following modification helps us to prove that there are no two consecutive edges along a pseudoline and that no comb contains two free vertices.

Let ρ_i be the parts of $\mathcal{X}_i$ and $\mathcal{X}_{i+1}$ that are on the boundary of the cell $\mathcal{Q}_i$, see Fig. 5. We modify ρ_i as follows. We first, join the endpoints of ρ_i in the infinity such that it becomes a simple closed curve. Let u be a vertex that lies on ρ_i.

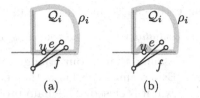

(a) (b)

Fig. 5. The curve ρ_i (a) and its modification in (b).

We reroute ρ_i such that u now lies outside of ρ_i. Since G is triangulated and ρ_i only intersects edges, ρ_i corresponds to a cycle in G^* and therefore to a cut C_i in G. Note, each edge of a connected component in $G - C_i$ is a free edge.

Now assume that there are two distinct edges e, f that consecutively cross a pseudoline $\mathcal{X}_i \in \mathcal{X}$. By the premises of the lemma there is a vertex that lies on the origin $\mathcal{O}$. Hence both e and f cross $\mathcal{X}_i$ on the same side with respect to $\mathcal{O}$. Since e and f are distinct and $(G, \mathcal{X})$ is ccw-aligned, there is a cell $\mathcal{Q}_j$ such that $\mathcal{Q}_j$ contains two distinct vertices u and w incident to e and f, respectively. Since G is triangulated and e and f are consecutive along $\mathcal{X}_i$, u and w are vertices in the same connected component of $G - C_j$. Therefore, $(G, \mathcal{X})$ contains a free edge. A contradiction.

Consider a comb $\mathcal{C}$ in a cell $\mathcal{Q}_i$ that contains two distinct vertices u and v in its interior. Since G is triangulated and $\mathcal{C}$ is inclusion-minimal (it does not contain another edge-region), u and v belong to the same connected component of $G - C_i$. Therefore $(G, \mathcal{X})$ contains a free edge. □

We call a comb *closed* if its two separating edges have the same source vertex.

Lemma 4. *For every reduced aligned triangulation $(G_R, \mathcal{X})$ there is a reduced aligned triangulation $(G''_R, \mathcal{X})$ where no closed comb contains a vertex such that $(G_R, \mathcal{X})$ has an aligned drawing if $(G''_R, \mathcal{X})$ has an aligned drawing.*

Proof. By Lemma 3 we know that each comb contains at most one vertex. We apply induction over the number of closed combs that contain a vertex. Let v be a free vertex in a closed comb with separating edges uw_1, uw_2. Then we obtain an aligned graph $(G'_R, \mathcal{X})$ by contracting edge uv in the embedding. Since $(G_R, \mathcal{X})$ is reduced ccw-aligned, all edges outgoing from the free vertex v are 1-anchored 0-crossed or 0-anchored 1-crossed. In $(G'_R, \mathcal{X})$ they are now 2-anchored 0-crossed or 1-anchored 1-crossed with free target vertex. Since there is no other vertex in the comb and the comb is closed, v only has uv as incoming edge which is contracted. Therefore $(G'_R, \mathcal{X})$ is ccw-aligned. Assume that $(G'_R, \mathcal{X})$ has an aligned drawing. Since v is a free vertex, we obtain an aligned drawing of $(G, \mathcal{X})$ by placing v close to u within in its closed comb. By Lemma 2 we obtain a reduced aligned triangulation $(G''_R, \mathcal{X})$ from $(G'_R, \mathcal{X})$ such that $(G'_R, \mathcal{X})$ has an aligned drawing if $(G''_R, \mathcal{X})$ has an aligned drawing. In the construction the number of closed combs that contain a vertex is not increased. □

We can now show that each reduced instance has an aligned drawing.

Lemma 5. *Every reduced ccw-aligned triangulation has an aligned drawing.*

Proof. By Lemma 4 we can assume that in our triangulation $(G, \mathcal{X})$ the closed combs contain no vertices. By Lemma 3 we know that each comb contains at most one vertex and no vertex if it is closed. The main problem is to draw the 1-crossed edges. For those, we place each free vertex v close to the right boundary of its comb. This allows to draw the incoming edges. Since $(G, \mathcal{X})$ is ccw-aligned, the target of each 1-crossed edge vu is free and allows to draw vu.

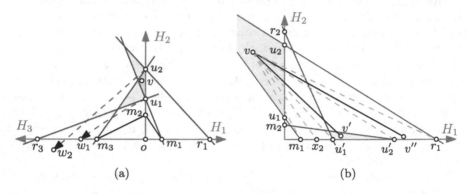

Fig. 6. (a) Placement of a free vertex v in cell $\mathcal{Q}_2$. It may be placed within the gray triangle. (b) Example for the observations with $u_1' = x_3$ and $u_2' = x_4$. (Color figure online)

We construct the aligned drawing (Γ, X) as follows. Let o be the vertex on the origin. We call the sources of separating edges *corners*. First place o and all corners on X in the order induced from $\mathcal{X}$. For $i = 1, \ldots, 2|X|$, let $\mathcal{H}_i$ be the half-pseudoline that is the right boundary of cell $\mathcal{Q}_i$. Let m_i denote the vertex on $\mathcal{H}_i$ that is adjacent to o and let r_i denote the vertex incident to the outer face on $\mathcal{H}_i$. Note that m_i, r_i are corners. We write $u <_i v$ if u lies between o and v on $\mathcal{H}_i$ where u, v may be vertices and intersections of edges with $\mathcal{H}_i$. Note that $<_i$ is a linear order. Define H_i correspondingly for X; see Fig. 7. The indices for m_i, $\mathcal{Q}_i$, etc. are considered mod $2|X|$. In the following, we denote by $\overline{uv}$ the line through two distinct points u, v. Now consider a free vertex v in some cell $\mathcal{Q}_i$; see Fig. 6a. It lies in a comb that is bounded by two separating edges u_1w_1, u_2w_2 with $u_1 <_i u_2$ on $\mathcal{H}_i$. Note that we have $u_1 \neq u_2$ since the comb contains v and is thus not closed. We place v within the triangle bounded by $\overline{m_{i+1}u_2}$, $\overline{r_{i+1}u_1}$, H_i and between $\overline{m_{i-1}u_1}$, $\overline{r_{i-1}u_2}$ (if these lines cross within $\mathcal{Q}_i$, then this means within the triangle bounded by $\overline{m_{i-1}u_1}$, $\overline{r_{i-1}u_2}$, H_i). Note that v lies in $\mathcal{Q}_i$. We will show that the intersections of 1-crossed edges with H_i and the corners on H_i respect the order $<_i$. Finally, we place for $i = 1, \ldots, 2|X|$ the vertices on $\mathcal{H}_i$ that are neither o nor a corner arbitrarily on H_i respecting the order $<_i$. This finishes the construction (edges are placed accordingly).

We next show that the vertices and edges of G appear for $1 \leq i \leq |X|$ along X_i and $\mathcal{X}_i$ in the same order. Consider the free vertex v and the separating edges $u_1 w_1$, $u_2 w_2$ as defined above. Let $m_{i-1} = x_1 <_{i-1} \cdots <_{i-1} x_k = r_{i-1}$ denote the corners on H_{i-1}. The following three observations imply that all 1-crossed edges with target v cross H_i in the correct order between u_1 and u_2; refer to Fig. 6b.

1. $\overline{m_{i-1}v}$ and $\overline{r_{i-1}v}$ cross H_i between u_1 and u_2.
2. $\overline{x_1 v}, \ldots, \overline{x_k v}$ intersect H_i in the same order as $x_1, \ldots, x_k$ lie on $\mathcal{H}_{i-1}$.
3. Let v' be a free vertex in Q_{i-1}. Let $u_1' w_1'$, $u_2' w_2'$ be the separating edges of the comb containing v'. Then $v'v$ crosses H_i between $\overline{u_1' v} \cap H_i$ and $\overline{u_2' v} \cap H_i$.

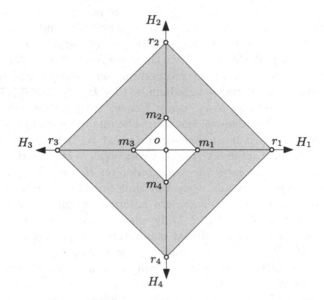

Fig. 7. The vertex o and the half-lines H_i and the vertices m_i, r_i for $i = 1, \ldots, 4$. All remaining edges and vertices lie in the green area. (Color figure online)

For Observation 1, note that v lies between $\overline{m_{i-1} u_1}$, $\overline{r_{i-1} u_2}$. For Observation 2, note that $\overline{x_1 v}, \ldots, \overline{x_k v}$ cross pairwise in v and thus not in section Q_{i-1}. These two observations imply that $\overline{x_1 v}, \ldots, \overline{x_k v}$ cross H_{i-1} between u_1 and u_2. For Observation 3 note now that v' lies in the triangle bounded by H_{i-1}, $\overline{u_2' m_i}$ and $\overline{u_1 r_i'}$. Observation 3 follows from v and this triangle lying between $\overline{u_1 m_{i-1}}$ and $\overline{u_2 r_{i-1}}$.

We now show that all 1-crossed edges with target v cross H_i in the correct order between u_1 and u_2. By Observations 2, 3 the 1-crossed edges with target v cross H_i between $\overline{m_{i-1}v} \cap H_i$ and $\overline{r_{i-1}v} \cap H_i$. With Observation 1, they cross H_i between u_1 and u_2. By Observation 2, we know that the 1-anchored 1-crossed edges with target v cross H_i in the correct order. By Observations 2, 3, we obtain that each pair of a 0-anchored 1-crossed and a 1-anchored 1-crossed edge cross H_i in the correct order. Since the sources of 0-anchored 1-crossed edges with target v lie in different combs, they lie pairwise on different sides of some edge $x_j v$ by Observation 3. Observation 2 then yields their correct ordering.

Since the corners on H_i respect $<_i$ and all 1-crossed edges have free target vertices (as the triangulation is ccw-aligned), this implies that the intersections of 1-crossed edges with H_i and the corners on H_i respect the order $<_i$. By construction, we placed the vertices on $\mathcal{H}_i$ that are not corners such that they also respect order $<_i$. Thus the lines X_j intersect the vertices and edges in the same order as $\mathcal{X}_j$.

We next show that our embedding is planar by showing that there is no location where edges cross. Since the order of intersections with lines in X is correct, there are no crossings on X. This leaves us with the cells. Since the separating edges of $\mathcal{Q}_i$ appear in the same order on $\mathcal{H}_i$ and $\mathcal{H}_{i+1}$, they also appear in the same order on H_i and H_{i+1}. Thus, separating edges of the same cell do not cross each other. We further obtain the same combs for (Γ, XY). Consider again a free vertex v in Q_i and the corresponding separating edges $u_1 w_1$, $u_2 w_2$; see Fig. 6a. Since v lies in the triangle bounded by H_i, T_1 and $\overline{m_{i+1}u_2}$, it also lies in the comb bounded by $u_1 w_1$, $u_2 w_2$. Hence, every free vertex lies in the correct comb. Let e be an edge incident to v. Then its other end vertex does not lie within the comb of v. It must therefore intersect $\mathcal{H}_i$ between u_1 and u_2 if it is incoming, and it must intersect $\mathcal{H}_{i+1}$ between $u_1 w_1 \cap \mathcal{H}_{i+1}$ and $u_2 w_2 \cap \mathcal{H}_{i+1}$ if it is outgoing. Since we have the same order on H_i and H_{i+1} respectively, edge e crosses neither $u_1 w_1$ nor $u_2 w_2$ and thus not the interior of any other comb in Q_i. This means that 1. There are no crossings on separating edges in the corresponding cells. And that 2. Only edges incident to the free vertex v in a comb intersect the interior of that comb. Those edges are all adjacent in v and do not cross. We obtain that there are no crossings on X, no crossings on separating edges in the corresponding cells and no crossings within combs. Hence, our embedding is planar.

Since there are no free edges and the order of intersections with lines in X is fixed, the order of incident edges around a free vertex is also fixed. For a vertex u on X we note that each adjacent free vertex is in another comb and therefore the order of incident edges around u is also fixed. Therefore, our embedding Γ induces the same combinatorial embedding as the embedding of G. □

From Lemma 2 and Lemma 5 we directly obtain our main theorem.

Theorem 6. *Every ccw-aligned graph $(G, \mathcal{X})$ has an aligned drawing.*

4   Parallel Lines

In this section, we prove that every aligned graph $(G, \mathcal{A})$ has an aligned drawing, if $\mathcal{A}$ is intersection free, i.e., the line arrangement A is a set of parallel lines.

Our result uses a result of Eades at al. [6], and of Pach and Toth [10]. Eades et al. consider hierarchical plane graphs. A graph $G = (V, E)$ with a mapping of the vertices to a layer L_i is a *hierarchical graph*, where a set of *layers* $\mathcal{L}$ is a set of ordered parallel horizontal lines $L_i \in \mathcal{L}$. A hierarchical plane drawing of a hierarchical graph is a planar drawing where each vertex is on its desired layer and each edge is drawn as a y-monotone curve. Two hierarchical drawings are *equivalent* if each layer, directed from $-\infty$ to ∞, crosses the same set of edges and vertices in the same order. Eades et. al. [6] proved that for every hierarchical planar drawing of a graph there is an equivalent hierarchical planar straight-line drawing. Pach and Toth [10] proved a similar result stating that for every y-monotone drawing where no two vertices have the same y-coordinate there is an equivalent y-monotone straight-line drawing such that each vertex keeps its y-coordinate. In contrast to these two results, we have that the y-coordinate is only prescribed for a subset of the vertices, i.e., there are some (free) vertices that have to be positioned between two layers (lines). The proof strategy is to extend the initial pseudoline arrangement with an additional set of intersection-free pseudolines such that there are no free vertices.

Due to [9] (compare Lemma 2), we can assume that there are neither free nor aligned edges. For the purpose of this section, a *reduced aligned graph* is an aligned graph that has no aligned edges and no free vertices. Note that previously only free edges were forbidden. Thus, the current definition is more restrictive. The following theorem is an immediate corollary from the results of Eades et al. [6], and Pach and Toth [10].

Theorem 7. *For every intersection-free pseudoline arrangement, every reduced aligned graph $(G, \mathcal{A})$ has an aligned drawing.*

Lemma 8. *Let $\mathcal{A}$ be an intersection-free pseudoline arrangement and let A be a line arrangement homeomorphic to $\mathcal{A}$. For every aligned graph $(G, \mathcal{A})$ there is a reduced aligned graph $(G, \mathcal{A}')$ such that $\mathcal{A} \subset \mathcal{A}'$ and $(G, \mathcal{A})$ has an aligned drawing if $(G, \mathcal{A}')$ has an aligned drawing.*

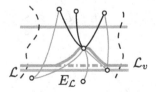

Fig. 8. Construction of the new pseudoline $\mathcal{L}_v$ (red) that contains v. The red-dotted pseudoline L'_v indicates the copy of L (bottom blue) that crossed the edges in E_L (green) in the same order as L (Color figure online)

Proof. We first insert for each free vertex v a new pseudoline $\mathcal{L}_v$ to $\mathcal{A}$ such that v is on $\mathcal{L}$. Thus, the aligned graph $(G, \mathcal{A}')$ does not have free vertices.

Let $\mathcal{L}$ be a pseudoline that is on the boundary the region R_v of $\mathcal{A}$ that contains v. Let $E_{\mathcal{L}}$ be the set of edges of G that are (partially) routed through R_v and that are either crossed by $\mathcal{L}$ or that have an endpoint on $\mathcal{L}$. We assume that $\mathcal{L}$ is directed. Then the direction of $\mathcal{L}$ induces a total order of the edges in $E_{\mathcal{L}}$. We obtain a curve $\mathcal{L}_v'$ that crosses the edges in $E_{\mathcal{L}}$ in this order and in their interior. Since v is free, G is triangulated and $(G, \mathcal{A})$ contains neither free nor aligned edges, there is at-least one edge $e \in E_{\mathcal{L}}$ that is incident to v. Denote by e_f and e_l in $E_{\mathcal{L}}$ the first and last edge incident to v. We obtain a pseudoline $\mathcal{L}_v$ that contains v from $\mathcal{L}_v'$ by rerouting $\mathcal{L}_v'$ along e_f and e_l such that it is does not cross these edges in their interior and such that v is on the line (Fig. 8).

Now, let $(G, \mathcal{A}')$ be the aligned graph that is obtained by the previous procedure for each free vertex v. Let A' be any set of parallel lines that contains A and corresponds to $\mathcal{A}'$. Clearly, (Γ, A) is an aligned drawing of $(G, \mathcal{A})$ if (Γ, A') is an aligned drawing of $(G, \mathcal{A}')$. This finishes the proof.   □

Theorem 7 and Lemma 8 together prove the following theorem.

Theorem 9. *Let $\mathcal{A}$ be an intersection-free pseudoline arrangement and let A be a (parallel) line arrangement homeomorphic to $\mathcal{A}$. Then every aligned graph $(G, \mathcal{A})$ has an aligned drawing (G, A).*

5   Conclusion

In the paper, we showed that every aligned graph $(G, \mathcal{A})$ has an aligned drawing if $(G, \mathcal{A})$ is either a ccw-aligned graph or if $\mathcal{A}$ is intersection-free. Further, we provided a non-trivial example of a 2-star aligned graph that does not admit an aligned drawing. Thus, in our opinion the most intriguing open question is whether every aligned graph of alignment complexity $(1, 0, 0)$ has an aligned

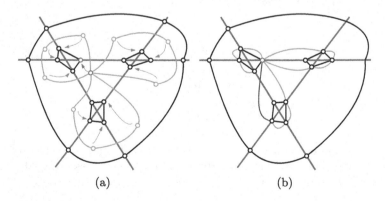

(a)   (b)

Fig. 9. There is no mapping of free vertices to aligned vertices on the boundary of the same cell such that moving the free vertices onto their image results in an aligned graph of alignment complexity $(1, 0, 0)$.

drawing, for general stretchable pseudoline arrangements $\mathcal{A}$. Our counter example shows that this statement is not true for aligned graphs of alignment complexity $(1, 1, 0)$. Our stretchability proof of counterclockwise aligned graphs uses the fact that we can move each free vertex v to an aligned vertex u on the cell of v. Performing this operation for all free vertices at once ensures that we do not introduce edges of a forbidden alignment complexity. Figure 9 indicates that for general aligned graphs of alignment complexity $(1, 0, 0)$ there is not always a consistent mapping of free vertices to aligned vertices such that that the resulting graph has the same alignment complexity. Thus it is unclear whether the techniques used in the paper can be used to decide whether every aligned graph of alignment complexity $(1, 0, 0)$ has an aligned drawing.

References

1. Biedl, T.C.: Drawing planar partitions I: LL-drawings and LH-drawings. In: Proceedings of the 14th Annual Symposium on Computational Geometry (SoCG'98), pp. 287–296. ACM (1998). https://doi.org/10.1145/276884.276917
2. Biedl, T., Kaufmann, M., Mutzel, P.: Drawing planar partitions II: HH-drawings. In: Hromkovič, J., Sýkora, O. (eds.) WG 1998. LNCS, vol. 1517, pp. 124–136. Springer, Heidelberg (1998). https://doi.org/10.1007/10692760_11
3. Chaplick, S., Fleszar, K., Lipp, F., Ravsky, A., Verbitsky, O., Wolff, A.: Drawing graphs on few lines and few planes. In: Hu, Y., Nöllenburg, M. (eds.) GD 2016. LNCS, vol. 9801, pp. 166–180. Springer, Cham (2016). https://doi.org/10.1007/978-3-319-50106-2_14
4. Da Lozzo, G., Dujmovic, V., Frati, F., Mchedlidze, T., Roselli, V.: Drawing planar graphs with many collinear vertices. J. Comput. Geom. **9**(1), 94–130 (2018). https://doi.org/10.20382/jocg.v9i1a4
5. Dujmovic, V., Frati, F., Gonçalves, D., Morin, P., Rote, G.: Every collinear set in a planar graph is free. In: Proceedings of the 30th Annual ACM-SIAM Symposium on Discrete Algorithms (SODA'19), pp. 1521–1538 (2019). https://doi.org/10.1137/1.9781611975482.92
6. Eades, P., Feng, Q., Lin, X., Nagamochi, H.: Straight-line drawing algorithms for hierarchical graphs and clustered graphs. Algorithmica **44**(1), 1–32 (2006). https://doi.org/10.1007/s00453-004-1144-8
7. Eppstein, D.: Cubic planar graphs that cannot be drawn on few lines. CoRR abs/1903.05256 (2019). http://arxiv.org/abs/1903.05256
8. Fáry, I.: On straight line representation of planar graphs. Acta Universitatis Szegediensis **11**, 229–233 (1948). Sectio Scientiarum Mathematicarum
9. Mchedlidze, T., Radermacher, M., Rutter, I.: Aligned drawings of planar graphs. J. Graph Algorithm. Appl. **22**(3), 401–429 (2018). https://doi.org/10.7155/jgaa.00475
10. Pach, J., Tóth, G.: Monotone drawings of planar graphs. J. Graph Theory **46**(1), 39–47 (2004). https://doi.org/10.1002/jgt.10168
11. Tutte, W.T.: How to draw a graph. Proc. Lond. Math. Soc. **s3–13**(1), 743–767 (1963)

Exploring the Design Space of Aesthetics with the Repertory Grid Technique

David Baum[(✉)]

Leipzig University, Grimmaische Straße 12, 04109 Leipzig, Germany
david.baum@uni.leipzig.de

Abstract. By optimizing aesthetics, graph diagrams can be generated that are easier to read and understand. However, the challenge lies in identifying suitable aesthetics. We present a novel approach based on repertory grids to explore the design space of aesthetics systematically. We applied our approach with three independent groups of participants to systematically identify graph aesthetics. In all three cases, we were able to reproduce the aesthetics with positively evaluated influence on readability without any prior knowledge. We also applied our approach to two- and three-dimensional domain-specific software visualizations to demonstrate its versatility. In this case, we were also able to acquire several aesthetics that are relevant for perceiving the visualization.

Keywords: Aesthetics · Graph · Repertory grid technique · Software visualization · Visual analytics

1 Introduction

Making visualizations easier to read and to understand is a challenging task and has been researched for decades [8]. Aesthetics are a suitable method to address this problem [31]. They represent heuristics to predict human perception of the visualization. Aesthetics are visual metrics that must be both objectively measurable and perceptible to the observer [1]. They are independent of the semantic context of a visualization and refer only to visual properties.

For graph layouts consisting of nodes and edges, aesthetics are well researched. Typical aesthetics are, e.g., edge crossings and cutting angles of edges [31]. These criteria are used as optimization goals, e.g., minimizing the number of edge crossings or maximizing the average cutting angle to generate perceivable and comprehensible graph layouts. Aesthetics have been adapted to other visualizations, e.g., different sorts of diagrams [9,34] as well as complex graphical user interfaces such as websites [27]. Each type of visualization has its own aesthetics. Therefore, the state of the art research process has to be repeated for every type of visualization. The process is always similar and comprises the following steps.

1. **Define one or multiple aesthetics.** Every aesthetic must be measurable. There is no established way to derive aesthetics. Many aesthetics are only chosen because they seem to be plausible, so this step is subjective.

© Springer Nature Switzerland AG 2020
D. Auber and P. Valtr (Eds.): GD 2020, LNCS 12590, pp. 308–323, 2020.
https://doi.org/10.1007/978-3-030-68766-3_24

2. **Evaluate impact of proposed aesthetics empirically.** In this step, participants solve tasks using different visualizations, measuring error rate and time to complete the task. It is necessary to be able to trace possible differences in solving the tasks back to different aesthetics. This can be achieved, for example, by changing one aesthetic while keeping all others approximately constant. This is often only possible to a limited extent due to dependencies between different aesthetics.
3. **Implement layout algorithm.** To make the positively evaluated aesthetics usable in practice, it is necessary to provide a suitable layout algorithm. It should have a reasonable runtime behavior and take care of conflicting optimization goals.

The whole process is iterative. Depending on the procedure, step 3 might be performed before step 2. When new aesthetics are defined, the subsequent steps have to be repeated accordingly. However, this approach leads to significant problems. Without being aware of all relevant aesthetics, interactions between them cannot be considered. Unknown but relevant aesthetics might distort the outcome of empirical evaluations significantly [18]. In addition, some aesthetics are not obvious, especially for complex visualizations with many different visual primitives. Hence, there is a risk that important aesthetics may be overlooked. The whole process is very tedious because aesthetics are also defined and examined that have no measurable effect on readability.

In this paper, we want to improve the identification of aesthetics by making the process more reproducible and less based on the researcher's intuition. We use a novel approach based on the repertory grid technique (RGT). This is an interview technique that triggers the participants' creativity to describe verbally the differences between certain elements. These descriptions then serve as a basis for the definition of aesthetics. Therefore, more relevant aesthetics are known when it comes to conducting the evaluations. This will simplify the outlined research process and help to overcome the mentioned problems.

2 Related Work

Several models and guidelines exist for designing and evaluating visualizations [23,24,26]. However, these process models do not use any aesthetics. The only framework known to us that takes aesthetics into account is [22]. It assumes that aesthetics and its effects are already known. Most aesthetics are selected based on intuition without giving an explicit rationale. Bennett et al. [4] justify established aesthetics with Gestalt principles. However, they do not show how new aesthetics can be derived from Gestalt principles.

We are only aware of one approach to improve the iterative process by making it less subjective and more efficient: drawings [28]. The participants are asked to draw visualizations with a given structure, often node-link diagrams. Subsequently, it is examined by statistical means which aesthetics the respondent applied to their drawing. Drawings can help to some extent to weigh aesthetics or identify any irrelevant aesthetics. However, the capabilities of this approach to explore the aesthetics design space are limited. This approach requires well-defined aesthetics to check if they have been used by the subject or not. Also, drawings will not work for complex or three-dimensional visualizations, since most participants will be unable to express their mental model adequate in a drawing of such visualizations.

We see drawings as a step towards improving the described research process. Nevertheless, some problems remain unsolved, which we address within this paper. Our approach is based on the RGT. We are not aware that this method has already been used in the context of aesthetics. In our previous work [1] we used RGT to identify neglected and overemphasized information in visualizations.

3   Repertory Grid Technique

The RGT is an empirical and qualitative research method. Its basic assumption is that everybody describes and evaluates elements based on a large set of personal constructs that can be expressed by using *bipolar constructs* [10, p. 15]. Elements are for example objects, persons, experiences, or even products. A construct is defined as "a way in which two or more things are alike and thereby different from a third or more things" [20, p. 61]. These constructs consist of two opposite poles, e.g., "clear" and "confusing" as well as a construct continuum in between, i.e., different degrees of clarity. The RGT is an approach to make these constructs explicit and visible. The process is reproducible and facilitates the structured exploration of an unknown domain. To apply the RGT, multiple design decisions have to be made, e.g., how elements and constructs are selected. In the following, we will discuss the research design that corresponds to our research questions. We will not discuss variants that are not reasonable for exploring design spaces such as constructs provided by the researcher.

3.1   Element Selection

Every interview is done with the same set of elements. They are selected by the researcher and should represent as much breadth of the domain as possible. The RGT helps to recognize differences between those elements. Something that all elements have in common will most likely not be taken into account by the participants. For example, if all visualizations only consist of black entities, no constructs for color mapping can be expected. The constructs obtained in this way are still valid, but it is possible that they only describe a subset of the domain. This threat can be reduced by asking the subject to provide additional elements that differ from those given [13]. Placeholder elements such as "ideal visualization" or "worst visualization" can also be used to ensure adequate coverage of the domain [11]. The elements can be based on real or artificially generated data.

3.2   Construct Elicitation

The constructs are not predefined. It will be investigated which constructs the participants use to describe the elements shown to them. For this purpose, three elements are randomly selected and presented at once to the participant (cf. 1. The participant has to answer the following question: "How are any two of these alike in some way?", complemented by "What is the opposite of that?" [12]. The answers to both questions are the respective poles. For example, a participant might describe those visualizations with two bipolar constructs, "helpful – unhelpful" and "ugly – beautiful". For them, these are the relevant attributes in which the two visualizations differ. Constructs differ in their level of abstraction. Some constructs are abstract, e.g., "ugly – beautiful", others are very concrete, e.g., "no edge crossings – many edge crossings". Abstract constructs are less helpful for our research question since they are subjective and hard to measure. These abstract constructs might lead to furtheer constructs if they are investigated in depth. It is not uncommon that a construct implies another construct. They only vary in their level of abstraction. The process of using a construct to attain a more concrete construct is called *laddering* and is a common part of the repertory grid interview [11, 13]. This can be done by asking "Why does this visualization appear more beautiful to you?". For example, the answer could lead to the construct "symmetrical – asymmetrical". The whole procedure is repeated by using other randomly selected elements as long as the participant creates new constructs to distinguish between the elements. It is not feasible to use all possible combinations during the interview, hence a reasonable stop criterion is necessary. We advise stopping the interview when three times in a row the participant did not use any new constructs. This will lead to enough constructs and does not prolong the interview unnecessarily. During the interview, the participants have no access to any constructs they used before. Otherwise, participants may try to avoid repetition or use synonyms to find as many constructs as possible. Further, there is no restriction on how many constructs may be named.

The interviewer must understand what the participant describes with a construct. For this reason, informal communication between both persons is a regular and intended part of the RGT. This may include further explanations by the participant, showing examples or simple drawings. The RGT demands high standards of the interviewer and the research design. The interviewer should be familiar with the established guidelines for conducting the interviews. We have mainly followed the recommendation of Kurzhals et al. [21] and Fransella [11].

Normally, a repertory grid interview also includes the creation of the name-giving grids. The participant evaluates for each element and each construct which pole is more appropriate. For us, however, this information is of little value as we are interested in the constructs used. For this reason, we have skipped this step.

3.3   Analysis

The output of the interview is a list of constructs, that has to be further analyzed. Some constructs will represent aesthetics directly, but many constructs are not interesting to us. This is expected and cannot be avoided. Kurzhals et al. propose the following categorization to analyze the constructs of repertory grid interviews to explore the design space [21]:

- **Visual Mapping** This category covers all constructs, that refer to the use of visual primitives (e.g., straight edges – bent edges) and color mapping.
- **Composition** This category consists of constructs that refer to the composition of visualization elements, i.e., layout, alignment, and visual density.
- **Data-related** Constructs are data-related and therefore belong to the third category, if they depend on the underlying data, such as "few nodes – many nodes".
- **Visual experience** The last category describes the hedonistic qualities of visualizations, such as "ugly – beautiful" [15].

For our research question, only the first two categories are interesting since they represent aesthetics. Data-related constructs do not describe the properties of the visualization but of the underlying data. Constructs of the last category are often vague and used as a starting point for laddering during the interview. In the process, more concrete constructs can be revealed that refer to visual mapping or composition. The last step is to reformulate the constructs as aesthetics. "straight edges – bend edges" becomes "edge curve" and so on. This step is straight forward and should not cause any problems. If ambiguities should arise here, the laddering was not sufficiently performed. The final result is a comprehensive list of aesthetics. The method of extraction ensures that all aesthetics are perceivable for human beings. However, there is no guarantee that all of them will have a significant influence on the readability of the visualization.

4   Evaluation

Many graph aesthetics have been proposed, and some of them have been evaluated in empirical studies [6]. We define positively evaluated aesthetics as aesthetics for which a significant influence on readability has already been empirically demonstrated. We applied the RGT to the domain of graph visualization to check the following hypotheses:

- **H1:** With RGT all positively evaluated aesthetics can be reproduced.
- **H2:** The results of RGT can be reproduced when using different elements and different participants.

H1 is used to check whether the RGT provides valid results. With **H2**, we check whether the results are reproducible or depend on the selected elements or participants. We are not aware of any other approach to systematically explore the design space of aesthetics. A comparative evaluation with other approaches is therefore not possible.

4.1 Ground Truth

To verify the results of our evaluation, we have conducted an extensive literature study following the guideline from vom Brocke et al. [39] to establish a ground truth for **H1**. It contains all the aesthetics proposed in the literature and whether a significant influence on readability could be empirically evaluated. We have searched the databases available to us with the search terms listed in Table 1.

Table 1. In- and exclusion criteria for literature study

Database	Search term	Inclusion (+) and Exclusion (–) Criteria
ScienceDirect	graph aesthetics	+ Publication Type: *Research Article*
		+ Journal: *Computer Aided Design*
		+ Journal: *Journal of Visual Languages*
ACM	(+graph +aesthetics)	
IEEE	graph aesthetics	– Publication Type: *Book*
SpringerLink	graph aesthetics	+ Publication Type: Conference Papers
		+ Discipline: *Computer Science*
		+ Subdiscipline: *Information Systems Appl.*
		+ Subdiscipline: *User Interfaces and HCI*

The additional inclusion and exclusion criteria are necessary because the term aesthetics is used in many different disciplines with different meanings. In total, we received 519 hits, 47 from ScienceDirect, 69 from ACM, 42 from IEEE, and 373 from SpringerLink. Two entries had to be removed due to duplicates, leaving 517 entries. We then manually sorted out the publications where the term *aesthetics* is not used in the sense mentioned here. Then, we performed a backward search on the 95 remaining publications. This was necessary because many publications use aesthetics, but it was not the original source in which the metric was proposed. We also included the summaries from Taylor [37] and Bennett [4], who did a similar literature study with a smaller focus. The first three columns of Table 2 summarize the results of our literature study. All in all, we identified 29 different graph aesthetics proposed in 14 different publications. For 13 aesthetics we could find an empirical evaluation that showed a significant influence on readability. For some aesthetics, we were not able to trace them back to exactly one source. In such a case we listed all found publications.

Most aesthetics refer to the position of the nodes, edge intersections, the length and curvature of the edges, and the angles between them. Some aesthetics refer to paths, i.e., combinations of edges. For example, path bendiness describes how straight a path is or how many bends it has. Most aesthetics can be sorted into the "Composition" category since they refer to layouting. Only a few aesthetics belong to the "Visual Mapping" category, they are highlighted in Table 2.

Table 2. List of all aesthetics derived from literature. Entries of the category "Visual Mapping" are highlighted.

Name	Source	Evaluation	Group A	Group B	Group C
Angular resolution	[8, 32, 37]	[18]	4	3	3
Area	[36, 37]	[33]	10	8	8
Aspect ratio	[8]		3	4	3
Cluster similar nodes	[36, 37]	[16]	5	5	4
Convex faces	[36]		–	–	–
Consistent flow direction	[32]		3	4	6
Crossing angle	[17, 18, 40]	[18, 40]	8	9	7
Degree of edge bends	[7, 32, 36]	[30, 31, 33]	9	9	10
Difference between angles	[19]		–	–	–
Distribute nodes evenly	[36, 37]		6	8	8
Edge orthogonality	[32]	[33]	5	4	4
Global symmetry	[5, 36]	[31]	4	3	4
Keep nodes apart from edges	[7]		3	6	7
Local symmetry	[5, 36]	[33]	8	10	8
Maximum bends	[8]		9	9	8
Maximum edge length	[8, 36, 37]		6	4	4
Node orthogonality	[32]		–	3	–
Nodes should not overlap	[35]		4	3	3
Number of bends	[8]		3	3	4
Number of branches	[40]	[40]	5	3	5
Number of edge crossings	[5, 7, 32, 36, 37]	[29–31, 33]	6	3	8
Path bendiness	[40]	[40]	3	3	5
Shortest path length	[40]	[40]	4	3	3
SD of crossing angles	[18]		–	–	–
SD of angular resolution	[18]		–	–	–
Total edge length	[36, 37]		–	–	–
Uniform edge bends	[37]		3	3	4
Uniform edge lengths	[5, 7, 14, 36]		4	3	3
Whitespace to ink ratio	[29, 38]	[29]	3	3	6

4.2   Study Design

Elements. For each group, we used 12 undirected graphs as elements. They can be found in the full version [2]. They consist only of black nodes and black undirected edges. We did not use any text labels or color mappings to keep the graphs as simple as possible. The graphs are not based on real but on artificially generated data. We used the *igraph* library for R[1] to generate random graphs. The smallest graph contains 5 edges, the largest graph contains 69 edges. Each node position was assigned randomly, i.e., overlaps could and did occur.

---

[1] https://igraph.org/r/.

For each edge, the degree and direction of edge curvature were determined randomly as well as which nodes the edge connects. No other properties were taken into account. Figure 1 shows three of the used graphs.

Participants. In total, we interviewed 30 participants. Initially, these participants were divided into three groups to check **H2**. We decided on a group size of 10 because it has proven to be sufficient in many studies. If the method is widely applied, it may be possible to find a convergence point at which additional participants do not add any value. All participants were bachelor or master students of economics and have received an expense allowance. They were all native speakers of German, which was also the language of the interviews. In group A, the students were between 19 and 40 years old (mean: 23.3 years). 50% were female, 50% male. In group B, the students were between 18 and 29 years old (mean: 21.9 years). 40% were female, 60% male. In group C, the students were between 19 and 25 years old (mean: 21.5 years). 60% were female, 40% male. Participants of the same group have worked with the same elements.

Interview. The complete evaluation was done using the evaluation server of Getaviz [3]. It displays three random graphs at the same time (see Fig. 1). The participant cannot interact with the visualizations, i.e., there are no tooltips and it is not possible to navigate or zoom in and out. In the prestudy, we noticed that sometimes rather vague terms such as "simple" or "complex" were used as constructs. To improve the laddering, we asked the participants to draw for instance a "very simple" or "very complex" graph and used it as an additional element. Having additional elements with extreme properties helps the participant to name differences between the elements [20]. Besides that, we conducted the interview as described in the method section.

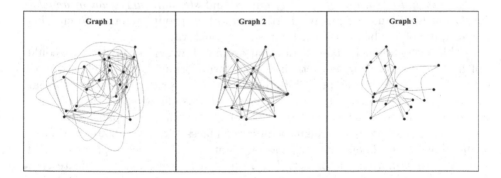

Fig. 1. User interface for repertory grid interview showing three graphs

Table 3. List of all novel aesthetics elicited in the evaluation

Name	Group A	Group B	Group C
Face area	2	3	3
Uniform faces	4	3	3

4.3   Results and Discussion

The interview procedure led to a set of 56 different constructs from all participants. These constructs are divided into the four categories as follows: Visual Mapping (4 constructs), Composition (21 constructs), Data-related (11 constructs), Visual Experience (20 constructs). The distribution of the categories is similar to previous studies but with fewer constructs referring to visual mapping [21]. That was to be expected since the visual mapping was given by using node-link diagrams and corresponds to the distribution of published graph aesthetics, which refer to the composition in most cases as well. The further analysis will focus on the 25 constructs from the first two categories since the other constructs are not relevant concerning aesthetics. For each aesthetic in Table 2 it is indicated which groups have used it. We can fully confirm hypothesis **H1**. An aesthetic was used by 51.7% of the participants on average (min: 33.3%, max: 93.3%) With a softer stop criterion, some aesthetics might have been used by more participants. It is neither necessary nor likely that all participants use identical constructs.

We were able to reproduce all published graph aesthetics that have an empirically verified impact on readability with all three groups. Group A reproduced 82%, Group B reproduced 86%, and Group C reproduced 82% of published graph aesthetics. The five aesthetics not mentioned were not positively evaluated without exception. In the case of *differences between smallest and optimal crossing angle, standard deviation of crossing angles*, and *standard deviation of angular resolution* this is not surprising. Participants of all groups referred to crossing angles quite often, but not in such a mathematical way.

Table 3 lists all elicited aesthetics that are novel, which means that we could not find a corresponding aesthetic in our literature study. Both novel aesthetics refer to faces, i.e., the empty white areas that are bordered by edges. So far in the literature, it has only been suggested to consider whether the faces are convex or concave. This was not relevant for any of the participants. However, participants of all groups distinguished between faces with a small area and faces with a huge area. They also took into account, whether the graph consists of faces with a similar shape or not. The results of our evaluation indicate that the area and shape of the faces might influence how a graph is perceived. It has to be verified empirically whether these aesthetics have a significant impact on understandability and readability.

With one exception, the used aesthetics are consistent among all three groups. Only participants of Group 2 used node orthogonality to differentiate between the elements. Therefore, we can accept hypothesis **H2** conditionally.

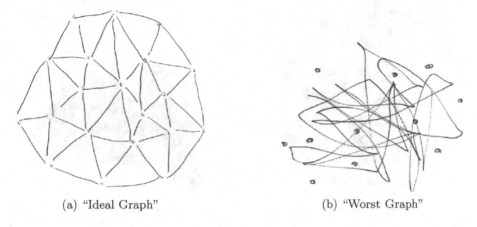

(a) "Ideal Graph"   (b) "Worst Graph"

Fig. 2. Two example graphs drawn by participants

4.4 Threats to Validity

For the interviews, we have specified the elements and deliberately used random values for different properties of the graphs. There is a risk that thereby the aesthetics are predetermined and reflect only our assumptions. However, the participants mentioned aesthetics that have no direct connection to the randomized graph properties. For example, all groups used *global symmetry* as a construct. None of the given elements was symmetric or designed with respect to symmetry. However, many self-drawn graphs were symmetrical as shown in 2a, making them different from the elements provided.

All interviews were conducted by the same person, therefore there is a risk of confirmation bias. Other potential confounding factors are the background and degree of experience of the participants.

5   Application to Software Visualization

Software visualization is a subdomain of information visualization about visualizing the structure, behavior, and evolution of software systems. These visualizations are used in visual analytics tools to support software developers, project managers, and other stakeholders to improve their understanding of development artifacts and corresponding activities. Software visualizations are complex domain-specific diagrams that might contain multiple thousand data points, various relationships between them, and a multitude of different visual primitives. Presenting this amount of information in such a way that it can be processed well by a human being is a central challenge of this domain. The Recursive Disk (RD) Metaphor (Fig. 3a) [25] and the City Metaphor (Fig. 3b) [41] are two approaches to adequately visualize these data.

Both metaphors are hierarchical visualizations that represent the internal structure of a software system. The RD Metaphor is an abstract two-dimensional

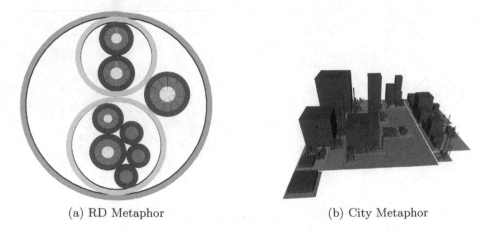

(a) RD Metaphor   (b) City Metaphor

Fig. 3. Software visualizations generated by Getaviz (Color figure online)

metaphor. It consists of two different kinds of disks (gray and purple) as well as two different kinds of disk segments (blue and yellow). The disks can be nested to represent contains-relationships between the elements as shown in Fig. 3a. The area of the disks and disk segments is also used to visualize the properties of the software system. The City metaphor is a three-dimensional real-world metaphor. It consists of gray districts and purple buildings as shown in Fig. 3b. The building's height and base area also represent the properties of the software system.

A high degree of readability and comprehensibility is a central requirement for these kinds of diagrams. To improve them, however, no aesthetics have been considered to date, i.e., there are no known aesthetics at all for this kind of visualization. One of the reasons for this is that the described problems of the current research process are even greater with such complex visualizations. Therefore, we apply our approach to software visualizations to elicit aesthetics that will help improve readability and comprehensibility in the future. We have conducted one study on RD visualizations and one on City visualizations. Both studies are independent of each other. However, since the study design is very similar, we will describe both studies together.

5.1 Study Design

For each study, we used 12 visualizations as elements. We chose 12 different software systems based on software metrics (number of packages, number of classes, number of methods, number of attributes, and number of statements) to cover a wide range. We used Getaviz to generate the corresponding visualizations for each system. For RD, we conducted interviews with ten participants (50% male, 50% female). Their age varies between 19 and 52 years (mean: 22.8 years). For City, we conducted interviews with ten different participants (70% male, 30% female). Their age varies between 18 and 38 years (mean: 24 years). During the construct elicitation we gave participants the possibility to navigate, i.e., rotate the visualization as well as zoom in and out, so they could view the visualization as they liked. Otherwise, it would not be possible to perceive all visual entities since some entities might be occluded or too small to perceive. Apart from that, the interviews were the same as described in Sect. 4.2.

5.2 Results and Discussion

We elicited 53 constructs during the RD interviews, 19 of them qualified as aesthetics. Each participant used 15.5 constructs on average. To describe the city metaphor, 45 constructs were used, 15 of them are aesthetics. Each participant used 13.5 constructs on average. Table 4 lists all elicited aesthetics for both metaphors. The aesthetics for the RD visualizations are mostly about color distribution and nesting, i.e., many aesthetics refer to a local context. This makes sense considering the recursive structure of the visualizations. City visualizations have a similar structure, but buildings are clearly dominant since most aesthetics refer to buildings. Some aesthetics refer to the three-dimensionality of the visualization, where the height of the buildings plays a major role.

Table 4. Elicited aesthetics for RD and city metaphor

RD aesthetics	City aesthetics
Area	Area
Blue segments evenly distributed (global)	Aspect ratio (global)
Blue segments evenly distributed (local)	Aspect ratio of districts (local)
Centered focus	Buildings in a row
Edge thickness	Building density
Face area	Clustering of similar buildings
Global symmetry	Empty district area
Length of spiral windings	Gap between buildings
Local symmetry	Largest difference in building height
Nesting depth	Nesting depth
Number of spiral turns	Share of empty area
Share of empty area	Sort buildings by height
Sorting of purple disks (local)	Uniform base area of buildings
Uniform size of gray disks	Uniform buildings
Uniform size of purple disks	Uniform faces
Uniform structure of gray disks	
Uniform structure of purple disks	
Yellow segments evenly distributed (global)	
Yellow segments evenly distributed (local)	

It is particularly noticeable that fewer constructs were used compared to graphs, both per interview and overall. This is most likely because the edges of the graphs have many degrees of freedom that are not present in the RD and City visualizations. Due to the semantic constraints, e.g., a building must always be located in a district, the design space is not as extensive as it is for graphs.

The elicited aesthetics serve as a starting point to design better layout algorithms. In previous work, only density and area were considered. The elicited aesthetics must now be empirically evaluated to find out which of them have a significant influence on readability.

6   Conclusion

Our approach to explore aesthetics design space using repertory grids has been effective. We have evaluated the approach as far as possible and were able to show in an empirical study that with only 10 participants all published and positively evaluated aesthetics can be identified. We could also show that our approach delivers reproducible results and can be applied to diverse visualizations. The quality and validity of the results depend above all on the selection of the suitable elements. The inclusion of drawings and placeholder elements was particularly helpful. However, the assessment of a domain expert is still necessary to create and select suitable elements. Nevertheless, the process is much less subjective and intuition-based than before.

The analysis of the repertory grid data applied in this paper is rather simple and could be enhanced in the future. For example, we did not analyze how often certain aesthetics have been used by participants. In our future work, we will evaluate the derived aesthetics from software visualizations to further validate the results.

References

1. Baum, D.: Introducing aesthetics to software visualization. In: 23rd International Conference in Central Europe on Computer Graphics, Visualization and Computer Vision, WSCG 2015 - Short Papers Proceedings, vol. 23, pp. 65–73 (2015). http:// wscg.zcu.cz/WSCG2015/!_2015_WSCG_SHORT_proceedings.pdf
2. Baum, D.: Exploring design space of aesthetics with repertory grids. In: 28th International Symposium on Graph Drawing and Network Visualization (GD 2020) (2020). https://arxiv.org/abs/2008.07862
3. Baum, D., Schilbach, J., Kovacs, P., Eisenecker, U., Müller, R.: GETAVIZ: generating structural, behavioral, and evolutionary views of software systems for empirical evaluation. In: Proceedings - 2017 IEEE Working Conference on Software Visualization, IEEE VISSOFT 2017, pp. 114–118 (2017). https://doi.org/10.1109/VISSOFT.2017.12
4. Bennett, C., Ryall, J., Spalteholz, L., Gooch, A.: The aesthetics of graph visualization. In: Proceedings of the 2007 Computational Aesthetics in Graphics, Visualization, and Imaging, pp. 57–64 (2007). https://doi.org/10.2312/COMPAESTH/COMPAESTH07/057-064

5. Biedl, T., Marks, J., Ryall, K., Whitesides, S.: Graph multidrawing: finding nice drawings without defining nice. In: Whitesides, S.H. (ed.) GD 1998. LNCS, vol. 1547, pp. 347–355. Springer, Heidelberg (1998). https://doi.org/10.1007/3-540-37623-2_26

6. Chen, C.: Top 10 unsolved information visualization problems. IEEE Comput. Graph. Appl. **25**(4), 12–16 (2005). https://doi.org/10.1109/MCG.2005.91

7. Davidson, R., Harel, D.: Drawing graphs nicely using simulated annealing. ACM Trans. Graph. **15**(4), 301–331 (1996). https://doi.org/10.1145/234535.234538

8. Brandenburg, F.J. (ed.): GD 1995. LNCS, vol. 1027. Springer, Heidelberg (1996). https://doi.org/10.1007/BFb0021783

9. Effinger, P., Jogsch, N., Seiz, S.: On a study of layout aesthetics for business process models using BPMN. In: Mendling, J., Weidlich, M., Weske, M. (eds.) BPMN 2010. LNBIP, vol. 67, pp. 31–45. Springer, Heidelberg (2010). https://doi.org/10.1007/978-3-642-16298-5_5

10. Fransella, F.: International Handbook of Personal Construct Psychology. John Wiley & Sons Ltd., Chichester (2005). https://doi.org/10.1002/0470013370

11. Fransella, F.: Some skills and tools for personal construct practitioners. In: International Handbook of Personal Construct Psychology, pp. 105–122 (2005)

12. Fransella, F., Neimeyer, R.A.: George Alexander Kelly: the man and his theory. International Handbook of Personal Construct Psychology, pp. 21–31 (2005). https://doi.org/10.1002/0470013370.ch2

13. Fried, R., Mayer, M.F.: Grid technique as tool for improving health services to institutionalized children: a ten-year experience. J. Am. Med. Assoc. **161**(1), 1–5 (1956). https://doi.org/10.1001/jama.1956.02970010003001

14. Gansner, E.R., Koren, Y., North, S.: Graph drawing by stress majorization. In: Pach, J. (ed.) GD 2004. LNCS, vol. 3383, pp. 239–250. Springer, Heidelberg (2005). https://doi.org/10.1007/978-3-540-31843-9_25

15. Hassenzahl, M., Wessler, R.: Capturing design space from a user perspective: the repertory grid technique revisited. Int. J. Hum. Comput. Interact. **12**(3–4), 441–459 (2000)

16. Huang, W., Eades, P., Hong, S.H.: Layout effects: Comparison of sociogram drawing conventions (575) (2005)

17. Huang, W., Eades, P., Hong, S.H.: Larger crossing angles make graphs easier to read. J. Vis. Lang. Comput. **25**(4), 452–465 (2014). https://doi.org/10.1016/j.jvlc.2014.03.001

18. Huang, W., Eadesy, P., Hongy, S.H., Linz, C.C.: Improving force-directed graph drawings by making compromises between aesthetics. In: Proceedings - 2010 IEEE Symposium on Visual Languages and Human-Centric Computing, VL/HCC 2010, pp. 176–183 (2010). https://doi.org/10.1109/VLHCC.2010.32

19. Hutchison, D., Mitchell, J.C.: Graph Drawing, p. 536 (2004). https://doi.org/10.1007/3-540-68339-9_34

20. Kelly, G.: A Theory of Personality: The Psychology of Personal Constructs (1963)

21. Kurzhals, K., Weiskopf, D.: Exploring the visualization design space with repertory grids. Comput. Graph. Forum **37**(3), 133–144 (2018). https://doi.org/10.1111/cgf.13407

22. Lau, A., Moere, A.V.: Towards a model of information aesthetics in information visualization. In: Proceedings of the International Conference on Information Visualisation, pp. 87–92 (2007). https://doi.org/10.1109/IV.2007.114

23. McKenna, S., Mazur, D., Agutter, J., Meyer, M.: Design activity framework for visualization design. IEEE Trans. Visual. Comput. Graphics **20**(12), 2191–2200 (2014). https://doi.org/10.1109/TVCG.2014.2346331

24. Meyer, M., Sedlmair, M., Quinan, P.S., Munzner, T.: The nested blocks and guidelines model. Inf. Visual. **14**(3), 234–249 (2015). https://doi.org/10.1177/1473871613510429

25. Müller, R., Zeckzer, D.: The recursive disk metaphor: a glyph-based approach for software visualization. In: IVAPP 2015–6th International Conference on Information Visualization Theory and Applications. VISIGRAPP, Proceedings, pp. 171–176. SciTePress, Setúbal (2015). https://doi.org/10.5220/0005342701710176

26. Munzner, T.: A nested model for visualization design and validation. IEEE Trans. Visual. Comput. Graphics **15**(6), 921–928 (2009). https://doi.org/10.1109/TVCG.2009.111

27. Pajusalu, M., Torres, R., Lamas, D.: The Evaluation of User Interface Aesthetics, p. 74 (2012)

28. Papadopoulos, C., Voglis, C.: Untangling graphs representing spatial relationships driven by drawing aesthetics. In: ACM International Conference Proceeding Series, pp. 158–165 (2013). https://doi.org/10.1145/2491845.2491853

29. Polisciuc, E., Cruz, A., Machado, P., Arrais, J.P.: On the role of aesthetics in genetic algorithms applied to graph drawing. In: GECCO 2017 - Proceedings of the Genetic and Evolutionary Computation Conference Companion, pp. 1713–1720 (2017). https://doi.org/10.1145/3067695.3082552

30. Purchase, H.C., James, M.I., Cohen, R.F.: An experimental study of the basis for graph drawing algorithms. ACM J. Exp. Algorithm. **2**, 4 (1997). https://doi.org/10.1145/264216.264222

31. Purchase, H.: Which aesthetic has the greatest effect on human understanding? In: DiBattista, G. (ed.) GD 1997. LNCS, vol. 1353, pp. 248–261. Springer, Heidelberg (1997). https://doi.org/10.1007/3-540-63938-1_67

32. Purchase, H.C.: Metrics for graph drawing aesthetics. J. Visual Lang. Comput. **13**(5), 501–516 (2002). https://doi.org/10.1016/S1045-926X(02)90232-6

33. Purchase, H.C., Carrington, D., Allder, J.A.: Empirical evaluation of aesthetics-based graph layout. Empir. Softw. Eng. **7**(3), 233–255 (2002). https://doi.org/10.1023/A:1016344215610

34. Purchase, H.C., Mcgili, M., Colpoys, L., Carrington, D.: Graph drawing aesthetics and the comprehension of UML class diagrams: an empirical study, pp. 129–137 (2001)

35. Shannon, A.: Tidy drawings of trees. IEEE Trans. Softw. Eng. **5**(5), 514–520 (1979). https://doi.org/10.1109/TSE.1979.234212

36. Tamassia, R., Battista, G.D., Batini, C.: Automatic graph drawing and readability of diagrams. IEEE Trans. Syst. Man Cybern. **18**(1), 61–79 (1988). https://doi.org/10.1109/21.87055

37. Taylor, M., Rodgers, P.: Applying graphical design techniques to graph visualisation. In: Proceedings of the International Conference on Information Visualisation, pp. 651–656 (2005). https://doi.org/10.1109/IV.2005.19

38. Tullis, T.S.: Evaluation of alphanumeric, graphic, and color information displays. Hum. Factors **23**(5), 541–550 (1981). https://doi.org/10.1177/001872088102300504

39. Vom Brocke, J., Simons, A., Niehaves, B., Riemer, K., Plattfaut, R., Cleven, A.: Reconstructing the giant: on the importance of rigour in documenting the literature search process. In: 17th European Conference on Information Systems, ECIS 2009 (2009)

40. Ware, C., Purchase, H., Colpoys, L., McGill, M.: Cognitive measurements of graph aesthetics. Inf. Visual. **1**(2), 103–110 (2002). https://doi.org/10.1057/palgrave.ivs. 9500013
41. Wettel, R., Lanza, M.: Code city. In: Proceedings of WASDeTT 2008 (1st International Workshop on Advanced Software Development Tools and Techniques), pp. 1–13 (2008). http://www.inf.usi.ch/phd/wettel/publications.html

Storyline Visualizations with Ubiquitous Actors

Emilio Di Giacomo⬛, Walter Didimo⬛, Giuseppe Liotta⬛,
Fabrizio Montecchiani⬛, and Alessandra Tappini$^{(\boxtimes)}$⬛

Engineering Department, University of Perugia, Perugia, Italy
{emilio.digiacomo,walter.didimo,giuseppe.liotta,fabrizio.montecchiani,
alessandra.tappini}@unipg.it

Abstract. Storyline visualizations depict the temporal dynamics of social interactions, as they describe how groups of actors (individuals or organizations) change over time. A common constraint in storyline visualizations is that an actor cannot belong to two different groups at the same time instant. However, this constraint may be too severe in some application scenarios, thus we generalize the model by allowing an actor to simultaneously belong to distinct groups at any point in time. We call this model *Storyline with Ubiquitous Actors (SUA)*. Essential to our model is that an actor is represented as a tree rather than a single line. We describe an algorithmic pipeline to compute storyline visualizations in the SUA model and discuss case studies on publication data.

Keywords: Storyline visualization · Ubiquitous actors

1 Introduction

Storyline visualizations have been the focus of intense research in the last decade. Originally introduced to describe the narrative of a movie [12], this visualization paradigm has been successfully used to represent the temporal dynamics of the interactions between actors (individuals or organizations) in a social network or in a working environment [10,14,16–20]. In a storyline visualization, the narrative unfolds from left to right, each actor is represented as a line, and two lines may converge or diverge at a time instant based on whether the two corresponding actors interact or not at that instant; see Fig. 1(a). Since a group of lines bundled together usually reflects an in-person meeting, a common constraint in a storyline visualization is that an actor cannot belong to two different groups at the same point in time. However, this constraint represents a severe limitation for some application scenarios, for example when groups model associations that are

---

This work is partially supported by: (*i*) MIUR, grant 20174LF3T8 "AHeAD: efficient Algorithms for HArnessing networked Data", (*ii*) Dipartimento di Ingegneria - Università degli Studi di Perugia, grant RICBA19FM: "Modelli, algoritmi e sistemi per la visualizzazione di grafi e reti".

D. Auber and P. Valtr (Eds.): GD 2020, LNCS 12590, pp. 324–332, 2020.
https://doi.org/10.1007/978-3-030-68766-3_25

not in-person meetings (e.g., paper co-authorships) or when each point in time of the storyline corresponds to a relatively long time interval (e.g., one year).

In this paper we generalize the classical storyline model by allowing an actor to simultaneously belong to distinct groups. We call this model *Storyline with Ubiquitous Actors (SUA)*; see Fig. 1(b). Essential to our model is that an actor is represented as a tree rather than a single line. Our contribution is: (*i*) We propose a visualization paradigm for the SUA model and identify quality metrics for it. (*ii*) We define an algorithmic pipeline for storyline visualizations in the SUA model. (*iii*) We provide a proof-of-concept implementation and apply it to produce visualizations in real-life scenarios.

Related Work. Tanahashi and Ma [18] present a general framework for generating aesthetically pleasing storyline visualizations. Subsequent papers focus on specific optimization problems like crossing minimization [5,7,8] and wiggle minimization [6]. Padia et al. [15,16] consider storyline visualizations with multiple timelines. Efficient approaches that compute storyline visualizations with hierarchical relationships or with streaming data are described by Liu et al. [10] and by Tanahashi et al. [19], respectively. Qiang and Bingjie [17] present a system that embeds storyline visualizations into a radial layout. For a broader dissertation on storytelling and visualization refer to the survey of Tong et al. [20]. We remark that our scenario is strongly related to the dynamic sets visualization; see, e.g., [11,13], and [3] for a survey. In this regard, it is worth mentioning a recent work by Agarwal and Beck [2], who adopt storylines for visualizing dynamic sets.

2 Storyline Visualizations and Ubiquitous Actors

We first recall basic definitions and principles of classical storyline visualizations and then define our visualization for the SUA model.

Classical Storyline Visualizations. A *storyline* $S = (\mathcal{A}, \mathcal{G})$ consists of a set $\mathcal{A} = \{a_1, a_2, \ldots, a_n\}$ of *actors* and a set $\mathcal{G} = \{G_1, G_2, \ldots, G_k\}$ of *groups*. Each group $G_i \in \mathcal{G}$ is a triple $\langle \mathcal{A}(G_i), b_i, e_i \rangle$, where $\mathcal{A}(G_i) \subseteq \mathcal{A}$ is a subset of actors, b_i is the *begin-time* of G_i and e_i is the *end-time* of G_i. We say that G_i is *active* at any time instant in the interval $[b_i, e_i]$, and that each actor $a_j \in \mathcal{A}(G_i)$ *participates* to G_i. A common assumption is that an actor cannot participate to two distinct groups at the same point in time, i.e., if $\mathcal{G}_i$ and $\mathcal{G}_j$ are two distinct groups such that $[b_i, e_i] \cap [b_j, e_j] \neq \emptyset$ then $\mathcal{A}(G_i) \cap \mathcal{A}(G_j) = \emptyset$.

In a *storyline visualization*, each actor a_j is represented as a line ℓ_j that flows from left to right; see Fig. 1(a). Some basic *principles* are considered: (*i*) For each group G_i, the lines representing the actors in $\mathcal{A}(G_i)$ are *adjacent*, i.e., they run close together from the begin-time b_i to the end-time e_i of G_i; (*ii*) lines of actors that are not in the same group at the same time are depicted relatively far from one another; (*iii*) a line should not deviate unless it converges or diverges with another line. In addition, common *quality metrics* for the readability of storyline visualizations are: (a) *Line or block crossings* – a line crossing occurs when two lines intersect while a block crossing is caused by two blocks of parallel lines that pairwise intersect. (b) *Line wiggles* – line deviations that, when frequent, negatively affect the visual flow of the layout. (c) *White space gaps* – white areas used to separate lines of actors that do not participate to the same group.

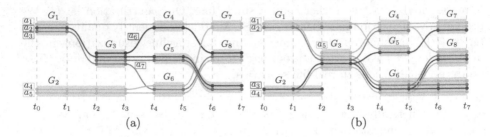

Fig. 1. Storyline visualization: (a) Classical model. (b) SUA model.

Visualizations with Ubiquitous Actors. To support the visualization of ubiquitous actors, we represent an actor a_j as a tree τ_j rather than as a line (see Sect. 3 for a formal definition of τ_j). Informally speaking, when an actor simultaneously participates to different groups, the line of the actor branches out and forms a tree. For example, in Fig. 1(b) we see the trees of five actors $a_1, \ldots, a_5$. At time t_1 actor a_2 participates to group G_1 while at time t_2 it simultaneously participates to groups G_1 and G_3. As a consequence, the line of a_2 at time t_1 is split into two branches. The choice of a tree is motivated by the fact that we want to represent each actor by a connected geometric feature (avoiding discontinuities); at the same time, we want to keep such geometric feature as simple as possible, since the addition of edges may increase the number of crossings. Such tree representations add new quality metrics:

- *Actor planarity.* It is natural to require that each tree representing an actor is not self-intersecting. While this is trivially guaranteed when an actor is a line, it requires an algorithmic effort in the SUA model.
- *Branch continuity.* To avoid interruptions in the continuity of the story, the number of branches of an actor at time t_h that continue at time t_{h+1} should be maximized. If an actor participates to m groups at time t_h and to $m' \geq m$ groups at time t_{h+1}, all branches at time t_h should continue at time t_{h+1}.
- *Branch degree.* When an actor tree needs new branches at some time instant t_h, it is desirable that the maximum number of branches that emanates from a common branch at time t_{h-1} is minimized.

We note that such new metrics may be in conflict with classical ones (see Fig. 2).

3   The SUA Algorithmic Pipeline

We compute storyline visualizations in the SUA model by means of an algorithmic pipeline based on the concept of *actor-tree* τ_j associated with an actor a_j.

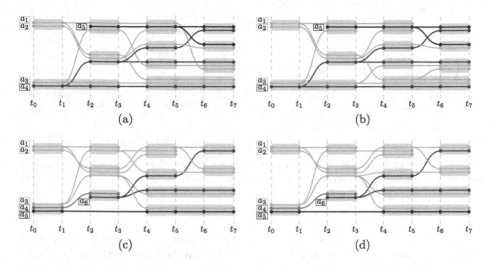

Fig. 2. (a) A layout with optimal branch continuity. (b) Violating branch continuity for a_3 at t_3 and for a_1 at t_5 removes 9 crossings and reduces line wiggles. (c) A layout with optimal branch degree. (d) Violating branch degree for a_3 at t_3 reduces line wiggles and removes 2 crossings.

The *life-time* of actor a_j is the interval between the first and the last time instant at which a_j belongs to some group. Tree τ_j is defined as follows. **Node set** – Tree τ_j has a root r_j. For each time instant t_h in the life-time of a_j: If a_j participates to at least one group G_i $(i > 0)$ active at t_h, τ_j has a node $u_{h,i}$ for each such group; otherwise τ_j has a single node $u_{h,0}$ that is not associated with any group. **Edge set** – The parent of a node $u_{h,i}$ $(i \geq 0)$ is assigned as follows: If t_{h-1} is not in the life-time of a_j, the parent of $u_{h,i}$ is the root r_j. Else, if $i > 0$ and G_i is active before time t_h, the parent of $u_{h,i}$ is $u_{h-1,i}$. Else, the parent of $u_{h,i}$ is one of the nodes $u_{h-1,l}$. Our algorithmic pipeline consists of four steps:

1. **Actor-tree Initialization.** It defines an initial actor-tree τ_j for each actor a_j. Namely, given the nodes of τ_j, it assigns the parent to each node of τ_j.

2. **Branch Permutation.** For each time instant t_h this step computes a permutation (i.e., a vertical order) of all nodes at time t_h in the union of all actor-trees .

3. **Actor-tree Untangling.** For each actor-tree τ_j, it redefines the parent of some nodes, so to reduce self-intersections of τ_j without changing its node degrees.

4. **Branch-coordinate Assignment.** It assigns the y-coordinates to actor-tree nodes.
   In Step 1 we aim to optimize branch continuity and branch degree. Step 2 aims to minimize block or line crossings. Step 3 tries to enforce actor planarity. Step 4 aims to reduce line wiggles and space gaps. Different algorithmic strategies are applicable to each step. We briefly describe our solution; see also Fig. 4.

Actor-Tree Initialization. For any actor-tree τ_j, let V_{h-1} and V_h be the sets of nodes at time instant t_{h-1} and t_h. The parents of the nodes in V_h are chosen

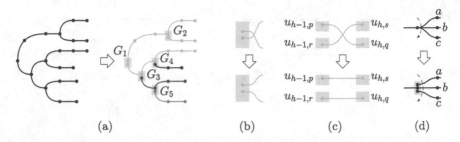

Fig. 3. (a) Transformation of an actor tree into a set of disjoint paths. (b) Crossing removal when merging two copies of the same node. (c) Actor-tree untangling. (d) Preservation of the edge order around a node when splitting it.

among the nodes in V_{h-1} so that the distribution of the degrees in V_{h-1} is as uniform as possible. For each node $u_{h,i}$ in V_h that belongs to the same group of a node $u_{h-1,i}$ in V_{h-1}, the parent of $u_{h,i}$ is $u_{h-1,i}$. For the remaining nodes of V_h, we adopt a round robin policy to assign children to the nodes in V_{h-1} so that the difference of the degrees of any two nodes of V_{h-1} is at most one.

Branch Permutation. We exploit a state-of-the-art algorithm for classical storyline visualizations, namely the algorithm by van Dijk et al. [5] based on a SAT formulation, which optimally solves the problem of minimizing block crossings. To this aim, we transform the output of Step 1 into an instance for a classical storyline visualization: Each actor tree is partitioned into a set of edge-disjoint paths by duplicating each node with $k \geq 2$ children into k nodes each having one child (see Fig. 3(a)). Each path is processed by the algorithm in [5] as a distinct actor. All copies of the same node are then recombined into a single node to restore the tree. However, if disjoint paths originating from two copies of the same node are treated independently, they can create many crossings when recombined back into the tree. To alleviate this drawback, we let the initial node of each path belong to the same group of its original duplicate, unless this operation makes the path belonging to multiple groups at some other point in time. Moreover, when copies of the same node are recombined into a single node, two edges incident to this node may create a crossing, which is easily removed as depicted in Fig. 3(b). Hence, a crossing in an actor tree only involves independent edges.

Actor-Tree Untangling. For each actor-tree τ_j, we redefine the parent of some nodes to reduce the number of crossings between the edges of τ_j. If two edges $(u_{h-1,p}, u_{h,q})$ and $(u_{h-1,r}, u_{h,s})$ ($p \neq q$, $r \neq s$) of τ_j cross, we replace them with two new edges $(u_{h-1,p}, u_{h,s})$ and $(u_{h-1,r}, u_{h,q})$ (see Fig. 3(c)). This operation removes at least one self-intersection and does not create any new one. Also, the degree of $u_{h-1,p}$, $u_{h,q}$, $u_{h-1,r}$, and $u_{h,s}$ does not change. We repeat this procedure until it is no longer possible to remove self-intersections from τ_j.

Branch-Coordinate Assignment. As in the Branch-permutation step, we consider the set of paths that decompose the tree and make them disjoint by duplicating each node with $k \geq 2$ children into k nodes each having one child. The cyclic

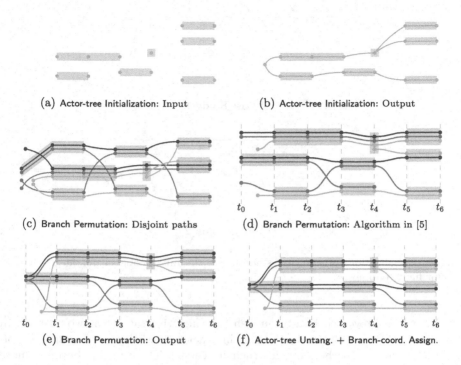

(a) Actor-tree Initialization: Input   (b) Actor-tree Initialization: Output

(c) Branch Permutation: Disjoint paths   (d) Branch Permutation: Algorithm in [5]

(e) Branch Permutation: Output   (f) Actor-tree Untang. + Branch-coord. Assign.

Fig. 4. Illustration of the algorithmic pipeline.

order of the edges around each node defines the permutation of the lines that correspond to these edges (see Fig. 3(d)). Any technique that assigns coordinates to the paths, while reducing line wiggles and white space gaps can be applied (see, e.g., [6,18]). This assignment preserves the vertical permutations of the paths.

4   Implementation and Case Studies

We developed a prototype web application, STORYTREEVIEWER, which implements the algorithmic pipeline of Sect. 3, see https://bit.ly/2yS3Fvi. STORY-TREEVIEWER offers a simple interactive interface, which we used to evaluate effectiveness and limits of our model through two case studies on publication data extracted from DBLP [9] and Scopus [1].

Case Study 1. The first case study, see Fig. 5(a), describes scientific collaborations among the authors of this work in the various editions of the Graph Drawing Symposium (GD) since 1999. Each actor is an author and a group G_i is a subset of actors who co-authored some papers. G_i is active in $[b_i, e_i]$ if all their members co-authored at least one paper in each year from b_i to e_i. The layout reveals the following dynamic. In the first part of the story there is a strong collaboration between the three oldest actors (pink, green, and blue), in particular they form a group lasting from 2004 to 2009. In 2003, the pink actor was the chair

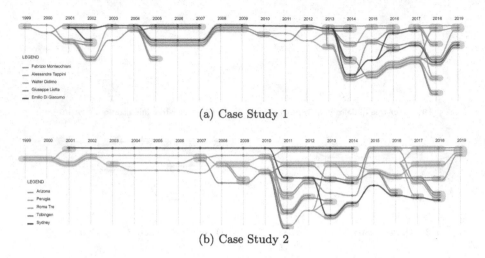

(a) Case Study 1

(b) Case Study 2

Fig. 5. Visualizations of our case studies. See the full version [4] for larger images. (Color figure online)

of GD, which prevented him to publish together with the other two authors. In 2010 and 2011 the collaboration of the three actors is weaker, as they mainly collaborated with researchers outside their university. The dynamic becomes more involved in the last years, when two new members joined the group (cyan and orange), and new theoretical and application research topics were activated.

Case Study 2. The second case study, see Fig. 5(b), describes scientific collaborations among five of the research teams (universities) with the highest number of papers published at GD. The actors are the teams and the groups are defined as in case study 1. Namely, a group G_i is a subset of teams that appear together in some papers (in terms of author affiliations); G_i is active in $[b_i, e_i]$ if all its teams appear together in at least one paper in each year from b_i to e_i. The layout shows some interesting facts. From 1999 to 2002 there is a strong collaboration between Roma Tre and Perugia, witnessing that the group in Perugia stems from researchers coming from Rome. The collaboration between the five research teams increases since 2007 and becomes stronger since 2011. This is partly explained by the series of workshops started around 2006 (BWGD, HOMONOLO, GNV, etc.) that increased international collaborations.

Limits. Working on the case studies, we observed some limits of our approach: (*i*) The implementation for the Branch Permutation step exploits the algorithm in [5], splitting each actor-tree into multiple disjoint paths. The size of this transformed instance raises some computational complexity issues. (*ii*) Our visualizations appear to be readable for relatively few actors and further work is needed to better evaluate the effectiveness of the SUA model on larger instances.

5   Conclusions and Future Work

We introduced the SUA model, which allows ubiquitous actors in storyline visualizations. This model extends the spectrum of applications for this type of representation and opens up to many intriguing research directions. Among them: (*i*) Are there more effective ways of modeling ubiquitous actors other than using trees? (*ii*) Design and experiment different algorithms for the SUA pipeline.

References

1. http://www.scopus.com . Accessed 03 June-2020
2. Agarwal, S., Beck, F.: Set streams: visual exploration of dynamic overlapping sets. Comput. Graph. Forum **39**(3), 383–391 (2020)
3. Alsallakh, B., Micallef, L., Aigner, W., Hauser, H., Miksch, S., Rodgers, P.J.: The state-of-the-art of set visualization. Comput. Graph. Forum **35**(1), 234–260 (2016)
4. Di Giacomo, E., Didimo, W., Liotta, G., Montecchiani, F., Tappini, A.: Storyline visualizations with ubiquitous actors. CoRR abs/2008.04125v2 (2020).http://arxiv.org/abs/2008.04125v2
5. van Dijk, T.C., Lipp, F., Markfelder, P., Wolff, A.: Computing storyline visualizations with few block crossings. In: Frati, F., Ma, K.-L. (eds.) GD 2017. LNCS, vol. 10692, pp. 365–378. Springer, Cham (2018). https://doi.org/10.1007/978-3-319-73915-1_29
6. Fröschl, T., Nöllenburg, M.: Minimizing wiggles in storyline visualizations. In: Frati, F., Ma, K.L. (eds.) Graph Drawing and Network Visualization. pp. 585–587. Springer International Publishing (2018)
7. Gronemann, M., Jünger, M., Liers, F., Mambelli, F.: Crossing minimization in storyline visualization. In: Hu, Y., Nöllenburg, M. (eds.) GD 2016. LNCS, vol. 9801, pp. 367–381. Springer, Cham (2016). https://doi.org/10.1007/978-3-319-50106-2_29
8. Kostitsyna, I., Nöllenburg, M., Polishchuk, V., Schulz, A., Strash, D.: On minimizing crossings in storyline visualizations. In: Di Giacomo, E., Lubiw, A. (eds.) GD 2015. LNCS, vol. 9411, pp. 192–198. Springer, Cham (2015). https://doi.org/10.1007/978-3-319-27261-0_16
9. Ley, M.: The DBLP computer science bibliography. https://dblp.uni-trier.de
10. Liu, S., Wu, Y., Wei, E., Liu, M., Liu, Y.: Storyflow: tracking the evolution of stories. IEEE Trans. Vis. Comput. Graph. **19**(12), 2436–2445 (2013)
11. Mizuno, K., Wu, H., Takahashi, S., Igarashi, T.: Optimizing stepwise animation in dynamic set diagrams. Comput. Graph. Forum **38**(3), 13–24 (2019)
12. Munroe, R.: Xkcd #657: Movie narrative charts, December 2009. http://xkcd.com/657
13. Nguyen, P.H., Xu, K., Walker, R., Wong, B.L.W.: Timesets: timeline visualization with set relations. Inf. Vis. **15**(3), 253–269 (2016)
14. Ogawa, M., Ma, K.: Software evolution storylines. In: Telea, A., Görg, C., Reiss, S.P. (eds.) Proceedings of the ACM 2010 Symposium on Software Visualization, pp. 35–42. ACM (2010)
15. Padia, K., Bandara, K.H., Healey, C.G.: Yarn: generating storyline visualizations using HTN planning. In: Graphics Interface, pp. 26–33. ACM (2018)

16. Padia, K., Bandara, K.H., Healey, C.G.: A system for generating storyline visualizations using hierarchical task network planning. Comput. Graph. **78**, 64–75 (2019)
17. Qiang, L., Chai, B.: Storycake: a hierarchical plot visualization method for storytelling in polar coordinates. In: CW, pp. 211–218. IEEE Computer Society (2016)
18. Tanahashi, Y., Ma, K.: Design considerations for optimizing storyline visualizations. IEEE Trans. Vis. Comput. Graph. **18**(12), 2679–2688 (2012)
19. Tanahashi, Y., Hsueh, C., Ma, K.: An efficient framework for generating storyline visualizations from streaming data. IEEE Trans. Vis. Comput. Graph. **21**(6), 730–742 (2015)
20. Tong, C., et al.: Storytelling and visualization: an extended survey. Information **9**(3), 65 (2018)

Drawing Shortest Paths
in Geodetic Graphs

Sabine Cornelsen[1]([✉])[ID], Maximilian Pfister[2][ID], Henry Förster[2][ID],
Martin Gronemann[3][ID], Michael Hoffmann[4][ID], Stephen Kobourov[5][ID],
and Thomas Schneck[2][ID]

[1] University of Konstanz, Konstanz, Germany
`sabine.cornelsen@uni-konstanz.de`
[2] University of Tübingen, Tübingen, Germany
`{pfister,foersth,schneck}@informatik.uni-tuebingen.de`
[3] University of Osnabrück, Osnabrück, Germany
`martin.gronemann@uni-osnabrueck.de`
[4] Department of Computer Science, ETH Zürich, Zürich, Switzerland
`hoffmann@inf.ethz.ch`
[5] Department of Computer Science, University of Arizona, Tucson, USA
`kobourov@cs.arizona.edu`

Abstract. Motivated by the fact that in a space where shortest paths
are unique, no two shortest paths meet twice, we study a question posed
by Greg Bodwin: Given a geodetic graph G, i.e., an unweighted graph in
which the shortest path between any pair of vertices is unique, is there
a *philogeodetic* drawing of G, i.e., a drawing of G in which the curves
of any two shortest paths meet at most once? We answer this question
in the negative by showing the existence of geodetic graphs that require
some pair of shortest paths to cross at least four times. The bound on
the number of crossings is tight for the class of graphs we construct.
Furthermore, we exhibit geodetic graphs of diameter two that do not
admit a philogeodetic drawing.

Keywords: Edge crossings · Unique shortest paths · Geodetic graphs

1 Introduction

Greg Bodwin [1] examined the structure of shortest paths in graphs with edge
weights that guarantee that the shortest path between any pair of vertices is
unique. Motivated by the fact that a set of unique shortest paths is *consistent* in

This research began at the Graph and Network Visualization Workshop 2019 (GNV'19)
in Heiligkreuztal. S. C. is funded by the German Research Foundation DFG – Project-
ID 50974019 – TRR 161 (B06). M. H. is supported by the Swiss National Science
Foundation within the collaborative DACH project *Arrangements and Drawings* as
SNSF Project 200021E-171681. S. K. is supported by NSF grants CCF-1740858, CCF-
1712119, and DMS-1839274.

D. Auber and P. Valtr (Eds.): GD 2020, LNCS 12590, pp. 333–340, 2020.
https://doi.org/10.1007/978-3-030-68766-3_26

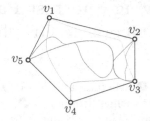

Fig. 1. A drawing of the geodetic graph K_5. It has a crossing formed by edges v_1v_3 and v_2v_5. In addition, edges v_1v_4 and v_2v_4 meet but do not cross since their meet includes vertex v_4. Finally, edges v_2v_5 and v_3v_5 meet twice violating the property of philogeodetic drawings.

the sense that no two such paths can "intersect, split apart, and then intersect again", he conjectured that if the shortest path between any pair of vertices in a graph is unique then the graph can be drawn so that any two shortest paths meet at most once. Formally, a *meet* of two Jordan curves $\gamma_1, \gamma_2 : [0,1] \to \mathbb{R}^2$ is a pair of maximal intervals $I_1, I_2 \subseteq [0,1]$ for which there is a bijection $\iota : I_1 \to I_2$ so that $\gamma_1(x) = \gamma_2(\iota(x))$ for all $x \in I_1$. A *crossing* is a meet with $(I_1 \cup I_2) \cap \{0,1\} = \emptyset$. Two curves *meet k times* if they have k pairwise distinct meets. For example, shortest paths in a simple polygon (geodesic paths) have the property that they meet at most once [6].

A *drawing* of a graph G in $\mathbb{R}^2$ maps the vertices to pairwise distinct points and maps each edge to a Jordan arc between the two end-vertices that is disjoint from any other vertex. Drawings extend in a natural fashion to paths: Let φ be a drawing of G, and let $P = v_1, \ldots, v_n$ be a path in G. Then let $\varphi(P)$ denote the Jordan arc that is obtained as the composition of the curves $\varphi(v_1v_2), \ldots, \varphi(v_{n-1}v_n)$. A drawing φ of a graph G is *philogeodetic* if for every pair P_1, P_2 of shortest paths in G the curves $\varphi(P_1)$ and $\varphi(P_2)$ meet at most once.

An unweighted graph is *geodetic* if there is a unique shortest path between every pair of vertices. Trivial examples of geodetic graphs are trees and complete graphs. Observe that any two shortest paths in a geodetic graph are either disjoint or they intersect in a path. Thus, a planar drawing of a planar geodetic graph is philogeodetic. Also every straight-line drawing of a complete graph is philogeodetic. Refer to Fig. 1 for an illustration of a drawing of a complete graph that is not philogeodetic; this example also highlights some of the concepts discussed above. It is a natural question to ask whether every (geodetic) graph admits a philogeodetic drawing.

Results. We show that there exist geodetic graphs that require some pair of shortest paths to meet at least four times (Theorem 1). The idea is to start with a sufficiently large complete graph and subdivide every edge exactly twice. The Crossing Lemma [8] can be used to show that some pair of shortest paths must cross at least four times. By increasing the number of subdivisions per edge, we can reduce the density and obtain sparse counterexamples. The bound on the number of crossings is tight because any uniformly subdivided K_n can be drawn so that every pair of shortest paths meets at most four times (Theorem 2).

On one hand, our construction yields counterexamples of diameter five. On the other hand, the unique graph of diameter one is the complete graph, which is geodetic and admits a philogeodetic drawing (e.g., any straight-line drawing since all unique shortest paths are single edges). Hence, it is natural to ask what is the largest d so that every geodetic graph of diameter d admits a philogeodetic drawing. We show that $d = 1$ by exhibiting an infinite family of geodetic graphs of diameter two that do not admit philogeodetic drawings (Theorem 3). The construction is based on incidence graphs of finite affine planes. The proof also relies on the crossing lemma.

Geodetic Graphs. Geodetic graphs were introduced by Ore who asked for a characterization as Problem 3 in Chap. 6 of his book "Theory of Graphs" [7, p. 104]. An asterisk flags this problem as a research question, which seems justified, as more than sixty years later a full characterization is still elusive.

Stemple and Watkins [14,15] and Plesník [10] resolved the planar case by showing that a connected planar graph is geodetic if and only if every block is (1) a single edge, (2) an odd cycle, or (3) stems from a K_4 by iteratively choosing a vertex v of the K_4 and subdividing the edges incident to v uniformly. Geodetic graphs of diameter two were fully characterized by Scapellato [12]. They include the Moore graphs [3] and graphs constructed from a generalization of affine planes. Further constructions for geodetic graphs were given by Plesník [10,11], Parthasarathy and Srinvasan [9], and Frasser and Vostrov [2].

Plesník [10] and Stemple [13] proved that a geodetic graph is homeomorphic to a complete graph if and only if it is obtained from a complete graph K_n by iteratively choosing a vertex v of the K_n and subdividing the edges incident to v uniformly. A graph is geodetic if it is obtained from any geodetic graph by uniformly subdividing each edge an even number of times [9,11]. However, the graph G obtained by uniformly subdividing each edge of a complete graph K_n an odd number of times is not geodetic: Let u, v, w be three vertices of K_n and let x be the middle subdivision vertex of the edge uv. Then there are two shortest x-w-paths in G, one containing v and one containing u.

2 Subdivision of a Complete Graph

The complete graph K_n is geodetic and rather dense. However, all shortest paths are very short, as they comprise a single edge only. So despite the large number of edge crossings in any drawing, every pair of shortest paths meets at most once, as witnessed, for instance, by any straight-line drawing of K_n. In order to lengthen the shortest paths it is natural to consider subdivisions of K_n.

As a first attempt, one may want to "take out" some edge uv by subdividing it many times. However, Stemple [13] has shown that in a geodetic graph every path where all internal vertices have degree two must be a shortest path. Thus, it is impossible to take out an edge using subdivisions. So we use a different approach instead, where all edges are subdivided uniformly.

Theorem 1. *There exists an infinite family of sparse geodetic graphs for which in any drawing in $\mathbb{R}^2$ some pair of shortest paths meets at least four times.*

Proof. Take an even number t and a complete graph K_s for some $s \in \mathbb{N}$. Subdivide each edge t times. The resulting graph $K(s,t)$ is geodetic. See Fig. 4 for a drawing of $K(8,2)$. Note that $K(s,t)$ has $n = s + t\binom{s}{2}$ vertices and $m = (t+1)\binom{s}{2}$ edges, with $m \in O(n)$, for s fixed and t sufficiently large. Consider a drawing Γ of $K(s,t)$.

Let B denote the set of s *branch vertices* in $K(s,t)$, which correspond to the vertices of the original K_s. For two distinct vertices $u, v \in B$, let $[uv]$ denote the shortest uv-path in $K(s,t)$, which corresponds to the subdivided edge uv of the underlying K_s. As t is even, the path $[uv]$ consists of $t + 1$ (an odd number of) edges. For every such path $[uv]$, with $u, v \in B$, we charge the crossings in Γ along the $t + 1$ edges of $[uv]$ to one or both of u and v as detailed below; see Fig. 2 for illustration.

- Crossings along an edge that is closer to u than to v are charged to u;
- crossings along an edge that is closer to v than to u are charged to v; *and*
- crossings along the single central edge of $[uv]$ are charged to both u and v.

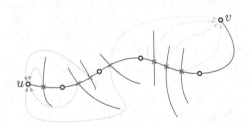

Fig. 2. Every crossing is charged to at least one endpoint of each of the two involved (independent) edges. Vertices are shown as white disks, crossings as red crosses, and charges by dotted arrows. The figure shows an edge uv that is subdivided four times, splitting it into a path with five segments. A crossing along any such segment is assigned to the closest of u or v. For the central segment, both u and v are at the same distance, and any crossing there is assigned to both u and v. (Color figure online)

Let Γ_s be the drawing of K_s induced by Γ: every vertex of K_s is placed at the position of the corresponding branch vertex of $K(s,t)$ in Γ and every edge of K_s is drawn as a Jordan arc along the corresponding path of $K(s,t)$ in Γ. Assuming $\binom{s}{2} \geq 4s$ (i.e., $s \geq 9$), by the Crossing Lemma [8], at least

$$\frac{1}{64}\frac{\binom{s}{2}^3}{s^2} = \frac{1}{512}s(s-1)^3 \geq c \cdot s^4$$

pairs of independent edges cross in Γ_s, for some constant c. Every crossing in Γ_s corresponds to a crossing in Γ and is charged to at least two (and up to four) vertices of B. Thus, the overall charge is at least $2cs^4$, and at least one vertex $u \in B$ gets at least the average charge of $2cs^3$.

Each charge unit corresponds to a crossing of two independent edges in Γ_s, which is also charged to at least one other vertex of B. Hence, there is a vertex $v \neq u$ so that at least $2cs^2$ crossings are charged to both u and v. Note that there are only $s - 1$ edges incident to each of u and v, and the common edge uv is not involved in any of the charged crossings (as adjacent rather than independent edge). Let E_x, for $x \in B$, denote the set of edges of K_s that are incident to x.

We claim that there are two pairs of mutually crossing edges incident to u and v, respectively; that is, there are sets $C_u \subset E_u \setminus \{uv\}$ and $C_v \subset E_v \setminus \{uv\}$ with $|C_u| = |C_v| = 2$ so that e_1 crosses e_2, for all $e_1 \in C_u$ and $e_2 \in C_v$.

Before proving this claim, we argue that establishing it completes the proof of the theorem. By our charging scheme, every crossing $e_1 \cap e_2$ happens at an edge of the path $[e_1]$ in Γ that is at least as close to u as to the other endpoint of e_1. Denote the three vertices that span the edges of C_u by u, x, y. Consider the two subdivision vertices x' along $[ux]$ and y' along $[uy]$ that form the endpoint of the middle edge closer to x and y, respectively, than to u; see Fig. 3 for illustration.

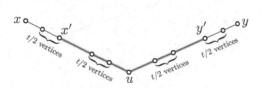

Fig. 3. Two adjacent edges ux and uy, both subdivided t times, and the shortest path between the "far" endpoints x' and y' of the central segments of $[ux]$ and $[uy]$.

The triangle uxy in K_s corresponds to an odd cycle of length $3(t + 1)$ in $K(s,t)$. So the shortest path between x' and y' in $K(s,t)$ has length $2(1+t/2) = t+2$ and passes through u, whereas the path from x' via x and y to y' has length $3(t+1) - (t+2) = 2t+1$, which is strictly larger than $t+2$ for $t \geq 2$. It follows that the shortest path between x' and y' in $K(s,t)$ is crossed by both edges in C_v. A symmetric argument yields two subdivision vertices a' and b' along the two edges in C_v so that the shortest $a'b'$-path in $K(s,t)$ is crossed by both edges in C_u. By definition of our charging scheme (that charges only "nearby" crossings to a vertex), the shortest paths $x'y'$ and $a'b'$ in $K(s,t)$ have at least four crossings.

It remains to prove the claim. To this end, consider the bipartite graph X on the vertex set $E_u \cup E_v$ where two vertices are connected if the corresponding edges are independent and cross in Γ_s. Observe that two sets C_u and C_v of mutually crossing pairs of edges (as in the claim) correspond to a 4-cycle C_4 in X. So suppose for the sake of a contradiction that X does not contain C_4 as a subgraph. Then by the Kővári-Sós-Turán Theorem [5] the graph X has $O(s^{3/2})$ edges. But we already know that X has at least $2cs^2 = \Omega(s^2)$ edges, which yields a contradiction. Hence, X is not C_4-free and the claim holds. $\square$

The bound on the number of crossings in Theorem 1 is tight.

Theorem 2. *A graph obtained from a complete graph by subdividing the edges uniformly an even number of times can be drawn so that every pair of shortest paths crosses at most four times.*

Proof (Sketch). Place the vertices in convex position. Draw the subdivided edges along straight-line segments. For each edge, put half of the subdivision vertices very close to one endpoint and the other half very close to the other endpoint (Fig. 4). As a result, all crossings fall into the central segment of the path.   □

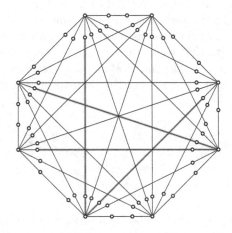

Fig. 4. A drawing of $K(8,2)$, the complete graph K_8 where every edge is subdivided twice, so that every pair of shortest paths meets at most four times. Two shortest paths that meet four times are shown bold and orange. (Color figure online)

3   Graphs of Diameter Two

In this section we give examples of geodetic graphs of diameter two that cannot be drawn in the plane such that any two shortest paths meet at most once.

An *affine plane* of order $k \geq 2$ consists of a set of lines and a set of points with a containment relationship such that (i) each line contains k points, (ii) for any two points there is a unique line containing both, (iii) there are three points that are not contained in the same line, and (iv) for any line ℓ and any point p not on ℓ there is a line ℓ' that contains p, but no point from ℓ. Two lines that do not contain a common point are *parallel*. Observe that each point is contained in $k + 1$ lines. Moreover, there are k^2 points and $k + 1$ classes of parallel lines each containing k lines. The 2-dimensional vector space $\mathbb{F}^2$ over a finite field $\mathbb{F}$ of order k with the lines $\{(x, mx + b);\ x \in \mathbb{F}\}$, $m, b \in \mathbb{F}$ and $\{(x_0, y);\ y \in \mathbb{F}\}$,

$x_0 \in \mathbb{F}$ is a finite affine plane of order k. Thus, there exists a finite affine plane of order k for any k that is a prime power (see, e.g., [4]).

Scapellato [12] showed how to construct geodetic graphs of diameter two as follows: Take a finite affine plane of order k. Let L be the set of lines and let P be the set of points of the affine plane. Consider now the graph G_k with vertex set $L \cup P$ and the following two types of edges: There is an edge between two lines if and only if they are parallel. There is an edge between a point and a line if and only if the point lies on the line; see Fig. 5. There are no edges between points. It is easy to check that G_k is a geodetic graph of diameter two.

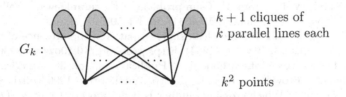

$G_k:$

$k+1$ cliques of
k parallel lines each

k^2 points

Fig. 5. Structure of the graph G_k.

Theorem 3. *There are geodetic graphs of diameter two that cannot be drawn in the plane such that any two shortest paths meet at most once.*

Proof. Let $k \geq 129$ be such that there exists an affine plane of order k (e.g., the prime $k = 131$). Assume there was a drawing of G_k in which any two shortest paths meet at most once. Let G be the bipartite subgraph of G_k without edges between lines. Observe that any path of length two in G is a shortest path in G_k. As G has $n = 2k^2 + k$ vertices and $m = k^2(k+1) > kn/2$ edges, we have $m > 4n$, for $k \geq 8$. Therefore, by the Crossing Lemma [8, Remark 2 on p. 238] there are at least $m^3/64n^2 > k^3n/512$ crossings between independent edges in G.

Hence, there is a vertex v such that the edges incident to v are crossed more than $k^3/128$ times by edges not incident to v. By assumption, (a) any two edges meet at most once, (b) any edge meets any pair of adjacent edges at most once, and (c) any pair of adjacent edges meets any pair of adjacent edges at most once. Thus, the crossings with the edges incident to v stem from a matching. It follows that there are at most $(n-1)/2 = (2k^2 + k - 1)/2$ such crossings. However, $(2k^2 + k - 1)/2 < k^3/128$, for $k \geq 129$. $\square$

4   Open Problems

We conclude with two open problems: (1) Are there diameter-2 geodetic graphs with edge density $1 + \varepsilon$ that do not admit a philogeodetic drawing? (2) What is the complexity of deciding if a geodetic graph admits a philogeodetic drawing?

References

1. Bodwin, G.: On the structure of unique shortest paths in graphs. In: Proceedings of the Thirtieth Annual ACM-SIAM Symposium on Discrete Algorithms (SODA 2019), pp. 2071–2089. SIAM (2019). https://doi.org/10.1137/1.9781611975482.125
2. Frasser, C.E., Vostrov, G.N.: Geodetic graphs homeomorphic to a given geodetic graph. CoRR abs/1611.01873 (2016). http://arxiv.org/abs/1611.01873
3. Hoffman, A.J., Singleton, R.R.: Moore graphs with diameter 2 and 3. IBM J. Res. Dev. **5**(4), 497–504 (1960). https://doi.org/10.1147/rd.45.0497
4. Hughes, D.R., Piper, F.C.: Projective Planes, Graduate Texts in Mathematics, vol. 6. Springer, New York (1973)
5. Kővári, T., Sós, V.T., Turán, P.: On a problem of K. Zarankiewicz. Colloq. Math. **3**, 50–57 (1954). https://doi.org/10.4064/cm-3-1-50-57
6. Lee, D.T., Preparata, F.P.: Euclidean shortest paths in the presence of rectilinear barriers. Networks **14**, 393–410 (1984). https://doi.org/10.1002/net.3230140304
7. Ore, Ø.: Theory of Graphs, Colloquium Publications, vol. 38. American Mathematical Society, Providence, RI (1962). https://doi.org/10.1090/coll/038
8. Pach, J., Tóth, G.: Which crossing number is it anyway? J. Combin. Theory Ser. B **80**(2), 225–246 (2000). https://doi.org/10.1006/jctb.2000.1978
9. Parthasarathy, K.R., Srinivasan, N.: Some general constructions of geodetic blocks. J. Combin. Theory Ser. B **33**(2), 121–136 (1982). https://doi.org/10.1016/0095-8956(82)90063-6
10. Plesník, J.: Two constructions of geodetic graphs. Math. Slovaca **27**(1), 65–71 (1977). https://dml.cz/handle/10338.dmlcz/136134
11. Plesník, J.: A construction of geodetic graphs based on pulling subgraphs homeomorphic to complete graphs. J. Combin. Theory Ser. B **36**(3), 284–297 (1984). https://doi.org/10.1016/0095-8956(84)90034-0
12. Scapellato, R.: Geodetic graphs of diameter two and some related structures. J. Combin. Theory Ser. B **41**(2), 218–229 (1986). https://doi.org/10.1016/0095-8956(86)90045-6
13. Stemple, J.G.: Geodetic graphs homeomorphic to a complete graph. Ann. N. Y. Acad. Sci. **319**(1), 512–517 (1979). https://doi.org/10.1111/j.1749-6632.1979.tb32829.x
14. Stemple, J.G., Watkins, M.E.: On planar geodetic graphs. J. Comb. Theory Ser. B **4**(2), 101–117 (1968). https://doi.org/10.1016/S0021-9800(68)80035-3
15. Watkins, M.E.: A characterization of planar geodetic graphs. J. Comb. Theory **2**(1), 102–103 (1967). https://doi.org/10.1016/S0021-9800(67)80118-2

Limiting Crossing Numbers for Geodesic Drawings on the Sphere

Marthe Bonamy[1] , Bojan Mohar[2] , and Alexandra Wesolek[2(✉)]

[1] CNRS, LaBRI, Université de Bordeaux, Bordeaux, France
`marthe.bonamy@u-bordeaux.fr`
[2] Department of Mathematics, Simon Fraser University, Burnaby, BC, Canada
`{mohar,agwesole}@sfu.ca`

Abstract. We introduce a model for random geodesic drawings of the complete bipartite graph $K_{n,n}$ on the unit sphere $\mathbb{S}^2$ in $\mathbb{R}^3$, where we select the vertices in each bipartite class of $K_{n,n}$ with respect to two non-degenerate probability measures on $\mathbb{S}^2$. It has been proved recently that many such measures give drawings whose crossing number approximates the Zarankiewicz number (the conjectured crossing number of $K_{n,n}$). In this paper we consider the intersection graphs associated with such random drawings. We prove that for any probability measures, the resulting random intersection graphs form a convergent graph sequence in the sense of graph limits. The edge density of the limiting graphon turns out to be independent of the two measures as long as they are antipodally symmetric. However, it is shown that the triangle densities behave differently. We examine a specific random model, blowups of antipodal drawings D of $K_{4,4}$, and show that the triangle density in the corresponding crossing graphon depends on the angles between the great circles containing the edges in D and can attain any value in the interval $\left(\frac{83}{12288}, \frac{128}{12288} \right)$.

Keywords: Crossing number · Graph limits · Geodesic drawing · Random drawing · Triangle density

1 Introduction

The crossing number $cr(G)$ of a graph G is the minimum number of crossings obtained by drawing G in the plane (or the sphere). In this paper we consider the *(spherical) geodesic crossing number* $cr_0(G)$, for which we minimize the number of crossings taken over all drawings of G in the unit sphere $\mathbb{S}^2$ in $\mathbb{R}^3$ such that

M. Bonamy—Supported in part by the ANR Project DISTANCIA (ANR-17-CE40-0015) operated by the French National Research Agency (ANR).
B. Mohar—Supported in part by the NSERC Discovery Grant R611450 (Canada), by the Canada Research Chairs program, and by the Research Project J1-8130 of ARRS (Slovenia).
A. Wesolek—Supported by the Vanier Canada Graduate Scholarships program.

© Springer Nature Switzerland AG 2020
D. Auber and P. Valtr (Eds.): GD 2020, LNCS 12590, pp. 341–355, 2020.
https://doi.org/10.1007/978-3-030-68766-3_27

each edge uv is a geodesic segment joining points u and v in $\mathbb{S}^2$. Recall that *geodesic segments* (or *geodesic arcs*) in $\mathbb{S}^2$ are arcs of great circles whose length is at most π. Also note that $cr(G) \leq cr_0(G)$ for every graph G.

Crossing number minimization has a long history and is used both in applications and as a theoretical tool in mathematics. We refer to [14] for an overview about the history and the use of crossing numbers. Despite various breakthrough results about crossing numbers, some of the very basic questions remain open as of today, two of the most intriguing being what are the crossing numbers of the complete graphs K_n and what are the crossing numbers of the complete bipartite graphs $K_{n,n}$ (the Turán Brickyard Problem). The asymptotic versions of both problems are strongly related [13] and a lower bound for the limiting crossing number of $K_{n,n}$ gives a related lower bound for K_n. The asymptotic version of the rectilinear crossing number of K_n is related to Sylvester's Four point problem in the plane [15,16], see also [14] for recent results. The geodesic version on the sphere, which we discuss in this paper, is a spherical version of Sylvester's problem.

1.1   Outline

In this paper we initiate the study of limiting properties of intersection graphs associated with drawings of complete and complete bipartite graphs. We limit ourselves to geodesic drawings on the unit sphere in $\mathbb{R}^3$ in which case the drawings are determined by the choice of the placements of the vertices on the sphere. The first main result of this work shows that whenever the vertices in each bipartite class of $K_{n,n}$ are selected according to some (non-degenerate) probability measure on $\mathbb{S}^2$ (where the two measures used for each class can be different), then, with probability 1, the intersection graphs form a convergent sequence of graphs in the sense of graph limits [7]. See Theorem 2.

The basic combinatorial property of convergent graph sequences is that of subgraph densities. The density of edges in the crossing graphs corresponds to the asymptotic crossing number. In addition to this, we examine one particular related basic question: what is the density of triangles. We show that their density can be substantially different among different randomized models. Although this result may be seen as "expected", it is still somewhat surprising. Indeed, it shows that there is a large variety of drawings of $K_{n,n}$, all attaining the Zarankiewicz bound, in which the number of triples of mutually crossing edges varies significantly, and can attain any value in the interval $\left(\frac{83}{12288}, \frac{128}{12288}\right)$. See Theorems 4 and 6. We believe that further exploring of subgraph densities in crossing graphons may give a deeper insight into the basic Turán's Brickyard Problem for geodesic drawings on the sphere.

1.2   Asymptotic Zarankiewicz Conjecture

During World War II, Hungarian mathematician Pál Turán worked in a brick factory near Budapest. There the bricks were transported on wheeled trucks from kilns to storage yards. It was difficult to push the trucks past the rail crossings and it would result in extra work if bricks fell of the trucks. Therefore Turán wondered if there was a way of arranging the rails such that there would be less crossings between them. Seeing the kilns and storage yards as parts of a bipartite graph, this led to the more general question of the minimum number of crossings in drawings of complete bipartite graphs $K_{n,n}$. Zarankiewicz [19] and Urbanik [17] suggested drawings that involved

$$Z(m,n) = \lfloor \tfrac{n}{2} \rfloor \lfloor \tfrac{n-1}{2} \rfloor \lfloor \tfrac{m}{2} \rfloor \lfloor \tfrac{m-1}{2} \rfloor = \begin{cases} \tfrac{1}{16} n(n-2)m(m-2), & n,m \text{ are even}; \\ \tfrac{1}{16} n(n-2)(m-1)^2, & n \text{ is even}, m \text{ is odd}; \\ \tfrac{1}{16}(n-1)^2(m-1)^2, & n,m \text{ are odd} \end{cases}$$

(1)

crossings. Whether this value is the best possible remains unanswered to this day despite numerous attacks using powerful machinery in trying to resolve this conjecture.

A general construction of drawings of complete bipartite graphs attaining the Zarankiewicz bound was recently exhibited [10]. All of them are geodesic drawings in $\mathbb{S}^2$ and they show that

$$cr(K_{n,n}) \le cr_0(K_{n,n}) \le Z(n,n) \quad \text{for every } n \ge 1.$$

(2)

It is not hard to see that the following limits exist:

$$\lambda := \lim_{n \to \infty} n^{-4} cr(K_{n,n}) \quad \text{and} \quad \lambda_0 := \lim_{n \to \infty} n^{-4} cr_0(K_{n,n}).$$

Clearly, (2) implies that $\lambda \le \lambda_0 \le \tfrac{1}{16}$. The *asymptotic Zarankiewicz conjecture* for the usual and the geodesic crossing number is also open.

Conjecture 1. $\lambda = \lambda_0 = \tfrac{1}{16}$.

1.3 Random Drawings of Complete Bipartite Graphs

In 1965, Moon [12] proved that a random set of n points on the unit sphere $\mathbb{S}^2$ in $\mathbb{R}^3$ joined by geodesics gives rise to a drawing of K_n whose number of crossings asymptotically approaches the conjectured value. It was proved recently [11] that the same phenomenon appears in a much more general random setting. These results can also be extended to random drawings of the complete bipartite graphs $K_{n,n}$ where it was shown that under a symmetry condition on the probability measures the crossings in such drawings converge to the Zarankiewicz value.

A probability distribution μ on $\mathbb{S}^2$ is *nondegenerate* if for every great circle $Q \subset \mathbb{S}^2$, $\mu(Q) = 0$. It is *antipodally-symmetric* if for every measurable set $A \subseteq \mathbb{S}^2$ the measure of its antipodal set $\overline{A}$ is the same, $\mu(A) = \mu(\overline{A})$.

Theorem 1 ([11]). *Let μ_1, μ_2 be nondegenerate antipodally-symmetric probability distributions on the unit sphere $\mathbb{S}^2$. Then a μ_1-random set of n points on $\mathbb{S}^2$ joined by geodesics (segments of great circles) to a μ_2-random set of n points gives rise to a drawing D_n of the complete bipartite graph $K_{n,n}$ such that $cr(D_n)/Z(n,n) = 1 + o(1)$ a.a.s.*

The random drawing model in the theorem will be referred to as (μ_1, μ_2)-*random drawing* of the complete bipartite graph $K_{n,n}$.

1.4   Crossing Graphon

Let $N = \{n_1, n_2, n_3, \dots\}$ be an infinite set of positive integers, where $n_1 < n_2 < n_3 < \cdots$. Suppose that for each $n \in N$, we have a drawing D_n of $K_{n,n}$. To each such drawing we associate the *crossing graph* $X_n = X_n(D_n)$, whose vertices are all n^2 edges in D_n, and two of them are adjacent in X_n if they cross in D_n. Then we can consider what may be the *limit* of the sequence $(X_n)_{n \in N}$. The notion of *graph limits* has been introduced by Lovász et al. [3,4,9], see [7]. The basic setup is described below.

Let $(X_n)_{n \in N}$ be a sequence of graphs. For any fixed graph H, let $k = |H|$ be its order, and let $hom(H, X_n)$ denote the number of graph homomorphisms $H \to X_n$, i.e. the number of maps $\phi : V(H) \to V(X_n)$ such that for each edge $uv \in E(H)$, $\phi(u)\phi(v) \in E(X_n)$. Then we define the *homomorphism density* for H as

$$t(H, X_n) = \frac{hom(H, X_n)}{|X_n|^k}.$$

Note that this is the probability that a random mapping $V(H) \to V(X_n)$ is a homomorphism. If the sequence $t(H, X_n)$ converges, we denote its limit by $t(H)$. If $t(H)$ exists for every H, then we say that (X_n) is a *convergent sequence of graphs*. In that case there is a well-defined object W, called a *graphon*, and the graphon W is called the *limit* of this convergent sequence [3,4]. We define the *homomorphism densities* of W by setting $t(H, W) = \lim_{n \to \infty} t(H, X_n) = t(H)$.

The space of all graphons is a compact metric space [7,9]. Given any graphon W, one can define W-*random graphs* [8]. A sequence (R_n) of W-random graphs is convergent with probability 1, and its limit is W.

In this paper we consider nondegenerate probability measures on $\mathbb{S}^2$. For each pair of such probability measures μ_1 and μ_2, we have a (μ_1, μ_2)-random sequence of drawings D_n of complete bipartite graphs $K_{n,n}$ and we consider their crossing graphs X_n. We prove that these sequences are convergent with probability 1 and discuss their homomorphism densities with the goal to better understand Conjecture 1.

Theorem 2. *Let μ_1 and μ_2 be nondegenerate probability measures on $\mathbb{S}^2$. Let A_n and B_n be a μ_1-random and a μ_2-random set of n points in $\mathbb{S}^2$, respectively, let D_n be the corresponding (μ_1, μ_2)-random geodesic drawing of $K_{n,n}$ on parts A_n and B_n, and let X_n be its crossing graph. The sequence of graphs (X_n) is convergent with probability 1 and there is a graphon $W = W(\mu_1, \mu_2)$ that is the limit of this convergent sequence.*

Since the number of edges in the crossing graph corresponds to the number of crossings in D_n, we have

$$t(K_2, X_n) = \frac{2|E(X_n)|}{|X_n|^2} = \frac{2cr(D_n)}{n^4}.$$

Thus, Theorem 1 shows a tight relationship with the asymptotic Zarankiewicz conjecture and can be expressed as follows.

Theorem 3. *Let μ_1, μ_2 be nondegenerate antipodally-symmetric probability measures on $\mathbb{S}^2$. Let $W = W(\mu_1, \mu_2)$ be the corresponding graphon of the sequence (X_n) as defined above. Then*

$$t(K_2, W(\mu_1, \mu_2)) = \frac{1}{8}.$$

1.5   Definitions

We follow standard terminology from [2,5] for graph theory and from [14] for drawings of graphs. A drawing of a graph is *good* if any two edges cross at most once, no two edges with a common endvertex cross, and no three edges cross at the same point. The first two conditions are clear when we consider geodesic drawings, and the third condition can always be satisfied if we make an infinitesimal perturbation.

We say that a set of points on the unit sphere $\mathbb{S}^2$ is *in general position* if no two of the points are antipodal to each other, no three of them lie on the same great circle and no three geodesic arcs joining pairs of points cross at the same point. If μ is a nondegenerate probability distribution on $\mathbb{S}^2$, then randomly chosen vertices will be in general position with probability 1.

2   The Proof of Theorem 2

In the following we want to draw a comparison of subgraph densities of the crossing graphs X_n to a concept similar to the Buffon Needle Problem (see, e.g. [6] or [18]). We pick endpoints of segments randomly w.r.t. some probability distribution and consider the crossings formed by the segments. If the probability distribution is uniform on the sphere, it is equivalent as throwing a (bended) needle onto the sphere, where the needle length varies. Now considering a small number of such segments on the sphere we ask how they will cross each other.

Let μ_1 and μ_2 be nondegenerate probability measures on $\mathbb{S}^2$. A (μ_1, μ_2)-*random geodesic segment* is a geodesic segment uv whose endpoints u, v are chosen randomly w.r.t. μ_1 and μ_2, respectively. For a given graph H of order $k = |H|$, we pick k (μ_1, μ_2)-random geodesic segments on the sphere and look at the probability that H is a subgraph of their intersection graph. Let $A = \{a_1, \ldots, a_k\}$ be a μ_1-random set of points in $\mathbb{S}^2$ and $B = \{b_1, \ldots, b_k\}$ be a μ_2 random set of points in $\mathbb{S}^2$. The segments we are considering are $a_1b_1, \ldots, a_kb_k$. Note that the probability that H is a subgraph of the intersection graph of $a_1b_1, \ldots, a_kb_k$ depends on μ_1 and μ_2 only.

Definition 1. *Let X be the intersection graph of k (μ_1, μ_2)-random geodesic segments $a_1b_1, \ldots, a_kb_k$ and let H be a graph of order k. For a bijection $\phi : V(H) \to V(X)$ we define*

$$p_H := Pr[\phi \text{ is a graph homomorphism}].$$

Observe that p_H is independent of ϕ, since the segments $a_i b_i$ $(i = 1, \ldots, k)$ are selected independently.

We want to compare the above model with another model where we pick $n \gg k$ points with respect to μ_1 and μ_2 each, and consider the corresponding crossing graph X_n of a drawing D_n of $K_{n,n}$. We will show that the models are closely related: with growing n, picking k vertices from X_n, they will with high probability come from k independent geodesic segments and therefore represent (μ_1, μ_2)-random geodesic segments. In the following we fix a graph H and a mapping $\phi : V(H) \to V(X_n)$.

Definition 2. *For given X_n, let $\phi : V(H) \to V(X_n)$ and we define the random variable $y_{H,\phi}$ on X_n to be*

$$y_{H,\phi}(X_n) = \begin{cases} 1 & \text{if } \phi \text{ is a graph homomorphism } H \to X_n \\ 0 & \text{otherwise} \end{cases}$$

and denote its expectation by

$$E_\phi := \mathbb{E}[y_{H,\phi}].$$

Note that E_ϕ is not the same for every ϕ. For example, if H is a complete graph, then $E_\phi = 0$ whenever $im(\phi)$ contains edges that share a vertex, as those edges never cross and hence are not adjacent in the crossing graph.

Lemma 1. *Let (X_n) be a sequence of the crossing graphs of (μ_1, μ_2)-random geodesic drawings D_n of $K_{n,n}$ for $n = 1, 2, \ldots$, and let H be a fixed graph of order k. Then*

$$\lim_{n \to \infty} \frac{1}{|X_n|^k} \sum_{\phi : V(H) \to V(X_n)} E_\phi = p_H.$$

Proof. Let $im(\phi) = \{v_1 w_1, \ldots, v_k w_k\}$. Then if $|\{v_1, \ldots, v_k, w_1, \ldots, w_k\}| = 2n$ we are in the setup of Definition 1 and $\mathbb{E}[y_{H,\phi}] = p_H$. Moreover, there are $O(n^{2k-1})$ choices for ϕ for which $|\{v_1, \ldots, v_k, w_1, \ldots, w_k\}| < 2n$ and the result follows. $\square$

Let us now consider the sum of the above defined random variables

$$Y_H := \sum_{\phi : V(H) \to V(X_n)} y_{H,\phi}, \tag{3}$$

and note that $Y_H(X_n) = hom(H, X_n)$ and $\mathbb{E}[Y_H] = \sum_{\phi : V(H) \to V(X_n)} E_\phi$. The aim is to show that Y_H is in general not far from its expectation. This then gives us the tool to show the existence of $\lim_{n \to \infty} \frac{|Y_H|}{|X_n|^k} = t(H)$ with probability 1.

Proposition 1. *Let Y_H be defined as in (3). Then we have*

$$var(Y_H) = O(n^{4k-2}).$$

The proof of the proposition can be found in the full version [1].

Proof (of Theorem 2). By Proposition 1 and Chebyshev's inequality there exists a constant C such that

$$Pr\left[|Y_H - \mathbb{E}[Y_H]| \geq kCn^{2k-1}\right] \leq \frac{1}{k^2}.$$

Now if we choose $k = k(n)$ appropriately such that $k(n)n^{-1}$ converges to zero and the sum $\sum_{n=1}^{\infty} \frac{1}{k(n)^2}$ is finite we can use the Borel-Cantelli Lemma. For example, we can choose $k = n^{3/4}$ and using Lemma 1 we get

$$Pr\left[\left|\frac{|Y_H|}{|X_n|^k} - p_H\right| - \left|p_H - \frac{\mathbb{E}[Y_H]}{|X_n|^k}\right| \geq \frac{Cn^{2k-1/4}}{|X_n|^k}\right] \leq \frac{1}{n^{3/2}}$$

$$\implies Pr\left[\left|\frac{|Y_H|}{|X_n|^k} - p_H\right| \geq \frac{C'}{n^{1/4}}\right] \leq \frac{1}{n^{3/2}}.$$

for some constant C'. Then the Borel-Cantelli Lemma implies the following.

Claim. For each fixed H, $\frac{|Y_H|}{|X_n|^k} \to p_H := t(H)$ with probability 1.

Given that for each H, $t(H, X_n) \to t(H)$ with probability 1, and since the probabilities are countably additive, it follows with probability 1 that $t(H, X_n) \to t(H)$ for every H. Consequently, the sequence of random crossing graphs (X_n) is convergent with probability 1. □

3   Blowup of an Antipodal Drawing of $K_{4,4}$

In the previous sections, we have established the existence of crossing graphons and determined densities $t(H)$ for $H = K_2$ if our measures μ_1, μ_2 are antipodally symmetric. Somewhat surprisingly, these edge densities are the same for any "suitable" measures μ_1, μ_2. It is natural to ask what happens with other homomorphism densities in these crossing graphons. The purpose of this section is to show that the homomorphism densities of triangles behave differently. To us, this was not *a priori* clear. We study a particular case of (μ_1, μ_2)-random drawings of complete bipartite graphs and determine $t(K_3)$ for the corresponding graphon $W(\mu_1, \mu_2)$.

In the following we fix a drawing D_4 of the complete bipartite graph $K_{4,4}$ where each part consists of two antipodal pairs of vertices on $\mathbb{S}^2$ as in Fig. 1.

We will be considering a *blowup drawing* $D_4^{(n)}$ of D_4 for which we replace each vertex from D_4 with a circle of some small radius $r = r(n)$ that is centered at that vertex, and position n evenly spaced vertices on that circle. These n vertices will be referred to as the *node* of the corresponding vertex of $K_{4,4}$. We also assume that all $8n$ vertices obtained in this way are in general position. In that way, each edge of $K_{4,4}$ is replaced by a complete bipartite graph between the

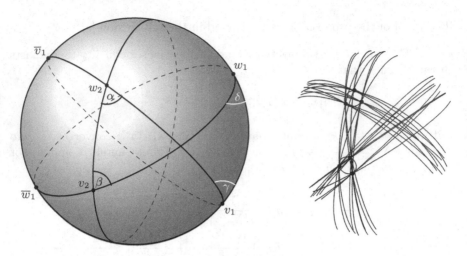

Fig. 1. The left part shows a drawing D_4 of a $K_{4,4}$ on parts $\{v_1, \overline{v}_1, v_2, \overline{v}_2\}$ and $\{w_1, \overline{w}_1, w_2, \overline{w}_2\}$. The angles α and β are in the triangle formed by w_2, v_2 and a crossing, whereas γ and δ are in a triangle formed by v_1, w_1 and the same crossing. The right-hand side shows part of a $D_4^{(3)}$ drawing with the circles of w_2 and v_2 each containing 3 vertices and with nine edges for each incident bundle emanating from these two nodes.

corresponding nodes which we call the *edge bundle*. This means for $N = 4n$ that $D_4^{(n)}$ is a drawing of $K_{N,N}$. In what follows, we discuss the number of triangles in the intersection graph (of edges in $D_4^{(n)}$) when n grows large. To simplify our discussion about triangles, we first classify the crossings in $D_4^{(n)}$.

3.1   Types of Crossings in $D_4^{(n)}$

In the blowup drawing $D_4^{(n)}$, we distinguish three types of crossings, depending on what they stem from, as depicted in Fig. 2.

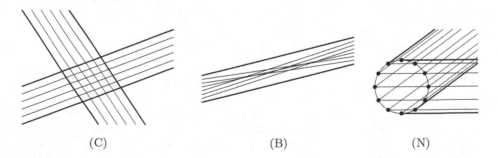

(C)   (B)   (N)

Fig. 2. Possible crossings in the blow up: Bundle-bundle crossings (C), bundle crossings (B) and node crossings (N).

Let us define these types (B), (C), and (N) more precisely and state their count. The corresponding counting process is described in the full version [1].

(C) Two edge-bundles cross in a small neighborhood of a previous crossing in D_4. We call these *bundle-bundle crossings* (C). Since each edge-bundle consists of n^2 edges, this gives n^4 bundle-bundle crossings for each crossing in D_4.

(B) Two edges cross within a bundle. We call these *bundle crossings* (B). Here we have $\binom{n}{2}^2$ crossings per bundle assuming $r(n) \ll n^{-1}$ and a suitable rotation of the circles.

(N) Two edge-bundles cross at a node. We call these *node crossings* (N). Let $\alpha \in (0, \pi)$ be the angle between two incident edges e, f in D_4 which were blown up to the edge-bundles, and let cr_α be the resulting number of node crossings between the edges in the corresponding edge-bundles. Then we have: $\mathrm{cr}_\alpha + \mathrm{cr}_{\pi-\alpha} = \frac{n^3(n-1)}{2}$.

3.2   Triangle Densities in $D_4^{(n)}$

The crossings in a triangle need to stem from bundle-bundle crossings (C), bundle crossings (B) or node crossings (N) as specified above. We first prove the following lemma.

Lemma 2. *Let D_4 be a spherical drawing of a $K_{4,4}$ where each part consists of two pairs of antipodal vertices. Then no edge in D_4 is crossed twice.*

Proof. Let the parts of the $K_{4,4}$ be $A = \{v_1, \overline{v_1}, v_2, \overline{v_2}\}$ and $B = \{w_1, \overline{w_1}, w_2, \overline{w_2}\}$. Note that the edge $v_1 w_1$ can only be crossed by an edge between the other antipodal pairs, i.e. $v_2 w_2, v_2 \overline{w_2}, \overline{v_2} w_2, \overline{v_2 w_2}$. All of them lie on the great circle defined by $v_2 w_2$ so in fact only one of these edges can cross $v_1 w_1$. By symmetry the same holds for the other edges. □

We classify the triangles in the intersection graph of the blowup drawing $D_4^{(n)}$ as follows. We assign each crossing (which is an edge in the intersection graph) a *type* (C), (B), or (N) depending on whether it is a bundle-bundle, within bundle or a node crossing. We say a triangle $c_1 c_2 c_3$ is of *type* $(l(c_1) l(c_2) l(c_3))$ where $l(c_i)$ is the type of crossing c_i.

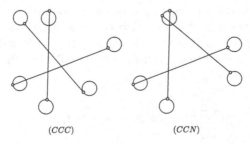

(CCC)   (CCN)

Fig. 3. Triangles of type (CCC) and (CCN).

The above lemma shows that there are no (CCC) or (CCN) triangles in $D_4^{(n)}$ (Fig. 3). Also note that (CBB), (BBN) and (CBN) are not possible in general since BB suggests that all edges are from the same bundle and the bundled edges in (CBN) cross the third edge either at a node or at a bundle-bundle crossing but not at both. Triangles of type (NNN) either appear at three different nodes or at one node. However, we can not have (NNN) triangles at three different nodes since $K_{4,4}$ is bipartite and hence triangle-free. By the following lemma, the number of (NNN) triangles with all three crossings at one node is only of order rn^6 and can therefore be neglected.

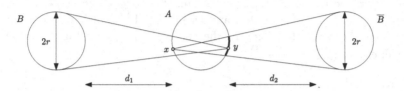

Fig. 4. Two edges from node A leading to antipodal nodes B and $\overline{B}$ can cross. If $d = \min\{d_1, d_2\}$ and $r \leq d$, then $|L_x| = O(rn)$.

Lemma 3. *The number of (NNN) triangles in $D_4^{(n)}$ that correspond to three edges at the same node is $O(rn^6)$. Moreover, if $r(n) \ll n^{-1}$, there are no such triangles.*

Proof. Let us refer to Fig. 4 and consider the possibility that an edge incident with a vertex y and leading to a node B crosses an edge incident with a vertex x that leads to the antipodal node $\overline{B}$. If the geodesics from x to $\overline{B}$ intersect the circle C_A corresponding to A, we denote by L_x the set of vertices in A that are on the smallest circular arc that contains those intersections.

Then it is easy to see that either $x \in L_y$ or $y \in L_x$ (or both as shown in the figure). It can be shown (details can be found in the full paper) that the number of cases where $y \in L_x$ or $x \in L_y$ is $O(rn)$. In particular, if $r \ll n^{-1}$ then L_x is empty. For each such pair x, y, the number of vertices z whose incident edges leading to a node different from B and $\overline{B}$ make an (NNN) crossing triangle with two edges incident with x and y, respectively, is $O((t+r)n)$, where t is the number of vertices on the arc between x and y. We define the parameter l which is the number of vertices in the node A between x and the lowest point on the circle of A (assuming that x is in the lower half of the circle and on the left side). Then $t \in [2l - \Theta(rn), 2l + \Theta(rn)]$. This gives the following upper bound for the number of such triples (x, y, z):

$$4 \sum_{l=1}^{n/4} O(rn)O(2l + rn) = O(rn^3).$$

Finally, since each such triple involves $O(n^3)$ triples of mutually crossing edges incident with x, y, z, we confirm that the number of considered (NNN) triangles is $O(rn^6)$.   □

We are left with the following four cases.

(CNN) We consider pairwise crossings of three edges such that two cross at a bundle-bundle crossing and the third edge crosses one edge each at one node each. These crossings depend on the angles $\alpha, \beta, \gamma, \delta$ as depicted in Fig. 1. By Section (N) in the full version [1] the number of pairs of vertices x, y such that all edges at angle α incident to x cross all horizontal edges incident to y is $\frac{\pi-\alpha}{2\pi}n^2 + O(rn^2 + n)$. It is easy to see that the number of crossings we get in the triangle including α and β is $\left(\frac{\pi-\alpha}{2\pi}\right)\left(\frac{\pi-\beta}{2\pi}\right)n^6 + O(rn^6 + n^5)$. We have a similar count for the angles γ and δ. Then we have to add three other contributions corresponding to other crossings in D_4. The antipodal crossing involves a triangles with α, β and γ, δ, whereas the other two crossings involve triangles with α, γ and β, δ. Overall, this gives $\frac{2}{n^2}(\mathrm{cr}_\alpha + \mathrm{cr}_\delta)(\mathrm{cr}_\gamma + \mathrm{cr}_\beta) + O(rn^6 + n^5)$ triangles of this kind.

(BBB) We consider pairwise crossings of three edges such that all edges are from one bundle. For each bundle we get $\binom{n}{3}^2 + O(rn^6)$ such triangles by Section (B) in the full version [1]. There are 16 bundles so in total we have $\sim \frac{4}{9}n^6 + O(rn^6)$ triangles of the type (BBB).

(CCB) We consider pairwise crossings of three edges such that two edges are in one bundle and cross the third edge at a bundle-bundle crossing. There are $2\binom{n}{2}^2 n^2 + O(rn^6)$ triangles per each crossing in D_4. We have 4 crossings so in total $\sim 2n^6 + O(rn^6)$ triangles of this kind.

(BNN) We consider pairwise crossings of three edges such that two are in the same bundle and cross the third edge at a node. The argument is analogous to the one for crossings of type (N). Starting at the top vertex, we enumerate the vertices clockwise along the cycle. We consider an edge at angle α which ends in the i-th vertex in part (A) and its crossings to horizontal edges. From Section (N) in the full version [1], we know that $|S_i| = 2i + O(rn)$, where S_i is as defined there. We can choose from $\binom{2i+O(rn)}{2}$ pairs of left endpoints and $\binom{n}{2}$ pairs of right endpoints for a triangle. The number of triangles with an edge ending in i and another edge ending in a vertex in $W_x = \{y \in A \mid x \in L_y\}$ is of order $O(rn^4)$, where L_y is defined as in the proof of Lemma 3. We consider now edges at angle α ending in a vertex x in (B). Note that $|S_x| = \frac{\pi-\alpha}{\pi}n + O(rn)$. We can choose for any one of $\binom{\frac{\pi-\alpha}{\pi}n+O(rn)}{2}$ pairs of left endpoints $\binom{n}{2}$ pairs of right endpoints for a triangle. The number of triangles with another edge ending in a vertex in W_x is of order $O(rn^4)$. The contribution of triangles from edges in (C) is the same as for edges in (A). Hence the number of triangles of type (BNN) is

$$2\left(n\sum_{i=1}^{(\pi-\alpha)n/2\pi}\binom{2i}{2}\cdot\binom{n}{2}\right) + \left(\frac{\alpha}{2\pi}n^2\right)\cdot\binom{\frac{\pi-\alpha}{\pi}n}{2}\binom{n}{2} + O(rn^6 + n^5).$$

For α and $\pi - \alpha$ added together, this gives

$$\frac{1}{12}n^6 - \frac{\alpha(\pi - \alpha)}{8\pi^2}n^6 + O(rn^6 + n^5).$$

Now note that for two bundles at angle α we can choose one of the bundles to contain the bundled edges. This gives two options. At each node we have two pairs of bundles meeting at angle α and two pairs of bundles meeting at angle $\pi - \alpha$. (In addition to these possibilities we get further (BNN) triangles from two bundles at the same node that lead to antipodal nodes and correspond to the value of $\alpha = \pi$. They give only $O(rn^6 + n^5)$ triangles.) If $\alpha, \beta, \gamma, \delta$ are the angles as in Fig. 1, the overall number of (BNN) triangles is

$$\frac{\alpha(\alpha - \pi) + \beta(\beta - \pi) + \gamma(\gamma - \pi) + \delta(\delta - \pi)}{\pi^2}n^6 + \frac{8}{3}n^6 + O(rn^6 + n^5).$$

If we leave out smaller order terms, the total number of triangles in the intersection graph by summing up the number of (CNN), (BBB), (CCB) and (BNN) triangles is

$$\frac{\alpha^2 + \beta^2 + \gamma^2 + \delta^2 - \pi(\alpha + \beta + \gamma + \delta)}{\pi^2}n^6 + \frac{(2\pi - \alpha - \delta)(2\pi - \gamma - \beta)}{2\pi^2}n^6 + \frac{46}{9}n^6.$$

Theorem 4. *Given a drawing D_4 of a $K_{4,4}$ where each part has two antipodal pairs, let $D_4^{(n)}$ be the blowup drawing, and let $\alpha, \beta, \gamma, \delta$ be the angles defined above. Then the limiting triangle density $t(K_3)$ of the sequence $D_4^{(1)}, D_4^{(2)}, \ldots$ is equal to*

$$\frac{3}{2^{12}\pi^2}\left((2\pi - \alpha - \delta)(2\pi - \gamma - \beta) + 2(\alpha^2 + \beta^2 + \gamma^2 + \delta^2) - 2\pi(\alpha + \beta + \gamma + \delta)\right)$$

$$+ \frac{23}{3 \cdot 2^{10}} + O(r).$$

Proof. We have determined the number of triangles in the intersection graphs. Dividing by the number of possible triangles in the intersection graph, $\binom{16n^2}{3} = \frac{16^3}{6}n^6 + O(n^5)$, gives the triangle density. □

4   Blowups as Graphons

Finally, let us show that the crossing graphs of drawings $D_4^{(n)}$ can be interpreted as certain graphons.

Theorem 5. *For fixed $r > 0$ let μ_1 and μ_2 be uniform distributions over two pairs of antipodal circles on $\mathbb{S}^2$ of radius r each and let $W(\mu_1, \mu_2)$ be the crossing graph limit of corresponding drawings. If we consider blowup drawings $D_4^{(n)}$ w.r.t. the centers of the circles of radius r, then the crossing graphs of $D_4^{(n)}$ converge and their limit is the graphon $W(\mu_1, \mu_2)$.*

Proof. All we need to show is that the density $t_1(H)$ in the random case limit and the density $t_2(H)$ of the blowup drawing limit are the same for each graph H. Let $k = |H|$ be the number of vertices of H and let $\phi : V(H) \to [k]$ be a bijection. For distinct points $x_1, \ldots, x_k, y_1, \ldots, y_k$ in $\mathbb{S}^2$, let $X(x_1, \ldots, x_k, y_1, \ldots, y_k)$ be the intersection graph of the geodesic segments $x_1 y_1, \ldots, x_k y_k$. Consider the following function

$$f(x_1, \ldots, x_k, y_1, \ldots, y_k) = \begin{cases} 1, & \text{if } v \mapsto x_{\phi(v)} y_{\phi(v)} \text{ is a hom. } H \to X(x_1, \ldots, y_k) \\ 0, & \text{otherwise.} \end{cases}$$

Let S_1 and S_2 be the two circles on which μ_1 and μ_2 are defined, respectively. Since f as defined above is measurable because $f^{-1}(1)$ is open, we can represent $t_1(H)$ as

$$t_1(H) = \frac{1}{(8\pi r)^k} \int_{x \in S_1^n \times S_2^n} f(x)\, dx.$$

In order to approximate $t_1(H)$ consider a set C_n which consists of n equidistant points on each of the cycles from S_1, S_2. Let $\pi_n : S_1 \cup S_2 \to C_n$ be the function that maps a points from $S_1 \cup S_2$ to its closest point in X. Let g_n be a function $g_n : (S_1 \cup S_2)^{2n} \to (C_n)^{2n}$ that applies π_n componentwise. Then $f_n = f \circ g_n$ converges pointwise to f on $S_1^n \times S_2^n$. By the bounded convergence theorem

$$t_1(H) = \frac{1}{(8\pi r)^k} \int_{x \in S_1^n \times S_2^n} f(x)\, dx = \frac{1}{(8\pi r)^k} \lim_{n \to \infty} \int_{x \in S_1^n \times S_2^n} f_n(x)\, dx = t_2(H). \qquad \square$$

The theorem shows that the same values for triangle densities in the (μ_1, μ_2)-random setting hold as for the blowup limit in Theorem 4.

Theorem 6. *For fixed $r > 0$ let μ_1 and μ_2 be uniform distributions over two pairs of antipodal circles on $\mathbb{S}^2$ of radius r each and let $W(\mu_1, \mu_2)$ be the crossing graph limit of the corresponding drawings. Then*

$$\frac{83}{3 \cdot 2^{12}} + O(r) \leq t(K_3, W(\mu_1, \mu_2)) \leq \frac{1}{3 \cdot 2^5} + O(r),$$

and these bounds are best possible. The limiting triangle density $t(K_3)$ depends on the angles $\alpha, \beta, \gamma, \delta$, and any value in the interval $\left(\frac{83}{12288}, \frac{128}{12288}\right)$ is possible.

The proof can be found in the full version [1].

5   Conclusion

It should be noted that the proofs of Theorem 2 and Theorem 3 also extend to the case of the complete graph K_n where we choose n points from the sphere with respect to some antipodally symmetric probability measure μ. (Let us observe that antipodal symmetry is needed for such a result.) In Theorem 4 the value

$\frac{23}{3 \cdot 2^{10}} = 0.00748$ appears which is included in the interval given by Theorem 6. Numerical experiments show that the triangle density with respect to the uniform distribution is close to 0.0075. This matches the mentioned special value from the blowup setting. It would be of interest to study the crossing graph limit for drawings on the sphere of the complete graph or the complete bipartite graph when we restrict our probability measure to a uniform measure on the sphere. As Moon already showed in 1965 [12], it holds asymptotically almost surely that $t(K_2) = \frac{1}{8}$, so it would be of interest to find a closed expression for $t(K_3)$.

References

1. Bonamy, M., Mohar, B., Wesolek, A.: Limiting crossing numbers for geodesic drawings on the sphere (2020). https://arxiv.org/abs/2008.10459v1
2. Bondy, J.A., Murty, U.S.R.: Graph Theory, Graduate Texts in Mathematics, vol. 244. Springer, New York (2008). https://doi.org/10.1007/978-1-84628-970-5
3. Borgs, C., Chayes, J.T., Lovász, L., Sós, V.T., Vesztergombi, K.: Convergent sequences of dense graphs I. Subgraph frequencies metric properties and testing. Adv. Math. **219**(66), 1801–1851 (2008). https://doi.org/10.1016/j.aim.2008.07.008
4. Borgs, C., Chaye, J., Lovász, L., Sós, V.T., Vesztergomb, K.: Counting graph homomorphisms. In: Klazar, M., Klazar, J., Kratochvíl, J., Loebl, M., Matoušek, J.J., Valtr, P., Thomas, ThomaR (eds.) Topics in Discrete Mathematics. Algorithms and Combinatori, vol. 26. Springer, Heidelberg (2006). https://doi.org/10.1007/3-540-33700-8_18
5. Diestel, Reinhard: Graph Theory. GTM, vol. 173. Springer, Heidelberg (2017). https://doi.org/10.1007/978-3-662-53622-3
6. Isokawa, Y.: Buffon's short needle on the sphere. Bull. Fac. Ed. Kagoshima Univ. Natur. Sci. **51**, 17–36 (2000)
7. Lovász, L.: Large Networks and Graph Limits, vol. 60. American Mathematical Society Colloquium Publications. American Mathematical Society, Providence, RI (2012). https://doi.org/10.1090/coll/060
8. Lovász, L., Sós, V.T.: Generalized quasirandom graphs. J. Comb. Theory Ser. B **98**(1), 146–163 (2008). https://doi.org/10.1016/j.jctb.2007.06.005
9. Lovász, L., Szegedy, B.: Limits of dense graph sequences. J. Comb. Theory Ser. B **96**(6), 933–957 (2006). https://doi.org/10.1016/j.jctb.2006.05.002
10. Mohar, B.: On a conjecture by Anthony Hill (2020). https://arxiv.org/abs/2009.03418
11. Mohar, B., Wesolek, A.: Random geodesic drawings. In preparation
12. Moon, J.W.: On the distribution of crossings in random complete graphs. J. Soc. Ind. Appl. Math. **13**, 506–510 (1965)
13. Richter, R.B., Thomassen, C.: Relations between crossing numbers of complete and complete bipartite graphs. Am. Math. Mon. **104**(2), 131–137 (1997)
14. Schaefer, M.: Crossing numbers of graphs. Discrete Mathematics and its Applications (Boca Raton). CRC Press, Boca Raton (2018)
15. Scheinerman, E.R., Wilf, H.S.: The rectilinear crossing number of a complete graph and Sylvester's "four point problem" of geometric probability. Am. Math. Mon. **101**(10), 939–943 (1994). https://doi.org/10.2307/2975158
16. Sylvester, J.J.: On a special class of questions on the theory of probabilities, pp. 8–9. Birmingham British Assoc. Report (1865)

17. Urbanik, K.: Solution du problème posé par P. Turán. In: Colloq. Math. **3**, 200–201 (1955)
18. Wegert, E., Trefethen, L.N.: From the Buffon needle problem to the Kreiss matrix theorem. Am. Math. Mon. **101**(2), 132–139 (1994). https://doi.org/10.2307/2324361
19. Zarankiewicz, K.: On a problem of P. Turán concerning graphs. Fundam. Math. **1**(41), 137–145 (1955)

Crossings, k-Planar Graphs

Crossings Between Non-homotopic Edges

János Pach[1,2,4], Gábor Tardos[1,3,4], and Géza Tóth[1,5(✉)]

[1] Rényi Institute, Budapest, Hungary
{pach,tardos,geza}@renyi.hu
[2] IST Austria, Vienna, Austria
[3] Department of Mathematics, Central European University, Budapest, Hungary
[4] Moscow Institute of Physics and Technology, Moscow, Russia
[5] Budapest University of Technology and Economics, SZIT, Budapest, Hungary

Abstract. We call a multigraph *non-homotopic* if it can be drawn in the plane in such a way that no two edges connecting the same pair of vertices can be continuously transformed into each other without passing through a vertex, and no loop can be shrunk to its end-vertex in the same way. It is easy to see that a non-homotopic multigraph on $n > 1$ vertices can have arbitrarily many edges. We prove that the number of crossings between the edges of a non-homotopic multigraph with n vertices and $m > 4n$ edges is larger than $c\frac{m^2}{n}$ for some constant $c > 0$, and that this bound is tight up to a polylogarithmic factor. We also show that the lower bound is not asymptotically sharp as n is fixed and $m \longrightarrow \infty$.

Keywords: Crossing number · Loop · Homotopic

1 Introduction

A standard parameter for measuring the non-planarity of a graph G is its *crossing number*, which is defined as the smallest number $\mathrm{cr}(G)$ of crossing points in any drawing of G in the plane. For many interesting variants of the crossing number, see [11,13,14,16]. Computing $\mathrm{cr}(G)$ is an NP-complete problem [5].

Perhaps the most useful result on crossing numbers, is the so-called *crossing lemma*, proved independently by Ajtai, Chvátal, Newborn, Szemerédi [3] and Leighton [8], according to which the crossing number of any graph with n vertices and $m > 4n$ edges is at least $c\frac{m^3}{n^2}$, for a suitable constant $c > 0$. For the best known value of the constant c, see [1,10]. This result, which is tight up to the constant factor, has been successfully applied to a variety of problems in discrete and computational geometry, additive number theory, algebra, and elsewhere [4,

Supported by the National Research, Development and Innovation Office, NKFIH, KKP-133864, K-131529, K-116769, K-132696, by the Higher Educational Institutional Excellence Program 2019 NKFIH-1158-6/2019, the Austrian Science Fund (FWF), grant Z 342-N31, by the Ministry of Education and Science of the Russian Federation MegaGrant No. 075-15-2019-1926, and by the ERC Synergy Grant "Dynasnet" No. 810115. A full version can be found at https://arxiv.org/abs/2006.14908.

D. Auber and P. Valtr (Eds.): GD 2020, LNCS 12590, pp. 359–371, 2020.
https://doi.org/10.1007/978-3-030-68766-3_28

15]. In some applications, it was the bottleneck that one needed a lower bound on the crossing number of a *multigraph* rather than a graph. Obviously, the crossing lemma does not hold in this case, as stated. Indeed, one can connect a pair of vertices ($n = 2$) with m parallel edges without creating any crossing. However, for multigraphs G with maximum edge *multiplicity* k and $m > 4kn$ edges, Székely [15] established the lower bound $\mathrm{cr}(G) > c'\frac{m^3}{kn^2}$, where $c' > 0$ is another constant. This bound is also tight, up to the constant factor. Ágoston and Pálvölgyi [2] observed that c' can be chosen to be the same as the best known constant c in the crossing lemma (presently, $\frac{1}{29}$).

As the multiplicity k increases, Székely's bound gets weaker and weaker. Luckily, the term k in the denominator can be eliminated in several special cases; see [7,12]. That is, the result holds without putting any upper bound on the edge multiplicity. However, in all of these cases, we have to assume (among other things) that no two adjacent edges cross.

In this paper, we study the analogous question under the weakest possible assumption. Obviously, we need to assume that no pair of parallel edges or loops are *homotopic*, i.e., they cannot be continuously deformed into each other so that their interiors do not pass through any vertex. As we have noted above, without this assumption, a multigraph can have arbitrarily many non-crossing edges. For simplicity, we will also assume that there are no *trivial* loops, that is, no loop can be transformed into a point. Clearly, this latter assumption can be eliminated as the first condition already implies that there is at most a single trivial loop at any vertex.

To state our results, we need to agree about the definitions.

A *multigraph* is a graph in which parallel edges and loops are permitted. A *topological graph (or multigraph)* is a graph (multigraph) $G = (V, E)$ drawn in the plane with the property that every vertex is represented by a distinct point and every edge $e \in E$ is represented by a continuous curve, i.e., a continuous function $f_e \colon [0, 1] \to \mathbb{R}^2$ with $f_e(0)$ and $f_e(1)$ being the endpoints of e. In terminology, we do not distinguish between the vertices and the points representing them. In the same spirit, if there is no danger of confusion, we often use the term edge instead of the curve f_e representing it or the image of f_e. As we deal with non-oriented multigraphs, we treat the functions $f_e(t)$ and $f_e(1 - t)$ as being the same. We assume that no edge passes through any vertex (i.e., $f_e(t) \notin V$ for $0 < t < 1$).

The *crossing number* of a *topological multigraph* G is the number of crossings between its edges, i.e, the number of unordered pairs of distinct pairs $(e, t), (e', t') \in E \times (0, 1)$ with $f_e(t) = f_{e'}(t')$. With a slight abuse of notation, this number will be denoted also by $\mathrm{cr}(G)$.

Two parallel edges, e, e', connecting the same pair of vertices, $u, v \in V$ are *homotopic*, if there exists a continuous function (*homotopy*) $g \colon [0, 1]^2 \to \mathbb{R}^2$ satisfying the following three conditions.

$$g(0, t) = f_e(t) \text{ and } g(1, t) = f_{e'}(t) \text{ for all } t \in [0, 1],$$

$$g(s, 0) = u \text{ and } g(s, 1) = v \text{ for all } s \in [0, 1],$$

$$g(s, t) \notin V \text{ for all } s, t \in (0, 1).$$

Recall that we do not distinguish f_e from $f_e(1-t)$, so we call e and e' homotopic also if $f_e(1-t)$ and $f_{e'}(t)$ are homotopic in the above sense. A loop at vertex u is said to be *trivial* if it is homotopic to the constant function $f(t) = u$.

A topological multigraph $G = (V, E)$ is called *non-homotopic* if it does not contain two homotopic edges, and does not contain a trivial loop.

Obviously, if G is a simple topological graph (no parallel edges or loops), then it is non-homotopic. A non-homotopic multigraph with zero or one vertex has no edge. However, if the number of vertices n is at least 2, the number of edges can be arbitrarily large, even infinite. Our first result provides a lower bound on the crossing number of non-homotopic topological multigraphs in terms of the number of their vertices and edges.

Theorem 1. *The crossing number of a non-homotopic topological multigraph G with $n > 1$ vertices and $m > 4n$ edges satisfies $\mathrm{cr}(G) \geq \frac{1}{24}\frac{m^2}{n}$.*

This bound is tight up to a polylogarithmic factor.

Theorem 2. *For any $n \geq 2$, $m > 4n$, there exists a non-homotopic multigraph G with n vertices and m edges such that its crossing number satisfies $\mathrm{cr}(G) \leq 30\frac{m^2}{n}\log_2^2\frac{m}{n}$.*

The constant 30 in the theorem was chosen for the proof to work for all n and m, and we made no attempt to optimize it. However, it can be replaced by $1 + o(1)$ if both n and m/n go to infinity.

Define the function $\mathrm{cr}(n, m)$ as the minimum crossing number of a non-homotopic multigraph with n vertices and m edges. Theorems 1 and 2 can be stated as

$$\frac{1}{24}\frac{m^2}{n} \leq \mathrm{cr}(n, m) \leq 30\frac{m^2}{n}\log_2^2\frac{m}{n},$$

for any $n \geq 2$ and $m > 4n$. We have been unable to close the gap between these bounds. Our next theorem shows that the lower bound is not tight.

Theorem 3. *The minimum crossing number of a non-homotopic multigraph with $n \geq 2$ vertices and m edges is super-quadratic in m. That is, for any fixed $n \geq 2$, we have*

$$\lim_{m \to \infty} \frac{\mathrm{cr}(n, m)}{m^2} = \infty.$$

More precisely, we obtain $\frac{\mathrm{cr}(n,m)}{m^2} = \Omega(\log m^{1/(6n)}/n^7)$.

Let n, k be positive integers, and consider a set S obtained from the Euclidean plane by removing n distinct points. Fix a point $x \in S$. An oriented loop in S that starts and ends at x is called an *x-loop*. An x-loop may have self-intersections. Contrary to our convention for edges of a topological multigraph, we do distinguish between an x-loop and its reverse. We consider the homotopy type of x-loops in S, that is, we consider two loops *homotopic* if one can be continuously transformed to the other within S. When counting self-intersections of x-loops

362   J. Pach et al.

or intersections between two x-loops, we count points of multiple intersections with the appropriate multiplicity.

To establish Theorems 2 and 3, we study the following topological problem of independent interest [6].

Problem 1. Let $n, k \geq 1$ be integers, let S denote the set obtained from $\mathbb{R}^2$ by removing n distinct points, and let us fix $x \in S$. Determine or estimate the maximum number $f(n, k)$ of pairwise non-homotopic x-loops in S such that none of them passes through x, each of them has fewer than k self-intersections and every pair of them cross fewer than k times.

It is not at all obvious that $f(n, k)$ is finite. However, in the sequel we show that this is the case. This fact is crucially important for the proof of Theorem 3.

Theorem 4. *For any integers $n \geq 2$ and $k \geq 1$, we have*

$$f(n, k) < 2^{(2k)^{2n}}.$$

Our proof of Theorem 2 is based on a lower bound on $f(n, k)$. For this application, all we need is the $n = 2$ special case. Next we state a lower bound valid for all n.

Theorem 5. *Let $n \geq 2$ and $k \geq 1$ be integers. If $2 \leq n \leq 2k$, then*

$$f(n, k) \geq 2^{\sqrt{nk}/3}$$

holds. For $n \geq 2k$, we have

$$f(n, k) \geq (n/k)^{k-1}.$$

There is a huge gap between this bound and the upper bound in Theorem 4. We suspect that the truth is to the lower bound. More precisely, we conjecture that $\log f(n, k)$ can be bounded from above by a polynomial of k whose degree does not depend on n. For $n = 2$ we have $2^{\sqrt{k}/3} \leq f(2, k) \leq 2^{16k^4}$.

Our paper is organized as follows. In Sect. 2, we establish Theorem 1. In Sect. 3, we present some constructions proving Theorem 5, and apply them to deduce Theorem 2. In Sect. 4, we prove Theorem 4. The proof of Theorem 3 is omitted in this version.

2   Loose Multigraphs—Proof of Theorem 1

One can also define topological multigraphs and non-homotopic multigraphs on the sphere S^2. If we consider S^2 as the single point compactification of the plane with the *ideal point* p^*, then any topological multigraph H drawn in the plane remains a topological multigraph on the sphere. However, it may lose the non-homotopic property, as the addition of the ideal point p^* may turn a loop trivial or two parallel edges homotopic. This can be avoided by adding p^* as an isolated

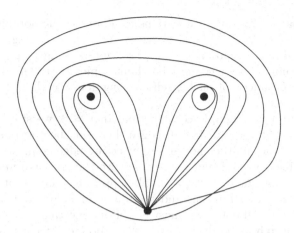

Fig. 1. A non-homotopic loose multigraph with 3 vertices and 6 edges (loops).

vertex to H: in this case, the resulting multigraph H^* is non-homotopic even on the sphere.

We say that a topological multigraph is *loose* if no pair of distinct edges cross each other. An edge (in particular, a loop) is allowed to cross itself. We start by finding the maximum number of edges in a loose non-homotopic multigraph on the sphere or in the plane, for a given number of vertices. We will see that despite allowing parallel edges, loops, and self-intersections, loose non-homotopic multigraphs with $n > 2$ vertices on the sphere cannot have more than $3n - 6$ edges, the maximum number of edges of a simple planar graph. However, there are many other nontrivial examples, for which this bound is tight. The interested reader can verify that, for all $n > 2$, there are extremal examples, all of whose edges are loops. See Fig. 1 for the case of three vertices in the plane.

Lemma 1. *On the sphere, any loose non-homotopic multigraph with $n > 2$ vertices has $m \leq 3n - 6$ edges. For $n = 2$, the maximum number of edges is 1.*

The proof of Lemma 1 is omitted in this version.

Lemma 2. *In the plane, any loose non-homotopic multigraph with $n \geq 1$ vertices has at most $3n - 3$ edges. This bound can be achieved for every n.*

Proof. Let H be a loose non-homotopic multigraph in the plane with $n \geq 1$ vertices and m edges. Consider the plane as the sphere S^2 with a point p^* removed. Add p^* to H as an isolated vertex, to obtain a topological multigraph H' on the sphere. Then H' is a loose non-homotopic multigraph with $n + 1$ vertices and m edges. If $n > 1$, applying Lemma 1 to H', we obtain that $m \leq 3n - 3$, as required. If $n = 1$, then H is a single-vertex topological multigraph in the plane, so all of its edges must be trivial loops. However, by definition, a non-homotopic multigraph cannot have any trivial loop. This completes the proof of the upper bound.

There are many different constructions for loose non-homotopic multigraphs for which the bound in the lemma is achieved. Such a topological multigraph may have several components and several self-intersecting loops. (However, all self-crossings of non-loop edges must be "homotopically trivial": the removal of the closed curve produced by such a self-crossing does not change the homotopy type of the edge.)

Here, we give a very simple construction. If $n > 2$, we start with a triangulation with n vertices and $3n - 6$ edges. Let uvw be the boundary of the unbounded face. Add another non-self-intersecting edge connecting u and v in the unbounded face, which is not homotopic with the arc uv of uvw. Finally, we add two further loops at u. First, a simple loop l that has all other edges and vertices (except u) in its interior, and then another loop l' outside of l, which goes twice around l. (Of course, l' must be self-intersecting.)

If $n = 1$, the graph with no edge achieves the bound of the lemma. For $n = 2$, draw an edge e connecting the two vertices, u and v. Then add two loops at u, as above: a simple loop l around e and another loop l' that winds around l twice. ☐

Proof of Theorem 1. Let G be a non-homotopic topological multigraph in the plane with $n > 1$ vertices and $m > 4n$ edges.

Let D denote the *non-crossing graph* of the edges of G, that is, let $V(D) = E(G)$ and connect two vertices of D by an edge if and only if the corresponding edges of G do not share an interior point. Any clique in D corresponds to a loose non-homotopic sub-multigraph of G. Therefore, by Lemma 2, D has no clique of size $3n - 2$. Thus, by Turán's theorem [17],

$$|E(D)| \le \frac{|V(D)|^2}{2}\left(1 - \frac{1}{3n - 3}\right) = \frac{m^2}{2}\left(1 - \frac{1}{3n - 3}\right).$$

The crossing number $\mathrm{cr}(G)$ is at least the number of crossing pairs of edges in G, which is equal to the number of non-edges of D. Since $m > 4n$, we have

$$\mathrm{crG} \ge \binom{m}{2} - \frac{m^2}{2}\left(1 - \frac{1}{3n - 3}\right) \ge \frac{1}{24}\frac{m^2}{n},$$

as claimed. ☐

The proof above gives a lower bound on the number of crossing pairs of edges in G, and in this respect it is tight up to a constant factor. To see this, suppose for simplicity that n is even and m is divisible by n. Let G_0 be a non-homotopic topological multigraph with two vertices and $\frac{2m}{n}$ non-homotopic loops on one of its vertices. Taking $\frac{n}{2}$ disjoint copies of G_0, we obtain a non-homotopic topological multigraph with n vertices, m edges, and $< \frac{m^2}{n}$ crossing pairs.

3   Two Constructions—Proofs of Theorems 5 and 2

The aim of this section is to demonstrate how to construct topological graphs with many edges and families consisting of many loops, without creating many crossings. The constructions are based on the description of the fundamental group of the plane from which a certain number of points have been removed.

Proof of Theorem 5. Let $S = \mathbb{R}^2 \setminus \{a_1, \ldots, a_n\}$, where $a_1, \ldots, a_n$ are distinct points in the plane, and let $x \in S$ be also fixed. Assume without loss of generality that $a_i = (i, 0)$, $1 \leq i \leq n$, and $x = (0, -1)$. Recall that an x-*loop* is a (possibly self-crossing) oriented path in S from x to x, i.e., a continuous function $f \colon [0, 1] \to S$ with $f(0) = f(1) = x$.

Note that the homotopy group of S is the free group F_n generated by $g_1, \ldots, g_n$, where g_i can be represented by a triangular x-loop around a_i, for example the one going from x to $(2i - 1, 1)$, from here to $(2i + 1, 1)$, and then back to x along three straight-line segments; see [9].

We define an *elementary loop* to be a polygonal x-loop with intermediate vertices

$$(1, \pm 1/2), (2, \pm 1/2), \ldots, (n, \pm 1/2), (n + 1, -1),$$

in this order. There are 2^n distinct elementary loops, depending on the choice of the signs. Each of them represents a distinct homotopy class of the form $g_{i_1} \cdots g_{i_t}$, where the indices form a strictly increasing sequence. By making infinitesimal perturbations on the interior vertices of the elementary loops, we can make sure that every pair of them intersect in at most $n - 1$ points. Thus, we have $f(n, n) \geq 2^n$.

We call $1 \leq i < n$ a *sign change* in the elementary loop l if l passes through both $(i, 1/2)$ and $(i + 1, -1/2)$, or both $(i, -1/2)$ and $(i + 1, 1/2)$. There are precisely $2\binom{n-1}{j}$ elementary loops with exactly j sign changes. The reader can easily verify that crossings between perturbed elementary loops are unavoidable only if a sign change occurs. More precisely, for $k \leq n$, one can perturb all elementary loops with at most $k - 1$ sign changes such that every pair cross at most $k - 1$ times. Hence, we have $f(n, k) \geq 2\sum_{j=0}^{k-1} \binom{n-1}{j} \geq 2\binom{n}{k-1} > (n/k)^{k-1}$, completing the proof of the theorem, whenever $n \geq 2k$.

If $k \leq n \leq 2k$, we have $f(n, k) \geq f(k, k) \geq 2^k < 2^{\sqrt{nk}/3}$. Similarly, if $n \leq k \leq 9n$, we have $f(n, k) \geq f(n, n) \geq 2^n \geq 2^{\sqrt{nk}/3}$, and we are done.

Finally, in the case $k > 9n$, we consider all x-loops which can be obtained as the product (concatenation) of $j = \lfloor \sqrt{\frac{k-1}{n}} \rfloor \geq 3$ elementary loops. Unfortunately, some of these concatenated x-loops will be homotopic. For example, if the elementary loops l_1, l_2, l_3, and l_4 represent the homotopy classes g_1, g_2g_3, g_1g_2, and g_3, respectively, then l_1l_2 and l_3l_4 are homotopic. To avoid this complication, we only use the 2^{n-1} elementary loops that represent homotopy classes involving g_1 (that is, the ones with $(1, +1/2)$ as their first intermediate vertex). Then no two of the resulting $2^{j(n-1)}$ x-loops will be homotopic. By infinitesimal perturbation of the interior vertices of these x-loops (including the $j - 1$ interior vertices at x), we can attain that they do not pass through x, and no two polygonal paths corresponding to a single elementary loop intersect more than n times. Therefore, any pair of perturbed concatenated loops cross at most $j^2n < k$ times, and the same bound holds for the number of self-intersections of any concatenated loop. This yields that $f(n, k) \geq 2^{j(n-1)} \geq 2^{\sqrt{nk}/3}$. $\qquad \square$

Proof of Theorem 2. We want to construct a non-homotopic topological multi-graph G with n vertices, m edges, and few crossings. We distinguish 3 cases.

Case A: If $n = 3$, we set $k = \lceil 2\log_2^2(2m) \rceil$. Theorem 5 guarantees that $f(2, k) \geq 2m$. Thus, there are $2m$ pairwise non-homotopic x-loops in $S = \mathbb{R}^2 \setminus \{a_1, a_2\}$ such that each of them has fewer than k self-intersections and any pair intersect fewer than k times. Regard this arrangement as a topological multigraph G with $2m$ edges on the vertex set $\{a_1, a_2, x\}$. All edges are x-loops. At most one of them is trivial, and for each loop edge there is at most one other loop edge homotopic to it (which must come from an x-loop with inverse orientation). Therefore, we can always select m edges that form a non-homotopic multigraph. Obviously, we have $\mathrm{cr}(G) < k(m + \binom{m}{2})$.

Case B: If $n > 3$, we set $n^* = \lfloor n/3 \rfloor$, $m_0 = \lceil m/n^* \rceil$. Take n^* disjoint copies of the non-homotopic multigraph G_0 with 3 vertices and m_0 edges constructed in Case A. We add at most 2 isolated vertices and remove a few edges if necessary to obtain a non-homotopic multigraph on n vertices and m edges. We clearly have $\mathrm{cr}(G) \leq n^*\mathrm{cr}(G_0)$.

Clearly, the crossing numbers of the graphs constructed in Cases A and B are within the bound stated in the theorem.

Case C: If $n = 2$, we cannot use Theorem 5 directly. Note that all edges of the non-homotopic multigraphs G constructed in Case A were loops at a vertex x, and these x-loops were pairwise non-homotopic even in the set obtained from the plane by keeping x, but removing every other vertex. Now we cannot afford this luxury without creating $\Omega(m^3)$ crossings. However, even for $n = 2$, we can construct a topological multigraph G with many pairwise non-homotopic edges and relatively few crossings, as sketched below.

Let $V(G) = \{a_1, a_2\}$, where a_1 and a_2 are distinct points in the plane, and set $S = \mathbb{R} \setminus V(G)$. Choose a base point $x \in S$ not on the line a_1a_2. Now the homotopy group of S is the free group generated by two elements, g_1 and g_2, that can be represented by triangular x-loops around a_1 and a_2, respectively. By the proof of Theorem 4, with the notation used there, we can construct 2^j pairwise non-homotopic x-loops in S with few crossings. Each of these x-loops, l, can be turned into either a loop edge at the vertex a_1 or into an a_1a_2 edge, as follows: we start with the straight-line segment a_1x, then follow l, finally add a straight-line segment from x to either a_1 (for a loop edge) or to a_2 (to obtain a non-loop edge). After infinitesimally perturbing the resulting edges, one can easily bound the crossing number. However, now we face a new complication: there may be a large number of pairwise homotopic edges. In Case A, when we regarded x-loops as loop edges in a topological multigraph having x as a vertex, two loop edges could only be homotopic if the corresponding x-loops represented the same or inverse homotopy classes. Now the situation is more complicated: a loop edge constructed from an x-loop representing an element g in the homotopy group is homotopic to an another edge constructed from another x-loop representing g' if and only if we have $g' = g_1^s g g_1^t$ or $g' = g_1^s g^{-1} g_1^t$ for some integers s and t. (For non-loop edges the corresponding condition is $g' = g_1^s g g_2^t$.) We may have constructed more than two (even an unbounded number of) homotopic edges, but a closer look at the 2^j x-loops constructed in the proof of the lower bound on $f(2, k)$ reveals that 2^{j-2} of them yield pairwise non-homotopic edges.   □

Remark. For $n \geq 3$, in our constructions all edges are loops. By splitting the base points of the loops, we can get constructions with no loops.

4 Loops with Bounded Number of Pairwise Intersections—Proof of Theorem 4

Consider a loop (oriented closed curve) l in the plane, and a point r not belonging to l. The *winding number* of l around r is the number of times the loop goes around r in the counter-clockwise direction. Going around r in the clockwise direction counts negatively.

Let S be obtained by removing a single point r from the plane. It is well known that two loops in S are homotopic if and only if their winding numbers around r are the same.

Lemma 3. *Let l be any loop in the plane with fewer than k self-intersections, and let x be a point that does not belong to l. Then the absolute value of the winding number of l around x is at most k.*

Proof. Removing the image of l from the plane, it falls into connected components, called *faces*. Obviously, the winding number of l is the same around any two points, x and y, that belong to the same face. Take a point in each face and connect two distinct points if the corresponding faces have a common boundary curve. We get a connected graph. If x and y are adjacent, then the winding number of l around x and y differs by precisely 1. As l has fewer than k self-intersections, the number of faces is at most $k + 1$. The winding number of l around any point of the unbounded face is zero. Therefore, the winding number of l around any point not belonging to l is between $-k$ and $+k$, as claimed.   □

Corollary 1. *For any integer $k > 0$, we have $f(1, k) \leq 2k + 1$.*

Proof. Let x and a be two distinct points in the plane $\mathbb{R}^2$. Any two x-loops in $S = \mathbb{R}^2 \setminus \{a\}$ are homotopic if they have the same winding number around a. For an x-loop with fewer than k self-crossing this is winding number is takes values between $-k$ and k. therefore, any collection of pairwise non-homotopic such loops has cardinality at most $2k + 1$.   □

In the rest of this section, we estimate the function $f(n, k)$ for $n > 1$. By the definition of $f(n, k)$, we have to consider a set S that can be obtained from $\mathbb{R}^2$ by removing n distinct points. As before, we consider the 2-sphere S^2 as the compactification of the plane with a single point p^*, the "ideal point". To simplify the presentation, we view S as a set obtained from S^2 by the removal of a set T of $n + 1$ points (including p^*). We also fix the common starting point $x \in S$ of all loops in S that we consider.

Let L be a collection of loops in S. The connected components of S^2 minus the set of all points of the elements of L are called L-*faces*. Obviously, all L-faces are homeomorphic to the plane and the points of T are scattered among them. We call L *balanced* if no L-face contains n or $n+1$ points of T.

Lemma 4. *Let k be a positive integer, let $x \in S$, and let H be a collection of pairwise non-homotopic nontrivial x-loops in S, each of which has fewer than k self-intersections.*

If $|H| > 2k+1$, then there is a balanced pair, L, of loops in H.

Proof. For $n = 1$, there is no balanced family. Nevertheless, formally the statement holds even in this case, because Lemma 3 implies that $|H| \leq 2k+1$. (In fact, now $|H| < 2k+1$ because of the non-triviality condition.)

Suppose that $n > 1$. Consider any loop $l \in H$. If $\{l\}$ is balanced, then any pair containing l is also balanced and we are done. Otherwise, there is an $\{l\}$-face F containing at least n points of T. It cannot contain all points of T, because then l would be contractible, that is, trivial.

Therefore, we can assume that there is a single point $t \in T$ outside F. We say that l *separates* t. If two loops, $l_1, l_2 \in H$, separate distinct points, $t_1, t_2 \in T$, respectively, then $L = \{l_1, l_2\}$ is a balanced pair, because t_1 and t_2 must lie in separate L-faces, distinct from all L-faces containing other points of T.

Hence, we may assume that all loops in H separate the same point $t \in T$. By symmetry, we may also assume $t \neq p^*$ (t is not the ideal point), so t is in the plane. By Lemma 3, the winding number of any loop $l \in H$ around t is between $-k$ and $+k$. If $|H| > 2k+1$, by the pigeonhole principle, there are two distinct loops, $l_1, l_2 \in H$, with the same winding number around t. This implies that $L = \{l_1, l_2\}$ is a balanced pair. Indeed, otherwise all points in $T \setminus \{t\}$ would be in the same L-face F. In this case, all points of $T \setminus \{t, r\}$ would lie in the unbounded face of the arrangement of the loops l_1 and l_2 in the plane. Since l_1 and l_2 have the same winding number around t, it would follow that they are homotopic, a contradiction. $\square$

Now we are in a position to establish the following recurrence relation for $f(n, k)$, which, together with Lemma 1, implies the upper bound in Theorem 4.

Lemma 5. *For any $n > 1$, $k > 0$, we have $f(n, k) \leq (6kf(n-1, k))^{2k}$.*

Proof. Consider a family H of loops for which in the definition of $f(n, k)$ the maximum is attained. That is, consider $S = S^2 \setminus T$ with $|T| = n + 1$, fix a point $x \in S$, and let H consist of $f(n, k)$ pairwise non-homotopic x-loops in S not passing through x, such that each loop has fewer than k self-intersections and every pair of loops intersect in fewer than k points. We may also assume, by infinitesimal perturbations, that there is no triple-intersection and that any intersection point of the loops is a *transversal* crossing, where one arc passes from one side of the other arc to the other side.

If $|H| \leq 2k+1$, we are done. Suppose that $|H| > 2k+1$. By Lemma 4, there exists a balanced two-element subset $L \subset H$. Fix such a subset L, and turn the arrangement of the two loops in L to a multigraph drawn on the sphere, as follows. Regard x and all intersection points as vertices, so that we obtain a planar drawing of a 4-regular connected multigraph G with at most $3k - 2$ vertices. Thus, the number of edges of G satisfies $|E(G)| \leq 6k - 4$. For any edge f of G, we designate an arbitrary curve in S starting at x and ending at an internal point of f, and we call it the *leash* of f. (For example, we may choose the leash to pass very close to the edges of G.) For any internal point a of f, the *standard path from x to a* is the leash of f followed by the piece of f between the endpoint of the leash and a. When referring to the standard path to x, we mean the single point curve.

Consider a loop $l \in H \setminus L$. Here l starts at x and has later some $j \leq 2k - 2$ further intersections with the edges of G, each time properly crossing an edge from one face of G to another. Define the *signature of l* as the sequence of these j edges of G, together with the information where the loop starts and ends in a tiny neighborhood of x. For the latter, we just record the cyclic order of the initial and final portions of l and the edges of G as they appear around x, so we have at most 20 possibilities. (For the initial portion, we have 4 possibilities and for the final one 5.) For the sequence of edges, we have at most $|E(G)|^j$ possibilities. Taking into account that $0 \leq j \leq 2k - 2$ and $|E(G)| \leq 6k - 4$, the number of different signatures of the loops in $H \setminus L$ is smaller than $(6k)^{2k}$.

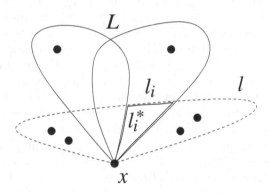

Fig. 2. The definition of the x-loops l_i^*.

Next, we fix a signature and bound the number of elements in the subset $H^* \subseteq H \setminus L$ of all loops that have this signature. Any element $l \in H^*$ that has j crossings with the edges of G, is divided into $j + 1$ *curve-segments* (or, simply, *segments*) $l_0, l_1, \ldots, l_j$. We extend each l_i into an x-loop l_i^* as follows. Let l_i^* start with the standard path from x to the initial point of l_i, followed by l_i, and then completed by the *reverse* of the standard path from x to the final point b of l_i. See Fig. 2. Note that the product $l_0^* l_1^* \ldots l_j^*$ is an x-loop homotopic to l.

This means that for any two distinct (and, therefore, non-homotopic) loops $l, l' \in H^*$, there must be an index $0 \leq i \leq j$ such that (the extension of) the ith segment of l is not homotopic to (the extension of) the ith segment of l'.

We claim that for any fixed i, the number of distinct homotopy classes of the ith segments of the loops in H^* is at most $f(n-1,k)$. If true, this would immediately imply that $|H^*| \leq (f(n-1,k))^{j+1}$. Summing this bound over all signatures would eventually imply that

$$f(n,k) = |H| \leq (6k)^{2k}(f(n-1,k))^{2k}.$$

It remains to prove the claim. We fix i and a subset $H_0 \subseteq H^*$ such that the ith segments of the loops in H_0 are pairwise non-homotopic. Let F be the L-face (i.e., face of the drawing of the graph G) that contains the ith segment l_i of a loop $l \in H^*$, and let f denote the edge at which l_i starts. Let us fix a point $x' \in F$ very close to f. (For $i = 0$, the segment l_i starts at x, between two edges of G, consecutive in the cyclic order. Then we pick x' very close to x, between these two consecutive edges.) Assign to each $l \in H_0$ an x'-loop l^* in $F \setminus T$, as follows. First, l^* follows f very closely till it reaches the ith curve-segment l_i of l close to its starting point. The second piece of l^* follows l_i almost to its endpoint, and then its third piece follows the boundary of F very closely to get back to p'. If l_i ends on the same edge f of G where it starts, the third piece of l^* follows f very closely. Otherwise, it follows the boundary of F in a fixed cyclic direction.

If the pieces of l^* that follow the boundary of F run closer to it than the distance of any point in $T \cap F$ from the boundary of F, then the homotopy type of l^* determines the homotopy type of l_i^*, and hence all the $|H_0|$ x'-loops will be pairwise non-homotopic. We can also choose these new loops in such a way that every self-intersection of l^* is also a self-intersection of l_i, and hence there are fewer than k such self-intersections. In a similar manner, we can make sure that every intersection between two new loops is actually an intersection between the corresponding loops in H_0, and hence any two new loops intersect fewer than k times. These new loops are pairwise non-homotopic in S. All of them lie in $F \setminus T \subset S$, therefore they are also non-homotopic there. Since F is homeomorphic to the plane, $F \setminus T$ can be obtained from the plane by discarding $|F \cap T|$ points. We know that $|F \cap T| \leq n - 1$, because F is an L-face and L is balanced. This completes the proof of the claim and, hence, the lemma. □

Fix any $k \geq 1$. According to Corollary 1, the upper bound in Theorem 4 holds for $n = 1$ and any $k \geq 1$. Let $n \geq 2$ and suppose that we have already verified the inequality $f(n-1,k) < 2^{(2k)^{2(n-1)}}$. By Lemma 5, we obtain

$$f(n,k) \leq (6kf(n-1,k))^{2k} < (6k2^{(2k)^{2(n-1)}})^{2k} < 2^{(2k)^{2n}},$$

completing the proof of Theorem 4.

References

1. Ackerman, E.: On topological graphs with at most four crossings per edge. Comput. Geom. **85**, 101574 (2019). 31 p

2. Ágoston, P., Pálvölgyi, D.: Improved constant factor for the unit distance problem. manuscript arXiv:2006.06285 (2020)
3. Ajtai, M., Chvátal, V., Newborn, M.N., Szemerédi, E.: Crossing-free subgraphs. In: Theory and Practice of Combinatorics. North-Holland Mathematics Studies, vol. 60, pp. 9–12. North-Holland, Amsterdam (1982)
4. Dey, T.L.: Improved bounds for planar k-sets and related problems. Discrete Comput. Geom. **19**(3), 373–382 (1998)
5. Garey, M.R., Johnson, D.S.: Crossing number is NP-complete. SIAM J. Algebraic Discrete Methods **4**(3), 312–316 (1983)
6. Juvan, M., Malnič, A., Mohar, B.: Systems of curves on surfaces. J. Comb. Theory Ser. B **68**, 7–22 (1996)
7. Kaufmann, M., Pach, J., Tóth, G., Ueckerdt, T.: The number of crossings in multigraphs with no empty lens. In: Biedl, T., Kerren, A. (eds.) GD 2018. LNCS, vol. 11282, pp. 242–254. Springer, Cham (2018). https://doi.org/10.1007/978-3-030-04414-5_17
8. Leighton, T.: Complexity Issues in VLSI. Foundations of Computing Series. MIT Press, Cambridge (1983)
9. Lyndon, R.C., Schupp, P.E.: Combinatorial Group Theory. Ergebnisse der Mathematik und ihrer Grenzgebiete, Band 89. Springer, New York (1977)
10. Pach, J., Radoičić, R., Tardos, G., Tóth, G.: Improving the crossing lemma by finding more crossings in sparse graphs. Discrete Comput. Geom. **36**(4), 527–552 (2006)
11. Pach, J., Tóth, G.: Thirteen problems on crossing numbers. Geombinatorics **9**(4), 199–207 (2000)
12. Pach, J., Tóth, G.: A crossing lemma for multigraphs. In: Proceedings of the 34th Annual Symposium on Computational Geometry (SoCG 2018), vol. 65, pp. 1–13 (2018)
13. Schaefer, M.: The graph crossing number and its variants: a survey. Electron. J. Comb. **1000**, DS21 (2013). Dynamic Survey
14. Schaefer, M.: Crossing Numbers of Graphs. CRC Press, Boca Raton (2018)
15. Székely, L.A.: Crossing numbers and hard Erdős problems in discrete geometry. Comb. Probab. Comput. **6**(3), 353–358 (1997)
16. Székely, L.A.: A successful concept for measuring non-planarity of graphs: the crossing number. In: 6th International Conference on Graph Theory (2004). Discrete Math. 276, no. 1–3, 331–352
17. Turán, P.: On an extremal problem in graph theory. Matematikai és Fizikai Lapok **48**, 436–452 (1941). (in Hungarian)

Improvement on the Crossing Number of Crossing-Critical Graphs

János Barát[1] and Géza Tóth[2,3]([✉])

[1] Department of Mathematics, University of Pannonia, Veszprém, Hungary
barat@mik.pannon.hu
[2] Alfréd Rényi Institute of Mathematics, Budapest, Hungary
geza@renyi.hu
[3] Budapest University of Technology and Economics, SZIT, Budapest, Hungary

Abstract. The crossing number of a graph G is the minimum number of edge crossings over all drawings of G in the plane. A graph G is k-crossing-critical if its crossing number is at least k, but if we remove any edge of G, its crossing number drops below k. There are examples of k-crossing-critical graphs that do not have drawings with exactly k crossings. Richter and Thomassen proved in 1993 that if G is k-crossing-critical, then its crossing number is at most $2.5k + 16$. We improve this bound to $2k + 6\sqrt{k} + 47$.

Keywords: Crossing critical · Crossing number · Graph drawing

1 Introduction

The crossing number $\mathrm{CR}(G)$ of a graph G is the minimum number of edge crossings over all drawings of G in the plane. In the optimal drawing of G, crossings are not necessarily distributed uniformly on the edges. Some edges can be more "responsible" for the crossing number than others. For any positive integer k, there exists a graph G whose crossing number is k, but it has an edge e such that $G - e$ is planar.

On the other hand, Richter and Thomassen [6] (Sect. 3) conjectured that if $\mathrm{CR}(G) = k$, then G contains an edge e such that $\mathrm{CR}(G - e) \geq k - c\sqrt{k}$ for some constant c. They observed that this bound would be optimal, as shown, e.g., by the graph $K_{3,n}$. They managed to prove a much weaker bound, namely, if $\mathrm{CR}(G) = k$, then G contains an edge e such that $\mathrm{CR}(G - e) \geq 2k/5 - 8$.

A graph G is k-crossing-critical if $\mathrm{CR}(G) \geq k$, but $\mathrm{CR}(G - e) < k$ for any edge e of G.

The structure and properties of crossing-critical graphs are fundamental in the study of crossing numbers. It is easy to describe 1-crossing-critical graphs,

Supported by National Research, Development and Innovation Office, NKFIH, K-131529 and the Higher Educational Institutional Excellence Program 2019, the grant of the Hungarian Ministry for Innovation and Technology (Grant Number: NKFIH-1158-6/2019).

D. Auber and P. Valtr (Eds.): GD 2020, LNCS 12590, pp. 372–381, 2020.
https://doi.org/10.1007/978-3-030-68766-3_29

and there is an almost complete description of 2-crossing-critical graphs [3]. For $k > 2$, a description of k-crossing-critical graphs seems hopeless at the moment.

It has been proved recently, that the bounded maximum degree conjecture for k-crossing-critical graphs holds for $k \leq 12$ and does not hold for $k > 12$ [2]. More precisely, there is a constant D with the property that for every $k \leq 12$, every k-crossing-critical graph has maximum degree at most D, and for every $k > 12$, $d \geq 1$, there is a k-crossing-critical graph with maximum degree at least d.

We rephrase the result and conjecture of Richter and Thomassen [6] as follows. They conjectured that if G is k-crossing-critical, then $\mathrm{CR}(G) \leq k + c'\sqrt{k}$ for some $c' > 0$ and this bound would be optimal. They proved that if G is k-crossing-critical, then $\mathrm{CR}(G) \leq 2.5k + 16$. This result has been improved in two special cases.

Lomelí and Salazar [5] proved that for any k there is an $n(k)$ such that if G is k-crossing-critical and has at least $n(k)$ vertices, then $\mathrm{CR}(G) \leq 2k + 23$.

Salazar [7] proved that if G is k-crossing-critical and all vertices of G have degree at least 4, then $\mathrm{CR}(G) \leq 2k + 35$.

It is an easy consequence of the Crossing Lemma [1] that if the average degree in a k-crossing-critical graph is large, then its crossing number is close to k [4]. More precisely, if G is k-crossing critical and it has at least cn edges, where $c \geq 7$, then $\mathrm{CR}(G) \leq kc^2/(c^2 - 29)$.

In this note, we obtain a general improvement.

Theorem 1. *For any $k > 0$, if G is a k-crossing-critical multigraph, then* $\mathrm{CR}(G) \leq 2k + 6\sqrt{k} + 47$.

We need a few definitions and introduce now several parameters for the proof. We also list them at the end of the paper.

Let G be a graph. We call a pair (C, v), where C is a cycle of G and v is a vertex of C, the *cycle C with special vertex v* . (The special vertex meant to be a vertex with large degree.) When it is clear from the context, which one is the special vertex, we just write C instead of (C, v).

Suppose C is a cycle with special vertex v. Let x be a vertex of C. An edge, adjacent to x but not in C, is *hanging from x* in short. Let $l(C) = l(C, v)$ be the *length of C* , that is, the number of its edges. For any vertex x, let $d(x)$ denote the degree of x. Let $h(C) = h(C, v) = \sum_{u \in C, u \neq v}(d(u) - 2)$, that is, the total number of hanging edges from all non-special vertices of C (with multiplicity).

A set of edges is *independent* if no two of them have a common endvertex.

2   The Proof of Richter and Thomassen

In [6], the most important tool in the proof was the following technical result. In this section we review and analyze its proof. The algorithmic argument finds a cycle C recursively such that $h(C)$ is small.

Theorem 0. [6] *Let H be a simple graph with minimum degree at least* 3. *Assume that H has a set E of t edges such that H − E is planar. Then H has a cycle K with special vertex v such that* $h(K) \leq t + 36$.

Proof of Lemma 0. The proof is by induction on t. The induction step can be considered as a process, which constructs a graph H^* from graph H, and cycle K of H, either directly, or from cycle K^* in H^*. For convenience, for any planar graph H, define $H^* = \emptyset$. In the rest of the paper we refer to this as the Richter-Thomassen procedure. The statement of Theorem 0 for $t = 0$ is the following.

Lemma 0. [6] *Let H be a simple planar graph with minimum degree at least* 3. *Then H has a cycle K with special vertex v such that* $l(K) \leq 5$ *and* $h(K) \leq 36$.

Here we omit the proof of Lemma 0. Suppose now that $t > 0$ and we have already shown Theorem 0 for smaller values of t. Let H be a simple graph with minimum degree at least 3. Assume that H has a set E of t edges such that $H − E$ is planar and let $e = uw \in E$. Let $H' = H − e$. We distinguish several cases.

<u>1.</u> H' has no vertex of degree 2. By the induction hypothesis, H' has a cycle K^* with a special vertex v such that $h(K^*) \leq t + 35$. If e is not a chord of K^*, then $K = K^*$ with the same special vertex satisfies the conditions for H. Let $H^* = H'$.

If e is a chord of K^*, then $K^* + e$ determines two cycles, and it is easy to see that either one satisfies the conditions. So, let K be one of them. If K, contains v, then v remains the special vertex. If K does not contain v, then we can choose the special vertex of K arbitrarily. Let $H^* = H'$.

<u>2.</u> H' has a vertex of degree 2. Clearly, only u and w can have degree 2. Suppress vertices of degree 2. That is, for each vertex of degree 2, remove the vertex and connect its neighbors by an edge. Let H'' be the resulting graph. It can have at most two sets of parallel edges.

> <u>2.1.</u> H'' has no parallel edges. By the induction hypothesis, H'' contains a cycle K^* with a special vertex v such that $h(K^*) \leq t + 35$. It corresponds to a cycle K' in H. Let $H^* = H''$.
>
> > <u>2.1.1.</u> The edge e is not incident with K'. In this case, $K = K'$ satisfies the conditions, with the same special vertex as H^*.
> >
> > <u>2.1.2.</u> The edge e has exactly one endvertex on K'. In this case, let $K = K'$ with the same special vertex. Now $h(K) = h(K^*)+1 \leq t+36$ and we are done.
> >
> > <u>2.1.3.</u> The edge e has both endvertices on K'. Now, just like in Case 1, $K' + e$ determines two cycles and it is easy to see that either one satisfies the conditions. If the new cycle contains v, then it will remain the special vertex, if not, then we can choose the special vertex arbitrarily.
>
> <u>2.2.</u> H'' has one set of parallel edges. Let x and y be the endvertices of the parallel edges. We can assume that one of the xy edges in H'' corresponds to the path xuy in H and H'. Clearly, $d(u) = 3$.

<u>2.2.1.</u> Another xy edge in H'' corresponds to the path xwy in H and H'. In this case $d(u) = d(w) = 3$, so for the cycle $K = uxw$ with special vertex x we have $h(K) \leq 2$ and we are done. Let $H^* = \emptyset$. We do not define K^* in this case.

<u>2.2.2.</u> No xy edge in H'' corresponds to the path xwy in H and H' and either $d(x) \leq 37 + t$ or $d(y) \leq 37 + t$. Assume that $d(x) \leq 37 + t$, the other case is treated analogously. Since there were at least two xy edges in H'', H contains the edge xy. For the cycle $K = uxy$, with special vertex y, we have $h(K) \leq 35 + t + 1$, so we are done. Let $H^* = \emptyset$.

<u>2.2.3.</u> No xy edge in H'' corresponds to the path xwy in H and H' and both $d(x), d(y) > 37 + t$. Replace the parallel edges by a single xy edge in H''. In the resulting graph H^*, we can apply the induction hypothesis and get a cycle K^* with special vertex v such that $h(K^*) \leq 35 + t$. Now K^* cannot contain both x and y and if it contains either one, then it has to be the special vertex. Therefore, the cycle K in H, corresponding to K^*, with the same special vertex, satisfies the conditions, since the only edge that can increase $h(K)$ is e, and e is not a chord of K.

<u>2.3.</u> H'' has two sets of parallel edges, xy and ab say. Now H contains the edges xy and ab. We can assume by symmetry that H contains the paths xuy and awb. Also $d(u) = d(w) = 3$ in H.

<u>2.3.1.</u> At least one of a, b, x, y has degree at most $37 + t$ in H. Assume that $d(x) \leq 37 + t$, the other cases are treated analogously. For the cycle $K = uxy$ with special vertex y, we have $h(K) \leq 35 + t + 1$, so we are done. Let $H^* = \emptyset$.

<u>2.3.2.</u> $d(x)$, $d(y)$, $d(a)$, $d(b) > 37 + t$. Replace the parallel edges by single edges xy and ab in H''. In the resulting graph H^*, we can apply the induction hypothesis and get a cycle K^* with special vertex v such that $h(K^*) \leq 35 + t$. However, K^* can contain at most one of x, y, a, and b, and if it contains one, that has to be the special vertex. Therefore, the cycle K in H, corresponding to K^* with the same special vertex satisfies the conditions.

This finishes the proof of Theorem 0. □

3   Proof of Theorem 1

The main idea in the proof of Richter and Thomassen [6] is the following. Suppose that G is k-crossing-critical. Then it has at most k edges whose removal makes G planar. Then by Theorem 0, we find a cycle C with special vertex v such that $h(C) \leq k + 36$. Let e be an edge of C, adjacent to v. We can draw $G - e$ with at most $k - 1$ crossings. Now we add the edge e, along $C - e$, on the "better" side. We get additional crossings from the crossings on $C - e$, and from the hanging edges and we can bound both.

Our contribution is the following. Take a "minimal" set of edges, whose removal makes G planar. Clearly, this set would contain at most k edges. However, we have to define "minimal" in a slightly more complicated way, but still, our set contains at most $k + \sqrt{k}$ edges. We carefully analyze the proof of Richter and Thomassen, extend it with some operations, and find a cycle C with special vertex v such that (roughly) $l(C) + h(C)/2 \leq k + 6\sqrt{k}$. Now, do the redrawing step. If $h(C)$, or the number of crossings on $C - e$ is small, then we get an improvement immediately. If both of them are large, then $l(C)$ is much smaller than the number of crossings on $C - e$. But in this case, we can remove the edges of C, and get rid of many crossings. This way, we can get a bound on the "minimal" set of edges whose removal makes G planar.

As we will see, for the proof we can assume that G is simple and all vertices have degree at least 3. But if we want to prove a better bound, say, $\mathrm{CR}(G) \leq (2 - \varepsilon)k + o(k)$, then we cannot prove that the result for simple graphs implies the result for multigraphs. Therefore, the whole proof collapses. Moreover, even if we could assume without loss of generality that G is simple, we still cannot go below the constant 2 with our method. We cannot rule out the possibility that all (or most of the) $k - 1$ crossings are on $C - e$.

Proof of Theorem 1. Suppose that G is k-crossing-critical. Just like in the paper of Richter and Thomassen [6], we can assume that G is simple and all vertices have degree at least 3. We sketch the argument.

If G has an isolated vertex, we can remove it from G. Suppose that a vertex v of G has degree 1. Then $\mathrm{CR}(G) = \mathrm{CR}(G - v)$, contradicting crossing criticality. Suppose now that v has degree 2. We can suppress v (remove it and connect its neighbors by an edge). The resulting (multi)graph is still k-crossing-critical and has the same crossing number as G.

Clearly, G cannot contain loops, as adding or removing a loop does not change the crossing number. Finally, suppose that e and f are parallel edges, both connecting x and y. Since G is k-crossing-critical, we have $\mathrm{CR}(G-e) \leq k-1$. Take a drawing of the graph $G - e$ with at most $k - 1$ crossings. Add the edge e, drawn very close to f. The obtained drawing of G has at most $2k - 2$ crossings.

So, we assume in the sequel, that G is simple and all vertices have degree at least 3.

Let k' be the smallest integer with the property that we can remove k' edges from G so that the remaining graph is planar. Define the function $f(x, y) = \sqrt{k}x + y$.

Let (t, t') be the pair of numbers that minimizes the function $f(t, t') = \sqrt{k}t + t'$ subject to the following property: There exists a set E of t edges such that $G - E$ is planar, and the set E contains at most t' independent edges. In the next lemma, part (i) is from [6], we repeat it here for completeness.

Lemma 1. *The following two statements hold.*

(i) [6] $k' \leq k$, *and*
(ii) $t \leq k' + \sqrt{k}$.

Proof of Lemma 1. (i) Since G is k-crossing-critical, $G - e$ can be drawn with at most $k - 1$ crossings for any edge e. Remove one of the edges from each crossing in such a drawing. We removed at most k edges in total and got a planar graph. (ii) Let E' be a set of k' edges such that $G - E'$ is planar. Suppose that E' contains at most k'' independent edges. Now $k'' \leq k'$. By the choice of (t, t'), $f(t, t') \leq f(k', k'')$. Consequently, $\sqrt{kt} \leq \sqrt{kt} + t' = f(t, t') \leq f(k', k'') \leq \sqrt{kk'} + k'$. Therefore, $t \leq k' + k'/\sqrt{k} \leq k' + \sqrt{k}$.   □

Now set $E = \{e_1, e_2, \ldots, e_t\}$, where E contains at most t' independent edges, and $G - E$ is planar. Apply the Richter-Thomassen procedure recursively starting with $H_0 = G$. We obtain a sequence of graphs $H_0, H_1, \ldots, H_s$, $(s \leq t)$ such that for $0 \leq i \leq s - 1$, $H_i^* = H_{i+1}$, and $H_s^* = \emptyset$. The procedure stops with graph H_s, where we obtain a cycle C_s either directly, in cases 2.2.1, 2.2.2, and 2.3.1, or by Lemma 0, when H_s is planar. In all cases, $l(C_s) \leq 5$. Following the procedure again, we also obtain cycles $C_{s-1}, \ldots, C_0$ of $H_{s-1}, \ldots, H_0$ respectively such that $0 \leq i \leq s - 1$, $C_i^* = C_{i+1}$. Let $C_0 = C$ with special vertex v.

Lemma 2. *There is a cycle K of G such that $l(K) + h(K)/2 \leq t + 5\sqrt{k} + 48$.*

Proof of Lemma 2. The cycle K will be either C, or a slightly modified version of C. It is clear from the procedure that C does not have a chord in G since we always choose C as a minimal cycle.

Consider the moment of the procedure, when we get cycle K from K^*. All hanging edges of K correspond to a hanging edge of K^*, with the possible exception of $e = uw$. Therefore, if we get a new hanging edge e, then $e \in E$. Taking into account the initial cases in the procedure, that is, when we apply Lemma 0, or we have Cases 2.2.1, 2.2.2, or 2.3.1, we get the following easy observations. We omit the proofs.

Observation 1. (i) *All but at most 36 edges of $G - C$ adjacent to a non-special vertex of C are in E.*
(ii) *For all but at most 4 non-special vertices z' of C, all edges of $G - C$ incident to z', are in E.*   □

Suppose that $l(C) > t' + 6$. Consider $t' + 5$ consecutive vertices on C, none of them being the special vertex v. By Observation 1 (ii), for at least $t' + 1$ of them, all hanging edges are in E. Consider one of these hanging edges at each of these $t' + 1$ vertices. By the definition of t', these $t' + 1$ edges cannot be independent, at least two of them have a common endvertex, which is not on C. Suppose that $x, y \in C$, $z \notin C$, $xz, yz \in E$. Let a be the xy arc of C, which does not contain the special vertex v. Take two consecutive neighbors of z. Assume for simplicity, that they are x and y. Let the cycle (C', z) be formed by arc a of C, together with the path xzy. See Fig. 1. The cycle C' does not have a chord in G. We have $l(C') \leq t' + 6$, $h(C') \leq h(C)$. The edges zx and zy are the only new hanging edges of C' from a non-special vertex. They might not be in E, therefore, the statement of Observation 1 holds in a slightly weaker form.

Observation 2. (i) *All but at most 38 edges of $G-C'$ adjacent to a non-special vertex of C' are in E.*
(ii) *For all but at most 6 non-special vertices z' of C', all edges of $G-C'$ incident to z', are in E.*   □

Let cycle $K = C$, if $l(C) \le t' + 6$, and let $K = C'$, if $l(C) > t' + 6$. In both cases, for the rest of the proof, let v denote the special vertex of K. Let $h = h(K)$, $l = l(K)$. We have

$$l \le t' + 6. \tag{1}$$

The cycle K does not have a chord. In particular, none of $e_1, e_2, \dots, e_t$ can be a chord of K. Now we partition E into three sets, $E = E_p \cup E_q \cup E_m$, where E_p is the subset of edges of E, which have exactly one endvertex on K (these are the hanging edges in E), $E_q = E \cap K$, E_m is the subset of edges of E, which do not have an endvertex on K. Let $p = |E_p|$, $q = |E_q|$, $m = |E_m|$. Let p' denote the number of edges of E_p hanging from the special vertex v. By definition,

$$t = p + q + m \tag{2}$$

and $p \ge p'$.

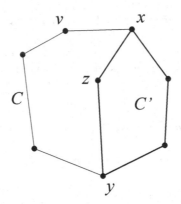

Fig. 1. Cycles C and C'.

It follows from Observations 1 (i) and 2 (i) that

$$h \ge p - p' \ge h - 38. \tag{3}$$

Therefore,

$$h + q + m \le p + q + m + 38 = t + 38.$$

Since all vertices have degree at least 3, and K does not have a chord,

$$h \ge l - 1.$$

Now, at each vertex x of C, where *all* hanging edges belong to E_p, take one such edge. The set of these edges is E'. By Observations 1 (ii) and 2 (ii), $|E'| \geq l - 7$. See Fig. 2. Let $F = E_p \cup E(K) \cup E_m - E'$ where $E(K)$ is the set of edges of K. Since $F \cup E' \supseteq E$, $G' = G - (F \cup E')$ is a planar graph. Let $G'' = G' \cup E' = G - F$. In G'', each edge of E' has an endvertex of degree one. Therefore, we can add all edges of E' to G' without losing planarity. Consequently, the graph $G'' = G' \cup E' = G - F$ is planar.

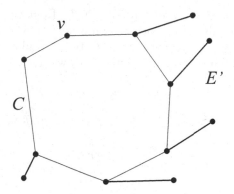

Fig. 2. The edge set E'.

The set F has at most $|F| \leq p + l + m - (l - 7) = p + 7 + m \leq t + 7$ edges by (2). That is

$$|F| \leq t + 7 \qquad (4)$$

Let $F' \subseteq F$ be a maximal set of independent edges in F. To estimate $|F'|$, observe that apart from the edges in $E_m \cap F'$, all edges in F' are adjacent to a vertex of C. Moreover, at most $p - p' - (l - 7) + 1$ of them have one vertex on K, the rest have two. Therefore,

$$|F'| \leq p - p' - (l - 7) + 1 + (l - (p - p' - l + 8))/2 + m = (p - p')/2 + 4 + m.$$

By the choice of the pair (t, t'), $\sqrt{k}t + t' \leq \sqrt{k}|F| + |F'|$.
Therefore, $(p - p')/2 + 4 + m \geq |F'| \geq \sqrt{k}t + t' - \sqrt{k}|F| \geq t' - 5\sqrt{k}$, using (4).

Now evoking (3): $h/2 + 4 + m \geq (p - p')/2 + 4 + m \geq t' - 5\sqrt{k}$.
Therefore, $h/2 + m \geq t' - 4 - 5\sqrt{k} \geq l - 10 - 5\sqrt{k}$ by (1).
Summarizing, we have

$$h + m \leq t + 38, \quad h/2 + m \geq l - 10 - 5\sqrt{k},$$

which implies

$$m \leq t - h + 38, \quad h/2 + t - h + 38 \geq l - 10 - 5\sqrt{k},$$

and finally
$$t + 5\sqrt{k} + 48 \geq l + h/2.$$

This concludes the proof of Lemma 2. □

Now we can finish the proof of Theorem 1. By Lemma 2, we have a cycle K in G with special vertex v such that $h(K)/2 + l(K) \leq t + 5\sqrt{k} + 48$. Let e be an edge of K adjacent to v. Since G was k-crossing-critical, the graph $G - e$ can be drawn with at most $k - 1$ crossings. Let us consider such a drawing D. Let $h = h(K)$, $l = l(K)$.

Suppose the path $K - e$ has cr crossings in D. Remove the edges of K from the drawing, and one edge from each crossing not on $K - e$. Together with e, we removed at most $k + l - cr$ edges from G to get a planar graph. Therefore, $l + k - cr \geq k'$. Combining it with Lemma 1 (ii) we have

$$l + k - cr \geq t - \sqrt{k}.$$

Consequently, $l + k - cr \geq t - \sqrt{k} \geq l + h/2 - 48 - 6\sqrt{k}$ by Lemma 2. That is

$$k + 6\sqrt{k} + 48 \geq cr + h/2.$$

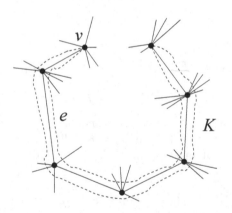

Fig. 3. Adding the missing edge e.

Consider the drawing D of $G - e$. We can add the missing edge e drawn along the path $K - e$ on either side. See Fig. 3. The two possibilities *together* create at most $h + 2cr$ crossings. Choose the one which creates fewer crossings. That makes at most $h/2 + cr$ crossings.

Since $k + 6\sqrt{k} + 48 \geq cr + h/2$, we can add e with at most $k + 6\sqrt{k} + 48$ additional crossings. Hence $\mathrm{CR}(G) \leq 2k + 6\sqrt{k} + 47$. □

Notations

Here we give a list of the parameters and their definitions, used in the proof.

k: G is k-crossing-critical

k': the smallest integer with the property that we can remove k' edges from G so that the remaining graph is planar.

(t, t'): the pair of numbers that minimizes the function $f(t, t') = \sqrt{k}t + t'$ subject to the following property: There exists a set E of t edges such that $G - E$ is planar, and the set E contains at most t' independent edges.

$p = |E_p|$: the number of edges in E that have exactly one endvertex on C.

$q = |E_q|$: the number of edges in $E \cap C$.

$m = |E_m|$: the number of edges in E that do not have an endvertex on C.

p': the number of edges of E_p hanging from the special vertex v of C.

$h = h(C) = h(C, v) = \sum_{u \in C, u \neq v} (d(u) - 2)$, the total number of hanging edges from all non-special vertices of C (with multiplicity).

$l = l(C)$: the length of C.

References

1. Ackerman, E.: On topological graphs with at most four crossings per edge. Comput. Geom. **85**, 1–31 (2019)
2. Bokal, D., Dvořák, Z., Hlinený, P., Leaños, J., Mohar, B., Wiedera, T.: Bounded degree conjecture holds precisely for c-crossing-critical graphs with $c \leq 12$. In: 35th International Symposium on Computational Geometry (SoCG 2019), pp. 14:1–14:15. Schloss Dagstuhl-Leibniz-Zentrum für Informatik (2019)
3. Bokal, D., Oporowski, B., Richter, R.B., Salazar, G.: Characterizing 2-crossing-critical graphs. Adv. Appl. Math. **74**, 23–208 (2016)
4. Fox, J., Tóth, C.D.: On the decay of crossing numbers. J. Comb. Theory Ser. B **98**(1), 33–42 (2008)
5. Lomelí, M., Salazar, G.: Nearly light cycles in embedded graphs and crossing-critical graphs. J. Graph Theory **53**(2), 151–156 (2006)
6. Richter, B.R., Thomassen, C.: Minimal graphs with crossing number at least k. J. Comb. Theory Ser. B **58**, 217–224 (1993)
7. Salazar, G.: On a crossing number result of Richter and Thomassen. J. Comb. Theory Ser. B **79**, 98–99 (2000)

On the Maximum Number of Crossings in Star-Simple Drawings of K_n with No Empty Lens

Stefan Felsner[1] ⓘ, Michael Hoffmann[2] ⓘ, Kristin Knorr[3](✉) ⓘ, and Irene Parada[4] ⓘ

[1] Institute of Mathematics, Technische Universität Berlin, Berlin, Germany
felsner@math.tu-berlin.de
[2] Department of Computer Science, ETH Zürich, Zürich, Switzerland
hoffmann@inf.ethz.ch
[3] Department of Computer Science, Freie Universität Berlin, Berlin, Germany
knorrkri@inf.fu-berlin.de
[4] Department of Mathematics and Computer Science, TU Eindhoven, Eindhoven, The Netherlands
i.m.de.parada.munoz@tue.nl

Abstract. A star-simple drawing of a graph is a drawing in which adjacent edges do not cross. In contrast, there is no restriction on the number of crossings between two independent edges. When allowing empty lenses (a face in the arrangement induced by two edges that is bounded by a 2-cycle), two independent edges may cross arbitrarily many times in a star-simple drawing. We consider star-simple drawings of K_n with no empty lens. In this setting we prove an upper bound of $3((n-4)!)$ on the maximum number of crossings between any pair of edges. It follows that the total number of crossings is finite and upper bounded by $n!$.

Keywords: Star-simple drawings · Topological graphs · Edge crossings

1 Introduction

A *topological drawing* of a graph G is a drawing in the plane where vertices are represented by pairwise distinct points, and edges are represented by Jordan arcs with their vertices as endpoints. Additionally, edges do not contain any

This research started at the 3rd Workshop within the collaborative DACH project *Arrangements and Drawings*, August 19–23, 2019, in Wergenstein (GR), Switzerland, supported by the German Research Foundation (DFG), the Austrian Science Fund (FWF), and the Swiss National Science Foundation (SNSF). We thank the participants for stimulating discussions. S.F. is supported by DFG Project FE 340/12-1. M.H. is supported by SNSF Project 200021E-171681. K.K. is supported by DFG Project MU 3501/3-1 and within the Research Training Group GRK 2434 *Facets of Complexity*. I.P. was supported by FWF project I 3340-N35.

© Springer Nature Switzerland AG 2020
D. Auber and P. Valtr (Eds.): GD 2020, LNCS 12590, pp. 382–389, 2020.
https://doi.org/10.1007/978-3-030-68766-3_30

other vertices, every common point of two edges is either a proper crossing or a common endpoint, and no three edges cross at a single point. A *simple drawing* is a topological drawing in which adjacent edges do not cross, and independent edges cross at most once.

We study a broader class of topological drawings, which are called *star-simple* drawings, where adjacent edges do not cross, but independent edges may cross any number of times; see Fig. 1 for illustration. In such a drawing, for every vertex v the induced substar centered at v is simple, that is, the drawing restricted to the edges incident to v forms a plane drawing. In the literature (e.g., [1,2]) these drawings also appear under the name *semi-simple*, but we prefer star-simple because the name is much more descriptive.

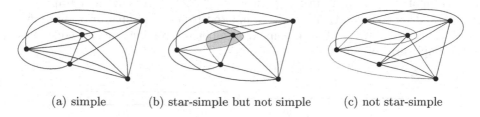

    (a) simple   (b) star-simple but not simple   (c) not star-simple

Fig. 1. Topological drawings of K_6 and a (nonempty) lens (shaded in (b)).

In contrast to simple drawings, star-simple drawings can have regions or cells whose boundary consists of two continuous pieces of (two) edges. We call such a region a *lens*; see Fig. 1b. A lens is *empty* if it has no vertex in its interior. If empty lenses are allowed, the number of crossings in star-simple drawings of graphs with at least two edges is unbounded (twisting), as illustrated in Fig. 2a. We restrict our attention to star-simple drawings with no empty lens. This restriction is—in general—not sufficient to guarantee a bounded number of crossings (spiraling), as illustrated in Fig. 2b. However, we will show that star-simple drawings *of the complete graph K_n* with no empty lens have a bounded number of crossings.

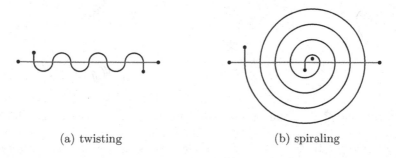

        (a) twisting   (b) spiraling

Fig. 2. Constructions to achieve an unbounded number of crossings.

Empty lenses also play a role in the context of the crossing lemma for multigraphs [5]. This is because a group of arbitrarily many parallel edges can be drawn without a single crossing. Hence, for general multigraphs there is no hope to get a lower bound on the number of crossings as a function of the number of edges. However, if we forbid empty lenses, we cannot draw arbitrarily many parallel edges.

Kynčl [3, Section 5], "Picture hanging without crossings"] proposed a construction of two edges in a graph on n vertices with an exponential number (2^{n-4}) of crossings and no empty lens; see Fig. 3. This configuration can be completed to a star-simple drawing of K_n, cf. [6]. For $n = 6$ it is possible to have one more crossing while maintaining the property that the drawing can be completed to a star-simple drawing of K_6; see Fig. 4. Repeated application of the doubling construction of Fig. 3 leads to two edges with $2^{n-4} + 2^{n-6}$ crossings in a graph on n vertices. This configuration can be completed to a star-simple drawing of K_n. We suspect that this is the maximum number of crossings of two edges in a star-simple drawing of K_n.

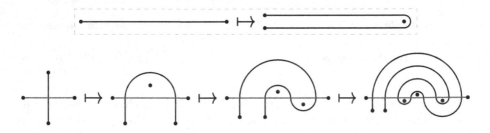

Fig. 3. The doubling construction yields an exponential number of crossings.

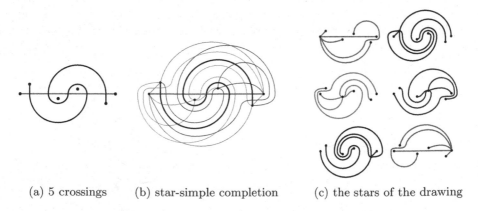

(a) 5 crossings   (b) star-simple completion   (c) the stars of the drawing

Fig. 4. Two edges with $2^{n-4} + 2^{n-6}$ crossings in a star-simple drawing of K_n, for $n = 6$.

2   Crossing Patterns

In this section we study the induced drawing $D(e, e')$ of two independent edges e and e' in a star-simple drawing D of the complete graph. We start by observing that the endpoints of e and e' must lie in the same region of $D(e, e')$. This fact was also used in earlier work by Aichholzer et al. [1] and by Kynčl [4].

Lemma 1. *The four vertices incident to e and e' belong to the same region of $D(e, e')$.*

Proof. Assuming that the two edges cross at least two times, the drawing $D(e, e')$ has at least two regions. Otherwise, the statement is trivial. If the four vertices do not belong to the same region of $D(e, e')$, then there is a vertex u of e and a vertex v of e' that belong to different regions. Now consider the edge uv in the drawing D of the complete graph. This edge has ends in different regions of $D(e, e')$, whence it has a crossing with either e or e'. This, however, makes a crossing in the star of u or v. This contradicts the assumption that D is a star-simple drawing.

Lemma 1 implies that the deadlock configurations as shown in Fig. 5a do not occur in star-simple drawings of complete graphs. Formally, a *deadlock* is a pair e, e' of edges such that not all incident vertices lie in the same region of the drawing $D(e, e')$.

Now suppose that D is a star-simple drawing of a complete graph with no empty lens. In this case we can argue that e and e' do not form a configuration as the black edge e and the red edge e' in Fig. 5b. Indeed, that configuration has an interior lens L and by assumption this lens is non-empty, i.e., L contains a vertex x. Let e and e' be the black and the red edge in Fig. 5b, respectively, and let u be a vertex of e. The edge xu (the green edge in the figure) has no crossing with e, hence it follows the "tunnel" of the black edge. This yields a deadlock configuration of the edges xu and e'. Note that if in Fig. 5b instead of drawing the green edge xu we connect x with an edge f to one of the vertices of the red edge e' such that f and the red edge have no crossing, then f and the black edge e form a deadlock.

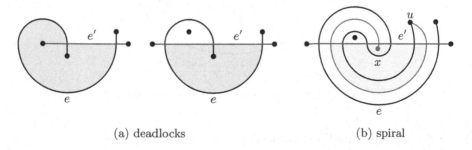

(a) deadlocks   (b) spiral

Fig. 5. Constructions to achieve an unbounded number of crossings.

We use this intuition to formally define a spiral. Two edges e, e' form a *spiral* if they form a lens L such that if we place a vertex x in L and draw a curve γ connecting x to a vertex u of e so that γ does not cross e, then γ and e' form a deadlock. The discussion above proves the following lemma:

Lemma 2. *A star-simple drawing of a complete graph with no empty lens has no pair e, e' of edges that form a spiral.*

3   Crossings of Pairs of Edges

In this section we derive an upper bound for the number of crossings of two edges in a star-simple drawing of K_n with no empty lens.

Theorem 1. *Consider a star-simple drawing of K_n with no empty lens. If $C(k)$ is the maximum number of crossings of a pair of edges that (a) form no deadlock and no spiral and such that (b) all lenses formed by the two edges can be hit by k points, then $C(k) \leq e \cdot k!$, where $e \approx 2.718$ is Euler's number.*

Proof. Due to Lemma 1 we can assume that all four vertices of e and e' are on the outer face of the drawing $D(e, e')$. We think of e' as being drawn red and horizontally and of e as being a black meander edge. Let $p_1, \ldots, p_k$ be points hitting all the lenses of the drawing $D(e, e')$. Let u be one of the endpoints of e. For each $i = 1, \ldots, k$ we draw an edge e_i connecting p_i to u such that e_i has no crossing with e and, subject to this, the number of crossings with e' is minimized. Figure 6 shows an example.

Note that we do not claim that all these edges $e_1, \ldots, e_k$ together with e and e' can be extended to a star-simple drawing of a complete graph. Therefore, we cannot use Lemma 2 directly but state the assumption (a) instead.

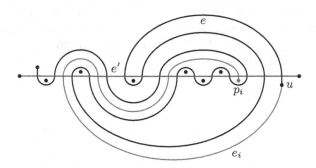

Fig. 6. The drawing $D(e, e')$ and an edge e_i connecting p_i to u.

We claim the following three properties:

(P1) The edges e_i and e' form no deadlock and no spiral.

(P2) All lenses of e_i and e' are hit by the $k-1$ points $p_1, \ldots, p_{i-1}, p_{i+1}, \ldots, p_k$.

(P3) Between any two crossings of e and e' from left to right, i.e., in the order along e', there is at least one crossing of e' with one of the edges e_i.

Before proving the properties, we show that they imply the statement of the theorem by induction on k. The base case $1 = C(0) \leq e \cdot 0! = e$ is obvious. From (P1) and (P2) we see that the number X_i of crossings of e_i and e' is upper bounded by $C(k-1)$. From (P3) we obtain that $C(k) \leq 1 + \sum_i X_i$. Combining these we get

$$C(k) \leq k \cdot C(k-1) + 1 \leq k! \cdot \sum_{s=0}^{k} \frac{1}{s!} \leq k! \cdot e. \qquad \square$$

For the proof of the three claims we need some notation. Let $\xi_1, \xi_2, \ldots, \xi_N$ be the crossings of e and e' indexed according to the left to right order along the horizontal edge e'. Let g_i and h_i be the pieces of e' and e, respectively, between crossings ξ_i and ξ_{i+1}. The bounded region enclosed by $g_i \cup h_i$ is the *bag* B_i and g_i is the *gap* of the bag. In the drawing $D(e, e')$ the bags B_i where h_i is a crossing free piece of e are exactly the inclusion-wise minimal lenses formed by e and e'. From now on when referring to a *lens* we always mean such a minimal lens. Indeed if there is no empty minimal lens, then there is no empty lens. The following observation is crucial.

Observation 2. *For two bags B_i and B_j the open interiors are either disjoint or one is contained in the other.*

Proof. Every bag is bounded by a closed Jordan curve, and the boundaries of two distinct bags do not cross (at most they may touch at a single point that is one of $\xi_1, \xi_2, \ldots, \xi_N$).

Observation 2 implies that the containment order on the bags is a downwards branching forest. The minimal elements in the containment order are the lenses. Consider a lens L and the point p_i inside L. Since the vertex u of e is in the outer face of $D(e, e')$, the edge e_i has to leave each bag that contains L. Furthermore, by definition e_i does not cross e and therefore it has to leave a bag B containing L through the gap g of B. We now reformulate and prove the third claim (P3).

(P3') For each pair ξ_i, ξ_{i+1} of consecutive crossings on e' there is a lens L and a point $p_j \in L$ such that e_j crosses e' between ξ_i and ξ_{i+1}.

Proof sketch for (P3'). The pair ξ_i, ξ_{i+1} is associated with the bag B_i. In the containment order of bags a minimal bag below B_i is a lens, let L be any of the minimal elements below B_i. By assumption, L contains a point p_j. Since $L \subseteq B_i$, we have that also $p_j \in B_i$. Thus, it follows that e_j has a crossing with the gap g_i, i.e., e_j has a crossing with e' between ξ_i and ξ_{i+1}.

Proof sketch for (P1). We have to show that e_i and e' form no deadlock and no spiral. The minimality condition in the definition of e_i implies that if $L = B_{i_1} \subset$

$B_{i_2} \subset \ldots \subset B_{i_t}$ is the maximal chain of bags with minimal element L, then e_i crosses the gaps of these bags in the given order and has no further crossings with e'. If γ is a curve from L to u that avoids e, then in the ordered sequence of gaps crossed by γ we find a subsequence that is identical to the ordered sequence of gaps crossed by e_i. Since e and e' form no spiral, there is such a curve γ that forms no deadlock with e'. Therefore, e_i forms no deadlock with e', either.

Now assume that e_i and e' form a spiral. Let B be the largest bag containing p_i. Think of B as a drawing of e_i with a broad pen, which may also have some extra branches that have no correspondence in e_i, see Fig. 7. The formalization of this picture is that for every bag β formed by e_i with e' there is a bag $B(\beta)$ formed by e and e' with $B(\beta) \subset \beta$. Now, if there is a lens λ formed by e_i with e' such that every e_i-avoiding[1] curve to u is a deadlock with e', then there is a lens $L(\lambda)$ formed by e and e' with $L(\lambda) \subset \lambda$ such that every e-avoiding curve from $L(\lambda)$ to u is also B-avoiding and hence e_i-avoiding. Thus, every such curve has a deadlock with e', whence e and e' form a spiral, contradiction.   □

Proof sketch for (P2). We know by P1 that e_i and e' form no deadlock. Therefore, by Lemma 1, the vertices of e_i and e' belong to the same region of $D(e_i, e')$. All crossings of e_i with e' correspond to bags of e and e', therefore the vertices of e and e' are in the outer face of $D(e_i, e')$. Together this shows that p_i is also in the outer face of $D(e_i, e')$. Since every lens of $D(e_i, e')$ contains a lens of $D(e, e')$, it also contains one of the points hitting all lenses of $D(e, e')$. Hence, all lenses of $D(e_i, e')$ are hit by the $k - 1$ points $p_1, \ldots, p_{i-1}, p_{i+1}, \ldots, p_k$.   □

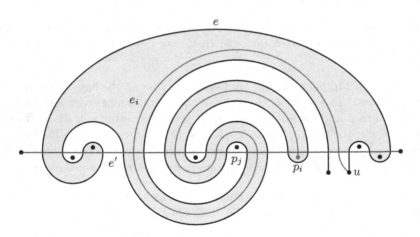

Fig. 7. An edge e_i(green) that forms a spiral with e'. The bag B in gray and the lens $L(\lambda)$ marked with the vertex p_j(blue). (Color figure online)

---

[1] That is, disjoint from e_i except for possibly a shared endpoint.

4   Crossings in Complete Drawings

Accounting for the four endpoints of the two crossing edges we have $k \leq n-4$ in Theorem 1. Therefore, we obtain that the number of crossings of a pair of edges in a star-simple drawing of K_n without empty lens is upper bounded by $3(n-4)!$. This directly implies that the drawing of K_n has at most $n!$ crossings. This is the first finite upper bound on the number of crossings in star-simple drawings of the complete graph K_n. We know drawings of K_n in this drawing mode that have an exponential number of crossings. Thus, it would be interesting to reduce the huge gap between the upper and the lower bound. Specifically, can a star-simple drawing of K_n have two edges with more than $2^{n-4} + 2^{n-6}$ crossings?

References

1. Aichholzer, O., Ebenführer, F., Parada, I., Pilz, A., Vogtenhuber, B.: On semi-simple drawings of the complete graph. In: Abstracts of the XVII Spanish Meeting on Computational Geometry (EGC 2017), pp. 25–28 (2017)
2. Balko, M., Fulek, R., Kynčl, J.: Crossing numbers and combinatorial characterization of monotone drawings of K_n. Discrete Comput. Geom. **53**(1), 107–143 (2014). https://doi.org/10.1007/s00454-014-9644-z
3. Kynčl, J.: Simple realizability of complete abstract topological graphs simplified (2016). arxiv.org/abs/1608.05867v1
4. Kynčl, J.: Simple realizability of complete abstract topological graphs simplified. Discrete Comput. Geom. **64**(1), 1–27 (2020). https://doi.org/10.1007/s00454-020-00204-0
5. Pach, J., Tóth, G.: A crossing lemma for multigraphs. Discrete Comput. Geom. **63**(4), 918–933 (2018). https://doi.org/10.1007/s00454-018-00052-z
6. Parada, I.: On straight-line and topological drawings of graphs in the plane. Ph.D. thesis, Graz University of Technology, Graz, Austria (2019)

Simple Topological Drawings of k-Planar Graphs

Michael Hoffmann[1] , Chih-Hung Liu[1] , Meghana M. Reddy[1(✉)] ,
and Csaba D. Tóth[2,3]

[1] Department of Computer Science, ETHZ, Zürich, Switzerland
{hoffmann,chih-hung.liu,meghana.mreddy}@inf.ethz.ch
[2] Department of Mathematics, Cal State Northridge, Los Angeles, CA, USA
csaba.toth@csun.edu
[3] Department of Computer Science, Tufts University, Medford, MA, USA

Abstract. Every finite graph admits a *simple (topological) drawing*, that
is, a drawing where every pair of edges intersects in at most one point.
However, in combination with other restrictions simple drawings do not
universally exist. For instance, *k-planar graphs* are those graphs that can
be drawn so that every edge has at most k crossings (i.e., they admit a k-
plane drawing). It is known that for $k \leq 3$, every k-planar graph admits
a k-plane simple drawing. But for $k \geq 4$, there exist k-planar graphs
that do not admit a k-plane simple drawing. Answering a question by
Schaefer, we show that there exists a function $f : \mathbb{N} \to \mathbb{N}$ such that every
k-planar graph admits an $f(k)$-plane simple drawing, for all $k \in \mathbb{N}$. Note
that the function f depends on k only and is independent of the size
of the graph. Furthermore, we develop an algorithm to show that every
4-planar graph admits an 8-plane simple drawing.

Keywords: Topological graphs · Local crossing number · k-planar
graphs

1 Introduction

A *topological drawing* of a graph G in the plane is a representation of G in
which the vertices are mapped to pairwise distinct points in the plane and edges
are mapped to Jordan arcs that do not pass through (the images of) vertices.
Moreover, no three Jordan arcs pass through the same point in the plane, and
every pair of Jordan arcs has finitely many intersection points, each of which is
either a common endpoint or a *crossing*, where the two arcs cross transversally.

This work was initiated at the 17th Gremo Workshop on Open Problems (GWOP)
2019. The authors thank the organizers of the workshop for inviting us and providing
a productive working atmosphere. M. H. and M. M. R. are supported by the Swiss
National Science Foundation within the collaborative DACH project *Arrangements
and Drawings* as SNSF Project 200021E-171681. Research by C. D. T. was supported
in part by the NSF award DMS-1800734.

© Springer Nature Switzerland AG 2020
D. Auber and P. Valtr (Eds.): GD 2020, LNCS 12590, pp. 390–402, 2020.
https://doi.org/10.1007/978-3-030-68766-3_31

A graph is *k-planar* if it admits a topological drawing in the plane where every edge is crossed at most k times, and such a drawing is called a *k-plane drawing*. A *simple topological drawing* of a graph refers to a topological drawing where no two edges cross more than once and no two adjacent edges cross. We study simple topological drawings of k-planar graphs.

It is well known that drawings of a graph G that attain the minimum number of crossings (i.e., the *crossing number* of G) are simple topological drawings [4, p. 18]. However, a drawing that minimizes the total number of crossings need not minimize the maximum number of crossings per edge; and a drawing that minimizes the maximum number of crossings per edge need not be simple. A *k-plane simple topological drawing* is a simple topological drawing where every edge is crossed at most k times. We study the simple topological drawings of k-planar graphs and prove that there exists a function $f : \mathbb{N} \to \mathbb{N}$ such that every k-planar graph admits an $f(k)$-plane simple topological drawing by designing an algorithm to obtain the simple topological drawing from a k-plane drawing. The function f in our bound is exponential in k, more precisely $f(k) \in O^*(3^k)$. It remains open whether this can be improved to a bound that is polynomial in k. We also present a significantly better bound for 4-planar graphs.

In a k-plane drawing adjacent edges may cross, and two edges may cross many times. To obtain a simple topological drawing, we need to eliminate crossings between adjacent edges and ensure that any two edges cross at most once.

Related Work. It is easy to see that every 1-planar graph admits a 1-plane simple topological drawing [3]. Pach et al. [2, Lemma 1.1] proved that every k-planar graph for $k \leq 3$ admits a k-plane simple topological drawing. However, these results do not extend to k-planar graphs, for $k > 3$. In fact, Schaefer [4, p. 57] constructed k-planar graphs that do not admit a k-plane simple topological drawing for $k = 4$. The construction idea can be extended to all $k > 4$. The *local crossing number* $\mathrm{lcr}(G)$ of a graph G is the minimum integer k such that G admits a drawing where every edge has at most k crossings. The *simple* local crossing number $\mathrm{lcr}^*(G)$ minimizes k over all simple topological drawings of G. Schaefer [4, p. 59] asked whether the $\mathrm{lcr}^*(G)$ can be bounded by a function of $\mathrm{lcr}(G)$. We answer this question in the affirmative and show that there exists a function $f : \mathbb{N} \to \mathbb{N}$ such that $\mathrm{lcr}^*(G) \leq f(\mathrm{lcr}(G))$.

The family of k-planar graphs, for small values of k, was instrumental in proving the current best bounds on the multiplicative constant in the Crossing Lemma and the Szemerédi-Trotter theorem on point-line incidences [1,2]. Ackerman [1] showed that every graph with $n \geq 3$ vertices that admits a simple 4-plane drawing has at most $m \leq 6n - 12$ edges, and claims that this bound holds for all 4-planar graphs. Pach et al. [2, Conjecture 5.4] conjectured that for all $k, n \geq 1$, the maximum number of edges in a k-planar n-vertex graph is attained by a graph that admits a simple k-plane drawing.

2   Preliminaries

Lenses in Topological Drawings. We start with definitions needed to describe the key operations in our algorithms. In a topological drawing, we define a structure called *lens*. Consider two edges, e and f, that intersect in two distinct points, α and β (each of which is either a common endpoint or a crossing). Let $e_{\alpha\beta}$ (resp., $f_{\alpha\beta}$) denote the portion of e (resp., f) between α and β. The arcs $e_{\alpha\beta}$ and $f_{\alpha\beta}$ together are called a *lens* if $e_{\alpha\beta}$ and $f_{\alpha\beta}$ do not intersect except at α and β. See Fig. 1 for examples. The lens is denoted by $L(e_{\alpha\beta}, f_{\alpha\beta})$. A lens $L(e_{\alpha\beta}, f_{\alpha\beta})$ is bounded by *independent arcs* if both α and β are crossings, else (if α or β is a vertex of G) it is bounded by *adjacent arcs*.

Lemma 1. *If a pair of edges e and f intersect in more than one point, then there exist arcs $e_{\alpha\beta} \subset e$ and $f_{\alpha\beta} \subset f$ that form a lens.*

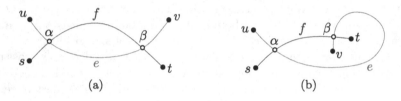

Fig. 1. Lenses formed by two edges.

Operations. We present algorithms that transform a k-plane drawing into a simple topological drawing by a sequence of elementary operations. Each operation modifies one or two edges that form a lens so that the lens is eliminated. We use two elementary operations, SWAP and REROUTE. Both have been used previously (e.g., in [2, Lemma 1.1]); we describe them here for completeness.

The common setup in both operations is the following. Let $e = uv$ and $f = st$ be edges that form a lens $L(e_{\alpha\beta}, f_{\alpha\beta})$, where α and β are each a crossing or a common endpoint. Assume that the Jordan arc of e visits u, α, β, v, and the Jordan arc of f visits s, α, β, and t in this order. Let $\overline{\alpha}$ and $\overline{\beta}$ be sufficiently small disks centered at α and β, resp., so that their boundary circles each intersect e and f twice, but do not intersect any other edge.

Swap Operation. We modify the drawing of e and f in three steps as follows. (1) Redraw e such that it follows its current arc from u to α, then continues along $f_{\alpha\beta}$ to β, and further to v along its original arc. Similarly, redraw f such that it follows its current arc from s to α, then continues along $e_{\alpha\beta}$ to β, and further to t along its original arc. (2) Replace the portion of e and f in $\overline{\alpha}$ and $\overline{\beta}$ by straight line segments. (3) Eliminate self-crossings, if any is introduced, by removing any loops from the modified arcs of e and f. The swap operation is denoted by SWAP($e_{\alpha\beta}, f_{\alpha\beta}$); see Fig. 2 for illustrations. The swap operation for a lens bounded by adjacent arcs is defined similarly.

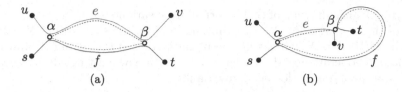

Fig. 2. SWAP($e_{\alpha\beta}, f_{\alpha\beta}$) applied to the two lenses in Fig. 1

Observation 1. *Let D be a topological drawing of a graph G, and let $L(e_{\alpha\beta}, f_{\alpha\beta})$ be a lens. Operation* SWAP($e_{\alpha\beta}, f_{\alpha\beta}$) *produces a topological drawing that has at least one fewer crossing than D.*

Reroute Operation. We modify the drawing of f in three steps as follows. (1) Redraw f such that it follows its current arc from s to the first intersection with $\overline{\alpha}$, it does not cross e in $\overline{\alpha}$, and then it closely follows arc $e_{\alpha\beta}$ to $\overline{\beta}$, and further follows its original arc from $\overline{\beta}$ to t. (2) Replace the portion of f in the interior of $\overline{\alpha}$ and $\overline{\beta}$ by straight line segments. (3) Eliminate self-crossings, if any are introduced, by removing any loops from the modified arc of f. The reroute operation is denoted by REROUTE($e_{\alpha\beta}, f_{\alpha\beta}$); see Fig. 3 for illustrations. The reroute operation for a lens bounded by adjacent arcs is defined similarly.

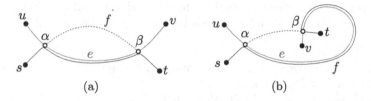

Fig. 3. REROUTE($e_{\alpha\beta}, f_{\alpha\beta}$) operation on the two lenses in Fig. 1

Observation 2. *Let D be a topological drawing of a graph G, and let $L(e_{\alpha\beta}, f_{\alpha\beta})$ be a lens. Operation* REROUTE($e_{\alpha\beta}, f_{\alpha\beta}$) *produces a topological drawing.*

While a REROUTE($e_{\alpha\beta}, f_{\alpha\beta}$) operation modifies only the edge f, it may increase the total number of crossings, as well as the number of crossings on f.

Planarization. Let D be a topological drawing of a graph G. Denote by N the planarization of D (i.e., we introduce a vertex of degree four at every crossing in D). We call this graph a *network*. We refer to the vertices and edges of N as *nodes* and *segments*, respectively, so as to distinguish them from the corresponding entities in G. Our algorithms in Sect. 3–4 use the planarization N of a drawing D, then successively modify the drawing D, and ultimately return a simple topological drawing of G. We formulate invariants for these algorithms in terms of the planarization N of the initial drawing. In other words, N remains fixed (in particular, N will not be the planarization of the modified drawings). As REROUTE operations redraw edges to closely follow existing edges, our algorithms will maintain the following invariants:

(I1) Every edge in D closely follows a path in the network N;
(I2) every pair of edges in D cross only in a small neighborhood of a node of N;
(I3) every pair of edges crosses at most once in each such neighborhood.

Length of an Arc and Number of Crossings. Let a be a Jordan arc that closely follows a path in N such that its endpoints are in the small neighborhoods of nodes of N. The *length* of a, denoted by $\ell(a)$, is the graph-theoretic length of the path of N that a closely follows. Let $x(a)$ denote the number of crossings on the arc a in a drawing D. Note that the length $\ell(a)$ is measured in terms of the (fixed) network N, and $x(a)$ is measured in terms of the (varying) drawing D. For instance, in Fig. 3(b) we have $\ell(f) = 3$ both before and after rerouting, whereas $x(f) = 2$ before and $x(f) = 1$ after rerouting.

3   General Bound for k-Planar Graphs

In this section we describe and analyze an algorithm to transform a topological drawing into a simple topological drawing whose local crossing number is bounded by a function of the local crossing number of the original drawing.

Algorithm 1
Let D_0 be a topological k-plane drawing of a graph $G = (V, E)$. Let N be the planarization of D_0. Let $D := D_0$.
While there exists a lens in D, do the following.
Let $L(e_{\alpha\beta}, f_{\alpha\beta})$ be a lens so that w.l.o.g. $\ell(e_{\alpha\beta}) < \ell(f_{\alpha\beta})$, or $\ell(e_{\alpha\beta}) = \ell(f_{\alpha\beta})$ and $x(e_{\alpha\beta}) \leq x(f_{\alpha\beta})$. Modify D by applying REROUTE$(e_{\alpha\beta}, f_{\alpha\beta})$.
When the while loop terminates, return the drawing D.

Observation 3. *Algorithm 1 maintains invariants (I1)–(I3), and the length of every edge decreases or remains the same.*

Corollary 1. *Algorithm 1 maintains the following invariant:*

(I4) The length of every edge in D is at most $k + 1$.

Lemma 2. *Algorithm 1 terminates and transforms a k-plane topological drawing into a simple topological drawing of G.*

Proof. Let the sum of lengths of all edges in the drawing be defined as the *total length* of the drawing (recall that the length of an edge is the length of the corresponding path in N). By Observation 3, the total length of the drawing monotonically decreases. If the total length remains the same in one iteration of the while loop, then $\ell(e_{\alpha\beta}) = \ell(f_{\alpha\beta})$ and $x(e_{\alpha\beta}) \leq x(f_{\alpha\beta})$. Since $\text{REROUTE}(e_{\alpha\beta}, f_{\alpha\beta})$ eliminates a crossing at α or β, the total number of crossings strictly decreases in this case. Thus, the algorithm terminates. By Observations 1–2, the algorithm maintains a topological drawing. The drawing D' returned by the algorithm does not contain lenses. By Lemma 1, any two edges in D' intersect in at most one point. Consequently, D' is a simple topological drawing of G. □

Lemma 3 (Crossing Lemma [1, Theorem 6]). *Let G be a graph with n vertices and m edges and D be a topological drawing of G. Let $cr(D)$ be defined as the total number of crossings in D, and $cr(G)$ be defined as the minimum of $cr(D)$ over all drawings D of G. If $m \geq 6.95n$, then $cr(G) \geq \frac{1}{29}\frac{m^3}{n^2}$.*

Theorem 1. *There exists a function $f(k)$ such that every k-planar graph admits an $f(k)$-plane simple topological drawing, and there exists an algorithm to obtain an $f(k)$-plane simple topological drawing from a given k-plane drawing of a graph.*

Proof. The statement holds for $k \leq 3$ with $f(k) = k$ [2, Lemma 1.1]. Hence we may suppose that $k \geq 4$. Consider the drawing D' returned by Algorithm 1, and a node γ of the network N that corresponds to a crossing. We analyse the subgraph G_γ of G formed by the edges of G that in D' pass through a small neighborhood $\overline{\gamma}$ of γ. Let n_γ and m_γ be the number of vertices and edges of G_γ, respectively. By (I4), every edge in D' corresponds to a path of length at most $k + 1$ in N. If an edge uv passes through $\overline{\gamma}$ in D', then N contains a path of length at most k from γ to u (resp., v) in which internal vertices correspond to crossings in D_0. Every node in N that corresponds to a crossing has degree 4. Hence the number of vertices reachable from γ on such a path is $n_\gamma \leq 4 \cdot 3^{k-1}$.

We apply Lemma 3 to the graph G_γ, and distinguish between two cases: Either $m_\gamma < 6.95n$, otherwise $m_\gamma \geq 6.95n$ and then $cr(G_\gamma) \geq \frac{1}{29}m_\gamma^3/n_\gamma^2$. Since G_γ has m_γ edges and each edge has at most k crossings in D, we obtain $\frac{1}{29}m_\gamma^3/n_\gamma^2 \leq m_\gamma k/2$, which implies $m_\gamma \leq \sqrt{29k/2}\,n_\gamma$. The combination of both cases yields an upper bound $m_\gamma \leq \max\{6.95n_\gamma, \sqrt{29k/2}\,n_\gamma\}$. So, for $k \geq 4$ we have $m_\gamma \leq \sqrt{29k/2}\,n_\gamma$.

Since m_γ edges pass through $\overline{\gamma}$, by invariant (I3) every edge passing through $\overline{\gamma}$ has at most $m_\gamma - 1$ crossings at $\overline{\gamma}$. By invariant (I4), every edge in G passes through (the neighborhood of) at most k nodes of N. By (I2), an edge passing through $\overline{\gamma_1}, \ldots, \overline{\gamma_k}$ crosses at most $\sum_{i=1}^{k}(m_{\gamma_i} - 1)$ edges in D'. Combining the upper bounds on m_γ and n_γ, we obtain that every edge in the output drawing D' has at most $\sqrt{29k/2} \cdot 4k \cdot 3^{k-1} = \frac{2}{3}\sqrt{58} \cdot k^{3/2} \cdot 3^k$ crossings, for $k \geq 4$. □

4 An Upper Bound for 4-Planar Graphs

The function f from our proof of Theorem 1 yields

$$f(4) = \frac{2}{3}\sqrt{58} \cdot 4^{3/2} \cdot 3^4 \approx 3290.01$$

and so every 4-plane drawing can be transformed into a 3290-plane simple topological drawing. In this section we improve this upper bound and show that 8 crossings per edge suffice.

Theorem 2. *Every 4-planar graph admits an 8-plane simple topological drawing. Given a 4-plane drawing of a graph with n vertices, an 8-plane simple topological drawing can be computed in $O(n)$ time.*

The proof of Theorem 2 is constructive: Let D_0 be a 4-plane drawing of a 4-planar graph $G = (V, E)$ with $n = |V|$ vertices and $m = |E|$ edges. The Crossing Lemma implies that a k-planar graph on n vertices has at most $3.81\sqrt{k}n$ edges. For $k = 4$, this implies $m \le 7.62n$. (We note that Ackerman [1] proved a bound $m \le 6n - 12$ for 4-plane simple topological drawings with $n \ge 3$ vertices; this bound is not applicable here.)

We want to eliminate all lenses using swap and reroute operations. We define three types of special lenses that we handle separately. A lens $L(e_{\alpha\beta}, f_{\alpha\beta})$ is

- a *0-lens* if $e_{\alpha\beta}$ has no crossings;
- a *quasi-0-lens* if the arc $e_{\alpha\beta}$ has exactly one crossing γ, where e crosses an edge h, the edges h and f have a common endpoint s, and the arcs $f_{s\alpha}$ and $h_{s\alpha}$ cross the same edges in the same order (see Fig. 5(a) for an example);
- a *1-3-lens* if $x(e) = 4$, $x(e_{\alpha\beta}) = 1$, and $x(f_{\alpha\beta}) = 3$; see Fig. 4(a).

We show that all lenses other than 0-lenses and 1-3-lenses can be eliminated by swap operations while maintaining a 4-plane drawing (Lemma 6). And 0-lenses can easily be eliminated by reroute operations (Lemma 4). The same holds for quasi-0-lenses (Lemma 5), which are of no particular concern in the initial drawing but are important for the analysis of the last phase of our algorithm. The main challenge is to eliminate 1-3-lenses, which we do by rerouting the arc with 3 crossings along the arc with 1 crossing.

Our algorithm proceeds in three phases: Phase 1 eliminates all lenses other than 1-3-lenses. We show that it maintains a 4-plane drawing (Lemma 8). Phase 2 eliminates every 1-3-lens using reroute operations. We show that this phase produces an 8-plane drawing. Phase 2 may also create new lenses, but only 0- and quasi-0-lenses, which are eliminated in Phase 3 without creating any new lenses.

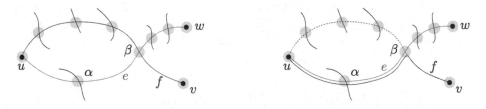

Fig. 4. REROUTE($e_{u\beta}, f_{u\beta}$) applied to a 1-3-lens $L(e_{u\beta}, f_{u\beta})$.

The initial 4-plane drawing has $O(n)$ crossings since the graph has $O(n)$ edges and each edge has at most four crossings. The set of lenses in the initial drawing can be identified in $O(n)$ time. Due to the elimination of a single lens, a constant number of other lenses can be affected, which can be computed in constant time. Further, each elimination operation strictly decreases the total number of crossings in the drawing. Consequently, Algorithm 2 performs $O(n)$ elimination operations and can be implemented in $O(n)$ time.

Lemma 4. *Let $L(e_{\alpha\beta}, f_{\alpha\beta})$ be a 0-lens. Then operation* REROUTE$(e_{\alpha\beta}, f_{\alpha\beta})$ *decreases the total number of crossings and does not create any new crossing. Further, if any two edges have at most two points in common, then it does not create any new lens.*

Proof. The operation REROUTE$(e_{\alpha\beta}, f_{\alpha\beta})$ modifies only the edge f, by rerouting the arc $f_{\alpha\beta}$ to closely follow $e_{\alpha\beta}$. Since the arc $e_{\alpha\beta}$ is crossing-free, the edge f loses one of its crossings and no edge gains any new crossing. Overall, the total number of crossings decreases, as claimed.

Assume that any two edges have at most two points in common before the operation. Consider a lens $L(g_{\gamma\delta}, h_{\gamma\delta})$ in the drawing after the operation. As no new crossings are created, γ and δ are already common points of g and h before the operation. Since g and h have no other common points by assumption, the lens $L(g_{\gamma\delta}, h_{\gamma\delta})$ is already present before the operation.   □

For quasi-0-lenses we define the operation QUASI-0-REROUTE$(e_{\alpha\beta}, f_{\alpha\beta})$ as follows; see Fig. 5. Let h be the edge that crosses $e_{\alpha\beta}$ at γ and shares an endpoint s with f. Redraw f such that it closely follows h from s to $\overline{\gamma}$, it does not cross e in $\overline{\gamma}$, and then it closely follows arc $e_{\alpha\beta}$ to $\overline{\beta}$, and further follows its original arc from $\overline{\beta}$ to t. The analogue of Lemma 4 reads as follows.

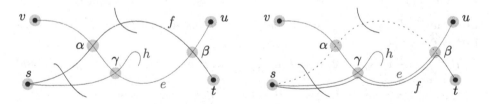

Fig. 5. QUASI-0-REROUTE$(e_{\alpha\beta}, f_{\alpha\beta})$ applied to a quasi-0-lens $L(e_{\alpha\beta}, f_{\alpha\beta})$.

Lemma 5. *Let $L(e_{\alpha\beta}, f_{\alpha\beta})$ be a quasi-0-lens, where h denotes the edge that crosses $e_{\alpha\beta}$ at γ and shares an endpoint s with f. Then operation* QUASI-0-REROUTE$(e_{\alpha\beta}, f_{\alpha\beta})$ *decreases the total number of crossings, and does not increase the number of crossings between any pair of edges. Further, if any two edges have at most two points in common, then it does not create any new lens.*

Proof. The operation QUASI-0-REROUTE$(e_{\alpha\beta}, f_{\alpha\beta})$ modifies only the edge f, by rerouting the arc $f_{s\beta}$ to closely follow first h from s to $\overline{\gamma}$ and then $e_{\alpha\beta}$ to $\overline{\beta}$. Let f' denote the new drawing of f. Since (1) e has at least one fewer crossing with f' than with f, and (2) every crossing of f' along the arc between s and $\overline{\gamma}$ corresponds to a crossing of f along the arc from s to $\overline{\alpha}$, the total number of crossings strictly decreases, and for each pair of edges the number of crossings between them does not increase, as claimed.

Assume that any two edges have at most two points in common before the operation. Suppose QUASI-0-REROUTE($e_{\alpha\beta}, f_{\alpha\beta}$) creates a new lens. This lens must be formed by f' and another edge, say g. Then f' and g must have at least two points in common, and g must cross $f_{\alpha\beta}$, implying that f and g have at least three points in common before the operation. However, by assumption, edges f and g have at most two points in common, which is a contradiction. Consequently, every lens in the resulting drawing corresponds to a lens in the original drawing, where the arc $f_{s\alpha}$ is shifted to the arc of f' from s to $\overline{\gamma}$.   □

Lemma 6. *Let $L(e_{\alpha\beta}, f_{\alpha\beta})$ be a lens either bounded by nonadjacent arcs with $x(e_{\alpha\beta}) \le x(f_{\alpha\beta}) \le x(e_{\alpha\beta}) + 2$, or by adjacent arcs with $x(e_{\alpha\beta}) \le x(f_{\alpha\beta}) \le x(e_{\alpha\beta}) + 1$. Then the operation SWAP($e_{\alpha\beta}, f_{\alpha\beta}$) produces a drawing in which the total number of crossings on each edge does not increase, and the total number of crossings decreases.*

Proof. The operation SWAP($e_{\alpha\beta}, f_{\alpha\beta}$) modifies only the edges e and f, by exchanging arcs $e_{\alpha\beta}$ and $f_{\alpha\beta}$, and eliminating any crossing at the endpoints of these arcs. In particular, the number of crossings on other edges cannot increase. This already implies that the total number of crossings decreases.

Let e' and f' denote the new drawing of e and f. If both α and β are crossings, then both crossings are eliminated, hence $x(e') = x(e) - 2 + (x(f_{\alpha\beta}) - x(e_{\alpha\beta})) \le x(e)$ and $x(f') = x(f) - 2 + (x(e_{\alpha\beta}) - x(f_{\alpha\beta})) \le x(f) - 2$. If α or β is a vertex of G, then only one crossing is eliminated, hence $x(e') = x(e) - 1 + (x(f_{\alpha\beta}) - x(e_{\alpha\beta})) \le x(e)$ and $x(f') = x(f) - 1 + (x(e_{\alpha\beta}) - x(f_{\alpha\beta})) \le x(f) - 1$, as required.   □

Lemma 7. *Let D be a 4-plane drawing of a graph, and let $L(e_{\alpha\beta}, f_{\alpha\beta})$ be a lens with $x(e_{\alpha\beta}) \le x(f_{\alpha\beta})$.*

1. *If $x(f_{\alpha\beta}) - x(e_{\alpha\beta}) \ge 2$, then $L(e_{\alpha\beta}, f_{\alpha\beta})$ is either a 0-lens or $x(e_{\alpha\beta}) = 1$ and $x(f_{\alpha\beta}) = 3$.*
2. *If $x(e_{\alpha\beta}) = 1$ and $x(f_{\alpha\beta}) = 3$, then $e_{\alpha\beta}$ and $f_{\alpha\beta}$ are adjacent arcs.*

Proof. As D is 4-plane, we have $x(e) \le 4$ and $x(f) \le 4$. Assume first that both α and β are crossings, and so $x(f_{\alpha\beta}) \le x(f) - 2 \le 2$. Combined with $x(f_{\alpha\beta}) - x(e_{\alpha\beta}) \ge 2$, this implies $x(e_{\alpha\beta}) = 0$, hence $L(e_{\alpha\beta}, f_{\alpha\beta})$ is a 0-lens. Assume next that α or β is a vertex in G. Then $x(f_{\alpha\beta}) \le x(f) - 1 \le 3$. With $x(f_{\alpha\beta}) - x(e_{\alpha\beta}) \ge 2$, this implies $x(e_{\alpha\beta}) = 0$, or $x(e_{\alpha\beta}) = 1$ and $x(f_{\alpha\beta}) = 3$.   □

Algorithm 2
Input. Let D_0 be a 4-plane drawing of a graph $G = (V, E)$.
Phase 1. While there is a lens $L(e_{\alpha\beta}, f_{\alpha\beta})$ that is not a 1-3-lens, do:
If it is a 0-lens, then REROUTE($e_{\alpha\beta}, f_{\alpha\beta}$), else SWAP($e_{\alpha\beta}, f_{\alpha\beta}$).
Phase 2. Let $\mathcal{L}$ be the set of 1-3-lenses. For every $L(e_{\alpha\beta}, f_{\alpha\beta}) \in \mathcal{L}$, if neither $e_{\alpha\beta}$ nor $f_{\alpha\beta}$ has been modified in previous iterations of Phase 2 (regardless of whether $x(e_{\alpha\beta})$ or $x(f_{\alpha\beta})$ has changed), apply REROUTE($e_{\alpha\beta}, f_{\alpha\beta}$).

Phase 3. While there is a 0-lens $L(e_{\alpha\beta}, f_{\alpha\beta})$, do: REROUTE$(e_{\alpha\beta}, f_{\alpha\beta})$.
While there is a quasi-0-lens $L(e_{\alpha\beta}, f_{\alpha\beta})$, do: QUASI-0-REROUTE$(e_{\alpha\beta}, f_{\alpha\beta})$.

For $i \in \{1, 2, 3\}$, let D_i denote the drawing obtained at the end of Phase i. We analyse the three phases separately.

Lemma 8. *Phase 1 terminates, and D_1 is a 4-plane drawing in which every lens is a 1-3-lens, and any two edges have at most two points in common.*

Proof. By Lemma 4 and Observation 1, each iteration of the while loop reduces the total number of crossings. Since D_0 has at most $\frac{1}{2} \cdot 4m \in O(n)$ crossings, the while loop terminates after $O(n)$ iterations. By Lemma 7 all lenses satisfy the conditions of Lemma 6, except for 0-lenses and lenses $L(e_{\alpha\beta}, f_{\alpha\beta})$ with $x(e_{\alpha\beta}) = 1$ and $x(f_{\alpha\beta}) = 3$. Each lens of the latter type is either a 1-3-lens, which remains untouched, or $x(e) < 4$ and the lens is eliminated by a swap operation. In this case, though the number of crossings on the edge e increases, it does not exceed four and the total number of crossings in the drawing strictly decreases. In all other cases we can apply either Lemma 4 or Lemma 6 to conclude that each iteration maintains a 4-plane drawing. By the end condition of the while loop, all lenses other than 1-3-lenses are eliminated.

To prove the final statement, suppose to the contrary, two edges e and f in D_1 have three or more points in common. By Lemma 1, there exist arcs $e_{\alpha\beta} \subset e$ and $f_{\alpha\beta} \subset f$ such that $L(e_{\alpha\beta}, f_{\alpha\beta})$ is a lens, which is necessarily a 1-3-lens. We may assume without loss of generality that $x(e) = 4$, $x(e_{\alpha\beta}) = 1$, and $x(f_{\alpha\beta}) = 3$. Denote by γ a common point of e and f other than α and β. Since D_1 is a 4-plane drawing and $x(f_{\alpha\beta}) = 3$, we may assume that α is common endpoint of e and f, furthermore γ is a crossing in the interior of $f_{\alpha\beta}$. Since $e_{\alpha\beta}$ and $f_{\alpha\beta}$ form a lens, the arc $e_{\alpha\beta}$ cannot pass through γ. Hence γ is a crossing between $f_{\alpha\beta}$ and $e \setminus e_{\alpha,\beta}$. By Lemma 1, $e_{\beta\gamma}$ and $f_{\beta\gamma}$ form a lens, which is necessarily a 1-3-lens. However, $x(e_{\beta\gamma}) \leq 2$ and $x(f_{\beta\gamma}) \leq 2$, which is a contradiction. $\square$

For the analysis of Phases 2 and 3, we introduce some notation. Let N denote the planarization of D_1. Note that N is a simple graph, since a double edge would correspond to a lens whose arcs are crossing-free (i.e., a 0-lens). Phases 2 and 3 apply only REROUTE and QUASI-0-REROUTE operations. Hence the resulting drawings satisfy invariants (I1)–(I3). For a node α of N, we denote by $\overline{\alpha}$ a small neighborhood of α. Recall that the length $\ell(a)$ of an arc a along an edge of G is the combinatorial length of the path in N that the arc closely follows.

Lemma 9. *D_2 has the following properties: (i) the length of every edge is at most five; (ii) at most two edges of G pass along every segment of N; (iii) through every node ν of N, at most two rerouted edges of G pass through ν; and (iv) at each node α of N, an edge passing through $\overline{\alpha}$ crosses at most two edges in $\overline{\alpha}$; (v) any two edges have at most two points in common.*

Proof. **(i)** By Lemma 8, the drawing D_1 is a 4-plane drawing. Therefore, every edge in D_1 passes through at most 4 crossings, hence its length is at most 5.

Each REROUTE operation in Phase 2 replaces an edge of length 5 with an edge of length 3 (cf. Fig. 4(a)). Property (i) follows.

(ii) Each REROUTE($e_{\alpha\beta}, f_{\alpha\beta}$) operation in Phase 2 reroutes the longer arc along the shorter arc of a 1-3-lens in $\mathcal{L}$. Let $\mathcal{A}$ be the set paths of length 2 in N that correspond to shorter arcs $e_{\alpha\beta}$ in some 1-3-lens $L(e_{\alpha\beta}, f_{\alpha\beta}) \in \mathcal{L}$. By the definition of 1-3-lenses, $\ell(e) = 5$ and $e_{\alpha\beta}$ consists of the first two segments of N along e. Thus every segment $\gamma\delta$ of N is contained in at most one path in $\mathcal{A}$. Consequently, at most one new edge can pass along $\gamma\delta$ due to reroute operations.

(iii) Let γ be a node in N that corresponds to a crossing in the drawing D_1. Then γ is incident to at most two paths in $\mathcal{A}$ (at most one along each of the two edges that cross at γ). Hence at most two rerouted edges can pass through γ.

(iv) Let γ be a node of N, and let e be an edge that passes through $\overline{\gamma}$ in D_2. By property (ii), at most 4 edges pass through $\overline{\gamma}$. If at most 3 edges pass through $\overline{\gamma}$, then it is clear that e crosses at most two edges in $\overline{\gamma}$. Suppose that four edges pass through $\overline{\gamma}$. Then the four segments of N incident to γ are each contained in the shorter arc of some 1-3-lens in D_1. Consequently, γ is the middle vertex of two distinct arcs in $\mathcal{A}$. In the drawing D_2 (after REROUTE operations), two edges run in parallel in each of these shorter arcs. Hence each edge that passes through $\overline{\gamma}$ crosses at most two other edges in $\overline{\gamma}$, as claimed.

(v) Suppose f_1 and f_2 have three points in common in D_2. By Lemma 8, we may assume that f_1 has been rerouted in Phase 2, and f_1 follows a path (v, γ, δ, u) in N and $(v, \gamma, \delta) \in \mathcal{A}$. Since f_1 and f_2 have at most one common endpoint, they cross in both $\overline{\gamma}$ and $\overline{\delta}$. After the rerouting operation, f_1 does not cross any edge of D_1 in $\overline{\delta}$, which implies that f_2 has also been rerouted in Phase 2. Since both f_1 and f_2 have length three and pass through $\overline{\gamma}$ and $\overline{\delta}$, and N is a simple graph, both f_1 and f_2 pass along segment $\gamma\delta$, which contradicts the fact that at most one new edge can pass along $\gamma\delta$ (see the proof of (ii) above). □

Corollary 2. D_2 is an 8-plane drawing of G.

Proof. Every edge of G passes through the small neighborhood of at most four nodes of N by Lemma 9(i). In each such neighborhood, it crosses at most two other edges by Lemma 9(iv), and it has at most one crossing with each by (I3). Overall, every edge has at most eight crossings in D_2. □

Unfortunately, Phase 2 may create new lenses, but only of very specific types. We analyze these types and argue that all remaining lenses are removed.

Lemma 10. *Phase 3 terminates with an 8-plane simple topological drawing D_3.*

Proof. The while loops in Phase 3 terminate, as each iteration decreases the number of crossings by Lemmas 4 and 5. The drawing D_2 at the beginning of Phase 3 is 8-plane by Corollary 2, and remains 8-plane and no new lens is created by Lemmas 4 and 5. It remains to show that Phase 3 eliminates all lenses of D_2.

Every lens in D_1 is a 1-3-lens by Lemma 8, and they are all in $\mathcal{L}$. Phase 2 modifies an arc in every lens in $\mathcal{L}$. Thus the lenses of D_1 are no longer present in D_2. (The two edges that form a lens $L \in \mathcal{L}$ may still form a lens L' in D_2, but technically this is a new lens, that is, $L \neq L'$, which is created in Phase 2 and will be discussed next.)

We classify the new lenses created in Phase 2. Assume that edges e and $f = uw$ form a lens in D_2. Without loss of generality, the edge f was modified in Phase 2. Each iteration in Phase 2 applies a reroute operation on a 1-3-lens, which decreases the length of an edge from 5 to 3. Therefore Phase 2 modifies every edge at most once. The drawing of edge f in D_2 was produced by a REROUTE$(g_{u\beta}, f_{u\beta})$ operation, for some edge g, where u is a common endpoint of f and g. The resulting drawing of f in D_2 closely follows a path (u, α, β) in N and then the original arc (in D_1) from $\overline{\beta}$ to w. After operation REROUTE$(g_{u\beta}, f_{u\beta})$, edges f and g do not cross each other.

Suppose first that f crosses e in $\overline{\beta}$. Then e was redrawn in Phase 2 to closely follow f from w to $\overline{\beta}$ and beyond; as in Fig. 6. However, in this case, e and f have a common endpoint at w. No other edges follow segment βw in N by Lemma 9(ii), hence e and f form a 0-lens. All such 0-lenses are eliminated in Phase 3, without creating any new lenses (cf. Lemma 4). Therefore, we may assume that f does not cross any edge in $\overline{\beta}$.

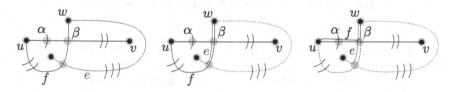

Fig. 6. New 0-lens formed by e and f crossing in $\overline{\beta}$.

By Lemma 9(iv), the edge f crosses at most two other edges in $\overline{\alpha}$. If it crosses exactly one other edge, and f forms a lens L with that edge, then this crossing in $\overline{\alpha}$ is the only crossing of f in D_2 and, thus, L is a 0-lens. Otherwise, f crosses two edges, denote them by e (for which we know that it crosses f) and h; one of them was redrawn in a REROUTE operation in Phase 2 to closely follow the other, which passes through $\overline{\alpha}$; see Fig. 7 and 8. Therefore, e and h are adjacent, and they do not cross at the end of that operation. Thus, they do not cross in D_2, either; otherwise, three rerouted edges would pass through $\overline{\alpha}$, contradicting Lemma 9(iii). As no new crossing is introduced in Phase 3, the edges e and h do not cross anytime during (and after) Phase 3, either.

If the common endpoint of e and h is u or w (see Fig. 7), then both e and h form a lens with f: One of these lenses is a 0-lens, and when this lens is eliminated, the other lens either disappears, or it becomes a 0-lens as well. Hence Phase 3 eliminates both crossings.

If e and h share distinct endpoints with f, without loss of generality e and f are adjacent at u and h and f are adjacent at w. As e and h do not cross, the crossing $e \cap f$ is closer to u and the crossing $h \cap f$ is closer to w along f. Hence, e and h each form a 0-lens with f, both of which are eliminated in Phase 3.

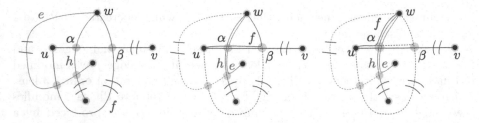

Fig. 7. f crosses two edges at $\bar{\alpha}$ and forms two 0-lenses.

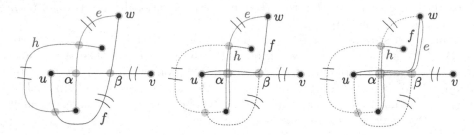

Fig. 8. f crosses two edges at $\bar{\alpha}$ and forms a quasi-0-lens.

It remains to consider the case that only e is adjacent to f (while h is not). Assume first that e and f are adjacent at w (see Fig. 8). If the crossing $e \cap f$ is closer to w along f than the crossing $h \cap f$, then the lens formed by e and f is a 0-lens; else it forms a quasi-0-lens. In any case, the lens is eliminated in Phase 3. The same argument works in case that e and f are adjacent at u.   □

References

1. Ackerman, E.: On topological graphs with at most four crossings per edge. Comput. Geom. **85**, 101574 (2019). https://doi.org/10.1016/j.comgeo.2019.101574
2. Pach, J., Radoičić, R., Tardos, G., Tóth, G.: Improving the crossing lemma by finding more crossings in sparse graphs. Discrete Comput. Geom. **36**(4), 527–552 (2006). https://doi.org/10.1007/s00454-006-1264-9
3. Ringel, G.: Ein Sechsfarbenproblem auf der Kugel. Abhandlungen aus dem Mathematischen Seminar der Universität Hamburg **29**, 107–117 (1965). https://doi.org/10.1007/BF02996313
4. Schaefer, M.: The graph crossing number and its variants: a survey. Electron. J. Comb. **20** (2013). https://doi.org/10.37236/2713. Version 4 (2020)

2-Layer k-Planar Graphs

Density, Crossing Lemma, Relationships, and Pathwidth

Patrizio Angelini[1] (ID), Giordano Da Lozzo[2] (ID), Henry Förster[3(✉)] (ID),
and Thomas Schneck[3] (ID)

[1] John Cabot University, Rome, Italy
pangelini@johncabot.edu
[2] Roma Tre University, Rome, Italy
giordano.dalozzo@uniroma3.it
[3] University of Tübingen, Tübingen, Germany
{foersth,schneck}@informatik.uni-tuebingen.de

Abstract. The 2-layer drawing model is a well-established paradigm to visualize bipartite graphs. Several beyond-planar graph classes have been studied under this model. Surprisingly, however, the fundamental class of k-planar graphs has been considered only for $k = 1$ in this context. We provide several contributions that address this gap in the literature. First, we show tight density bounds for the classes of 2-layer k-planar graphs with $k \in \{2, 3, 4, 5\}$. Based on these results, we provide a Crossing Lemma for 2-layer k-planar graphs, which then implies a general density bound for 2-layer k-planar graphs. We prove this bound to be almost optimal with a corresponding lower bound construction. Finally, we study relationships between k-planarity and h-quasiplanarity in the 2-layer model and show that 2-layer k-planar graphs have pathwidth at most $k + 1$.

Keywords: 2-layer graph drawing · k-planar graphs · Density · Crossing lemma · Pathwidth · Quasiplanar graphs

1 Introduction

Beyond-planarity is an active research area that studies graphs admitting drawings that avoid certain forbidden crossing configurations. Research on this subject has attracted considerable interest due to its theoretical appeal and due to the need of visualizing real-world non-planar graphs. A great deal of attention has been captured by two important graph families. The *k-planar* graphs, with $k \geq 1$, for which the forbidden configuration is an edge crossing more than k other edges, and the *h-quasiplanar* graphs, with $h \geq 3$, for which the forbidden configuration is a set of h pairwise crossing edges. The study of these two

G. Da Lozzo—The work of Giordano Da Lozzo was partially supported by MIUR grants 20157EFM5C *"MODE: MOrphing graph Drawings Efficiently"* and 20174LF3T8 *"AHeAD: efficient Algorithms for HArnessing networked Data"*.

D. Auber and P. Valtr (Eds.): GD 2020, LNCS 12590, pp. 403–419, 2020.
https://doi.org/10.1007/978-3-030-68766-3_32

families finds its origins in the 1960's [12,41], when the question arose about the *density* of these graphs, that is, the maximum number of edges of graphs in these families.

Many works have addressed this extremal graph theoretical question and established upper bounds for k-planar and h-quasiplanar graphs for various values of k and h. For small k and h, these upper bounds have been proven to be tight by lower bound constructions achieving the corresponding density. The most significant results include tight density bounds for 1-planar graphs [40] ($4n - 8$ edges), 2-planar graphs [40] ($5n - 10$ edges), 3-planar graphs [14,37] ($5.5n - 20$), and 4-planar graphs [2] ($6n - 12$). For general k, the currently best upper bound is $3.81\sqrt{k}\,n$, which can be derived from the result of Ackerman [2] on 4-planar graphs and from the renowned Crossing Lemma [5]. For h-quasiplanar graphs, despite considerable research efforts, a density upper bound that is linear in the number of vertices exists only for $h \leq 4$ [1,3,4,38]. In particular, a tight upper bound exists for simple 3-quasiplanar (for short, *quasiplanar*) graphs. Here, *simple* means that any two edges meet in at most one point, which is either a common endvertex or an internal point. For general h, only super-linear upper bounds are known [19,30,31,39,44,45] while a linear bound has been conjectured [39].

These two families have also been studied from other perspectives. A notable relationship is that every simple k-planar graph is also simple $(k + 1)$-quasiplanar [7], for every $k \geq 2$. It is also known that every *optimal* 3-planar graph, namely one with the maximum possible number of edges ($5.5n - 20$), is also 3-quasiplanar. This latter result follows from a characterization of the optimal 3-planar graphs [15], which also exists for the optimal 1- and 2-planar graphs [15,40]. Note that these characterizations do not directly yield recognition algorithms; in fact, recognizing (non-optimal) k-planar graphs is NP-complete for every $k \geq 1$ [35]. The complexity of recognizing h-quasiplanar graphs is still open for any $h \geq 3$.

Aside these two major families, we mention the *fan-planar* graphs, in which no edge is crossed by two independent edges or by two adjacent edges from different directions [13,17,18,34], and the *RAC graphs*, in which the edges are poly-lines with few bends and crossings only happen at right angles [8,24,25,28]. These and other graph classes have been also investigated with respect to their density, recognition, and relationship with other classes; see also the recent survey [26].

Beyond-planar classes have also been studied under additional constraints on the placement of the vertices. In the *outer model* [11,13,20,21,23,32,33] every vertex is incident to the unbounded region of the drawing, while in the *2-layer model* [17,18,22,23] the vertices lie on two horizontal lines and every edge is a y-monotone curve. The latter model requires the graph to be bipartite, and the constraints on the placement of the vertices emphasize the bipartite structure. Beyond-planar bipartite graphs have also been considered in the general drawing model, without any additional restriction [9]. We remark that the 2-layer model lies at the core of the Sugiyama framework for general layered drawings [42,43].

In [22], it was shown that 2-layer RAC graphs have at most $\frac{3}{2}n - 2$ edges and that this bound is tight, exploiting a characterization which also leads to an efficient recognition algorithm. Later, Didimo [23] observed that 2-layer 1-planar graphs are 2-layer RAC graphs, and that the optimal graphs in these two classes coincide. Thus, the tight bound of $\frac{3}{2}n - 2$ edges extends to 2-layer 1-planar graphs. For h-quasiplanar graphs, Walczak [46] provided a density upper bound of $(h - 1)(n - 1)$ edges, following from the fact that *convex bipartite geometric h-quasiplanar graphs* can be $(h - 1)$-colored so that edges with the same color do not cross. For (3-)quasiplanar graphs, the $2n - 2$ bound can be improved to $2n-4$ by observing that they are planar bipartite graphs. Since fan-planar graphs are also quasiplanar, this density bound holds for 2-layer fan-planar graphs, as well. Further, this bound is tight for both classes, since the complete bipartite graph $K_{2,n}$ is 2-layer fan-planar. Note that 2-layer fan-planar graphs have been characterized [17] and can be recognized when the graph is biconnected [17] or a tree [16]. Another property that has been investigated in the 2-layer model is the pathwidth. Namely, 2-layer fan-planar graphs have pathwidth 2 [16], while 2-layer graphs with at most c crossings in total have pathwidth $2c + 1$ [27]; note that both results can be extended to general layered graphs.

Our Contribution. From the above discussion it is evident that, in the wide literature on the 2-layer model, the study of the central class of k-planar graphs is completely missing, except for the special case $k = 1$. In this paper, we make several contributions towards filling this gap. We provide tight density bounds for 2-layer k-planar graphs with $k \in \{2, 3, 4, 5\}$ in Sect. 3. Exploiting these bounds, we deduce a Crossing Lemma for 2-layer graphs in Sect. 4. This implies a density upper bound for general values of k. We then show a lower bound construction that is within a factor of $1/1.84$ from the upper bound. Finally, in Sect. 5, we investigate two additional properties. First, we prove that 2-layer 2-planar graphs are 2-layer quasiplanar, as in the case where the vertices are not restricted to two layers [7]. For larger k, we show a stronger relationship, namely, every 2-layer k-planar graph is 2-layer h-quasiplanar for $h = \left\lceil \frac{2}{3}k + 2 \right\rceil$. Second, we demonstrate that 2-layer k-planar graphs have pathwidth at most $k + 1$, which is the first result of this type, since they may have a linear number of crossings and may not be fan-planar [10].

2 Preliminaries

The 2-Layer Model. A *bipartite graph* $G = (U \dot\cup V, E)$ is a graph with vertex subsets U and V, so that $E \subseteq U \times V$. A *topological* 2-*layer graph* is a bipartite graph drawn in the plane so that the vertices in U and V are mapped to distinct points on two horizontal lines L_u and L_v, respectively, and the edges are mapped to y-monotone Jordan arcs. A topological 2-layer graph can be assumed to be simple, that is, no two adjacent edges cross each other, and every two independent edges cross each other at most once.

Let G be a topological 2-layer graph. We denote the vertices in U and in V as $u_1, \ldots, u_p$ and $v_1, \ldots, v_q$, respectively, in the order in which they appear in positive x-direction along L_u and L_v. We denote the number of vertices of G by $n = p + q$ and the number of edges in E by m. We call G k-*planar* if each edge is crossed at most k times, and h-*quasiplanar* if there is no set of h pairwise crossing edges. Further, we say that a bipartite graph G is 2-*layer* k-*planar* (h-*quasiplanar*) if there exists a topological 2-layer k-planar (resp. h-quasiplanar) graph whose underlying abstract graph is isomorphic to G.

The *maximum number of edges* of a graph class $\mathcal{C}$ is a function $m_{\mathcal{C}} : \mathbb{N} \to \mathbb{N}$ such that (i) every n-vertex graph in $\mathcal{C}$ has at most $m_{\mathcal{C}}(n)$ edges, and (ii) for every n, there is an n-vertex graph in $\mathcal{C}$ with $m_{\mathcal{C}}(n)$ edges. The *(maximum edge) density* of $\mathcal{C}$ is a function $d_{\mathcal{C}} : \mathbb{N} \to \mathbb{N}$ such that (i) for every n, it holds that $d_{\mathcal{C}}(n) \geq m_{\mathcal{C}}(n)$, and (i) there are infinitely many values of n such that $d_{\mathcal{C}}(n) = m_{\mathcal{C}}(n)$. We say that an n-vertex graph in $\mathcal{C}$ with $d_{\mathcal{C}}(n)$ edges is *optimal*.

Note that 2-layer quasiplanar graphs are equivalent to the *convex bipartite geometric quasiplanar graphs*, where vertices lie on a convex shape so that the two partition sets are well-separated [46]. Since these graphs are planar bipartite, as discussed in Sect. 1, and include $K_{2,n}$, their density can be established using the same argumentation as for convex bipartite geometric quasiplanar graphs in [46]:

Theorem 1. *An n-vertex 2-layer quasiplanar graph has at most $2n - 4$ edges for $n \geq 3$. Also, there exist infinitely many 2-layer quasiplanar graphs with n vertices and $2n - 4$ edges.*

Tree and Path Decomposition. A *tree decomposition* of a graph $G = (V, E)$ is a tree T on vertices $B_1, \ldots, B_n$ called *bags* such that the following properties hold: (P.1) each bag B_i is a subset of V, (P.2) $V = \bigcup_{i=1}^{n} B_i$, (P.3) for every edge $(u, v) \in E$, there exists a bag B_i such that $u, v \in B_i$, and (P.4) for every vertex v, the bags containing v induce a connected subtree of T. If T is a path, we call T a *path decomposition*. The *width* of a tree decomposition T is the maximum cardinality of any of its bags minus one, i.e., width$(T) = \max_{i \in \{1, \ldots, n\}}(|B_i| - 1)$. The *treewidth* of a graph G is the minimum width of any of its tree decompositions, whereas the *pathwidth* of G is the minimum width of any of its path decompositions.

3   Tight Density Results for Small Values of k

In this section, we establish the density of 2-layer k-planar graphs for small values of k. We start with a preliminary observation, which follows from the fact that the density of k-planar graphs can be upper bounded by a linear function in n [2,40] and that the density of 2-layer 1-planar graphs is lower bounded by $\frac{3}{2}n - 2$ [22]. This allows us to derive the following:

Lemma 1. *For $k \geq 1$, there exist positive rational numbers $a_k \geq \frac{3}{2}$ and $b_k \geq 0$ such that (i) every n-vertex 2-layer k-planar graph has at most $a_k n - b_k$ edges for $n \geq n_k$ with n_k a constant, and (ii) there is a 2-layer k-planar graph with n vertices and exactly $a_k n - b_k$ edges for some $n > 0$.*

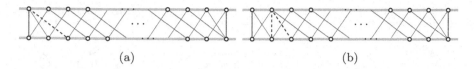

Fig. 1. (a) A maximal topological 2-layer 2-planar graph that is not optimal, as shown by the graph in (b). Differences between the two graphs are dashed blue. (Color figure online)

We then define a useful concept for the analysis of 2-layer k-planar graphs:

Definition 1. *Let G be a topological 2-layer k-planar graph and let $G[i, j\,|\,x, y]$, with $1 \leq i \leq j \leq p$ and $1 \leq x \leq y \leq q$, be the topological subgraph of G induced by vertices $\{u_i, \ldots, u_j, v_x, \ldots, v_y\}$. $G[i, j\,|\,x, y]$ is a brick if it contains two distinct crossing-free edges, namely (u_i, v_x) and (u_j, v_y), that are also crossing-free in G.*

The smallest brick, called *trivial*, contains one vertex of one partition set, say $u_i = u_j$, and two consecutive vertices of the second one, say v_x and $v_y = v_{x+1}$.

Observation 1. *Every optimal topological 2-layer k-planar graph contains planar edges (u_1, v_1) and (u_p, v_q), and hence at least one brick.*

Regarding the connectivity we observe the following. If a topological 2-layer k-planar graph G is not connected, we can draw the connected components as consecutive bricks and connect two consecutive bricks with another edge. Hence, we conclude the following:

Observation 2. *Every optimal topological 2-layer k-planar graph is connected.*

Next, we establish a useful property of an optimal 2-layer k-planar graph G.

Lemma 2. *Let G be an optimal topological 2-layer k-planar graph with exactly $a_k n - b_k$ edges. Then G contains no vertex of degree 1 and no trivial brick.*

Proof. Assume that G contains a degree-1 vertex v and consider the graph G' obtained from G by removing v. This graph has $m' = m - 1$ edges and $n' = n - 1$ vertices. Then, $m' = a_k n - b_k - 1 = a_k(n - 1) - b_k + (a_k - 1)$, which is larger than $a_k(n - 1) - b_k$ since $a_k \geq \frac{3}{2}$, by Lemma 1; a contradiction.

Second, assume that G contains a trivial brick $G[i, i\,|\,x, x + 1]$. Then, consider the graph G' obtained from G by identifying vertices v_x and v_{x+1}. Clearly G' has $m' = m - 1$ edges (edges (u_i, v_x) and (u_i, v_{x+1}) coincide in G') and $n' = n - 1$ vertices. This leads to the same contradiction as in the previous case. $\square$

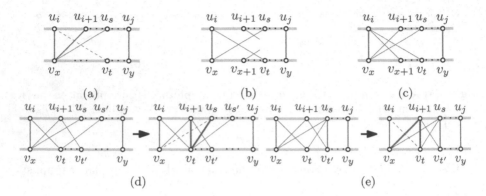

Fig. 2. Illustrations for the proof of Lemma 3.

3.1 2-Layer 2-Planar Graphs

We start with an observation about *maximal* topological 2-layer 2-planar graphs, that is, in which no edge may be inserted without violating 2-planarity.

Observation 3. *There exists a maximal topological 2-layer 2-planar graph that is not optimal; see Fig. 1.*

We now characterize the structure of bricks in optimal 2-layer 2-planar graphs.

Lemma 3. *Let G be an optimal topological 2-layer 2-planar graph with exactly $a_2n - b_2$ edges and let $G[i,j|x,y]$ be a brick of G. Then, $j \geq i+1$ and $y = x+1$, or $j = i+1$ and $y \geq x+1$.*

Proof. By Lemma 2, $G[i,j|x,y]$ is not a trivial brick. Assume, for a contradiction, that both $y \geq x+2$ and $j \geq i+2$. We first observe that u_i is connected to some $v_t \neq v_x$, while v_x is connected to some $u_s \neq u_i$. If this were not the case, say if u_i were only incident to v_x, then a crossing-free edge (v_x, u_{i+1}) could be inserted, contradicting the optimality of G; see Fig. 2a and recall that a brick has no crossing-free edge, except for (u_i, v_x) and (u_j, v_y). So in the following assume that (u_i, v_t) and (v_x, u_s) belong to $G[i,j|x,y]$, with $v_t \neq v_x$ and $u_s \neq u_i$, such that there exists no edge $(u_i, v_{t'})$ with $t' > t$ and no edge $(v_x, v_{s'})$ with $s' > s$. Next, we consider u_{i+1} and v_{x+1}. Assume first that $u_{i+1} \neq u_s$ and that $v_{x+1} \neq v_t$. Then, all edges incident to u_{i+1} and v_{x+1} have a crossing with (u_i, v_t) or (v_x, u_s). Since (u_i, v_t) and (v_x, u_s) cross each other, there can be at most two such edges, and thus u_{i+1} or v_{x+1} has degree one; see Figs. 2b and 2c. By Lemma 2, this contradicts the optimality of G. Hence, assume w.l.o.g. that $v_{x+1} = v_t$. Note that $u_s \neq u_{i+1}$, as otherwise the crossing-free edge (u_{i+1}, v_{x+1}) could be inserted, contradicting the optimality of G. In addition, $u_s = u_{i+2}$, since otherwise u_{i+1} and u_{i+2} could only be incident to a total of two edges, by

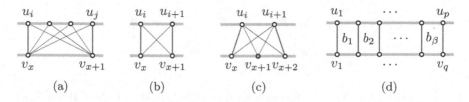

Fig. 3. The unique 2-layer drawings of (a) $K_{2,4}$; (b) $K_{2,2}$; (c) $K_{2,3}$. (d) An optimal 2-layer 2-planar graph is a sequence of bricks joint at planar edges.

the same argument as above, resulting in a degree-1 vertex, which contradicts the optimality of G. □

By Lemma 2, both u_{i+1} and v_{x+1} have degree at least 2. Let $u_{s'}$ and $v_{t'}$ denote the neighbors of v_{x+1} and u_{i+1} respectively, such that s' and t' are maximal. First assume that $t' \neq t$. If $s' = i + 1$, the crossing-free edge (u_s, v_t) can be inserted, contradicting the optimality of G. We observe that edge $(u_{i+1}, v_{t'})$ is crossed by edges (v_x, u_s) and $(v_t, u_{s'})$. If $u_s \neq u'_s$, we can obtain a topological 2-layer 2-planar graph G' by removing edge (v_x, u_s) and inserting edges (v_t, u_s) and (v_t, u_{i+1}); see Fig. 2d. This clearly contradicts the optimality of G. If $u_s = u'_s$, we can obtain a topological 2-layer 2-planar graph G' by removing edge (u_i, v_t) and inserting edges (v_x, u_{i+1}) and (v_t, u_{i+1}); see Fig. 2e. This again contradicts the optimality of G. We conclude that $t' = t$.

Since (v_x, u_s) is crossed by edges (u_i, v_t) and (u_{i+1}, v_t), we conclude that (u_s, v_t) can be inserted without crossings, contradicting the optimality of G. □

By Lemmas 2 and 3, we get that every brick must be a $K_{2,h}$ for some $h \geq 2$. The following observation shows that $h \leq 3$; see also Fig. 3a:

Observation 4. *The complete bipartite graph $K_{2,4}$ is not 2-layer 2-planar.*

We are ready to prove a tight bound for the density of 2-layer 2-planar graphs:

Theorem 2. *Any 2-layer 2-planar graph on n vertices has at most $\frac{5}{3}n - \frac{7}{3}$ edges. Moreover, the optimal 2-layer 2-planar graphs with exactly $\frac{5}{3}n - \frac{7}{3}$ edges are sequences of $K_{2,3}$'s such that consecutive $K_{2,3}$'s share one planar edge.*

Proof. Lemmas 2 and 3, and Observation 4 imply that G contains only $K_{2,2}$- and $K_{2,3}$-bricks; see Figs. 3b and 3c. Moreover, the planar edges separate G into a sequence of β bricks $(b_1, \ldots, b_\beta)$ such that b_i and b_{i+1} share one planar edge. Let β_2 denote the number of $K_{2,2}$-bricks. Then, G has $\beta - \beta_2$ $K_{2,3}$-bricks. Moreover, $n = 2\beta + 2 + (\beta - \beta_2) = 3\beta - \beta_2 + 2$ since each of the $\beta+1$ planar edges is incident to two distinct vertices while each $K_{2,3}$-brick contains an additional vertex; see Fig. 3c. Finally, $m = \beta + 1 + 2\beta_2 + 4(\beta - \beta_2) = 5\beta - 2\beta_2 + 1$ since every $K_{2,2}$-brick contains two non-planar edges while every $K_{2,3}$-brick contains four. For a fixed value of n, $\beta = \frac{1}{3}n + \frac{1}{3}\beta_2 - \frac{2}{3}$ and the density is $m = \frac{5}{3}n - \frac{1}{3}\beta_2 - \frac{7}{3}$. This is clearly maximized for $\beta_2 = 0$. Hence, the maximum density is $m = \frac{5}{3}n - \frac{7}{3}$ which is tightly achieved for graphs in which every brick is a $K_{2,3}$. □

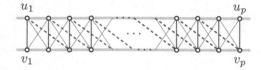

Fig. 4. A family of 3-planar graphs on $n = 2p$ vertices with $2n - 4$ edges. (Color figure online)

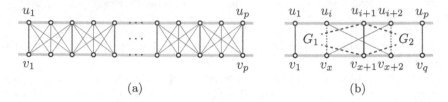

(a)   (b)

Fig. 5. (a) A family of 4-planar graphs on $n = 2p$ vertices with $2n - 3$ edges. (b) A triple of pairwise crossing edges and at most 4 additional edges separates an optimal 2-layer 4-planar graph into graphs G_1 and G_2.

3.2   2-Layer 3-Planar Graphs

Next, we give a tight bound on the density of 2-layer 3-planar graphs. We first present a lower bound construction:

Theorem 3. *There exist infinitely many 2-layer 3-planar graphs with n vertices and $2n - 4$ edges.*

Proof. We describe a family of graphs where $p = q$; refer to Fig. 4. Each graph has the following edges: (u_i, v_i) for $1 \leq i \leq p$ (red edges in Fig. 4); (u_i, v_{i+1}) for $1 \leq i \leq p - 1$, and (u_i, v_{i-1}) for $2 \leq i \leq p$ (green edges in Fig. 4); (u_i, v_{i+2}) for $1 \leq i \leq p - 2$ (dashed blue edges in Fig. 4). Vertices u_1, u_{p-1}, v_2 and v_p have degree 3, u_p and v_1 have degree 2, and all other vertices have degree 4, yielding $4n - 8$ for the sum of the vertex degrees and hence $2n - 4$ edges.   □

The following theorem provides the corresponding density upper bound:

Theorem 4. *Let G be a topological 2-layer 3-planar graph on n vertices. Then G has at most $2n-4$ edges for $n \geq 3$. Moreover, if G is optimal, it is quasiplanar.*

Proof (Sketch). We show that optimal 2-layer 3-planar graphs are quasiplanar, which implies the statement, by Theorem 1.   □

3.3   2-Layer 4-Planar Graphs

We first present a lower bound construction for this class of graphs:

Theorem 5. *There exist infinitely many 2-layer 4-planar graphs with n vertices and $2n - 3$ edges.*

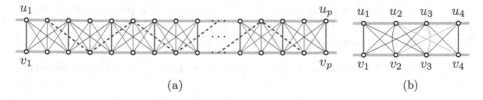

Fig. 6. (a) A family of 5-planar graphs on $n = 2p$ vertices with $\frac{9}{4}n - \frac{9}{2}$ edges. (b) Graph S with $n = 8$ vertices and $m = 14 > \frac{9}{4} \cdot 8 - \frac{9}{2} = 13.5$ edges.

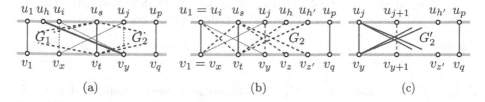

Fig. 7. (a) A triple (u_i, v_y), (u_s, v_t), (u_j, v_x) of pairwise crossing edges and at most six other edges separates an optimal 2-layer 5-planar graph into subgraphs G_1 and G_2. If G_1 consists of a single edge, (b) there can be edges (u_s, v_z), $(u_s, v_{z'})$, (v_t, u_h), $(v_t, u_{h'})$, in which case (c) G_2 consists of a graph G_2', vertices u_j, v_y and at most four of the green edges. (Color figure online)

Proof. We describe a family of graphs where $p = q$; see Fig. 5a. Each topological graph G consists of a sequence $(b_1, \ldots, b_\beta)$ of $K_{3,3}$-bricks such that b_i and b_{i+1} share a planar edge for $1 \leq i \leq \beta - 1$. Then G has $n = 4\beta + 2$ vertices and $m = 8\beta + 1 = 2n - 3$ edges. $\square$

Next, we provide a matching upper bound.

Theorem 6. *Any 2-layer 4-planar graph on n vertices has at most $2n - 3$ edges.*

Proof (Sketch). We first prove that in an optimal topological 2-layer 4-planar graph G, every triple of pairwise crossing edges is such that removing the triple and at most four other edges separates G into two subgraphs G_1 and G_2 as shown in Fig. 5b. Based on this observation, we apply induction on the number of such triples in G. Note that in the base case, i.e., no triples of pairwise crossing edges exist, the graph is quasiplanar. $\square$

3.4 2-Layer 5-Planar Graphs

We first provide a lower bound construction for this class of graphs:

Theorem 7. *There exist infinitely many 2-layer 5-planar graphs with n vertices and $\frac{9}{4}n - \frac{9}{2}$ edges.*

Proof. We augment the construction from Theroem 5 by a path of length $\beta - 1$, where β is the number of $K_{3,3}$ subgraphs; see the dashed blue edges in Fig. 6a. The obtained graph has $n = 4\beta + 2$ vertices and $m = 9\beta = \frac{9}{4}n - \frac{9}{2}$ edges.   □

For the specific value $n = 8$, we can provide a denser lower bound construction.

Observation 5. *There exists a topological 2-layer 5-planar graph $\mathcal{S}$ with $n = 8$ vertices and $m = 14 > \frac{9}{4}n - \frac{9}{2}$ edges; see Fig. 6b.*

We show that the graph $\mathcal{S}$ is in fact an exception, by demonstrating that the lower bound construction in Theorem 7 is tight for all other values of n.

Theorem 8. *Any 2-layer 5-planar graph on $n \geq 3$ vertices has at most $\frac{9}{4}n - \frac{9}{2}$ edges, except for graph $\mathcal{S}$ which has 8 vertices and 14 edges.*

Proof (Sketch). First observe that the theorem is clearly fulfilled if $G = \mathcal{S}$. Otherwise, we apply an argument similar to the proof of Theorem 6. Namely, we first prove that if there is a triple of pairwise crossing edges in an optimal topological 2-layer 5-planar graph, the removal of few edges separates the graph into two components G_1 and G_2; see Fig. 7a. We then apply induction on the number of such triples in G. In particular, we consider some special cases, namely G_1 could be $\mathcal{S}$ or a single edge; see also Fig. 7b. In the latter case, we also investigate the structure of graph G_2 in more careful detail to prove our result; see also Fig. 7c.   □

4   A Crossing Lemma and General Density Bounds

In this section we generalize the well-known Crossing Lemma [6, 29, 36] to a meta Crossing Lemma for general graphs (Theroem 9), which also yields a density upper bound for k-planar graphs. We denote by $\mathcal{R}$ a restriction on graphs, e.g., $\mathcal{R}$ can be "bipartite" or "2-layer". We assume that for a fixed $t > 0$, there are $\alpha_i, \beta_i \in \mathbb{R}$ for $i \in \{0, \ldots, t - 1\}$ such that $m \leq \alpha_i n - \beta_i$ is an upper bound for the number of edges in $\mathcal{R}$-restricted i-planar graphs. Let $\alpha := \sum_{i=0}^{t-1} \alpha_i$ and $\beta := \sum_{i=0}^{t-1} \beta_i$. The proof of the next theorem follows the probabilistic technique of Chazelle, Sharir and Welzl (see e.g. [5, Chapter 35]).

Theorem 9. *Let G be a simple $\mathcal{R}$-restricted graph with $n \geq 4$ vertices and $m \geq \frac{3\alpha}{2t}n$ edges. The following inequality holds for the crossing number $cr(G)$:*

$$cr(G) \geq \frac{4t^3}{27\alpha^2} \frac{m^3}{n^2}. \tag{1}$$

The meta Crossing Lemma is used to obtain the following theorem regarding the density. We follow closely the proof for corresponding statements for k-planar and bipartite k-planar graphs [2, 9].

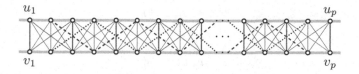

Fig. 8. A family of 6-planar graphs on $n = 2p$ vertices with $\frac{5}{2}n - 6$ edges.

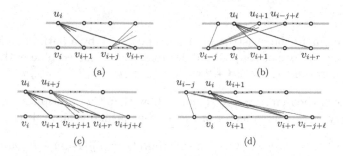

Fig. 9. Illustrations for the proof of Theroem 12.

Theorem 10. *Let G be a simple $\mathcal{R}$-restricted k-planar graph with $n \geq 4$ vertices for some $k \geq t$. Then*

$$m \leq \max\left\{1, \sqrt{\frac{3}{2t}}\sqrt{k}\right\} \cdot \frac{3\alpha}{2t}n.$$

We apply Theorem 9 and 10 to 2-layer k-planar graphs for $t = 6$. By [23], Theorems 2, 4, 6 and 8, we have $(\alpha_0, \alpha_1, \alpha_2, \alpha_3, \alpha_4, \alpha_5) = (1, \frac{3}{2}, \frac{5}{3}, 2, 2, \frac{9}{4})$, yielding $\alpha = \frac{125}{12}$. By substituting the numbers in Theroem 9 we obtain the following.

Corollary 1. *Let G be a simple 2-layer graph with $n \geq 4$ vertices and $m \geq \frac{125}{48}n$ edges. Then, the following inequality holds for the crossing number $cr(G)$:*

$$cr(G) \geq \frac{4.608}{15.625}\frac{m^3}{n^2} \approx 0.295\frac{m^3}{n^2}.$$

By plugging the result into Theorem 10 we obtain.

Corollary 2. *Let G be a simple 2-layer k-planar graph with $n \geq 4$ vertices for some $k > 5$. Then*
$$m \leq \max\left\{\frac{125}{48}, \frac{125}{96}\sqrt{k}\right\} \cdot n.$$

Note that for 2-layer 6-planar graphs, Corollary 2 certifies that $m \leq 3.19n$. We can show that there is only a gap of $0.69n$ towards an optimal solution:

Theorem 11. *There exist infinitely many 2-layer 6-planar graphs with n vertices and $\frac{5}{2}n - 6$ edges.*

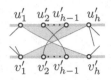

Fig. 10. A set of h pairwise crossing edges in a topological 2-layer graph.

Proof. We augment the construction from Theorem 7 by a path of length $\beta - 1$, where β is the number of $K_{3,3}$ subgraphs; refer to the dotted blue path in Fig. 8. The obtained graph has $n = 4\beta + 2$ vertices and $m = 10\beta - 1 = \frac{5}{2}n - 6$ edges. □

In the next theorem, we additionally show that the multiplicative constant from Colloary 2 is within a factor of 1.84 of the optimal achievable upper bound.

Theorem 12. *For any k, there exist infinitely many 2-layer k-planar graphs with n vertices and $m = \left\lfloor \sqrt{k/2} \right\rfloor n - \mathcal{O}(f(k)) \approx 0.707\sqrt{k}n - \mathcal{O}(f(k))$ edges.*

Proof (Sketch). We choose $p = q$ and a parameter $\ell = \lfloor \sqrt{k/2} \rfloor$. We connect vertex u_i to the ℓ vertices $v_{i+1} \ldots, v_{i+\ell}$ and vertex v_i to vertices $u_{i+1} \ldots, u_{i+\ell}$. Note that by symmetry, u_i is also incident to the ℓ vertices $v_{i-1} \ldots, v_{i-\ell}$ and vertex v_i to vertices $u_{i-1} \ldots, u_{i-\ell}$. Clearly, this gives the density bound in the statement of the theorem. Then, we consider an edge (u_i, v_{i+r}) and the crossings it forms with edges incident to some other vertices; see Fig. 9. This allows us to establish that each edge has at most k crossings. □

5   Properties of 2-Layer k-Planar Graphs

In this section, we present some properties of 2-layer k-planar graphs.

In Theorem 4, we have established that every optimal 2-layer 3-planar graph is (3-)quasiplanar, which is also the case in the general, non-layered, drawing model [15]. A more general relationship between the classes of k-planar and h-quasiplanar graphs was uncovered in [7], where it is proven that every k-planar graph is $(k + 1)$-quasiplanar, for every $k \geq 2$. Next, we show that for 2-layer drawings an even stronger relationship holds.

Theorem 13. *For $k \geq 3$, every 2-layer k-planar graph is 2-layer $\lceil \frac{2}{3}k + 2 \rceil$-quasiplanar. Further, every 2-layer 2-planar graph is 2-layer (3-)quasiplanar.*

Proof. Let G be a topological 2-layer k-planar graph, with $k \geq 3$, which we assume w.l.o.g. to be connected. Suppose for a contradiction that G contains $h := \lceil \frac{2}{3}k + 2 \rceil$ mutually crossing edges (u_i', v_{h+1-i}') for $1 \leq i \leq h$ in G, such that $u_1', \ldots, u_h'$ and $v_1', \ldots, v_h'$ appear in this order in $u_1, \ldots, u_p$ and $v_1, \ldots, v_q$, respectively. Observe that (u_1', v_h') and (v_1', u_h') have $h - 1$ crossings from this

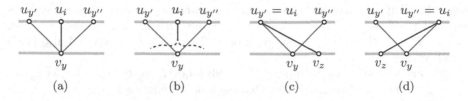

Fig. 11. Illustrations for the proof of Theorem 14.

h-tuple. Moreover, both endvertices of all the $h - 2$ edges (u_i', v_{h+1-i}'), for $i = 2, \ldots, h - 1$, are located in regions bounded by $e^{(1)} := (u_1', v_h')$ and $e^{(2)} := (v_1', u_h')$; see Fig. 10. Since G is connected, for each $2 \le i \le h - 1$, the edge (u_i', v_{h+1-i}') is adjacent to another edge e_i. Note that either $e_i = e_j$ for some $j \ne i$, and e_i crosses $e^{(1)}$ and $e^{(2)}$, or $e_i \ne e_j$ for all $j \ne i$, and e_i crosses one of $e^{(1)}$ and $e^{(2)}$. This implies $h - 2$ additional crossings for $\{e^{(1)}, e^{(2)}\}$, and, consequently, $e^{(1)}$ or $e^{(2)}$ is crossed by at least $h - 1 + \lceil (h - 2)/2 \rceil$ edges. We obtain $h - 1 + \lceil (h - 2)/2 \rceil \ge \frac{3}{2}h - 2 \ge \frac{3}{2}\left(\frac{2}{3}k + 2\right) - 2 = k + 1$ crossings for $e^{(1)}$ or $e^{(2)}$, a contradiction.

For the case $k = 2$, assume that G contains three mutually crossing edges $e_1 = (u_1', v_3')$, $e_2 = (u_2', v_2')$ and $e_3 = (u_3', v_1')$, such that u_1', u_2', u_3' and v_1', v_2', v_3'. appear in this order in $u_1, \ldots, u_p$ and $v_1, \ldots, v_q$, respectively. As e_1 and e_3 are already crossed twice, e_2 represents a connected component; contradiction.   □

Next, we show that the pathwidth of 2-layer k-planar graphs is bounded by $k + 1$. We point out that similar results are known for layered graphs with a bounded total number of crossings [27] and for layered fan-planar graphs [16], and that these bounds do not have any implication on 2-layer k-planar graphs.

Theorem 14. *Every 2-layer k-planar graph has pathwidth at most $k + 1$.*

Proof. Let G be a topological 2-layer k-planar graph with parts U and V. We first define a total ordering $\prec$ on the edges as follows: We say that edge $e_1 = (u_i, v_x)$ precedes edge $e_2 = (u_j, v_y)$, or $e_1 \prec e_2$, if $u_i, u_j \in U$ and either (i) $i < j$, or (i) $i = j$ and $x < y$. Let $E = (e_1, \ldots, e_m)$ be the set of edges ordered with respect to $\prec$. Let $e_i = (u_s, v_t)$ be an edge and let v_y be a vertex in V. Further let e_{y^-} and e_{y^+} be the first and the last edge incident to v_y in $\prec$, respectively. We call v_y *related* to e_i if v_y is incident to an edge crossing e_i and if $y^- < i < y^+$. For every edge $e_i = (u_s, v_t) \in E$, we construct a bag B_i that contains u_s, v_t and all the (at most k) related vertices of e_i. Then, we connect B_i to bags B_{i-1} and B_{i+1} (if they exist), obtaining a path of bags P.

In the following we show that P is a valid path decomposition of G. Since we assigned at most $k + 2$ vertices to each bag of P the width of P is at most $k + 1$. Properties P.1 and P.3 of a tree decomposition are fulfilled for P by construction. We may assume that G is connected, otherwise we compute a path decomposition for each connected component and link the obtained vertex disjoint paths. Hence also P.2 is fulfilled. Moreover, by the choice of $\prec$, all the edges incident to a vertex $u_i \in U$ occur in a consecutive sequence, i.e. u_i is incident to edges $e_j, \ldots, e_k$ for some $1 \le j \le k \le m$ and then u_i appears in all of bags $B_j, \ldots, B_k$, which is a subpath of P. Therefore, Property P.4 also holds for all vertices in U.

It remains to show that Property P.4 holds for every vertex $v_y \in V$. Let $e_{y^-} = (u_{y'}, v_y)$ and $e_{y^+} = (u_{y''}, v_y)$. Note that each of the edges $e_{y^-}, e_{y^-+1}, \ldots, e_{y^+}$ is either incident to v_y (see Fig. 11a), or it crosses one of e_{y^-} and e_{y^+}, since its endvertex in U is some u_i with $y' \le i \le y''$; see Figs. 11b to 11d. Note that for the endvertex v_z in V necessarily $z > y$ if $u_i = u_{y'}$ or $z < y$ if $u_i = u_{y''}$ by definition of $\prec$; see Fig. 11c or Fig. 11d, respectively. Hence v_y belongs to all bags $B_{y^-}, B_{y^-+1}, \ldots, B_{y^+}$ and P.4 holds. The statement follows. $\square$

6   Conclusions

We gave results for 2-layer k-planar graphs regarding their density, relationship to 2-layer h-quasiplanar graphs, and pathwidth. Tight density bounds for 2-layer k-planar graphs with $k = 6$ may be achievable following similar arguments to the proof of Theorem 8, which would also improve upon our results for the Crossing Lemma, and in turn on the density for general k. Moreover, a better lower bound for general k may exist. The relationship to other beyond-planar graph classes is also of interest. With respect to the pathwidth, we conjecture that our upper bound is tight. Finally, the recognition and characterization of 2-layer k-planar graphs remain important open problems.

References

1. Ackerman, E.: On the maximum number of edges in topological graphs with no four pairwise crossing edges. Discret. Comput. Geom. **41**(3), 365–375 (2009). https://doi.org/10.1007/s00454-009-9143-9
2. Ackerman, E.: On topological graphs with at most four crossings per edge. Comput. Geom. 85 (2019). https://doi.org/10.1016/j.comgeo.2019.101574
3. Ackerman, E., Tardos, G.: On the maximum number of edges in quasi-planar graphs. J. Comb. Theory Ser. A **114**(3), 563–571 (2007). https://doi.org/10.1016/j.jcta.2006.08.002
4. Agarwal, P.K., Aronov, B., Pach, J., Pollack, R., Sharir, M.: Quasi-planar graphs have a linear number of edges. Combinatorica **17**(1), 1–9 (1997). https://doi.org/10.1007/BF01196127
5. Aigner, M., Ziegler, G.M.: Probability makes counting (sometimes) easy. Proofs from THE BOOK, pp. 311–319. Springer, Heidelberg (2018). https://doi.org/10.1007/978-3-662-57265-8_45
6. Ajtai, M., Chvátal, V., Newborn, M., Szemerédi, E.: Crossing-free subgraphs. In: Hammer, P.L., Rosa, A., Sabidussi, G., Turgeon, J. (eds.) Theory and Practice of Combinatorics, North-Holland Mathematics Studies, vol. 60, pp. 9–12. North-Holland (1982). http://www.sciencedirect.com/science/article/pii/S0304020808734844
7. Angelini, P., et al.: Simple k-planar graphs are simple (k+1)-quasiplanar. J. Comb. Theory Ser. B **142**, 1–35 (2020). https://doi.org/10.1016/j.jctb.2019.08.006

8. Angelini, P., Bekos, M.A., Förster, H., Kaufmann, M.: On RAC drawings of graphs with one bend per edge. Theor. Comput. Sci. **828–829**, 42–54 (2020). https://doi.org/10.1016/j.tcs.2020.04.018
9. Angelini, P., Bekos, M.A., Kaufmann, M., Pfister, M., Ueckerdt, T.: Beyond-planarity: Turán-type results for non-planar bipartite graphs. In: ISAAC. LIPIcs, vol. 123, pp. 28:1–28:13. Schloss Dagstuhl (2018). https://doi.org/10.4230/LIPIcs.ISAAC.2018.28
10. Angelini, P., Da Lozzo, G., Förster, H., Schneck, T.: 2-layer k-planar graphs: Density, crossing lemma, relationships, and pathwidth (2020). https://arxiv.org/abs/2008.09329
11. Auer, C., et al.: Outer 1-Planar Graphs. Algorithmica **74**(4), 1293–1320 (2015). https://doi.org/10.1007/s00453-015-0002-1
12. Avital, S., Hanani, H.: Graphs. Gilyonot Lematematika **3**, 2–8 (1966)
13. Bekos, M.A., Cornelsen, S., Grilli, L., Hong, S., Kaufmann, M.: On the recognition of fan-planar and maximal outer-fan-planar graphs. Algorithmica **79**(2), 401–427 (2017). https://doi.org/10.1007/s00453-016-0200-5
14. Bekos, M.A., Kaufmann, M., Raftopoulou, C.N.: On the density of non-simple 3-planar graphs. In: Hu, Y., Nöllenburg, M. (eds.) GD 2016. LNCS, vol. 9801, pp. 344–356. Springer, Cham (2016). https://doi.org/10.1007/978-3-319-50106-2_27
15. Bekos, M.A., Kaufmann, M., Raftopoulou, C.N.: On optimal 2- and 3-planar graphs. In: Aronov, B., Katz, M.J. (eds.) 33rd International Symposium on Computational Geometry, SoCG 2017, July 4–7, 2017, Brisbane, Australia. LIPIcs, vol. 77, pp. 16:1–16:16. Schloss Dagstuhl - Leibniz-Zentrum für Informatik (2017). https://doi.org/10.4230/LIPIcs.SoCG.2017.16
16. Biedl, T.C., Chaplick, S., Kaufmann, M., Montecchiani, F., Nöllenburg, M., Raftopoulou, C.N.: Layered fan-planar graph drawings. In: Esparza, J., Král', D. (eds.) 45th International Symposium on Mathematical Foundations of Computer Science, MFCS 2020, August 24–28, 2020, Prague, Czech Republic. LIPIcs, vol. 170, pp. 14:1–14:13. Schloss Dagstuhl - Leibniz-Zentrum für Informatik (2020). https://doi.org/10.4230/LIPIcs.MFCS.2020.14
17. Binucci, C., et al.: Algorithms and characterizations for 2-layer fan-planarity: From caterpillar to stegosaurus. J. Graph Algorithms Appl. **21**(1), 81–102 (2017). https://doi.org/10.7155/jgaa.00398
18. Binucci, C., et al.: Fan-planarity: properties and complexity. Theor. Comput. Sci. **589**, 76–86 (2015). https://doi.org/10.1016/j.tcs.2015.04.020
19. Capoyleas, V., Pach, J.: A Turán-type theorem on chords of a convex polygon. J. Comb. Theory Ser. B **56**(1), 9–15 (1992)
20. Chaplick, S., Kryven, M., Liotta, G., Löffler, A., Wolff, A.: Beyond outerplanarity. In: Frati, F., Ma, K.-L. (eds.) GD 2017. LNCS, vol. 10692, pp. 546–559. Springer, Cham (2018). https://doi.org/10.1007/978-3-319-73915-1_42
21. Dehkordi, H.R., Eades, P., Hong, S., Nguyen, Q.H.: Circular right-angle crossing drawings in linear time. Theor. Comput. Sci. **639**, 26–41 (2016). https://doi.org/10.1016/j.tcs.2016.05.017
22. Di Giacomo, E., Didimo, W., Eades, P., Liotta, G.: 2-layer right angle crossing drawings. Algorithmica **68**(4), 954–997 (2012). https://doi.org/10.1007/s00453-012-9706-7
23. Didimo, W.: Density of straight-line 1-planar graph drawings. Inf. Process. Lett. **113**(7), 236–240 (2013). https://doi.org/10.1016/j.ipl.2013.01.013
24. Didimo, W., Eades, P., Liotta, G.: Drawing graphs with right angle crossings. Theor. Comput. Sci. **412**(39), 5156–5166 (2011). https://doi.org/10.1016/j.tcs.2011.05.025

25. Didimo, W., Liotta, G.: The crossing-angle resolution in graph drawing. In: Pach, J. (ed.)Thirty Essays on Geometric Graph Theory, pp. 167–184. Springer, Heidelberg (2013). https://doi.org/10.1007/978-1-4614-0110-0_10

26. Didimo, W., Liotta, G., Montecchiani, F.: A survey on graph drawing beyond planarity. ACM Comput. Surv. **52**(1), 4:1–4:37 (2019). https://doi.org/10.1145/3301281

27. Dujmović, V., et al.: On the parameterized complexity of layered graph drawing. Algorithmica **52**(2), 267–292 (2008). https://doi.org/10.1007/s00453-007-9151-1

28. Eades, P., Liotta, G.: Right angle crossing graphs and 1-planarity. Discret. Appl. Math. **161**(7–8), 961–969 (2013). https://doi.org/10.1016/j.dam.2012.11.019

29. Erdős, P., Guy, R.: Crossing number problems. Am. Math. Mon. **80**, 52–58 (1973). https://doi.org/10.2307/2319261

30. Fox, J., Pach, J.: Coloring K_k-free intersection graphs of geometric objects in the plane. In: Symposium on Computational Geometry, pp. 346–354. ACM (2008)

31. Fox, J., Pach, J., Suk, A.: The number of edges in k-quasi-planar graphs. SIAM J. Discrete Math. **27**(1), 550–561 (2013)

32. Hong, S.-H., Eades, P., Katoh, N., Liotta, G., Schweitzer, P., Suzuki, Y.: A linear-time algorithm for testing outer-1-planarity. Algorithmica **72**(4), 1033–1054 (2014). https://doi.org/10.1007/s00453-014-9890-8

33. Hong, S., Nagamochi, H.: A linear-time algorithm for testing full outer-2-planarity. Discret. Appl. Math. **255**, 234–257 (2019). https://doi.org/10.1016/j.dam.2018.08.018

34. Kaufmann, M., Ueckerdt, T.: The density of fan-planar graphs. CoRR **1403**, 6184 (2014)

35. Korzhik, V.P., Mohar, B.: Minimal obstructions for 1-immersions and hardness of 1-planarity testing. J. Graph Theory **72**(1), 30–71 (2013). https://doi.org/10.1002/jgt.21630

36. Leighton, F.T.: Complexity Issues in VLSI: Optimal Layouts for the Shuffle-exchange Graph and Other Networks. MIT Press, Cambridge, MA, USA (1983)

37. Pach, J., Radoicic, R., Tardos, G., Tóth, G.: Improving the crossing lemma by finding more crossings in sparse graphs. Discret. Comput. Geom. **36**(4), 527–552 (2006). https://doi.org/10.1007/s00454-006-1264-9

38. Pach, J., Radoicic, R., Tóth, G.: Relaxing planarity for topological graphs. JCDCG **2866**, 221–232 (2002). https://doi.org/10.1007/978-3-540-44400-8_24

39. Pach, J., Shahrokhi, F., Szegedy, M.: Applications of the crossing number. Algorithmica **16**(1), 111–117 (1996)

40. Pach, J., Tóth, G.: Graphs drawn with few crossings per edge. Combinatorica **17**(3), 427–439 (1997). https://doi.org/10.1007/BF01215922

41. Ringel, G.: Ein Sechsfarben problem auf der Kugel. Abh. Math. Sem. Univ. Hamb. **29**, 107–117 (1965). https://doi.org/10.1007/BF02996313

42. Sugiyama, K.: Graph drawing and applications for software and knowledge engineers. In: Series on Software Engineering and Knowledge Engineering, vol. 11. WorldScientific (2002). https://doi.org/10.1142/4902

43. Sugiyama, K., Tagawa, S., Toda, M.: Methods for visual understanding of hierarchical system structures. IEEE Trans. Syst. Man Cybern. SMC-11 **11**(2), 109–125 (1981). https://doi.org/10.1109/TSMC.1981.4308636

44. Suk, A., Walczak, B.: New bounds on the maximum number of edges in k-quasi-planar graphs. Comput. Geom. **50**, 24–33 (2015)

45. Valtr, P.: On geometric graphs with no k pairwise parallel edges. Discrete Comput. Geom. **19**(3), 461–469 (1998)
46. Walczak, B.: Old and new challenges in coloring graphs with geometric representations. In: Archambault, D., Tóth, C.D. (eds.) Graph Drawing and Network Visualization - 27th International Symposium, GD 2019, Prague, Czech Republic, 17–20 September 2019, Proceedings. Lecture Notes in Computer Science, vol. 11904. Springer, Heidelberg (2019). Invited talk

Planarity

Planar Rectilinear Drawings of Outerplanar Graphs in Linear Time

Fabrizio Frati[✉]

Roma Tre University, Rome, Italy
frati@dia.uniroma3.it

Abstract. We show how to test in linear time whether an outerplanar graph admits a planar rectilinear drawing, both if the graph has a prescribed plane embedding and if it does not. Our algorithm returns a planar rectilinear drawing if the graph admits one.

1 Introduction

Planar orthogonal graph drawings with a minimum number of bends have been studied for decades. In 1987, Tamassia [20] proved that, for an n-vertex planar graph with a prescribed plane embedding, a planar orthogonal drawing with the minimum number of bends can be constructed in polynomial time, thereby establishing a result that lies at the very foundations of the graph drawing research area. The running time of Tamassia's algorithm is $O(n^2 \log n)$, which has been improved to $O(n^{7/4}\sqrt{\log n})$ [11] and then to $O(n^{3/2})$ by Cornelsen and Karrenbauer [4]. However, achieving a linear running time is still an elusive goal.

Bend minimization in the variable embedding setting is a much harder problem; indeed, Garg and Tamassia [12] proved that testing whether a graph admits a planar orthogonal drawing with zero bends is NP-hard. However, some natural restrictions on the input make the problem tractable. A successful story is the one about n-vertex degree-3 planar graphs. Di Battista et al. [5] proved that, for such graphs, a planar orthogonal drawing with the minimum number of bends can be constructed in $O(n^5 \log n)$ time. After some improvements [3,9], a recent breakthrough result by Didimo et al. [8] has shown that $O(n)$ time is indeed sufficient. Di Battista et al. [5] also presented an $O(n^4)$-time algorithm for minimizing the number of bends in a planar orthogonal drawing of an n-vertex biconnected series-parallel graph. This result was first extended to not necessarily biconnected series-parallel graphs by Bläsius et al. [1] and then improved to an $O(n^3 \log n)$ running time by Di Giacomo et al. [6].

Evidence has shown that the bend-minimization problem is not much easier if one is only interested in the construction of planar orthogonal drawings with zero bends; these are also called *planar rectilinear drawings* (see Figs. 1(a)

Partially supported by MIUR Project "AHeAD" under PRIN 20174LF3T8, by H2020-MSCA-RISE Proj. "CONNECT" n° 734922, and by Roma Tre University Azione 4 Project "GeoView".

© Springer Nature Switzerland AG 2020
D. Auber and P. Valtr (Eds.): GD 2020, LNCS 12590, pp. 423–435, 2020.
https://doi.org/10.1007/978-3-030-68766-3_33

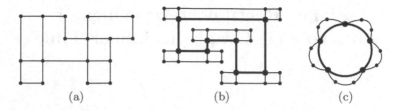

Fig. 1. (a) An outerplanar rectilinear drawing. (b) A planar rectilinear drawing of an outerplanar graph G. The graph G has no outerplanar rectilinear drawing. (c) An outerplanar graph G that has no 3-cycle and no planar rectilinear drawing.

and 1(b) for two such drawings). Namely, the cited NP-hardness proof of Garg and Tamassia [12] is designed for planar rectilinear drawings. Further, almost every efficient algorithm for testing the existence of planar rectilinear drawings [14,17,19] has been eventually subsumed by an algorithm in the more general bend-minimization scenario [8,18,21]. A notable exception is that of planar graphs with a fixed embedding, for which the fastest known algorithms for the bend-minimization problem and for the rectilinear-planarity testing problem run in $O(n^{3/2})$ time [4] and in $O(n \log^3 n)$ time [2,20], respectively.

In this paper, we show that the existence of a planar rectilinear drawing can be tested in $O(n)$ time for an n-vertex outerplanar graph. Our algorithm is constructive and covers both the fixed and the variable embedding scenarios, where the previously best known time bounds were $O(n \log^3 n)$ [2,20] and $O(n^3 \log n)$ [6], respectively; our algorithm also allows us to test in $O(n)$ time whether an n-vertex outerplanar graph admits an outerplanar rectilinear drawing. Given how common it is to study outerplanar graphs for a problem which is too difficult or too computationally expensive on general planar graphs, it is surprising that a systematic study of planar orthogonal and rectilinear drawings of outerplanar graphs has not been done before. The only result we are aware of that is tailored for outerplanar graphs is the one by Nomura et al. [16], which states that an outerplanar graph with maximum degree 3 admits a planar rectilinear drawing if and only if it does not contain any 3-cycle. This characterization is not true for outerplanar graphs with vertices of degree 4; see, e.g., Fig. 1(c).

We outline our algorithm for the variable embedding setting.

The first, natural, idea is to reduce the problem to the 2-connected case. This reduction builds on (an involved version of) a technique introduced by Didimo et al. [8] that, roughly speaking, allows us to perform postorder traversals of the block-cut-vertex tree of the graph in total linear time so that each edge is traversed in both directions; during these traversals, information is computed that allows us to decide whether solutions for the subproblems associated to the blocks of the graph can be combined into a solution for the entire graph. This reduction to the 2-connected case comes at the expense of having to solve a harder problem, in which some vertices of the graph have restrictions on their incident angles in the sought planar rectilinear drawing.

An analogous technique allows us to reduce the problem to the case in which the input 2-connected graph has a prescribed edge that is required to be incident to the outer face. The role that in the previous reduction is played by the block-cut-vertex tree is here undertaken by the "extended dual tree" of the outerplanar graph. Each edge of this tree is dual to an edge of the outerplanar graph; the latter edge splits the outerplanar graph into two smaller outerplanar graphs. These are the sub-instances whose solutions might be combined into a solution for the entire graph; whether this combination is possible is decided based on information that is computed during the traversals of the extended dual tree.

The core of our algorithm consists of an efficient solution for the problem of testing whether a 2-connected outerplanar graph admits a planar rectilinear drawing in which a prescribed edge is required to be incident to the outer face. Our starting point is a characterization of the positive instances in terms of the existence of a sequence of numerical values satisfying some conditions; these values represent certain geometric angles of a planar rectilinear drawing. Some of these numerical values can be chosen optimally, based on recursive solutions to smaller subproblems; further, a constant number of them have to be chosen in all possible ways; finally, we reduce the problem of finding the remaining numerical values to the one of testing for the existence of a set of integers, each of which is required to be in a certain interval, so that a linear equation on these integers is satisfied. We characterize the solutions to the latter problem so that not only it can be solved efficiently, but a solution can be modified in constant time if the interval associated to each integer changes slightly; this change corresponds to a different edge chosen to be incident to the outer face.

Together with our submission to GD 2020, another paper on the rectilinear-planarity testing problem was accepted to the same conference. Namely, Didimo et al. [7] presented an $O(n)$-time algorithm which tests whether an n-vertex 2-connected series-parallel graph with fixed embedding admits a planar rectilinear drawing; the techniques by Didimo et al. are different from ours.

In what follows, we assume w.l.o.g. that every considered graph is connected and has maximum degree 4. Because of space limitations, we only present our algorithm for outerplanar graphs with a variable embedding. We also neglect the construction of planar rectilinear drawings and focus on testing their existence. Finally, all proofs are omitted and deferred to the full version of the paper [10].

2   Preliminaries

A *block* of a connected graph G is a maximal 2-connected subgraph of G; it is *trivial* if it is a single edge and *non-trivial* otherwise. The *BC-tree* T of G [13,15] is the tree that has a *B-node* for each block of G and a *C-node* for each cut-vertex of G; a B-node b and a C-node c are adjacent in T if c lies in the block corresponding to b (we often identify a C-node and the corresponding cut-vertex).

A *drawing* of a graph maps each vertex to a point in the plane and each edge to a curve between its endpoints. A drawing is *planar* if no two edges cross and it is *rectilinear* if each edge is either a horizontal or a vertical segment. A planar

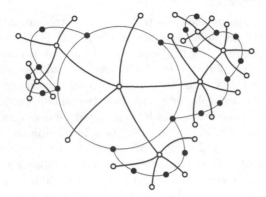

Fig. 2. The extended dual tree (represented by white disks and thick lines) of an outerplane embedding of a 2-connected outerplanar graph.

drawing divides the plane into topologically connected regions, called *faces*; the only unbounded face is the *outer face*, while all the other faces are *internal*.

Two planar drawings Γ_1 and Γ_2 of a connected planar graph G are *equivalent* if: (i) for each vertex w of G, the clockwise order of the edges incident to w is the same in Γ_1 and Γ_2; and (ii) the clockwise order of the edges incident to the outer face is the same in Γ_1 and Γ_2. A *plane embedding* is an equivalence class of planar drawings. Two drawings that correspond to the same plane embedding have faces delimited by the same walks, hence we often speak about *faces of a plane embedding*. We denote by $f_{\mathcal{E}}^*$ the outer face of a plane embedding $\mathcal{E}$.

Two planar rectilinear drawings of a 2-connected planar graph G are *equivalent* if they correspond to the same plane embedding $\mathcal{E}$ and if, for every face f of $\mathcal{E}$ and for every vertex w incident to f, the angle at w in f is the same in both drawings. A *rectilinear representation* of G is an equivalence class of planar rectilinear drawings of G. A rectilinear representation is hence a pair $(\mathcal{E}, \phi)$, where $\mathcal{E}$ is a plane embedding of G and ϕ is a function that assigns an *angle* $\phi(w, f) \in \{90°, 180°, 270°\}$ to every pair (w, f) such that w is a vertex incident to a face f of $\mathcal{E}$. For a rectilinear representation $(\mathcal{E}, \phi)$ and a vertex u incident to $f_{\mathcal{E}}^*$, we denote by $\phi^{\text{int}}(u)$ the sum of the *internal angles* incident to u, that is, $\phi^{\text{int}}(u) = \sum_f \phi(u, f)$, where the sum is over all the internal faces f of $\mathcal{E}$ incident to u. For planar graphs that are not 2-connected, the notions of equivalence between planar rectilinear drawings and of rectilinear representation are similar, however a vertex w might have several occurrences $w^1, \ldots, w^x$ on the boundary of a face f, hence ϕ assigns an angle to every pair (w^k, f), for $k \in \{1, \ldots, x\}$; further, the value 360° is admissible for $\phi(w^k, f)$.

An *outerplanar drawing* is a planar drawing such that all the vertices are incident to the outer face. An *outerplane embedding* is an equivalence class of outerplanar drawings. A graph is *outerplanar* if it admits an outerplanar drawing. The *extended dual tree* $\mathcal{T}$ of an outerplane embedding $\mathcal{O}$ of an n-vertex 2-connected outerplanar graph is obtained from the dual graph of $\mathcal{O}$ by replacing the vertex corresponding to $f_{\mathcal{O}}^*$ with n new degree-1 nodes. See Fig. 2.

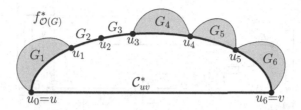

Fig. 3. A 2-connected outerplanar graph G rooted at uv. The uv-subgraphs of G are colored gray; G_2 and G_3 are trivial, while G_1, G_4, G_5, and G_6 are not.

3   Testing Algorithm for Outerplanar Graphs

In this section, we show how to test in $O(n)$ time whether an n-vertex outerplanar graph G with a variable embedding admits a planar rectilinear drawing. In Sect. 3.1, we assume that G is 2-connected and that an edge of G is prescribed to be incident to the outer face. In Sect. 3.2, we get rid of the second assumption. In Sect. 3.3, we get rid of the first assumption.

3.1   2-Connected Outerplanar Graphs with a Prescribed Edge

Let G be an n-vertex 2-connected outerplanar graph, let $\mathcal{O}$ be its outerplane embedding, let uv be an edge incident to $f_{\mathcal{O}}^*$, and let χ be a set of degree-2 vertices of G. A χ-*constrained* representation of G is a rectilinear representation $(\mathcal{E}, \phi)$ of G such that, for every vertex $x \in \chi$ and every face f of $\mathcal{E}$ incident to x, we have $\phi(x, f) \in \{90°, 270°\}$. We now show how to test in $O(n)$ time whether G admits a χ-constrained representation $(\mathcal{E}, \phi)$ in which uv is incident to $f_{\mathcal{E}}^*$.

The reason for introducing this seemingly artificial problem is the following. Consider an outerplanar graph H and assume that H contains a degree-4 cut-vertex v that belongs to two non-trivial blocks H_{b_1} and H_{b_2} of H. For $i = 1, 2$, in any plane embedding of H, the two edges of H_{b_i} incident to v are consecutive in clockwise order around v. Hence, in the restriction of a rectilinear representation of H to H_{b_i}, the angles incident to v are 90° and 270°. In Sect. 3.3, we will use the algorithm that tests whether a block H_{b_i} of a simply-connected outerplanar graph H admits a χ-constrained representation as one of the main ingredients for testing whether H admits a rectilinear representation.

We say that uv is the *root* of G. Let $\mathcal{C}_{uv}^* = (u = u_0, u_1, \ldots, u_k = v)$ be the cycle delimiting the internal face of $\mathcal{O}$ incident to uv; see Fig. 3. The blocks of the graph obtained from G by removing the edge uv are the uv-*subgraphs* of G. These are denoted by $G_1, \ldots, G_k$, where the root of G_i is $u_{i-1}u_i$. The assumption that uv is incident to the outer face of the desired plane embedding $\mathcal{E}$ ensures that $\mathcal{C}_{uv}^*$ lies in the outer face of each uv-subgraph G_i of G in $\mathcal{E}$. Conversely, each uv-subgraph G_i of G might lie inside or outside $\mathcal{C}_{uv}^*$ in $\mathcal{E}$.

A χ-constrained representation $(\mathcal{E}, \phi)$ in which the root uv is incident to $f_{\mathcal{E}}^*$ and the angles $\phi^{\mathrm{int}}(u)$ and $\phi^{\mathrm{int}}(v)$ are equal to μ and ν, respectively, is called a (χ, μ, ν)-*representation* of G. We show how to test, for any $\mu, \nu \in$

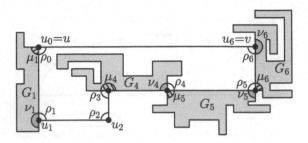

Fig. 4. A planar rectilinear drawing of the outerplanar graph G in Fig. 3.

$\{90°, 180°, 270°\}$, whether G admits a (χ, μ, ν)-representation. The following lemma is one of the main ingredients for our algorithm. For $i = 1, \ldots, k$, let $\chi_i := \chi \cap V(G_i)$. An *in-out assignment* is an assignment of each non-trivial uv-subgraph G_i of G either to the inside or to the outside of C^*_{uv}. Refer to Fig. 4.

Lemma 1. *For any $\mu, \nu \in \{90°, 180°, 270°\}$, we have that G admits a (χ, μ, ν)-representation if and only if there exist an in-out assignment $\mathcal{A}$ and a sequence of values $\rho_0, \rho_1, \mu_1, \nu_1, \rho_2, \mu_2, \nu_2, \ldots, \rho_k, \mu_k, \nu_k$ in $\{0°, 90°, 180°, 270°\}$ so that the following properties are satisfied.*

(V1): *for $i = 0, \ldots, k$, we have $\rho_i \geq 90°$; further, if $u_i \in \chi$, then either $\rho_i = 90°$ or $\rho_i = 270°$;*
(V2): *for $i = 1, \ldots, k$, if G_i is trivial then $\mu_i = \nu_i = 0°$, otherwise $\mu_i, \nu_i \in \{90°, 180°\}$ and G_i admits a (χ_i, μ_i, ν_i)-representation;*
(V3): *for $i = 1, \ldots, k - 1$, we have that $\nu_i + \rho_i + \mu_{i+1} \leq 270°$;*
(V4): *$\rho_0 + \mu_1 = \mu$ and $\rho_k + \nu_k = \nu$; and*
(V5): *for $i = 1, \ldots, k$, if G_i is trivial or is assigned by $\mathcal{A}$ to the outside of C^*_{uv}, then $\sigma_i = 0°$, else $\sigma_i = \mu_i + \nu_i$; then we have $\sum_{i=0}^{k} \rho_i + \sum_{i=1}^{k} \sigma_i = (k-1) \cdot 180°$.*

Let $(\mathcal{E}, \phi)$ be a rectilinear representation of G and let $f^{uv}_\mathcal{E}$ be the internal face of $\mathcal{E}$ incident to uv. Then, roughly speaking, ρ_i represents $\phi(u_i, f^{uv}_\mathcal{E})$; further, if G_i is non-trivial, then μ_i and ν_i represent the sums of the internal angles incident to u_{i-1} and u_i, respectively, in the restriction of $(\mathcal{E}, \phi)$ to G_i.

Due to Lemma 1, our goal becomes that of testing for the existence of a sequence $\rho_0, \rho_1, \mu_1, \nu_1, \ldots, \rho_k, \mu_k, \nu_k$ and of an in-out assignment $\mathcal{A}$ such that Properties **(V1)**–**(V5)** are satisfied. Property **(V2)** implies that, for every trivial uv-subgraph G_i of G, the values μ_i and ν_i can be set equal to $0°$ without loss of generality. The values μ_i and ν_i can also be chosen "optimally" for every non-trivial uv-subgraph G_i of G, except for G_1 and G_k. This choice selects one of the pairs (μ_i, ν_i) with $\mu_i, \nu_i \in \{90°, 180°\}$ such that G_i admits a (χ_i, μ_i, ν_i)-representation, as required by Property **(V2)**, and has to guarantee that Property **(V3)** is not violated. Subject to these constraints, the *optimal pair* (μ_i, ν_i) for G_i is the one for which $\mu_i + \nu_i$ is minimum. Choosing the optimal pair for G_i only requires to check information associated to G_{i-1}, G_i, and G_{i+1}, hence it can be done in $O(1)$ time, and thus in $O(k)$ time for all graphs $G_2, \ldots, G_{k-1}$.

If the *optimal sequence* of values $\mu_2, \nu_2, \mu_3, \nu_3, \ldots, \mu_{k-1}, \nu_{k-1}$ was established (otherwise we conclude that G admits no (χ, μ, ν)-representation), then we consider all the $3^6 \in O(1)$ tuples $(\mu_1, \nu_1, \mu_k, \nu_k, \rho_0, \rho_k)$ such that $\mu_1, \nu_1, \mu_k, \nu_k \in \{0°, 90°, 180°\}$ and $\rho_0, \rho_k \in \{90°, 180°, 270°\}$. For each of the tuples, we test in $O(1)$ time whether it violates Properties **(V1)**–**(V4)**, and in the positive case we discard the tuple. If we discarded all the tuples $(\mu_1, \nu_1, \mu_k, \nu_k, \rho_0, \rho_k)$, we conclude that G admits no (χ, μ, ν)-representation. Every tuple that was not discarded, together with the optimal sequence, forms a *promising sequence*. We process each promising sequence independently and check whether it is *extensible*, i.e., whether there exist an in-out assignment $\mathcal{A}$ and values $\rho_1, \rho_2, \ldots, \rho_{k-1}$ that, together with the promising sequence, satisfy Properties **(V1)**–**(V5)**.

In particular, the choice of $\mathcal{A}$ and $\rho_1, \rho_2, \ldots, \rho_{k-1}$ is done so as to satisfy Property **(V5)**, i.e., $\sum_{i=0}^{k} \rho_i + \sum_{i=1}^{k} \sigma_i = (k-1) \cdot 180°$, while complying with Properties **(V1)** and **(V3)**. For example, if a vertex u_i belongs to χ, then we need to set either $\rho_i = 90°$ or $\rho_i = 270°$. Recall that the sum $\mu_i + \nu_i$ is now fixed, for $i = 1, \ldots, k$. Suppose, for example, that $\mu_i + \nu_i = 360°$. If $\mathcal{A}$ assigns G_i to the inside of $\mathcal{C}_{uv}^*$, this contributes $360°$ to the sum $\sum_{i=1}^{k} \sigma_i$, otherwise it contributes $0°$. Let a be the number of uv-subgraphs for which $\mu_i + \nu_i = 360°$; then $\mathcal{A}$ can assign any $0 \le a' \le a$ of such graphs to the inside of $\mathcal{C}_{uv}^*$, and this will contribute $a' \cdot 360°$ to the sum $\sum_{i=1}^{k} \sigma_i$. Similar considerations allow us to reduce the problem of testing for the extensibility of a promising sequence to the problem of deciding whether integer values $0 \le a' \le a$, $0 \le b' \le b$, $0 \le c' \le c$, and $0 \le d' \le d$ exist such that $4a' + 3b' + 2c' + d' = t$, where a, b, c, d, and t are given integers inherent to the values of the promising sequence; the values a, b, c, d, and t can be computed in $O(k)$ time from the promising sequence.

If t is "very small" or "very large", then the existence of the values a', b', c', and d' can be decided in $O(1)$ time by means of exhaustive search. For values of t that are neither too small nor too large, we characterize the positive instances a, b, c, d, t as the ones satisfying a certain (constant size) boolean formula. For example, if $d \ge 3$, a solution to $4a' + 3b' + 2c' + d' = t$ subject to $0 \le a' \le a$, $0 \le b' \le b$, $0 \le c' \le c$, and $0 \le d' \le d$ always exists, while if $d = 2$ it exists if and only if $b > 0$, or $c > 0$, or $t \not\equiv 3 \mod 4$.

This concludes the description of our algorithm to test whether G admits a (χ, μ, ν)-representation. The algorithm runs in $O(k)$ time, assuming that the pairs (μ_i, ν_i) with $\mu_i, \nu_i \in \{90°, 180°\}$ such that each uv-subgraph G_i of G admits a (χ_i, μ_i, ν_i)-representation are known. In particular, once the values $\mu_1, \nu_1, \ldots, \mu_k, \nu_k, \rho_0, \rho_k, a, b, c, d, t$ have been computed, the algorithm concludes the test in $O(1)$ time (for each promising sequence).

We now use this algorithm inductively on the subgraphs of G. Namely, we root the extended dual tree $\mathcal{T}$ of the outerplane embedding of G at the leaf r^* such that the edge of $\mathcal{T}$ incident to r^* is dual to uv. Then, for any internal node s of $\mathcal{T}$, let G_s be the subgraph of G dual to the subtree of $\mathcal{T}$ rooted at s, let $\chi_s = \chi \cap V(G_s)$, and let $s_1, \ldots, s_k$ be the children of s in $\mathcal{T}$. We use the described algorithm to determine in $O(k)$ time the pairs (μ, ν) with $\mu, \nu \in \{90°, 180°, 270°\}$ such that G_s admits a (χ_s, μ, ν)-representation, starting from the pairs (μ_i, ν_i)

with $\mu_i, \nu_i \in \{0°, 90°, 180°\}$ such that G_{s_i} admits a $(\chi_{s_i}, \mu_i, \nu_i)$-representation, for $i = 1, \ldots, k$. This results in the following.

Theorem 1. *Let G be an n-vertex 2-connected outerplanar graph, uv be an edge incident to the outer face of the outerplane embedding of G, and χ be a subset of the degree-2 vertices of G. There is an $O(n)$-time algorithm which tests, for any values $\mu, \nu \in \{90°, 180°, 270°\}$, whether G admits a (χ, μ, ν)-representation.*

By independently considering all the pairs (μ, ν) with $\mu, \nu \in \{90°, 180°, 270°\}$, Theorem 1 also allows us to test whether a χ-constrained representation of G exists such that uv is incident to the outer face.

3.2   2-Connected Outerplanar Graphs

We now get rid of the assumption that there is a prescribed edge uv incident to the outer face of the rectilinear representation we seek, while maintaining the assumption that the input n-vertex outerplanar graph G is 2-connected. We are again required to look for χ-constrained representations.

Our $O(n)$-time algorithm to solve this problem will actually perform a more general task. Namely, our algorithm will label every vertex u of G whose degree is not larger than 3 with a set $\gamma(u)$ which contains all the values $\mu \in \{90°, 180°, 270°\}$ such that G admits a χ-constrained representation $(\mathcal{E}, \phi)$ in which u is incident to $f^*_{\mathcal{E}}$ and $\phi^{\mathrm{int}}(u) = \mu$.

First, we reduce the problem of computing the labels $\gamma(u)$ to the problem of computing labels $\mathcal{N}_{s \to t}$ and $\mathcal{N}_{t \to s}$ for each edge st of the extended dual tree $\mathcal{T}$ of the outerplane embedding $\mathcal{O}$ of G. The label $\mathcal{N}_{s \to t}$ is defined as follows (the label $\mathcal{N}_{t \to s}$ is defined symmetrically). Refer to Fig. 5. Let uv be the edge of G dual to st. The removal of the edge st splits $\mathcal{T}$ into two trees. Let $\mathcal{T}_{s \to t}$ be the one containing s. If s is an internal node of $\mathcal{T}$, then let $G_{s \to t}$ be the graph $\bigcup_{x \in \mathcal{T}_{s \to t}} C_x$, where C_x is the cycle of G delimiting the face of $\mathcal{O}$ dual to the node x of $\mathcal{T}$. If s is a leaf of $\mathcal{T}$, then let $G_{s \to t}$ be the edge uv. In both cases, $G_{s \to t}$ is rooted at uv. Let $\chi_{s \to t} = \chi \cap V(G_{s \to t})$. If s is an internal node of $\mathcal{T}$, then $\mathcal{N}_{s \to t}$ contains all the pairs (μ, ν) with $\mu, \nu \in \{90°, 180°, 270°\}$ such that $G_{s \to t}$ admits a $(\chi_{s \to t}, \mu, \nu)$-representation, while if s is a leaf, then $\mathcal{N}_{s \to t} = \{(0°, 0°)\}$.

The sets $\gamma(u)$ can be easily recovered from the labels $\mathcal{N}_{s \to t}$. Let $u_1 v_1, \ldots, u_n v_n$ be the edges of G incident to $f^*_{\mathcal{O}}$, in any order. The labels $\mathcal{N}_{s \to t}$ are computed by means of n postorder traversals of $\mathcal{T}$; during the h-th traversal, $\mathcal{T}$ is rooted at the leaf r^*_h such that the edge $r^*_h r_h$ incident to r^*_h is dual to $u_h v_h$.

When processing a node s with parent t during one of the traversals, we compute the label $\mathcal{N}_{s \to t}$. The computation of $\mathcal{N}_{s \to t}$ exploits the values of the already computed labels $\mathcal{N}_{s_1 \to s}, \ldots, \mathcal{N}_{s_k \to s}$, where $s_1, \ldots, s_k$ are the neighbors of s in $\mathcal{T}$ different from t. This is the problem we solved in Sect. 3.1! Namely, we want to compute the pairs (μ, ν) with $\mu, \nu \in \{90°, 180°, 270°\}$ such that $G_{s \to t}$ admits a $(\chi_{s \to t}, \mu, \nu)$-representation (these define $\mathcal{N}_{s \to t}$), starting from the pairs (μ_i, ν_i) with $\mu_i, \nu_i \in \{0°, 90°, 180°\}$ such that $G_{s_i \to s}$ admits a $(\chi_{s_i \to s}, \mu_i, \nu_i)$-representation (these define $\mathcal{N}_{s_1 \to s}, \ldots, \mathcal{N}_{s_k \to s}$). When $\mathcal{N}_{s_i \to s} = \emptyset$, in

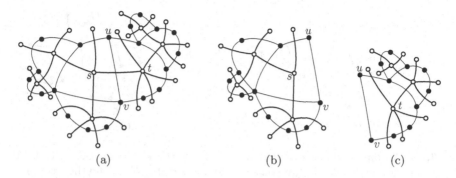

Fig. 5. (a) The graph G (represented with thin lines and black disks) and the extended dual tree $\mathcal{T}$ (represented with thick lines and white disks) of the outerplane embedding $\mathcal{O}$ of G. (b) The graph $G_{s \to t}$ and the tree $\mathcal{T}_{s \to t}$. (c) The graph $G_{t \to s}$ and the tree $\mathcal{T}_{t \to s}$.

particular, we also have $\mathcal{N}_{s \to t} = \emptyset$. That is, the non-existence of a $(\chi_{s_i \to s}, \mu_i, \nu_i)$-representation of G_{s_i} propagates towards the root of $\mathcal{T}$ in the current traversal.

Clearly, we cannot afford to perform each traversal independently of the other ones, as this would result in a quadratic running time. Then, as in [8], we re-use the already computed labels $\mathcal{N}_{s \to t}$; this implies that a postorder traversal is not invoked on a tree $\mathcal{T}_{s \to t}$ if the label $\mathcal{N}_{s \to t}$ has been computed by a previous traversal. As a result, during the traversals of $\mathcal{T}$, each edge is traversed at most once in each direction and each node with degree k is processed $O(k)$ times. Differently from [8], we need to handle the possibility that, when a node s of $\mathcal{T}$ is visited in a traversal after the first one, we might not have the sets $\mathcal{N}_{s_i \to s}$ ready, even for most children of s. This is a consequence of the propagation of the empty sets $\mathcal{N}_{s \to t}$ described above. Indeed, we cannot even afford to look at all the children s_i of s and see which sets $\mathcal{N}_{s_i \to s}$ have already been computed and which have not; if the degree of s is k, this would take $\Omega(k)$ time whenever we visit s (potentially k times), which would again result in a quadratic running time.

We cope with this problem by using, for each node of $\mathcal{T}$, some auxiliary labels that are dynamically computed during the traversals. For example, a label $\eta(s)$ points to a neighbor s_i of s for which $\mathcal{N}_{s_i \to s} = \emptyset$, two labels START$(s)$ and END(s) delimit the interval of neighbors of s for which an optimal pair has already been computed, and a label $a(s)$ stores the number of computed optimal pairs (μ_i, ν_i) such that $\mu_i + \nu_i = 360°$. The labels allow us to quickly determine which sets $\mathcal{N}_{s_i \to s}$ have already been computed and which have not, and to invoke a traversal recursively on the subtrees $\mathcal{T}_{s_i \to s}$ for which the sets $\mathcal{N}_{s_i \to s}$ have not been computed yet. Some labels (for example $a(s)$) store aggregate information on the values of the optimal pairs for the graphs $G_{s_i \to s}$. Thus, when the sets $\mathcal{N}_{s_i \to s}$ have been computed for all the children s_i of s, and we are hence in a position to apply the algorithm described in Sect. 3.1, we do not have to spend $O(k)$ time to compute the values a, b, c, d, and t, but we can extract them from the labels associated to s in $O(1)$ time, and then decide in $O(1)$ time whether a solution to the equation $4a' + 3b' + 2c' + d' = t$ subject to $0 \le a' \le a$, $0 \le b' \le b$,

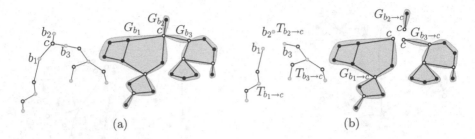

Fig. 6. (a) A graph G and its BC-tree T. The cut-vertices of G are empty disks; the blocks of G are surrounded by gray regions. (b) The graphs $G_{b_i \to c}$ and the trees $T_{b_i \to c}$, for $i = 1, 2, 3$, where c is a C-node and b_1, b_2, and b_3 are its adjacent B-nodes in T.

$0 \le c' \le c$, and $0 \le d' \le d$ exists; this ultimately determines whether a pair (μ, ν) belongs to $\mathcal{N}_{s \to t}$. We obtain the following.

Theorem 2. *Let G be an n-vertex 2-connected outerplanar graph and χ be a set of degree-2 vertices of G. There is an $O(n)$-time algorithm that labels every vertex u of G whose degree is not larger than 3 with a set $\gamma(u)$, which contains all the values $\mu \in \{90°, 180°, 270°\}$ such that G admits a χ-constrained representation $(\mathcal{E}, \phi)$ in which u is incident to $f^*_{\mathcal{E}}$ and $\phi^{\mathrm{int}}(u) = \mu$.*

3.3 General Outerplanar Graphs

In this section we remove the assumption that the input graph is 2-connected and show how to test whether an n-vertex outerplanar graph G admits a planar rectilinear drawing in $O(n)$ time. Consider the BC-tree T of G [13,15] and denote by G_b the block corresponding to a B-node b.

We now define a set χ_b for every non-trivial block G_b of G. We initialize $\chi_b = \emptyset$. Then, for every cut-vertex c that is shared by two non-trivial blocks G_{b_1} and G_{b_2} of G, we add c to both χ_{b_1} and χ_{b_2}. This concludes the construction of the sets χ_b. The next lemma justifies our study of χ-constrained representations.

Lemma 2. *For any non-trivial block G_b of G, the restriction to G_b of any rectilinear representation of G is a χ_b-constrained representation of G_b.*

Refer to Fig. 6. Consider any edge bc of T, where b is a B-node and c is a C-node. The removal of bc splits T into two trees. Let $T_{b \to c}$ be the one containing b. Let $G_{b \to c}$ be the subgraph of G composed of the blocks corresponding to B-nodes in $T_{b \to c}$. Let $\chi_{b \to c}$ be the restriction of χ to the vertices of $G_{b \to c}$.

We present an $O(n)$-time algorithm that computes, for every edge bc of T where b is a B-node and c is a C-node, a set $\mathcal{N}_{b \to c}$, which contains all the values $\mu \in \{0°, 90°, 180°, 270°\}$ such that $G_{b \to c}$ admits a rectilinear representation $(\mathcal{E}_{b \to c}, \phi_{b \to c})$ in which c is incident to $f^*_{\mathcal{E}_{b \to c}}$ and $\phi^{\mathrm{int}}_{b \to c}(c) = \mu$.

Let b be a B-node of T, let c_i be a C-node adjacent to b, and let $b_{i,1}, \dots, b_{i,m(i)}$ be the B-nodes adjacent to c_i and different from b. We say that c_i is a *friendly neighbor* of b if, for every $j = 1, \ldots, m(i)$, we have that $\mathcal{N}_{b_{i,j} \to c_i} \cap \{0°, 90°, 180°\} \neq \emptyset$ and G_b is trivial, or we have that $\mathcal{N}_{b_{i,j} \to c_i} \cap \{0°, 90°\} \neq \emptyset$ and G_b is non-trivial.

Determining the set $\mathcal{N}_{b \to c}$ for every edge bc of T is sufficient for determining whether G admits a rectilinear representation.

Lemma 3. *We have that G admits a rectilinear representation if and only if there exists a B-node b^* in T such that: (i) if G_{b^*} is non-trivial, then it admits a χ_{b^*}-constrained representation; and (ii) every neighbor of b^* in T is friendly.*

The first step of our algorithm for computing the sets $\mathcal{N}_{b \to c}$ labels, for each non-trivial block G_b of G, each vertex v of G_b whose degree in G_b is smaller than or equal to 3 with a set $\gamma_b(v)$; this set contains all the values $\mu \in \{90°, 180°, 270°\}$ such that G_b admits a χ_b-constrained representation $(\mathcal{E}_b, \phi_b)$ in which v is incident to $f^*_{\mathcal{E}_b}$ and $\phi_b^{int}(v) = \mu$. By Theorem 2, this can be done in $O(n_b)$ time for each non-trivial block G_b of G with n_b vertices, and hence in $O(n)$ time for all the non-trivial blocks of G. Further, for each trivial block G_b of G, we label each end-vertex c of G_b with a set $\gamma_b(c) = \{0°\}$.

For each leaf b of T, we then have $\mathcal{N}_{b \to c} = \gamma_b(c)$. If b is an internal node of T, in order to compute $\mathcal{N}_{b \to c}$, our algorithm exploits the following tool.

Lemma 4. *Let b be an internal B-node of T and let c be a C-node of T adjacent to b. Further, let $c_1, \ldots, c_h$ be the C-nodes adjacent to b and different from c; for $i = 1, \ldots, h$, let $b_{i,1}, \ldots, b_{i,m(i)}$ be the B-nodes adjacent to c_i and different from b. Finally, let $\mu \in \{0°, 90°, 180°, 270°\}$. We have that $\mu \in \mathcal{N}_{b \to c}$ if and only if $\mu \in \gamma_b(c)$ and c_i is a friendly neighbor of b, for every $i = 1, \ldots, h$.*

Similarly to Sect. 3.2, we construct the sets $\mathcal{N}_{b \to c}$ by performing several traversals of T. Some auxiliary labels are used also in this case, namely $\eta(b)$ points to a neighbor c_i of b that is not friendly and $\xi(b)$ tells us whether every neighbor, or almost every neighbor, of b is friendly. Lemma 4 is used in order to propagate the values $\mathcal{N}_{b \to c}$ in the tree. We obtain the following main theorem.

Theorem 3. *There is an $O(n)$-time algorithm that tests whether an n-vertex outerplanar graph admits a planar rectilinear drawing.*

4   Conclusions

In this paper, we proved that the existence of a planar rectilinear drawing of an outerplanar graph can be tested in linear time, both if the plane embedding of the outerplanar graph is prescribed and if it is not. We conclude with two natural generalizations of the questions we answered in this paper. Is it possible to determine in $O(n)$ time the minimum number of bends for a planar orthogonal drawing of an n-vertex outerplanar graph? Is it possible to test in $O(n)$ time whether an n-vertex series-parallel graph admits a planar rectilinear drawing? Didimo et al. [7] proved that the latter question has a positive answer for 2-connected series-parallel graphs with fixed embedding.

Acknowledgments. Thanks to Maurizio "Titto" Patrignani for many explanations about results and techniques from the state of the art.

References

1. Bläsius, T., Lehmann, S., Rutter, I.: Orthogonal graph drawing with inflexible edges. Comp. Geom. **55**, 26–40 (2016)
2. Borradaile, G., Klein, P.N., Mozes, S., Nussbaum, Y., Wulff-Nilsen, C.: Multiple-source multiple-sink maximum flow in directed planar graphs in near-linear time. SIAM J. Comput. **46**(4), 1280–1303 (2017)
3. Chang, Y., Yen, H.: On bend-minimized orthogonal drawings of planar 3-graphs. In: Aronov, B., Katz, M.J. (eds.) SoCG 2017. LIPIcs, vol. 77, pp. 29:1–29:15. Schloss Dagstuhl - Leibniz-Zentrum für Informatik (2017)
4. Cornelsen, S., Karrenbauer, A.: Accelerated bend minimization. J. Graph Algorithms Appl. **16**(3), 635–650 (2012)
5. Di Battista, G., Liotta, G., Vargiu, F.: Spirality and optimal orthogonal drawings. SIAM J. Comput. **27**(6), 1764–1811 (1998)
6. Di Giacomo, E., Liotta, G., Montecchiani, F.: Sketched representations and orthogonal planarity of bounded treewidth graphs. In: Archambault, D., Tóth, C.D. (eds.) GD 2019. LNCS, vol. 11904, pp. 379–392. Springer, Cham (2019). https://doi.org/10.1007/978-3-030-35802-0_29
7. Didimo, W., Kaufmann, M., Liotta, G., Ortali, G.: Rectilinear planarity testing of plane series-parallel graphs in linear time. CoRR abs/2008.03784 (2020)
8. Didimo, W., Liotta, G., Ortali, G., Patrignani, M.: Optimal orthogonal drawings of planar 3-graphs in linear time. In: Chawla, S. (ed.) SODA 2020, pp. 806–825. SIAM (2020)
9. Didimo, W., Liotta, G., Patrignani, M.: Bend-minimum orthogonal drawings in quadratic time. In: Biedl, T., Kerren, A. (eds.) GD 2018. LNCS, vol. 11282, pp. 481–494. Springer, Cham (2018). https://doi.org/10.1007/978-3-030-04414-5_34
10. Frati, F.: Planar rectilinear drawings of outerplanar graphs in linear time. CoRR abs/2006.06951 (2020)
11. Garg, A., Tamassia, R.: A new minimum cost flow algorithm with applications to graph drawing. In: North, S. (ed.) GD 1996. LNCS, vol. 1190, pp. 201–216. Springer, Heidelberg (1997). https://doi.org/10.1007/3-540-62495-3_49
12. Garg, A., Tamassia, R.: On the computational complexity of upward and rectilinear planarity testing. SIAM J. Comput. **31**(2), 601–625 (2001)
13. Harary, F.: Graph Theory. Addison-Wesley Pub. Co., Reading (1969)
14. Hasan, M.M., Rahman, M.S.: No-bend orthogonal drawings and no-bend orthogonally convex drawings of planar graphs (extended abstract). In: Du, D.-Z., Duan, Z., Tian, C. (eds.) COCOON 2019. LNCS, vol. 11653, pp. 254–265. Springer, Cham (2019). https://doi.org/10.1007/978-3-030-26176-4_21
15. Hopcroft, J.E., Tarjan, R.E.: Algorithm 447: efficient algorithms for graph manipulation. Commun. ACM **16**(6), 372–378 (1973)
16. Nomura, K., Tayu, S., Ueno, S.: On the orthogonal drawing of outerplanar graphs. IEICE Trans. **88–A**(6), 1583–1588 (2005)
17. Rahman, M.S., Egi, N., Nishizeki, T.: No-bend orthogonal drawings of series-parallel graphs. In: Healy, P., Nikolov, N.S. (eds.) GD 2005. LNCS, vol. 3843, pp. 409–420. Springer, Heidelberg (2006). https://doi.org/10.1007/11618058_37

18. Rahman, M.S., Nishizeki, T.: Bend-minimum orthogonal drawings of plane 3-graphs. In: Goos, G., Hartmanis, J., van Leeuwen, J., Kučera, L. (eds.) WG 2002. LNCS, vol. 2573, pp. 367–378. Springer, Heidelberg (2002). https://doi.org/10.1007/3-540-36379-3_32
19. Rahman, M.S., Nishizeki, T., Naznin, M.: Orthogonal drawings of plane graphs without bends. J. Graph Algorithms Appl. 7(4), 335–362 (2003)
20. Tamassia, R.: On embedding a graph in the grid with the minimum number of bends. SIAM J. Comput. 16(3), 421–444 (1987)
21. Zhou, X., Nishizeki, T.: Orthogonal drawings of series-parallel graphs with minimum bends. SIAM J. Discrete Math. 22(4), 1570–1604 (2008)

Rectilinear Planarity Testing of Plane Series-Parallel Graphs in Linear Time

Walter Didimo[1], Michael Kaufmann[2], Giuseppe Liotta[1], and Giacomo Ortali[1(✉)]

[1] Università degli Studi di Perugia, Perugia, Italy
{walter.didimo,giuseppe.liotta}@unipg.it,
giacomo.ortali@studenti.unipg.it
[2] University of Tübingen, Tübingen, Germany
mk@informatik.uni-tuebingen.de

Abstract. A plane graph is *rectilinear planar* if it admits an embedding-preserving straight-line drawing where each edge is either horizontal or vertical. We prove that rectilinear planarity testing can be solved in optimal $O(n)$ time for any plane series-parallel graph G with n vertices. If G is rectilinear planar, an embedding-preserving rectilinear planar drawing of G can be constructed in $O(n)$ time. Our result is based on a characterization of rectilinear planar series-parallel graphs in terms of intervals of orthogonal spirality that their components can have, and it leads to an algorithm that can be easily implemented.

Keywords: Orthogonal drawings · Rectilinear planarity testing · Series-parallel graphs

1 Introduction

A *planar orthogonal drawing* Γ of a planar graph G is a crossing-free drawing of G that maps each vertex to a distinct point of the plane and each edge to a sequence of horizontal and vertical segments between its end-points [4,10,15]. A graph is *rectilinear planar* if it admits a planar orthogonal drawing without bends.

Testing whether a graph is rectilinear planar is a fundamental question in graph drawing. The problem can be either studied for *plane* graphs, that is graphs that come with a fixed embedding, or in the variable embedding setting, where the algorithm can choose one of the planar embeddings of the input graph. Besides being an interesting topic on its own right, rectilinear planarity testing is at the core of efficient algorithms that compute orthogonal drawings with minimum number of bends. For example, Rahman et al. [18] characterize the rectilinear plane 3-graphs (i.e., graphs with vertex degree at most three) and then

Work partially supported by: (*i*) MIUR, grant 20174LF3T8 "AHeAD: efficient Algorithms for HArnessing networked Data", (*ii*) Engineering Dep., Univ. Perugia, grant RICBA19FM: "Modelli, algoritmi e sistemi per la visualizzazione di grafi e reti".

D. Auber and P. Valtr (Eds.): GD 2020, LNCS 12590, pp. 436–449, 2020.
https://doi.org/10.1007/978-3-030-68766-3_34

use this characterization to design linear time bend-minimization algorithms for these graphs in the fixed embedding setting [16,17]. On the other hand, Garg and Tamassia [12] prove that rectilinear planarity testing is NP-complete for planar 4-graphs in the variable embedding setting. Remarkably, the study of rectilinear plane 3-graphs has turned out to be an essential tool to design linear-time rectilinear planarity testing and bend-minimization algorithms for planar 3-graphs in the variable embedding setting [9,14].

In this paper we study rectilinear planarity testing in the fixed embedding setting. A seminal paper of Tamassia [19] implies that in this setting the problem can be solved in $O(n^2 \log n)$, where n is the number of vertices of the input graph; its approach is based on solving a min-cost flow network problem to compute a bend-minimum orthogonal drawing of the input graph. Since its time of publication, establishing a lower bound on the time complexity of computing bend-minimum orthogonal drawings of plane graphs has remained a fascinating open problem (see, e.g., [2,4,7]). Garg and Tamassia [13] improve the complexity to $O(n^{\frac{7}{4}}\sqrt{\log n})$ and then Cornelsen and Karrenbauer [3] further improve the upper bound to $O(n^{1.5})$. For rectilinear planarity testing, the approach in [19] reduces to compute a maximum flow in an n-vertex planar network with multiple sources and sinks; Borradaile et al. [1] prove that this problem can be solved in $O(n \log^3 n)$ time. Since, as already mentioned, an $O(n)$-time algorithm for rectilinear planarity testing is known when the input is a plane 3-graph, the challenge is to understand whether an $O(n)$-time bound exists for plane 4-graphs.

This paper sheds some light on this question by answering it for series-parallel graphs. An essential aspect of our approach is to tackle the problem without using any network-flow computation. Our results are as follows:

(i) We give a characterization of those plane series-parallel graphs (with two terminals s and t) that are rectilinear planar. This characterization is expressed in terms of values of *spirality* that each series or parallel component can have in a rectilinear drawing. Intuitively, the spirality of a component measures how much it can be "rolled-up" in a rectilinear drawing of the graph.

(ii) While the possible values of spirality for each component may be linear, we can encode them in constant space. This makes it possible to design a linear-time rectilinear planarity testing algorithm for a two-terminal series-parallel graph G based on a bottom-up visit of its decomposition tree T. If the test is positive, we compute in linear time a rectilinear drawing of G through a top-down visit of T. The algorithm is easy to implement.

The paper is organized as follows. Section 2 recalls basic concepts. Section 3 gives our characterization of rectilinear planar series-parallel graphs in terms of their orthogonal spirality. Section 4 describes the linear-time testing and drawing algorithm. Section 5 lists some open problems. For space restrictions some proofs are sketched or omitted and can be found in [8].

Together with our submission to GD 2020, another paper by Frati [11] was accepted to the same conference. The work of Frati is based on a different technique and it presents an $O(n)$-time algorithm for rectilinear planarity testing of outerplanar graphs. While the result of [11] does not apply to the family of graphs that are studied in this paper, it covers the variable embedding setting and the case of 1-connected outerplanar graphs.

2 Preliminaries

Orthogonal Representations. We focus on *orthogonal representations* rather than orthogonal drawings. An orthogonal representation H describes the shape of a class of orthogonal drawings in terms of sequences of bends along the edges and angles at the vertices. An (orthogonal) drawing Γ of H can be computed in linear time [19]. If H has no bend, it is a *rectilinear representation* (see Fig. 1(b)). The *degree* $\deg(v)$ of a vertex v denotes the number of edges incident to v.

Series-Parallel Graphs and Decomposition Trees. A *two-terminal series-parallel* graph, also called *series-parallel graph* in the rest of the paper, has two distinct vertices s and t, called its *source* and its *sink*, respectively, and it is inductively defined as follows: (*i*) A single edge (s,t) is a series-parallel graph with source s and sink t. (*ii*) Given $p \geq 2$ series-parallel graphs $G_1, \ldots, G_p$, each G_i with source s_i and sink t_i ($i = 1, \ldots, p$), a new series-parallel graph G can be obtained with any of these two operations: *Series composition* – It identifies t_i with s_{i+1} ($i = 1, \ldots, p - 1$); G has source $s = s_1$ and sink $t = t_p$. *Parallel composition* – It identifies all sources s_i together and all sinks t_i together; G has source $s = s_i$ and $t = t_i$ ($i = 1, \ldots, p$).

A series-parallel graph G is naturally associated with a *decomposition tree* T, which describes the series and parallel compositions that build G. Tree T has three types of nodes: S-, P-, and Q*-nodes. If G is the series composition of $p \geq 2$ graphs G_i that are not all single edges, the root of T is an S-node whose subtrees are the decomposition trees T_i of G_i. If G is the parallel composition of $p \geq 2$ graphs G_i, the root of T is a P-node whose subtrees are the decomposition trees T_i of G_i. If G is a series composition of $\ell \geq 1$ edges, its decomposition tree is a single Q*-node and for brevity we say that ℓ is the *length* of this node.

For a node ν of T, the *pertinent graph* G_ν of ν is the series-parallel subgraph of G formed by all edges associated with the Q*-nodes in the subtree rooted at ν. We also call G_ν a *component* of G. If u and v are the source and the sink of G_ν, respectively, we say that $\{u,v\}$ are the *poles* of G_ν and of ν: u is the *source pole* and v is the *sink pole*. If G is a biconnected plane series-parallel graph, for any edge $e = (s,t)$ on the external face of G, we can associate with G a decomposition tree T where the root is a P-node representing the parallel composition between e and the rest of the graph. Thus, the root of T is always a P-node with two children, one of which is a Q*-node corresponding to e. It will be called the (unique) P^r-node of T, to distinguish it by the other P-nodes. Edge e is the *reference edge* of T and T is the SPQ*-tree of G *with respect to* e. Also, it is always possible to make T such that each P-node (distinct from the root) has no P-node child and each S-node has no S-node child. Since we only deal with graphs of vertex-degree at most four, a P-node has either two or three

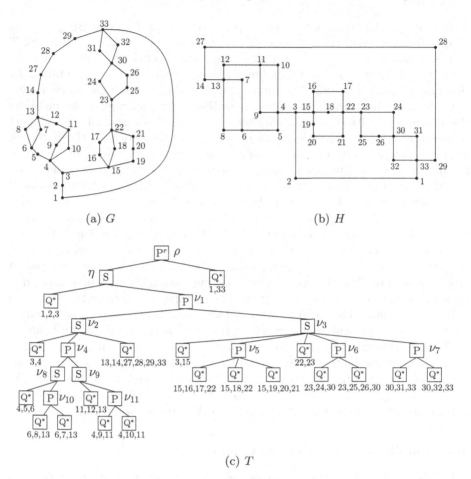

Fig. 1. (a) A biconnected series-parallel graph G. (b) A rectilinear planar representation H of G. (c) The SPQ*-tree T of G with reference edge $(1,33)$.

children. From now on we assume that T always satisfies the properties above for a biconnected series-parallel graph. Observe that the number of nodes of T is $O(n)$, where n is the number of vertices of G. Figure 1 shows a biconnected series-parallel graph G, a rectilinear planar representation H of G, and the SPQ*-tree T of G with respect to the reference edge $(1,33)$.

3   Characterizing Rectilinear Plane Series-Parallel Graphs

Let G be a plane series-parallel graph. If G is biconnected let $e = (s,t)$ be any edge on the external face of G; otherwise, by definition of two-terminal series-parallel graph, we can add a dummy edge e on the external face of G to make it biconnected. We assume that the external face of G is to the right of e while moving from s to t (as in Fig. 1(b)). Let T be an SPQ*-tree of G with respect to e. An overview of our algorithm is as follows. It visits T in post-order (a node is visited after its children). When the algorithm visits a node ν, it tests whether G_ν admits a planar rectilinear representation by checking whether a certain condition, which we call *representability condition*, is verified: In the negative case, the algorithm halts and rejects the instance; else it stores in ν its *representability interval* I_ν. Such an interval is a compact representation of the possible values of *orthogonal spirality* that the pertinent graph G_ν of ν may have in a rectilinear representation of G. Informally speaking, the orthogonal spirality is a measure of how much a rectilinear representation of pertinent graph G_ν is "rolled-up" in a rectilinear planar representation of G. As we shall see, the representability interval is such that for every value $k \in I_\nu$ graph G_ν admits a planar rectilinear representation with spirality k, while it does not for any value outside I_ν. If the testing algorithm does not halt and it reaches the root, two cases are considered: If e is a real edge of G, then the algorithm executes a final test to check whether a rectilinear planar representation of G can be obtained by merging a straight-line representation of e with a rectilinear representation of the child component of the root other than e. If e is a dummy edge added to make G biconnected this check is not required, because e is not present in the final representation and can arbitrarily bend.

We now present the characterization of the rectilinear planar components in terms of representability conditions and intervals that is at the base of the testing algorithm. We start in Sect. 3.1 with a formal definition of spirality. We characterize Q*-, S-, and P-components with three children in Sect. 3.2, and P-components with two children in Sect. 3.3. We summarize in Sect. 3.4.

3.1   Spirality of Series-Parallel Graphs

Let T be an SPQ*-tree of a biconnected plane series-parallel graph G for a given reference edge $e = (s, t)$. Let H be an embedding-preserving orthogonal representation of G. Also, let ν be a node of T with poles $\{u, v\}$, and let H_ν be the restriction of H to the pertinent graph G_ν of ν. We also say that H_ν is a *component* of H. For each pole $w \in \{u, v\}$, let $\text{indeg}_\nu(w)$ and $\text{outdeg}_\nu(w)$ be the degree of w inside and outside H_ν, respectively. Define two (possibly coincident) *alias vertices* of w, denoted by w' and w'', as follows: (i) if $\text{indeg}_\nu(w) = 1$, then $w' = w'' = w$; (ii) if $\text{indeg}_\nu(w) = \text{outdeg}_\nu(w) = 2$, then w' and w'' are dummy vertices, each splitting one of the two distinct edge segments incident to w outside H_ν; (iii) if $\text{indeg}_\nu(w) > 1$ and $\text{outdeg}_\nu(w) = 1$, then $w' = w''$ is a dummy vertex that splits the edge segment incident to w outside H_ν.

Let A^w be the set of distinct alias vertices of a pole w. Let P^{uv} be any simple path from u to v inside H_ν and let $u' \in A^u$ and $v' \in A^v$. The path $S^{u'v'}$ obtained concatenating (u', u), P^{uv}, and (v, v') is called a *spine* of H_ν. Denote by $n(S^{u'v'})$ the number of right turns minus the number of left turns encountered along $S^{u'v'}$ while moving from u' to v'. The *spirality* $\sigma(H_\nu)$ of H_ν is defined based on the following cases: (a) $A^u = \{u'\}$ and $A^v = \{v'\}$. Then $\sigma(H_\nu) = n(S^{u'v'})$.

(b) $A^u = \{u'\}$ and $A^v = \{v', v''\}$. Then $\sigma(H_\nu) = \frac{n(S^{u'v'}) + n(S^{u'v''})}{2}$. (c) $A^u = \{u', u''\}$ and $A^v = \{v'\}$. Then $\sigma(H_\nu) = \frac{n(S^{u'v'}) + n(S^{u''v'})}{2}$. (d) $A^u = \{u', u''\}$ and $A^v = \{v', v''\}$. Without loss of generality, assume that (u, u') precedes (u, u'') counterclockwise around u and that (v, v') precedes (v, v'') clockwise around v. Then $\sigma(H_\nu) = \frac{n(S^{u'v'}) + n(S^{u''v''})}{2}$. Notice that, by definition, the spirality of H_ν also depends on the angles at the poles of H_ν, not only on the shape of H_ν.

Di Battista et al. [5] showed that the spirality of H_ν does not vary with the choice of path P^{uv} and that two distinct representations of G_ν with the same spirality are interchangeable. Figure 2 reports the spiralities of some P- and S-components in the representation H of Fig. 1(b). For brevity, we shall denote by σ_ν the spirality of an orthogonal representation of G_ν.

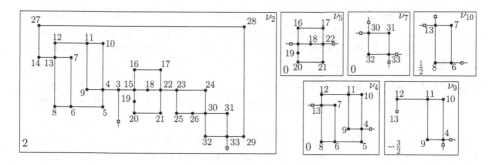

Fig. 2. Spiralities (left-bottom corners) of some components in the representation H of Fig. 1(b). Small squares indicate alias vertices.

Lemma 1 ([5]). *Let ν be an S-node of T with children $\mu_1, \ldots, \mu_h$. The following relationship holds: $\sigma_\nu = \sum_{i=1}^h \sigma_{\mu_i}$.*

If ν is a P-node with two children, we denote by μ_l and μ_r the left child and the right child of ν, respectively. If ν is a P-node with three children, we denote by μ_l, μ_c, and μ_r, the three children of ν from left to right. Also, for each pole $w \in \{u, v\}$ of ν, the *leftmost angle* at w in H is the angle formed by the leftmost external edge and the leftmost internal edge of H_ν incident to w. The *rightmost angle* at w in H is defined symmetrically. We define two binary variables α_w^l and α_w^r as follows: $\alpha_w^l = 0$ ($\alpha_w^r = 0$) if the leftmost (rightmost) angle at w in H is of $180°$, while $\alpha_w^l = 1$ ($\alpha_w^r = 1$) if this angle is of $90°$. Observe that if $\deg(w) = 4$, then $\alpha_w^l = \alpha_w^r = 1$. Also, if ν has two children, define two additional variables k_w^l and k_w^r as follows: $k_w^d = 1$ if $\text{indeg}_{\mu_d}(w) = \text{outdeg}_\nu(w) = 1$, while $k_w^d = 1/2$ otherwise, for $d \in \{l, r\}$.

For example, in Fig. 2 the P-component of ν_4 has poles $u = 4$ and $v = 13$, and we have $k_u^l = k_v^r = 1$, $k_u^r = k_v^l = \frac{1}{2}$, and $\alpha_u^l = \alpha_u^r = \alpha_v^l = \alpha_v^r = 1$. The P-component of ν_{10} has poles $u = 6$ and $v = 13$, and we have $k_u^l = k_u^r = 1$,

$k_v^l = k_v^r = \frac{1}{2}$, $\alpha_u^l = 0$, and $\alpha_u^r = \alpha_v^l = \alpha_v^r = 1$. Figure 3 reports all the values of k_w^d for the possible types of P-nodes with two children.

Lemma 2 ([5]). *Let ν be a P-node of T with two children μ_l and μ_r. The following relationships hold:* $\sigma_\nu = \sigma_{\mu_l} - k_u^l \alpha_u^l - k_v^l \alpha_v^l = \sigma_{\mu_r} + k_v^r \alpha_u^r + k_v^r \alpha_v^r$.

Lemma 3 ([5]). *Let ν be a P-node of T with three children μ_l, μ_c, and μ_r. The following relationships hold:* $\sigma_\nu = \sigma_{\mu_l} - 2 = \sigma_{\mu_c} = \sigma_{\mu_r} + 2$.

About the values of spirality σ_ν that a component H_ν can take, if ν is a Q*-node or a P-node with three children, σ_ν is always an integer. If ν is an S-node or a P-node with two children, σ_ν is either integer or semi-integer depending on whether the total number of alias vertices for the poles of ν is even or odd.

3.2   Q*-nodes, S-nodes, and P-nodes with Three Children

From now on, when we say that the spirality σ_ν of an orthogonal planar representation of G_ν can take *all* values in an interval $[a, b]$, we mean that such values are either all the integer numbers or all the semi-integer numbers in $[a, b]$, depending on the cases described above for ν.

Lemma 4. *Let ν be a Q*-node of length ℓ. Graph G_ν is always rectilinear planar (i.e., its representability condition is always true) and its representability interval is* $I_\nu = [-\ell + 1, \ell - 1]$.

Proof. G_ν is a path with $\ell - 1$ degree-2 vertices. For any integer $k \in [-\ell+1, 0]$, a rectilinear planar representation H_ν of G_ν with spirality k is obtained by making a left turn at k degree-2 vertices of G_ν (going from the source to the sink pole), and no turn at any remaining vertex of G_ν. Symmetrically, for any $k \in (0, \ell - 1]$, we realize H_ν with spirality k by making a right turn at exactly k degree-2 vertices of G_ν. It is clear that no values of spirality out of I_ν can be achieved. □

Lemma 5. *Let ν be an S-node with h children $\mu_1, \ldots, \mu_h$. Suppose that, for every $i \in [1, h]$, the representability interval of G_{μ_i} is $I_{\mu_i} = [m_i, M_i]$. Graph G_ν is always rectilinear planar (i.e., its representability condition is always true) and its representability interval is* $I_\nu = [\sum_{i=1}^{h} m_i, \sum_{i=1}^{h} M_i]$.

Proof. We use induction on the number of children of ν. In the base case $h = 2$. By hypothesis $I_{\mu_1} = [m_1, M_1]$ and $I_{\mu_2} = [m_2, M_2]$. By Lemma 1, a series composition of a rectilinear representation of G_{μ_1} with spirality σ_{μ_1} and of a rectilinear representation of G_{μ_2} with spirality σ_{μ_2} results in a rectilinear representation of G_ν with spirality $\sigma_\nu = \sigma_{\mu_1} + \sigma_{\mu_2}$. Hence, if $M_1 = m_1 + r_1$ and $M_2 = m_2 + r_2$, for two non-negative integers r_1 and r_2, then the possible values for σ_ν are exactly $m_1 + m_2, m_1 + 1 + m_2, \ldots, m_1 + r_1 + m_2, \ldots, m_1 + r_1 + m_2 + 1, \ldots, m_1 + r_1 + m_2 + r_2$, i.e., all values in the interval $[m_1 + m_2, M_1 + M_2]$. In the inductive case $h \geq 3$; consider the series composition G_1' of $G_{\mu_1}, \ldots, G_{\mu_{h-1}}$. Graph G_ν is the series composition of G_1' and G_{μ_2}. By inductive hypothesis the representability interval of G_1' is $[\sum_{i=1}^{h-1} m_i, \sum_{i=1}^{h-1} M_i]$ and by Lemma 1 applied to G_1' and G_{μ_2} we have $I_\nu = [\sum_{i=1}^{h} m_i, \sum_{i=1}^{h} M_i]$, using the same reasoning as for the base case. □

Lemma 6. *Let ν be a P-node with three children μ_l, μ_c, and μ_r. Suppose that G_{μ_l}, G_{μ_c}, and G_{μ_r} are rectilinear planar and that their representability intervals are $I_{\mu_l} = [m_l, M_l]$, $I_{\mu_c} = [m_c, M_c]$, and $I_{\mu_r} = [m_r, M_r]$, respectively. Graph G_ν is rectilinear planar if and only if $[m_l - 2, M_l - 2] \cap [m_c, M_c] \cap [m_r + 2, M_r + 2] \neq \emptyset$. Also, if this representability condition holds then the representability interval of G_ν is $I_\nu = [\max\{m_l - 2, m_c, m_r + 2\}, \min\{M_l - 2, M_c, M_r + 2\}]$.*

Proof. **Representability Condition.** Suppose first that G_ν is rectilinear planar and let H_ν be a rectilinear planar representation of G_ν with spirality σ_ν. By Lemma 3, the spiralities σ_{μ_l}, σ_{μ_c}, and σ_{μ_r} for the representations of G_{μ_l}, G_{μ_c}, and G_{μ_r} in H_ν are such that $\sigma_{\mu_l} = \sigma_\nu + 2$, $\sigma_{\mu_c} = \sigma_\nu$, and $\sigma_{\mu_r} = \sigma_\nu - 2$. Since $\sigma_{\mu_l} \in [m_l, M_l]$, $\sigma_{\mu_c} \in [m_c, M_c]$, $\sigma_{\mu_r} \in [m_r, M_r]$, we have $\sigma_\nu \in [m_l - 2, M_l - 2] \cap [m_c, M_c] \cap [m_r + 2, M_r + 2]$. Suppose vice versa that $[m_l - 2, M_l - 2] \cap [m_c, M_c] \cap [m_r + 2, M_r + 2] \neq \emptyset$, and let k be any value in such intersection. Setting $\sigma_{\mu_l} = k + 2$, $\sigma_{\mu_c} = k$, and $\sigma_{\mu_r} = k - 2$ we have $\sigma_{\mu_l} \in [m_l, M_l]$, $\sigma_{\mu_c} \in [m_c, M_c]$, and $\sigma_{\mu_r} \in [m_r, M_r]$. By Lemma 3, G_ν is rectilinear planar for a value of spirality $\sigma_\nu = k$.

Representability Interval. Assume that G_ν is rectilinear planar. Clearly $[\max\{m_l - 2, m_c, m_r + 2\}, \min\{M_l - 2, M_c, M_r + 2\}] = [m_l - 2, M_l - 2] \cap [m_c, M_c] \cap [m_r + 2, M_r + 2]$, and by the truth of the feasiblity condition we have $[\max\{m_l - 2, m_c, m_r + 2\}, \min\{M_l - 2, M_c, M_r + 2\}] \neq \emptyset$. Similarly to the first part of the proof of the representability condition, any rectilinear planar representation of G_ν has a value of spirality in the interaval $[\max\{m_l - 2, m_c, m_r + 2\}, \min\{M_l - 2, M_c, M_r + 2\}]$. On the other hand, let $k \in [\max\{m_l - 2, m_c, m_r + 2\}, \min\{M_l - 2, M_c, M_r + 2\}]$. Analogously to the second part of the proof of the representability condition, we can construct a rectilinear planar representation of G_ν with spirality $\sigma_\nu = k$, by combining in parallel rectilinear planar representations of G_{μ_l}, G_{μ_c}, and G_{μ_r} with spiralities $\sigma_{\mu_l} = \sigma_\nu + 2$, $\sigma_{\mu_c} = \sigma_\nu$, and $\sigma_{\mu_r} = \sigma_\nu - 2$, respectively. □

3.3 P-nodes with Two Children

For a P-node ν with two children μ_l and μ_r, the representability condition and interval depend on the indegree and outdegree of the poles of ν in G_ν, G_{μ_l}, and G_{μ_r}. We define the *type* of ν and of G_ν as follows (refer to Fig. 3):

- $I_2O_{\alpha\beta}$: Both poles of ν have indegree two in G_ν; also one pole has outdegree α in G_ν and the other pole has outdegree β in G_ν, for $1 \leq \alpha \leq \beta \leq 2$. This gives rise to the specific types I_2O_{11}, I_2O_{12}, and I_2O_{22}.
- $I_{3d}O_{\alpha\beta}$: One pole of ν has indegree two in G_ν, while the other pole has indegree three in G_ν and indegree two in G_{μ_d} for $d \in \{l, r\}$; also one pole has outdegree α in G_ν and the other has outdegree β in G_ν, for $1 \leq \alpha \leq \beta \leq 2$, where $\alpha = \beta = 2$ is not possible. This gives rise to the specific types $I_{3l}O_{11}$, $I_{3r}O_{11}$, $I_{3l}O_{12}$, $I_{3r}O_{12}$.
- $I_{3dd'}$: Both poles of ν have indegree three in G_ν; one of the two poles has indegree two in G_{μ_d} and the other has indegree two in $G_{\mu_{d'}}$, for $dd' \in \{ll, lr, rr\}$ (both poles have outdegree one in G_ν). Hence, the specific types are I_{3ll}, I_{3lr}, I_{3rr}.

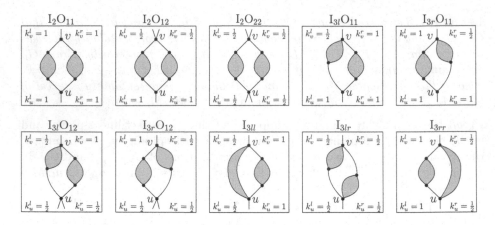

Fig. 3. Schematic illustration of the different types of P-nodes with two children.

To characterize P-nodes of type $I_2O_{\alpha\beta}$ we start with the following result.

Lemma 7. *Let G_ν be a P-node of type $I_2O_{\alpha\beta}$ with children μ_l and μ_r. G_ν is rectilinear planar if and only if G_{μ_l} and G_{μ_r} are rectilinear planar for values of spiralities σ_{μ_l} and σ_{μ_r} such that $\sigma_{\mu_l} - \sigma_{\mu_r} \in [2, 4 - \gamma]$, where $\gamma = \alpha + \beta - 2$.*

Sketch of Proof. We only give the proof for $\alpha = \beta = 2$. The other cases are treated similarly (see [8]). In this case G_ν is of type I_2O_{22} and we prove that G_ν is rectilinear planar if and only if G_{μ_l} and G_{μ_r} are rectilinear planar for values of spiralities σ_{μ_l} and σ_{μ_r} such that $\sigma_{\mu_l} - \sigma_{\mu_r} = 2$. We have $k_u^l = k_u^r = \frac{1}{2}$. If G_ν is rectilinear planar, we have that $\alpha_u^l + \alpha_u^r = \alpha_v^l + \alpha_v^r = 2$. By Lemma 2, $\sigma_{\mu_l} = \sigma_\nu + 1$ and $\sigma_{\mu_r} = \sigma_\nu - 1$; hence $\sigma_{\mu_l} - \sigma_{\mu_r} = 2$.

Suppose vice versa that $\sigma_{\mu_l} - \sigma_{\mu_r} = 2$. We show that G_ν admits a rectilinear planar representation H_ν. We obtain H_ν by combining in parallel the two rectilinear planar representations of G_{μ_l} and G_{μ_r} and by suitably setting α_u^d and α_v^d ($d \in \{l, r\}$). For any cycle C through u and v, the number of 90° angles minus the number of 270° angles in the interior of C can be expressed by $a_c = \sigma_{\mu_l} - \sigma_{\mu_r} + 1 + 1$ (both the angles at u and v inside C is always of 90° degrees). We then set $\alpha_u^l = \alpha_v^l = \alpha_u^r = \alpha_v^r = 1$, which guarantees $a_c = 4$. Also, any other cycle not passing through u and v is an orthogonal polygon because it belongs to a rectilinear planar representation of either G_{μ_l} or G_{μ_r}.

Lemma 8. *Let ν be a P-node of type $I_2O_{\alpha\beta}$ with children μ_l and μ_r. Suppose that G_{μ_l} and G_{μ_r} are rectilinear planar with representability intervals $I_{\mu_l} = [m_l, M_l]$ and $I_{\mu_r} = [m_r, M_r]$, respectively. Graph G_ν is rectilinear planar if and only if $[m_l - M_r, M_l - m_r] \cap [2, 4 - \gamma] \neq \emptyset$, where $\gamma = \alpha + \beta - 2$. Also, if this representability condition holds then the representability interval of G_ν is $I_\nu = [\max\{m_l - 2, m_r\} + \frac{\gamma}{2}, \min\{M_l, M_r + 2\} - \frac{\gamma}{2}]$.*

Sketch of Proof. We consider the case $\alpha = \beta = 2$. The other cases are treated similarly (see [8]). In this case G_ν is of type I_2O_{22} and we prove that $I_\nu = [\max\{m_l - 2, m_r\} + 1, \min\{M_l, M_r + 2\} - 1]$.

Assume first that G_ν is rectilinear planar and let H_ν be a rectilinear planar representation of G_ν with spirality σ_ν. Let H_{μ_l} and H_{μ_r} be the rectilinear planar representations of G_{μ_l} and G_{μ_r} contained in H_ν, and let σ_{μ_l} and σ_{μ_r} their spiralities. Since both u and v have outdegree two in G_ν we have that $\alpha_u^l + \alpha_u^r = \alpha_v^l + \alpha_v^r = 2$. By Lemma 2, $\sigma_{\mu_l} = \sigma_\nu + 1$ and $\sigma_{\mu_r} = \sigma_\nu - 1$. By the representability condition $\sigma_{\mu_r} = \sigma_{\mu_l} - 2$. Hence $\sigma_{\mu_r} \geq m_l - 2$ and $\sigma_{\mu_r} \geq \max\{m_l - 2, m_r\}$. Also by $\sigma_\nu = \sigma_{\mu_r} + 1$, $\sigma_\nu \geq \max\{m_l - 2, m_r\} + 1$. Similarly, by the representability condition $\sigma_{\mu_l} = \sigma_{\mu_r} + 2$. Hence $\sigma_{\mu_l} \leq M_r + 2$ and $\sigma_{\mu_l} \leq \max\{M_l, M_r + 2\}$. Since $\sigma_{\mu_l} = \sigma_\nu + 1$ we have $\sigma_\nu \leq \max\{M_l, M_r + 2\} - 1$.

Assume vice versa that k is an integer in the interval $I_\nu = [\max\{m_l - 2, m_r\} + 1, \min\{M_l, M_r + 2\} - 1]$. We show that there exists a rectilinear planar representation of G_ν with spirality $\sigma_\nu = k$. We have $k + 1 \in [\max\{m_l, m_r + 2\}, \min\{M_l, M_r + 2\}]$ and therefore $k + 1 \in [m_l, M_l]$. Hence there is a rectilinear planar representation H_{μ_l} of G_{μ_l} with spirality $\sigma_{\mu_l} = k + 1$. Similarly, $k - 1 \in [\max\{m_l - 2, m_r\}, \min\{M_l - 2, M_r\}]$ and therefore $k - 1 \in [m_r, M_r]$. Hence there is a rectilinear planar representation H_{μ_r} of G_{μ_r} with spirality $\sigma_{\mu_r} = k - 1$. By the representability condition, G_ν has a rectilinear planar representation H_ν; with the same construction as in Lemma 7, the spirality of H_ν is $\sigma_\nu = k$. □

The proofs of the next lemmas are similar to Lemma 8 (see [8]).

Lemma 9. *Let ν be a P-node of type $I_{3d}O_{\alpha\beta}$ with children μ_l and μ_r. Suppose that G_{μ_l} and G_{μ_r} are rectilinear planar with representability intervals $I_{\mu_l} = [m_l, M_l]$ and $I_{\mu_r} = [m_r, M_r]$, respectively. Graph G_ν is rectilinear planar if and only if $[m_l - M_r, M_l - m_r] \cap [\frac{5}{2}, \frac{7}{2} - \gamma] \neq \emptyset$, where $\gamma = \alpha + \beta - 2$. Also, if this representability condition holds then the representability interval of G_ν is $I_\nu = [\max\{m_l - \frac{3}{2}, m_r + 1\} + \frac{\gamma - \rho(d)}{2}, \min\{M_l - \frac{1}{2}, M_r + 2\} - \frac{\gamma + \rho(d)}{2}]$, where $\rho(\cdot)$ is a function such that $\rho(r) = 1$ and $\rho(l) = 0$.*

Lemma 10. *Let ν be a P-node of type $I_{3dd'}$ with children μ_l and μ_r. Suppose that G_{μ_l} and G_{μ_r} are rectilinear planar with representability intervals $I_{\mu_l} = [m_l, M_l]$ and $I_{\mu_r} = [m_r, M_r]$, respectively. Graph G_ν is rectilinear planar if and only if $3 \in [m_l - M_r, M_l - m_r]$. Also, if this representability condition holds then the representability interval of G_ν is $I_\nu = [\max\{m_l - 1, m_r + 2\} - \frac{\rho(d) + \rho(d')}{2}, \min\{M_l - 1, M_r + 2\} - \frac{\rho(d) + \rho(d')}{2}]$, where $\rho(\cdot)$ is a function such that $\rho(r) = 1$ and $\rho(l) = 0$.*

3.4 Characterization

Lemmas 4, 5, 6, 8, 9, and 10 give rise to the following characterization.

Theorem 1. *Let G be a plane series-parallel graph and let T be an SPQ*-tree of G. Let ν be any non-root node of T. The plane graph G_ν is rectilinear planar if and only if it satisfies the representability condition given in Table 1. Also, if such condition is satisfied, G_ν admits a rectilinear planar representation for all and only the values of spirality in the representability interval given in Table 1.*

Table 1. Representability conditions and intervals for the different types of nodes. In the formulas $\gamma = \alpha + \beta - 2$ and $\rho(\cdot)$ is such that $\rho(r) = 1$ and $\rho(l) = 0$.

Q*-node of length ℓ – Lemma 4	
Representability Condition	True
Representability Interval	$[-\ell + 1, \ell - 1]$
S-node with h children – Lemma 5	
Representability Condition	true
Representability Interval	$[\sum_{i=1}^{h} m_i, \sum_{i=1}^{h} M_i]$
P-node with three children – Lemma 6	
Representability Condition	$[m_l - 2, M_l - 2] \cap [m_c, M_c] \cap [m_r + 2, M_r + 2] \neq \emptyset$
Representability Interval	$[\max\{m_l - 2, m_c, m_r + 2\}, \min\{M_l - 2, M_c, M_r + 2\}]$
P-node with two children – $I_2O_{\alpha\beta}$ – Lemma 8	
Representability Condition	$[m_l - M_r, M_l - m_r] \cap [2, 4 - \gamma] \neq \emptyset$
Representability Interval	$[\max\{m_l - 2, m_r\} + \frac{\gamma}{2}, \min\{M_l, M_r + 2\} - \frac{\gamma}{2}]$
P-node with two children – $I_{3d}O_{\alpha\beta}$ – Lemma 9	
Representability Condition	$[m_l - M_r, M_l - m_r] \cap [\frac{5}{2}, \frac{7}{2} - \gamma] \neq \emptyset$
Representability Interval	$[\max\{m_l - \frac{3}{2}, m_r + 1\} + \frac{\gamma - \rho(d)}{2}, \min\{M_l - \frac{1}{2}, M_r + 2\} - \frac{\gamma + \rho(d)}{2}]$
P-node with two children – $I_{3dd'}$ – Lemma 10	
Representability Condition	$3 \in [m_l - M_r, M_l - m_r]$
Representability Interval	$[\max\{m_l - 1, m_r + 2\} - \frac{\rho(d) + \rho(d')}{2}, \min\{M_l - 1, M_r + 2\} - \frac{\rho(d) + \rho(d')}{2}]$

4 Rectilinear Planarity Testing Algorithm

Theorem 2. *Let G be an n-vertex plane series-parallel graph. There exists an $O(n)$-time algorithm that tests whether G admits a planar rectilinear representation and that constructs one in the positive case.*

Proof. If G is biconnected let e be an edge of G on the external face; otherwise, let e be a dummy edge added on the external face to make G biconnected. Let T be an SPQ*-tree of G with respect to e. We first show how to perform the test in linear time. If the test is positive, we show how to efficiently construct a rectilinear planar representation of G.

Testing Algorithm. Based on Theorem 1, the algorithm visits T in post-order and, for each non-root node ν of T, it checks the representability condition of ν and computes interval I_ν if the condition is positive. If the representability condition is violated for some node, the algorithm halts and returns a negative answer. Otherwise, the algorithm reaches the root ρ of T. If e is a dummy edge, the algorithm halts and returns a positive answer (since e will not appear in the representation, the algorithm does not need to check anything else). If e is real, let η be the child of ρ other than the child associated with e (see Fig. 1(c)). To complete the test, the algorithm must check the *root condition*, i.e., whether there exists a rectilinear planar representation H_η that can be merged with a straight-line representation of e. The spirality σ_η that H_η must have to be merged with e depends on the alias vertices for the poles of G_η: (*i*) If both these vertices coincide with the poles then $\sigma_\eta = 2$; (*ii*) if exactly one of them coincides with a pole of G_η then $\sigma_\eta = 3$ (one alias vertex subdivides e); (*iii*) if none of the alias vertices coincides with a pole of G_η (they both subdivide e) then $\sigma_\eta = 4$.

About the time complexity, each node of T is visited exactly once. By Theorem 1, for each non-root node ν, the representability condition for ν is checked in $O(1)$ time and the representability interval of ν is computed in $O(1)$ time. Also, the root condition is easily checked in $O(1)$ time. Finally, since T is computed in $O(n)$ time and has $O(n)$ nodes, the whole test is executed in $O(n)$ time.

Construction Algorithm. Suppose that the test is positive and that σ_η is the spirality required by a rectilinear planar representation of G_η (if e is dummy, we can set σ_η as any value in I_η). To construct a rectilinear planar representation H of G, the algorithm visits T top-down and determine the right value of spirality required by the component associated with each node of T. Once the spiralities for all nodes of T are determined, H is easily defined by fixing the vertex angles in each component as described in the proofs of Lemmas 4–6, 8–10. To compute the spiralities for the children of η we distinguish the following cases:

Case 1: η is an S-node, with children $\mu_1, \ldots, \mu_h$ ($i \in \{1, \ldots, h\}$). Let $I_{\mu_i} = [m_i, M_i]$ be the representability interval of μ_i. We must find a value $\sigma_{\mu_i} \in [m_i, M_i]$ for each $i = 1, \ldots, h$ such that $\sum_{i=i}^{h} \sigma_{\mu_i} = \sigma_\eta$. To this aim, initially set $\sigma_{\mu_i} = M_i$ for each $i = 1, \ldots, h$ and consider $\delta = \sum_{i=i}^{h} \sigma_{\mu_i} - \sigma_\eta$. By Theorem 1, $\delta \geq 0$. If $\delta = 0$ we are done. Otherwise, iterate over all $i = 1, \ldots, h$ and for each i decrease both σ_{μ_i} and δ by the value $\min\{\delta, M_i - m_i\}$, until $\delta = 0$.

Case 2: η is a P-node with three children, μ_l, μ_c, and μ_r. By Lemma 3, it suffices to set $\sigma_{\mu_l} = \sigma_\eta + 2$, $\sigma_{\mu_c} = \sigma_\eta$, and $\sigma_{\mu_r} = \sigma_\eta - 2$.

Case 3: η is a P-node with two children, μ_l and μ_r. Let u and v be the poles of η. By Lemma 2, σ_{μ_l} and σ_{μ_r} must be fixed in such a way that $\sigma_{\mu_l} = \sigma_\eta + k_u^l \alpha_u^l + k_v^l \alpha_v^l$ and $\sigma_{\mu_r} = \sigma_\eta - k_u^r \alpha_u^r - k_v^r \alpha_v^r$. The values of k_u^l, k_v^l, k_u^r, and k_v^r are fixed by the indegree and outdegree of u and v. Hence, it suffices to choose the values of α_u^l, α_v^l, α_u^r, α_v^r such that they are consistent with the type of η and they yield $\sigma_{\mu_l} \in I_{\mu_l}$ and $\sigma_{\mu_r} \in I_{\mu_r}$. Since each α_w^d ($w \in \{u, v\}, d \in \{l, r\}$) is either 0 or 1 there are at most four possible combinations of values to consider.

Once the spiralities for the children of η are computed, the algorithm continues its top-down visit, and for each node ν for which a spirality σ_ν has been fixed, it computes the spiralities of the children of ν with same procedure as for η. Concerning the time complexity, the procedure in Case 1 takes linear time in the number of children of the S-node, while the procedures in Case 2 and Case 3 take constant time. Therefore the whole visit requires $O(n)$ time.

Table 2 shows a running example based on Fig. 1. For each P- and S-component it reports the representability interval computed in the bottom-up visit of the tree and the spirality fixed in the top-down visit (see also Fig. 2).

Table 2. Running example based on Fig. 1.

Node Label	Node Type	Repres. Interval	Spirality in H
η	S-node	$[-3,3]$	3
ν_1	P-node (2 children) – $I_{3r}O_{11}$	$[-2,2]$	2
ν_2	S-node	$[-4,4]$	4
ν_3	S-node	$[-\frac{5}{2},\frac{1}{2}]$	$\frac{1}{2}$
ν_4	P-node (2 children) – I_{3lr}	$[0,0]$	0
ν_5	P-node (3 children)	$[-1,0]$	0
ν_6	P-node (2 children) – I_2O_{12}	$[-\frac{3}{2},\frac{1}{2}]$	$\frac{1}{2}$
ν_7	P-node (2 children) – I_2O_{22}	$[0,0]$	0
ν_8	S-node	$[-\frac{3}{2},\frac{3}{2}]$	$\frac{3}{2}$
ν_9	S-node	$[-\frac{3}{2},\frac{3}{2}]$	$-\frac{3}{2}$
ν_{10}	P-node (2 children) – I_2O_{12}	$[-\frac{1}{2},\frac{1}{2}]$	$\frac{1}{2}$
ν_{11}	P-node (2 children) – I_2O_{12}	$[-\frac{1}{2},\frac{1}{2}]$	$-\frac{1}{2}$

5   Conclusions and Open Problems

We proved that rectilinear planarity testing can be solved in linear time for series-parallel graphs with two terminals. Several open problems can be studied:

OP1. Can we extend Theorem 2 to 1-connected plane 4-graphs whose biconnected components are two-terminal series-parallel graphs (i.e., partial 2-trees)? The work in [11] solves the problem for 1-connected outerplanar graphs.

OP2. What is the time complexity of rectilinear planarity testing for general plane 4-graphs? The question is interesting even for triconnected plane 4-graphs. A linear-time solution exists for plane 3-graphs [16,17].

OP3. Testing rectilinear planarity is NP-complete in the variable embedding setting but it can be solved in $O(n^3 \log n)$-time for series-parallel graphs [6]. It is interesting to determine whether this complexity bound can be improved.

References

1. Borradaile, G., Klein, P.N., Mozes, S., Nussbaum, Y., Wulff-Nilsen, C.: Multiple-source multiple-sink maximum flow in directed planar graphs in near-linear time. SIAM J. Comput. **46**(4), 1280–1303 (2017)
2. Brandenburg, F., Eppstein, D., Goodrich, M.T., Kobourov, S., Liotta, G., Mutzel, P.: Selected open problems in graph drawing. In: Liotta, G. (ed.) GD 2003. LNCS, vol. 2912, pp. 515–539. Springer, Heidelberg (2004). https://doi.org/10.1007/978-3-540-24595-7_55
3. Cornelsen, S., Karrenbauer, A.: Accelerated bend minimization. J. Graph Algorithms Appl. **16**(3), 635–650 (2012)
4. Di Battista, G., Eades, P., Tamassia, R., Tollis, I.G.: Graph Drawing: Algorithms for the Visualization of Graphs. Prentice-Hall, Upper Saddle River (1999)
5. Di Battista, G., Liotta, G., Vargiu, F.: Spirality and optimal orthogonal drawings. SIAM J. Comput. **27**(6), 1764–1811 (1998)

6. Di Giacomo, E., Liotta, G., Montecchiani, F.: Sketched representations and orthogonal planarity of bounded treewidth graphs. CoRR, abs/1908.05015 (2019)
7. Di Giacomo, E., Liotta, G., Tamassia, R.: Graph drawing. In: Goodman, J.E., O'Rourke, J., Tóth, C.D. (eds.) Handbook of Discrete and Computational Geometry, Third Edition, chapter 55, pp. 1451–1478. Chapman and Hall/CRC (2017)
8. Didimo, W., Kaufmann, M., Liotta, G., Ortali, G.. Rectilinear planarity testing of plane series-parallel graphs in linear time. CoRR, 2008.03784 (2020). http://arxiv.org/abs/2008.03784v3
9. Didimo, W., Liotta, G., Ortali, G., Patrignani, M.: Optimal orthogonal drawings of planar 3-graphs in linear time. In: Chawla, S. (ed.) Proceedings of the 2020 ACM-SIAM Symposium on Discrete Algorithms, SODA 2020, Salt Lake City, UT, USA, 5–8 January 2020, pp. 806–825. SIAM (2020)
10. Duncan, C.A., Goodrich, M.T.: Planar orthogonal and polyline drawing algorithms. In: Tamassia, R. (ed.) Handbook on Graph Drawing and Visualization, pp. 223–246. Chapman and Hall/CRC (2013)
11. Frati, F.: Planar rectilinear drawings of outerplanar graphs in linear time. In: Auber, D., Valtr, P. (eds.) Graph Drawing, Symposium on Graph Drawing and Network Visualization, GD 2020, 16–18 September Proceedings, Lecture Notes in Computer Science. Springer (2020)
12. Garg, A., Liotta, G.: Almost bend-optimal planar orthogonal drawings of biconnected degree-3 planar graphs in quadratic time. In: Kratochvíyl, J. (ed.) GD 1999. LNCS, vol. 1731, pp. 38–48. Springer, Heidelberg (1999). https://doi.org/10.1007/3-540-46648-7_4
13. Garg, A., Tamassia, R.: A new minimum cost flow algorithm with applications to graph drawing. In: North, S. (ed.) GD 1996. LNCS, vol. 1190, pp. 201–216. Springer, Heidelberg (1997). https://doi.org/10.1007/3-540-62495-3_49
14. Hasan, M.M., Rahman, M.S.: No-bend orthogonal drawings and no-bend orthogonally convex drawings of planar graphs (extended abstract). In: Du, D.-Z., Duan, Z., Tian, C. (eds.) COCOON 2019. LNCS, vol. 11653, pp. 254–265. Springer, Cham (2019). https://doi.org/10.1007/978-3-030-26176-4_21
15. Nishizeki, T., Rahman, M.S.: Planar Graph Drawing. Lecture Notes Series on Computing, vol. 12. World Scientific (2004)
16. Rahman, M.S., Nakano, S., Nishizeki, T.: A linear algorithm for bend-optimal orthogonal drawings of triconnected cubic plane graphs. J. Graph Algorithms Appl. 3(4), 31–62 (1999)
17. Rahman, M.S., Nishizeki, T.: Bend-minimum orthogonal drawings of plane 3-graphs. In: Goos, G., Hartmanis, J., van Leeuwen, J., Kučera, L. (eds.) WG 2002. LNCS, vol. 2573, pp. 367–378. Springer, Heidelberg (2002). https://doi.org/10.1007/3-540-36379-3_32
18. Rahman, M.S., Nishizeki, T., Naznin, M.: Orthogonal drawings of plane graphs without bends. J. Graph Algorithms Appl. 7(4), 335–362 (2003)
19. Tamassia, R.: On embedding a graph in the grid with the minimum number of bends. SIAM J. Comput. 16(3), 421–444 (1987)

New Quality Metrics for Dynamic Graph Drawing

Amyra Meidiana[✉], Seok-Hee Hong, and Peter Eades

University of Sydney, Sydney, Australia
amei2916@uni.sydney.edu.au, {seokhee.hong,peter.eades}@sydney.edu.au

Abstract. In this paper, we present new quality metrics for dynamic graph drawings. Namely, we present a new framework for *change faithfulness* metrics for dynamic graph drawings, which compare the *ground truth change* in dynamic graphs and the *geometric change* in drawings.

   More specifically, we present two specific instances, *cluster change* faithfulness metrics and *distance change* faithfulness metrics. We first validate the effectiveness of our new metrics using deformation experiments. Then we compare various graph drawing algorithms using our metrics. Our experiments confirm that the best cluster (resp. distance) faithful graph drawing algorithms are also cluster (resp. distance) change faithful.

1 Introduction

Quality metrics (or *aesthetic criteria* [3]) for graph drawings play an important role in evaluating graph drawings as well as designing new algorithms to optimize the metrics. Traditional quality metrics for graph drawings mainly evaluate the *readability* of a drawing, such as edge crossings, edge bends, total edge length, and angular resolution [3]. Most of these metrics focus on *static* graphs.

   Network data are abundant in various domains, from social media to chemical pathways, and they are often changing with dynamics. Compared to static graph drawing, dynamic graph drawing brings its own challenges, such as the preservation of the user's mental map as the drawing evolves [12]. To evaluate dynamic graph drawing algorithms, we need quality metrics to measure how well a drawing of a dynamic graph reflects the changes in the graph.

   Faithfulness metrics measure how faithfully the ground truth about the data is displayed in the visualization [29]. For dynamic graphs, *change faithfulness* measures how proportional the change in the drawings of dynamic graphs is to the change in the graphs.

   However, existing work on quality metrics of dynamic graph drawings, such as preservation of the mental map [8,10,12], mainly focus on the *readability* metrics, which only measure the geometric change in the drawing without considering how well the change represents the change in the graph. Furthermore, recent

---

This work is supported by ARC DP grant.

D. Auber and P. Valtr (Eds.): GD 2020, LNCS 12590, pp. 450–465, 2020.
https://doi.org/10.1007/978-3-030-68766-3_35

qualitative studies have shown that mental map preservation alone may not be sufficient to aid users in understanding dynamic graphs [1].

In this paper, we present a new framework for *change faithfulness metrics* of dynamic graphs, quantitatively measuring how faithfully the ground truth change in dynamic graphs is proportionally displayed as the geometric change in dynamic graph drawings.

Based on the framework, we present two new quality metrics, *cluster change* faithfulness metrics and *distance change* faithfulness metrics. We validate the effectiveness of our new metrics using deformation experiments, and then compare various graph drawing algorithms using our metrics.

More specifically, we present the following contributions:

1. We present a general *change faithfulness metric* framework for dynamic graphs, which compares the ground truth change in dynamic graphs and the geometric change in the drawings.
2. We present the *cluster change faithfulness metrics CCQ* as an instance of the change faithfulness metrics, comparing the change in *ground truth clustering* of dynamic graphs to the change in *geometric clustering* of the drawing.
3. We present the *distance change faithfulness metrics DCQ* as another specific instance of the change faithfulness metrics, which compares the change in *graph theoretic distance* of dynamic graphs to the change in *geometric distance* of the drawing.
4. We validate the effectiveness of the cluster change faithfulness metrics and distance change faithfulness metrics using deformation experiments on drawings. Results of the experiments confirm that the *CCQ* and *DCQ* metrics decrease as the drawings are distorted such that the change between drawings are more disproportionate to the change in ground truth information.
5. We compare various graph drawing algorithms using the *CCQ* and *DCQ* metrics. Experiments confirm that the most cluster faithful layouts and distance faithful layouts indeed also obtain high cluster change faithfulness and distance faithfulness respectively. Interestingly, we also discover that in some cases, higher information faithfulness does not necessarily lead to higher change faithfulness.

2   Related Work

2.1   Quality Metrics for Graph Drawing

Traditional aesthetic criteria [3] for graph drawings are mainly concerned with the *readability* of graphs, such as the minimization of edge crossings, bends, total edge lengths and drawing area. They have been established as criteria to be optimized by graph drawing algorithms [3].

HCI studies have verified the correlation between aesthetic criteria with specific task performance on graphs. For example, few edge crossings [34,35] and large crossing angles [19] are important criteria for finding shortest paths between two vertices. However, these studies tend to focus on *small* graphs.

More recently, a new concept of *faithfulness* metrics have been introduced for *large* graphs, measuring how faithfully the ground truth information of graphs is displayed in graph drawings [29]. Subsequently, a series of new faithfulness metrics have been developed [11, 26–28].

Shape-based metrics [11] are introduced to evaluate *large* graph drawings, where traditional metrics such as edge crossings do not scale well. More specifically, the metrics compare the similarity between the original graph G with a shape graph (or proximity graph) G' computed from a drawing D of G.

The *cluster metrics* CQ [26, 27] measure how faithfully the ground truth clusters of a graph is displayed in a drawing, by comparing the ground truth clusters to the geometric clustering in a graph drawing.

The *symmetry metrics* [28] measure how faithfully the ground truth *automorphisms* of a graph (rotational or axial) and *automorphism groups* (cyclic or dihedral), are displayed as symmetries in a drawing, computed by approximate symmetry detection algorithms in $O(n \log n)$ time. A $O(n \log n)$ time algorithm for exact symmetry detection is also presented.

2.2 Quality Metrics for Dynamic Graph Drawing

A *dynamic* graph is defined by a sequence of static graphs $G_1, G_2, \ldots, G_k$ spanning k time steps, where G_i is a time slice of the graph at time step i [5]. Dynamic graphs are most commonly visualized using small multiples [39] or animation.

A long standing challenge with dynamic graph drawings is preserving the user's *mental map* [12], where dramatic changes in the positions of vertices can make it difficult for users to keep track of the state of a dynamic graph. The mental map can be modelled using e.g.. orthogonal ordering, clustering, or topology [12]. Related is the concept of *dynamic stability*, which aims to minimize the geometric distance between subsequent drawings [6, 37]. Stability has been shown to assist users in performing analytical tasks on dynamic graphs [2].

A recent survey on dynamic graph drawing [5] addresses that evaluation is one of the most important research questions on dynamic graph drawings. Quantitatively, dynamic graph drawings can be evaluated using *distance* metrics, including Euclidean distance, orthogonal distance, and edge routing, to measure the extent of mental map preservation [8, 10].

However, specific change faithfulness metrics for dynamic graph drawings have yet to be developed to measure how the ground truth change in dynamic graphs are proportionally displayed as geometric change in drawings.

3   Change Faithfulness Metric Framework

We propose the *change faithfulness metric* for measuring how well dynamic graph drawings show the structural changes in dynamic graphs. Roughly speaking, a drawing is *change faithful* if the extent of change in the drawing is proportional to the extent of (ground truth) change in the graph. Figure 1 illustrates the general framework for change faithfulness metrics.

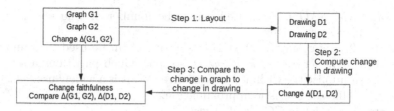

Fig. 1. Change faithfulness metric framework: The change faithfulness metric is computed by comparing the ground truth change $\Delta(G_1, G_2)$ between two graphs G_1 and G_2, and the geometric change $\Delta(D_1, D_2)$ in drawings of graphs.

In practice, the vertex set of a dynamic graph may change; in this paper we focus on cases where only the edge set changes. Let $G_1 = (V, E_1)$ and $G_2 = (V, E_2)$ be two time slices of a dynamic graph, with the change denoted as $\Delta(G_1, G_2)$. The change faithfulness metrics are computed as follows:

Step 1: Compute a drawing D_1 (resp. D_2) of G_1 (resp. G_2).
Step 2: Compute the geometric change $\Delta(D_1, D_2)$ between D_1 and D_2.
Step 3: Compute the change faithfulness metrics by comparing the ground truth change $\Delta(G_1, G_2)$ to $\Delta(D_1, D_2)$.

The framework in Fig. 1 is a general framework applicable to various types of change in dynamic graphs. The detailed definitions for $\Delta(G_1, G_2)$ and $\Delta(D_1, D_2)$, as well as how to compare them, depend on the nature of the considered change.

3.1 Cluster Change Faithfulness Metrics

We present the *cluster change faithfulness metric CCQ* as an example of a change faithfulness metric. *CCQ* measures how faithfully the change in *ground truth clustering* is reflected as a change in the *geometric clustering* between drawings of different time slices of a dynamic graph.

Let C_1 (resp. C_2) be the ground truth clustering of the vertices of G_1 (resp. G_2), with the change between the clusterings denoted as $\Delta(C_1, C_2)$. The cluster change faithfulness is defined as follows:

Step 1: Compute a drawing D_1 (resp. D_2) of G_1 (resp. G_2).
Step 2: Compute geometric clustering C_1' (resp. C_2') based on vertex positions in D_1 (resp. D_2), and compute the change in geometric clustering $\Delta(C_1', C_2')$.
Step 3: Compute *CCQ* by comparing $\Delta(C_1, C_2)$ to $\Delta(C_1', C_2')$.

To compute $\Delta(C_1, C_2)$ and $\Delta(C_1', C_2')$, any clustering comparison metrics can be used. In this paper, we use *ARI* (Adjusted Rand Index) [20,36] and *FMI* (Fowlkes-Mallows Index) [14], which showed superior performance in measuring cluster faithfulness in static graph drawing [27]. *ARI* is based on the number of pairs of elements classified into the same and different groups in two clusterings of the same set. FMI is computed using the number of true positives, false positives, and false negatives.

For Step 2, any geometric clustering algorithm can be used to compute C_1' and C_2'. In this paper we use *k-means clustering*, which partitions a set into k subsets that minimize the within-class variance [24]. It is a widely used clustering method with efficient heuristic approximation.

For Step 3, we define CCQ as follows:

$$CCQ = 1 - \frac{|\Delta(C_1, C_2) - \Delta(C_1', C_2')|}{max(\Delta(C_1, C_2), \Delta(C_1', C_2'))} \tag{1}$$

Specifically, we take the difference between $\Delta(C_1, C_2)$ to $\Delta(C_1', C_2')$, and express the difference as a fraction of the larger value between the two, as both are normalized to the same range by using the same clustering quality metrics. We then negate the result from 1, such that 1 represents completely change faithful drawings and less change faithful drawings obtain values closer to 0.

3.2 Distance Change Faithfulness Metrics

We also present the *distance change faithfulness metric DCQ* as another instance of change faithfulness metric. We define *distance faithfulness* as how faithfully graph theoretic distances between vertices in a graph are displayed as geometric distances between the positions of vertices in a drawing. Similarly, *distance change faithfulness* measures how faithfully the change in graph theoretic distances is reflected as a proportional change in the geometric distances.

Let $\Delta(SP_1, SP_2)$ be the change in graph theoretic distances between two time slices of a dynamic graph, G_1 and G_2. More specifically, the distance change faithfulness metric is defined as follows:

Step 1: Compute a drawing D_1 (resp. D_2) of G_1 (resp. G_2).
Step 2: Compute the change in geometric distance $\Delta(GD_1, GD_2)$.
Step 3: Compute DCQ by comparing $\Delta(SP_1, SP_2)$ to $\Delta(GD_1, GD_2)$.

One example measure of distance faithfulness is *stress* [3]. For each pair of vertices v_i, v_j in a graph G, the stress is defined by the difference between the graph theoretic distance (i.e., shortest path) between v_i and v_j, and the geometric distance between the positions of v_i and v_j in a drawing D of G.

Using stress measures, we present two types of distance change faithfulness metrics DCQ. The first metric DCQ_1 is based on the *target edge length* used in some stress-based layouts (e.g. [16]). Given a target edge length tl, we expect neighboring vertices (i.e. path length 1) to have a geometric distance of tl. We thus scale the geometric distance between each pair of vertices in D by tl.

Let $\Delta(v_i, v_j) = |\delta_1(v_i, v_j) - \delta_2(v_i, v_j)|/max(\delta_1(v_i, v_j), \delta_2(v_i, v_j))$ and $S(v_i, v_j) = |s_1(v_i, v_j) - s_2(v_i, v_j)|/max(s_1(v_i, v_j), s_2(v_i, v_j))$, where $\delta_1(v_i, v_j)$ (resp. $\delta_2(v_i, v_j)$) is the graph theoretic distance between vertices v_i, v_j in G_1 (resp. G_2) and $s_1(v_i, v_j)$ (resp. $s_2(v_i, v_j)$) is the geometric distance between vertices v_i, v_j in D_1 (resp. D_2). Scaling $S(v_i, v_j)$ by tl to ensure the change in geometric distance is scaled to the target edge length, we define DCQ_1 as follows:

$$DCQ_1 = 1 - \frac{2}{|V|^2} \sum_{i=0}^{|V|} \sum_{j=i+1}^{|V|} \left| \Delta(v_i, v_j) - \frac{S(v_i, v_j)}{tl} \right| \qquad (2)$$

In practice, not every layout algorithm takes an target edge length as input. Therefore, we instead use the *average* of all edge lengths as tl.

For the second type of distance change faithfulness metric DCQ_2, we scale both the graph theoretic and geometric distances by the *maximum* distance. For graph theoretic distances, it is the *diameter* of graph G, while for geometric distances, it is the largest distance between any pair of vertices in drawing D.

The scaled graph theoretic distance is given as $\delta'(i,j) = \delta(v_i, v_j)/diam(G)$, where $diam(G)$ is the diameter of G. The scaled geometric distance is given as $s'(i,j) = s(v_i, v_j)/max(s)$, where $max(s)$ is the maximum distance between any two vertices in D. We define DCQ_2 as follows:

$$DCQ_2 = 1 - \frac{2}{|V|^2} \sum_{i=0}^{|V|} \sum_{j=i+1}^{|V|} ||\delta_1'(i,j) - \delta_2'(i,j)| - |s_1'(i,j) - s_2'(i,j)|| \qquad (3)$$

4 Cluster Change Faithfulness Validation Experiment

To validate the cluster change faithfulness metrics, we design *deformation* experiments. Given two dynamic graph time slices G_1 and G_2 with ground truth clustering C_1 and C_2, we start with *cluster faithful drawings* D_1 and D_2, i.e. the geometric clustering C_1' of D_1 (resp. C_2' of D_2) is the same as C_1 (resp. C_2). This gives $\Delta(C_1, C_2) = \Delta(C_1', C_2')$, i.e. cluster change faithful.

We then progressively deform drawing D_2. In each experiment, we perform 10 steps of deformation, where in each step, the coordinates of each vertex from the previous step are perturbed by a value in the range $[0, \delta]$, where δ is the size of the drawing area multiplied by a value in the range [0.05,0.1]. We compute CCQ and compare the scores across all steps of the deformation.

We expect that CCQ will decrease with the deformation steps, as $\Delta(C_1', C_2')$ will grow further away from $\Delta(C_1, C_2)$. We formulate the following hypothesis:

Hypothesis 1. CCQ_{ARI} and CCQ_{FMI} decrease as D_2 is deformed.

We generate ten dynamic graph data sets for the CCQ validation experiment, with 200–1000 vertices each, as follows: First, we create a small graph (up to 30 vertices). We replace each vertex with a larger, denser graph, which becomes a cluster in G_1. We then replace each edge with inter-cluster edges between a randomly selected subset of vertices from each cluster. To create G_2, we change the cluster membership of vertices, either by merging clusters through randomly adding inter-cluster edges until a desired density for the new cluster is achieved, or splitting clusters by deleting edges between two partitions of the cluster until a desired lower intra-cluster edge density is reached.

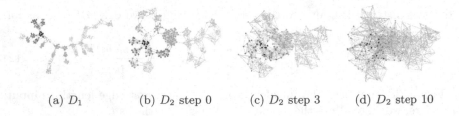

(a) D_1 (b) D_2 step 0   (c) D_2 step 3   (d) D_2 step 10

Fig. 2. Deformation experiment for $clusts - tree - 30$, showing deformation steps.

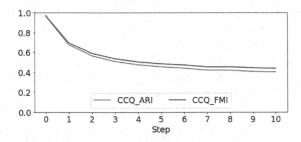

Fig. 3. Average of CCQ for all validation experiments. The decreasing trend for all versions of CCQ validates Hypothesis 1.

To compute the initial layouts, we use the Backbone layout from Visone [4], which produces cluster faithful layouts (i.e. $CQ = 1$) for our validation datasets. We use cluster comparison metric implementation from scikit-learn [33].

Figure 2 shows a deformation experiment example, where vertices are colored based on ground truth cluster membership. Figures 2 (a) and (b) show D_1 and D_2 at step 0. As the positions are perturbed in Figs. 2 (c) and (d), vertices in the same cluster grow further apart and mix with vertices from other clusters, making the drawing less cluster faithful and subsequently increasing the difference between the geometric clustering in D_1 and D_2.

Figure 3 shows the average CCQ scores for each deformation step, averaged for all data sets. Clearly, we can see that CCQ metrics decrease after each deformation step, validating Hypothesis 1.

4.1   Discussion and Summary

Figure 3 clearly shows a downward slope of the CCQ metrics, which validates the usage of both CCQ_{ARI} and CCQ_{FMI} metrics with our framework. Previous work on cluster faithfulness metrics CQ on static graphs [27] has shown that ARI is more sensitive to changes than FMI. To a lesser extent, a similar pattern can be seen here, where CCQ_{ARI} decreases to a lower score on latter perturbation steps compared to CCQ_{FMI}, indicating that it may be better in capturing changes in cluster change faithfulness as well.

In summary, the validation experiments have shown that the CCQ metrics effectively reflect the cluster change faithfulness of drawings of dynamic graphs *with dynamic clusters. Furthermore, we see that CCQ_{ARI} is slightly more effective in capturing cluster change faithfulness than CCQ_{FMI}.*

5   Cluster Change Faithfulness Layout Comparison

After validating the effectiveness of the cluster change faithfulness metrics, we use the CCQ metrics to compare the performance of various graph drawing algorithms. We select the following layout algorithms: *LinLog* [30], a force-directed layout emphasizing clusters; *Backbone* [31], which uses Simmelian backbones to extract communities; *tsNET* [22], which uses t-SNE [23] and aims to preserve point neighborhoods; and *sfdp* [18], a multi-level force-directed layout.

LinLog, Backbone, and tsNET are designed to display clusters, and sfdp was seen to be more cluster faithful than other non-cluster-focused layouts [27].

As LinLog was shown to be the most cluster faithful [27], we also expect it to be the most cluster change faithful, formulating the following hypothesis:

Hypothesis 2. *LinLog scores the highest CCQ metrics.*

We use Tulip [9] (LinLog), visone [4] (Backbone), Graphviz [13] (sfdp), and tsNET [21]. We use thirteen dynamic graphs including synthetic data created similarly as in Sect. 4, and real-world data the Social Evolution set [25]; the graph sizes range from around 80–1000 vertices.

Table 1 shows a layout comparison example, with a cluster split (yellow into yellow and pastel green). The CQ cell shows the cluster faithfulness metrics: green and orange show the CQ_{ARI} metric for G_1 and G_2 respectively, and purple and pink show the CQ_{FMI} metric for G_1 and G_2 respectively. The CCQ cell shows the CCQ metrics: red for CCQ_{ARI} and blue for CCQ_{FMI}.

LinLog obtains the highest CCQ score, supporting Hypothesis 2. We also see a case of higher CQ not always corresponding to higher CCQ: for example, tsNET obtains higher CQ than sfdp, however, it obtains lower CCQ than sfdp.

Figure 4 shows the average CCQ scores across all data sets used for the layout comparison experiment. On average, LinLog obtains the highest CCQ metrics, at 0.98 on CCQ_{ARI}, validating Hypothesis 2.

5.1   Discussion and Summary

Our experiments confirm that the LinLog layout, which was previously shown as the most cluster faithful layout for static graphs, also obtains the highest cluster change faithfulness for dynamic graphs.

We also find cases where better cluster faithfulness does not always correspond to better cluster change faithfulness, as seen in Table 1. This may be due to the clusters "moving around" between the drawings produced by tsNET, causing different misclassifications. For example, in D_1, some members of the pink cluster were misclassified to the dark purple or lime green clusters in D_1; however, they are misclassified into the lime green or orange instead in D_2.

Table 1. Layout comparison on *gnm_10_25*

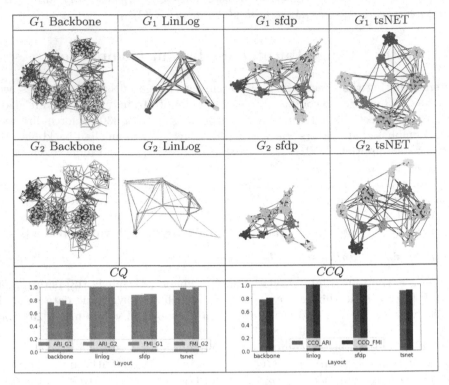

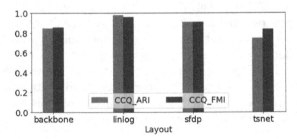

Fig. 4. Average of CCQ for layout comparison experiments. LinLog obtains the highest CCQ, validating Hypothesis 2. sfdp unexpectedly obtains the second highest CCQ.

Meanwhile, sfdp produces drawings where relative positions of the cluster are more stable, causing the misclassifications to be more "consistent", e.g. members of the pink cluster are misclassified only into the lime green and orange clusters in both D_1 and D_2. Stability alone does not always lead to high change faithfulness, however, as seen from Backbone in Fig. 1, where the cluster positions are stable yet CCQ is still low as CQ is lower compared to the other layouts.

In summary, our experiments confirm Hypothesis 2, showing that LinLog produces the most cluster change faithful drawings. We also show that cluster faithfulness does not always translate to cluster change faithfulness, in cases where subsequent drawings do not preserve the relative locations of the clusters.

6   Distance Change Faithfulness Validation Experiment

We also validate the distance change faithfulness metrics, using validation experiments. Given two graph time slices G_1 and G_2, we start with *stress faithful drawings* D_1 and D_2. We then perturb D_2 as follows: before perturbing, we divide the edges into two sets E_1' and E_2'. In each step, we select edges from E_1' to extend their lengths, and select edges from E_2' to shorten their lengths.

We expect that the DCQ scores decrease with the deformation steps. We therefore formulate the following hypothesis:

Hypothesis 3. *DCQ_1 and DCQ_2 decrease as the drawing D_2 is deformed, and DCQ_1 performs better than DCQ_2 in measuring distance change faithfulness.*

To create the validation data sets, we start with a randomly-generated graph G_1, typically with a long diameter. To create G_2, we add edges to G_1 that significantly reduces the diameter and introduces smaller cycles into the graph. We generate ten dynamic graphs with 20–300 vertices and draw them using the *Stress Majorization* layout from Tulip [9] to obtain low stress drawings.

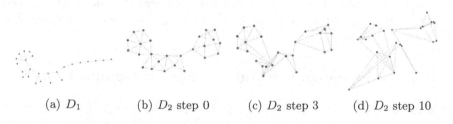

(a) D_1 (b) D_2 step 0   (c) D_2 step 3   (d) D_2 step 10

Fig. 5. Deformation experiment for *powertree_25_1*, showing deformation steps.

Figure 5 shows a deformation experiment example, where Figs. 5(a) and (b) show D_1 and D_2 at step 0 respectively, computed by the Stress Majorization layout to produce stress faithful drawings. As the positions are perturbed in Figs. 5(c) and (d), the geometric distances between the vertices are perturbed to be more disproportionate to their graph theoretic distance.

Figure 6 shows the average DCQ for each deformation step, averaged for all data sets. DCQ decreases with each deformation step, confirming Hypothesis 3.

We can also see that DCQ_1 decreases to a lower value in latter deformation steps compared to DCQ_2, which only decreases by about 0.1. Considering how far the drawings are from the initial distance faithful drawings at step 10, e.g. Fig. 5(d), the minor decrease with DCQ_2 does not capture the extent of change as closely as DCQ_1. This indicates that DCQ_1 is more effective at capturing the distance change faithfulness, also supporting Hypothesis 3.

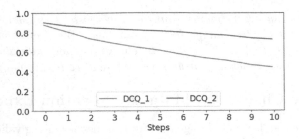

Fig. 6. Average of DCQ metrics for all validation experiments. The decreasing trend, especially with DCQ_1, validates Hypothesis 3.

6.1   Discussion and Summary

Our deformation experiment validates the effectiveness of DCQ metrics to measure the distance change faithfulness of drawings of dynamic graphs. We also observe that DCQ_1 is more effective at capturing differences in distance change faithfulness than DCQ_2. This may be due to the fact that scaling by maximum distance in DCQ_2 can be more susceptible to outliers, and may cause smaller distance changes to be underrepresented. Therefore, we will focus on DCQ_1 as the main comparison metric for the next experiments.

In summary, our experiments have validated Hypothesis 3, showing that DCQ effectively reflects the distance change faithfulness of dynamic graph drawing, and that DCQ_1 captures distance change faithfulness more effectively than DCQ_2.

7   Distance Change Faithfulness Layout Comparison

After validating the effectiveness of the distance change faithfulness metrics, we compare the performance of a number of graph drawing algorithms using the DCQ metrics. We select the following layout algorithms: *Stress-based* layouts *Stress Majorization* [17] and *Sparse Stress Minimization* [32]; *MDS (Multi-Dimensional Scaling)* layouts *Pivot MDS* [7] and *Metric MDS* [38]; *tsNET* [21]; *FR (Fruchterman-Reingold)* [15]; and *LinLog* [30].

Stress-based layouts aim to minimize stress (i.e. high distance faithfulness), therefore we expect them to be the most distance change faithful. As the concept of stress was adapted from MDS, we expect that MDS layouts will also perform quite well. Meanwhile, we expect force-directed layouts such as FR and LinLog to be less distance change faithful. We therefore formulate the following hypothesis:

Hypothesis 4. *Stress Majorization and Sparse Stress Minimization obtain the highest DCQ scores, while FR and LinLog obtain the lowest DCQ scores.*

We again use a mix of synthetic graphs and real-world graphs from the Social Evolution set [25], in total fifteen sets of dynamic graphs with 20–300 vertices.

Table 2. Layout comparison for *tree_100_1*

G_1 Stress Maj.	G_1 S. Stress Min.	G_1 Pivot MDS	G_1 Metric MDS	G_1 tsNET
G_2 Stress Maj.	G_2 S. Stress Min.	G_2 Pivot MDS	G_2 Metric MDS	G_2 tsNET
G_1 FR	G_1 LinLog	Stress		
G_2 FR	G_2 LinLog	DCQ		

Table 2 shows a layout comparison example. The stress of the drawings are shown in magenta (D_1) and cyan (D_2), and DCQ_1 and DCQ_2 are shown in red and blue respectively. Stress Majorization and Sparse Stress Minimization obtains the two highest DCQ, while FR and LinLog obtain notably higher stress and lower DCQ than other layouts, supporting Hypothesis 4.

Figure 7 shows the average stress and DCQ scores across all layout comparison experiment data sets. On average, Stress Majorization and Sparse Stress Minimization obtain the lowest stress and highest DCQ metrics, at around 0.86 on DCQ_1, and FR and LinLog obtain the highest stress and lowest DCQ metric, at around 0.7 and 0.66 respectively on DCQ_1, supporting Hypothesis 4.

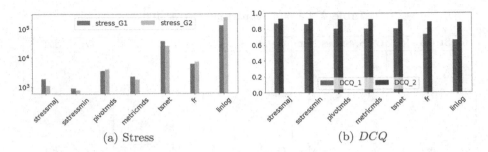

(a) Stress   (b) DCQ

Fig. 7. (a) Average stress scores; (b) average DCQ metrics for layout comparison experiments. Stress Majorization and Sparse Stress Minimization obtain the highest DCQ, while FR and LinLog obtain the lowest DCQ, supporting Hypothesis 4.

7.1 Discussion and Summary

Our experiments have supported Hypothesis 4, showing that the stress-based layouts, which explicitly aim to achieve low stress drawings, also obtain high DCQ, while FR and LinLog, which are not specifically designed to minimize stress, obtain lower DCQ.

While LinLog obtains the best results in the CCQ layout comparison, in this case, it obtains the lowest DCQ. This shows a case where a layout that is optimal for one metric may not perform as well on other metrics.

In summary, our experiments have supported Hypothesis 4 for stress-based layouts, which obtain the highest DCQ metrics on average. We also observe that a layout obtaining good performance on one change faithfulness metric may not perform as well on other change faithfulness metrics.

8   Conclusion and Future Work

We introduce a general framework for measuring change faithfulness in dynamic graph drawings. Based on the framework, we present cluster change faithfulness metrics CCQ and distance change faithfulness metrics DCQ, as specific instances of the framework.

We validate the effectiveness of both metrics using deformation experiments, and then compare various graph drawing layouts using the metrics. Our experiments confirm that LinLog obtains the highest cluster change faithfulness, while stress-based layouts obtain the highest distance change faithfulness.

Future work include designing other specific instances of the change faithfulness metric framework. More specifically, DCQ can be extended by using other notions of distance. As the general nature of the change faithfulness metric framework allows for the development of other specific metrics, this also presents the opportunity for designing new layout algorithms to optimize such new metrics.

References

1. Archambault, D., Purchase, H.C.: The "map" in the mental map: experimental results in dynamic graph drawing. Int. J. Hum Comput Stud. **71**(11), 1044–1055 (2013)
2. Archambault, D., Purchase, H.C.: Can animation support the visualisation of dynamic graphs? Inf. Sci. **330**, 495–509 (2016)
3. Battista, G.D., Eades, P., Tamassia, R., Tollis, I.G.: Graph Drawing: Algorithms for the Visualization of Graphs. Prentice Hall PTR, Upper Saddle River (1998)
4. Baur, M., et al.: Visone software for visual social network analysis. In: International Symposium on Graph Drawing, pp. 463–464. Springer (2001). https://doi.org/10.1007/3-540-45848-4_47
5. Beck, F., Burch, M., Diehl, S., Weiskopf, D.: A taxonomy and survey of dynamic graph visualization. In: Computer Graphics Forum, vol. 36, pp. 133–159. Wiley Online Library, Hoboken (2017)
6. Böhringer, K.F., Paulisch, F.N.: Using constraints to achieve stability in automatic graph layout algorithms. In: Proceedings of the SIGCHI Conference on Human Factors in Computing Systems, pp. 43–51 (1990)
7. Brandes, U., Pich, C.: Eigensolver methods for progressive multidimensional scaling of large data. In: Kaufmann, M., Wagner, D. (eds.) GD 2006. LNCS, vol. 4372, pp. 42–53. Springer, Heidelberg (2007). https://doi.org/10.1007/978-3-540-70904-6_6
8. Branke, J.: Dynamic graph drawing. In: Kaufmann, M., Wagner, D. (eds.) Drawing Graphs. LNCS, vol. 2025, pp. 228–246. Springer, Heidelberg (2001). https://doi.org/10.1007/3-540-44969-8_9
9. David, A.: Tulip. In: Mutzel, P., Jünger, M., Leipert, S. (eds.) GD 2001. LNCS, vol. 2265, pp. 435–437. Springer, Heidelberg (2002). https://doi.org/10.1007/3-540-45848-4_34
10. Diehl, S., Görg, C.: Graphs, they are changing. In: Goodrich, M.T., Kobourov, S.G. (eds.) GD 2002. LNCS, vol. 2528, pp. 23–31. Springer, Heidelberg (2002). https://doi.org/10.1007/3-540-36151-0_3
11. Eades, P., Hong, S.H., Nguyen, A., Klein, K.: Shape-based quality metrics for large graph visualization. J. Graph Algorithms Appl. **21**(1), 29–53 (2017)
12. Eades, P., Lai, W., Misue, K., Sugiyama, K.: Preserving the mental map of a diagram. Technical report, Technical Report IIAS-RR-91-16E, Fujitsu Laboratories (1991)
13. Ellson, J., Gansner, E., Koutsofios, L., North, S.C., Woodhull, G.: Graphviz— open source graph drawing tools. In: Mutzel, P., Jünger, M., Leipert, S. (eds.) GD 2001. LNCS, vol. 2265, pp. 483–484. Springer, Heidelberg (2002). https://doi.org/10.1007/3-540-45848-4_57
14. Fowlkes, E.B., Mallows, C.L.: A method for comparing two hierarchical clusterings. J. Am. Stat. Assoc. **78**(383), 553–569 (1983). https://doi.org/10.1080/01621459.1983.10478008
15. Fruchterman, T.M.J., Reingold, E.M.: Graph drawing by force-directed placement. Softw. Pract. Experience **21**(11), 1129–1164 (1991). https://doi.org/10.1002/spe.4380211102
16. Gansner, E.R., Hu, Y., North, S.: A maxent-stress model for graph layout. IEEE Trans. Visual Comput. Graph. **19**(6), 927–940 (2012)
17. Gansner, E.R., Koren, Y., North, S.: Graph drawing by stress majorization. In: Pach, J. (ed.) GD 2004. LNCS, vol. 3383, pp. 239–250. Springer, Heidelberg (2005). https://doi.org/10.1007/978-3-540-31843-9_25

18. Hu, Y.: Efficient, high-quality force-directed graph drawing. Math. J. **10**(1), 37–71 (2005)
19. Huang, W., Hong, S.H., Eades, P.: Effects of crossing angles. In: 2008 IEEE Pacific Visualization Symposium, pp. 41–46. IEEE (2008)
20. Hubert, L., Arabie, P.: Comparing partitions. J. Classif. **2**(1), 193–218 (1985). https://doi.org/10.1007/BF01908075
21. Kruiger, J.F.: tsNET (2017). https://github.com/HanKruiger/tsNET/
22. Kruiger, J.F., Rauber, P.E., Martins, R.M., Kerren, A., Kobourov, S., Telea, A.C.: Graph layouts by t-SNE. Comput. Graph. Forum **36**(3), 283–294 (2017). https://doi.org/10.1111/cgf.13187
23. Maaten, L.V.D., Hinton, G.: Visualizing data using t-SNE. J. Mach. Learn. Res. **9**, 2579–2605 (2008)
24. MacQueen, J., et al.: Some methods for classification and analysis of multivariate observations. In: Proceedings of the fifth Berkeley Symposium on Mathematical Statistics and Probability, vol. 1, pp. 281–297. University of California Press (1967)
25. Madan, A., Cebrian, M., Moturu, S., Farrahi, K., et al.: Sensing the "health state" of a community. IEEE Pervasive Comput. **11**(4), 36–45 (2011)
26. Meidiana, A., Hong, S.H., Eades, P., Keim, D.: A quality metric for symmetric graph drawings. arXiv preprint arXiv:1910.04974 (2019)
27. Meidiana, A., Hong, S.-H., Eades, P., Keim, D.: A quality metric for visualization of clusters in graphs. In: Archambault, D., Tóth, C.D. (eds.) GD 2019. LNCS, vol. 11904, pp. 125–138. Springer, Cham (2019). https://doi.org/10.1007/978-3-030-35802-0_10
28. Meidiana, A., Hong, S.H., Eades, P., Keim, D.: Quality metrics for symmetric graph drawings. In: 2020 IEEE Pacific Visualization Symposium (PacificVis), pp. 11–15. IEEE (2020)
29. Nguyen, Q., Eades, P., Hong, S.H.: On the faithfulness of graph visualizations. In: 2013 IEEE Pacific Visualization Symposium (PacificVis), pp. 209–216. IEEE (2013)
30. Noack, A.: An energy model for visual graph clustering. In: Liotta, G. (ed.) GD 2003. LNCS, vol. 2912, pp. 425–436. Springer, Heidelberg (2004). https://doi.org/10.1007/978-3-540-24595-7_40
31. Nocaj, A., Ortmann, M., Brandes, U.: Untangling the hairballs of multi-centered, small-world online social media networks. J. Graph Algorithms Appl. **19**(2), 595–618 (2015). https://doi.org/10.7155/jgaa.00370
32. Ortmann, M., Klimenta, M., Brandes, U.: A sparse stress model. In: Hu, Y., Nöllenburg, M. (eds.) GD 2016. LNCS, vol. 9801, pp. 18–32. Springer, Cham (2016). https://doi.org/10.1007/978-3-319-50106-2_2
33. Pedregosa, F., et al.: Scikit-learn: Machine learning in Python. J. Mach. Learn. Res. **12**, 2825–2830 (2011)
34. Purchase, H.: Which aesthetic has the greatest effect on human understanding? In: DiBattista, G. (ed.) GD 1997. LNCS, vol. 1353, pp. 248–261. Springer, Heidelberg (1997). https://doi.org/10.1007/3-540-63938-1_67
35. Purchase, H.C., Cohen, R.F., James, M.: Validating graph drawing aesthetics. In: Brandenburg, F.J. (ed.) GD 1995. LNCS, vol. 1027, pp. 435–446. Springer, Heidelberg (1996). https://doi.org/10.1007/BFb0021827
36. Rand, W.M.: Objective criteria for the evaluation of clustering methods. J. Am. Stat. Assoc. **66**(336), 846–850 (1971). https://doi.org/10.1080/01621459.1971.10482356

37. Tamassia, R., Di Battista, G., Batini, C.: Automatic graph drawing and readability of diagrams. IEEE Trans. Syst. Man Cybern. **18**(1), 61–79 (1988). https://doi.org/10.1109/21.87055
38. Torgerson, W.S.: Multidimensional scaling: I. theory and method. Psychometrika **17**(4), 401–419 (1952). https://doi.org/10.1007/BF02288916
39. Tufte, E.R., Goeler, N.H., Benson, R.: Envisioning information, vol. 126. Graphics press Cheshire, CT (1990)

The Turing Test for Graph Drawing Algorithms

Helen C. Purchase[1] [ID], Daniel Archambault[2]([✉])[ID], Stephen Kobourov[3][ID],
Martin Nöllenburg[4][ID], Sergey Pupyrev[5][ID], and Hsiang-Yun Wu[4][ID]

[1] University of Glasgow, Glasgow, UK
Helen.Purchase@glasgow.ac.uk
[2] Swansea University, Swansea, UK
d.w.archambault@swansea.ac.uk
[3] University of Arizona, Tucson, USA
kobourov@cs.arizona.edu
[4] TU Wien, Vienna, Austria
noellenburg@ac.tuwien.ac.at, hsiang.yun.wu@acm.org
[5] Facebook, Menlo Park, USA
spupyrev@gmail.com

Abstract. Do algorithms for drawing graphs pass the Turing Test? That
is, are their outputs indistinguishable from graphs drawn by humans? We
address this question through a human-centred experiment, focusing on
'small' graphs, of a size for which it would be reasonable for someone to
choose to draw the graph manually. Overall, we find that hand-drawn
layouts can be distinguished from those generated by graph drawing algorithms, although this is not always the case for graphs drawn by force-
directed or multi-dimensional scaling algorithms, making these good candidates for Turing Test success. We show that, in general, hand-drawn
graphs are judged to be of higher quality than automatically generated
ones, although this result varies with graph size and algorithm.

Keywords: Empirical studies · Graph drawing algorithms · Turing
test

1 Introduction

It is common practice to use node-link diagrams when presenting graphs to
an audience (e.g., online, in an article, to support a verbal presentation, or for
educational purposes), rather than the alternatives of adjacency matrices or edge
lists. Automatic graph layout algorithms replace the need for a human to draw
graphs; it is important to determine how well these algorithms fulfil the task of
replacing this human activity,

Such algorithms are essential for creating drawings of large graphs; it is less
clear that this is the case for drawing smaller graphs. In our experience as graph
drawing researchers, it is often preferable to draw a small graph ourselves, how we
wish to depict it, than be beholden to the layout criteria of automatic algorithms.

© Springer Nature Switzerland AG 2020
D. Auber and P. Valtr (Eds.): GD 2020, LNCS 12590, pp. 466–481, 2020.
https://doi.org/10.1007/978-3-030-68766-3_36

The question therefore arises: are automatic graph layout algorithms any use for small graphs? Indeed, for small graphs, is it even possible to tell the difference? If automatic graph layout algorithms were doing their job properly for small graphs, then they should produce drawings not dissimilar to those we would choose to create by hand.

Distinguishing human and algorithmic graph drawings can be considered a 'Turing Test'; as in Turing's 1950 'Imitation Game' [44], if someone cannot tell the difference between machine output and human output more than half the time, the machine passes the Turing Test. Thus, if someone cannot tell the difference between an algorithmically-drawn graph and a hand-drawn graph more than half the time, the algorithm passes the Turing Test: it is doing as good a job as human graph drawers. Of course, algorithms are useful for non-experts and for large graphs that cannot be drawn by humans effectively, but in the context of experts presenting a small graph, can their creations be distinguished from products from layout algorithms? Turing Tests have never yet been performed on graph layout algorithms.

This paper presents the results of an experiment where participants were asked to distinguish between small hand-drawn graphs and those created by four common graph layout algorithms. Using different algorithms and graphs of different size allows us to investigate under what conditions an algorithm might pass the Turing Test. Our Turing Test results led us to also ask, in common with the *Non-photorealistic rendering Turing Test* observational study [30], which of the two methods of graph drawing (by hand, or by algorithm) produce better drawings. We find that distinguishing hand-drawn layouts from automatically generated ones depends on the type of the layout algorithm, and that subjectively determined quality depends on graph size and the type of the algorithm.

2   Related Work

2.1   Automatic Graph Layout Algorithms

We focus on four popular families of layout algorithms [13,25]: force-directed, stress-minimisation, circular and orthogonal.

Most general-purpose graph layout algorithms use a force-directed (FD) [15, 19] or stress model [12,34]. FD works well for small graphs, but does not scale for large networks. Techniques to improve scalability often involve multilevel approaches, where a sequence of progressively coarser graphs is extracted from the graph, followed by a sequence of progressively finer layouts, ending with a layout for the entire graph [8,21,26,28,29].

Stress minimisation, introduced in the general context of multi-dimensional scaling (MDS) [36] is also frequently used to draw graphs [31,35]. Simple stress functions can be optimised by exploiting fast algebraic operations such as majorisation. Modifications to the stress model include the strain model (classical scaling) [43], PivotMDS [12], COAST [22], and MaxEnt [23].

Circular layout algorithms [41] place nodes evenly around a circle with edges drawn as straight lines. Layout quality (in particular the number of crossings) is influenced by the order of the nodes on the circle. Crossing minimisation in circular layouts is NP-hard [37], and various heuristics attempt to find good vertex orderings [9, 24, 33].

The orthogonal drawing style [16] is popular in applications requiring a clean and schematic appearance (e.g., in software engineering or database schema). Edges are drawn as polylines of horizontal and vertical segments only. Orthogonal layouts have been investigated for planar graphs of maximum degree four [42], non-planar graphs [10] and graphs with nodes of higher degree [11, 20].

We seek to understand if drawings produced by these types of algorithms can be distinguished from human-generated diagrams for small networks. We do this by asking experimental participants to identify the hand-drawn layout when it is paired with an algorithmically-created one.

2.2 Studies of Human-Created Graph Layouts

Early user studies [38, 39] confirmed that many of the aesthetic criteria incorporated in layout algorithms (e.g., uniform edge length, crossing minimisation) correlate with user performance in tasks such as path finding. Van Ham and Rogowitz [27] investigated how humans modified given small graph layouts so as to represent the structure of these graphs. They found that force-directed layouts were already good representations of human vertex distribution and cluster separation. Dwyer et al. [14] focused on the suitability of graph drawings for four particular tasks (identifying cliques, cut nodes, long paths and nodes of low degree), and found that the force-based automatic layout received the highest preference ratings, but the best manual drawings could compete with these layouts. Circular and orthogonal layouts were considerably less effective. Purchase et al. [40] presented graph data to participants as adjacency lists and asked them to create drawings by sketc.hing; their findings include that the participants preferred planar layouts with straight-line edges (except for some non-straight edges in the outer face), nodes aligned with an (invisible) grid, and somewhat similar edge lengths. Kieffer et al. [32] focused on orthogonal graph layouts, asking participants to draw a few small graphs (13 or fewer nodes) orthogonally by hand. The human drawings were compared to orthogonal layouts generated by yEd [46] and the best human layouts were consistently ranked better than automatic ones. They then developed an algorithm for creating human-like orthogonal drawings.

This paper builds on this prior work by considering drawings of small to medium-sized graphs (up to 108 nodes) and an example from each of four families of standard graph layout algorithms. We address the question of whether people can distinguish between algorithmic and human created drawings, and if so, is this the case for all layout algorithms?

Table 1. Characteristics of the experimental graphs. The *size* column indicates how the graphs were divided into sub-sets (small, medium, large) for the purposes of the experiment; (rw): real-world graphs; (ab): abstract graphs.

Graph	Nodes	Edges	Density	Mean shortest path	Clustering coefficient	Diam.	Planar	Size	Reference
G_1(rw)	108	156	0.03	5.03	0.11	11	N	L	Causes of obesity [7]
G_2(rw)	22	164	0.71	1.30	0.78	2	N	S	Causes of social problems in Alberta, Canada [4]
G_3(rw)	85	104	0.03	6.05	0.04	13	Y	L	Cross posting users on a newsgroup (final timeslice) [18]
G_4(rw)	34	77	0.14	2.45	0.48	5	N	M	Social network [47]
G_5(ab)	20	30	0.16	2.63	0.00	5	Y	S	Fullerene graph with 20 nodes [3]
G_6(ab)	24	38	0.14	3.41	0.64	6	N	S	A block graph (chordal, every biconnected component is a clique) [2]
G_7(ab)	42	113	0.13	2.55	0.48	5	Y	M	A maximal planar graph [6]
G_8(ab)	37	71	0.11	2.76	0.70	5	Y	M	A planar 2-tree [5]
G_9(ab)	18	27	0.18	2.41	0.00	4	N	S	Pappus graph (bipartite, 3-regular) [1]
Mean	43.3	86.7	0.18	3.18	0.36	6.2			
Median	34	77	0.14	2.63	0.48	5			

3 Experiment

3.1 Stimuli

The Graphs. Our experiment compares unconstrained hand-drawn graphs with the same graphs laid out using different algorithmic approaches. We considered 24 graphs, from which we selected 9, based on the following criteria:

- A balanced split between real-world graphs and abstract graphs, the abstract graphs being ones of graph-theoretic interest;
- A balanced split between planar and non-planar graphs;
- A range in the number of nodes between 15 and 108;
- A range in the number of edges (for our graphs, between 27 and 164);
- Connected and undirected graphs only: directionality was removed from the real-world graphs as necessary.

The diversity of our graphs is demonstrated by the range of values for other graph characteristics (diameter, density, average shortest path length, clustering coefficient) that they exhibit (Table 1).

The Algorithms. We included examples of major families of graph drawing algorithms (Table 2: force-directed, stress-based, circular, orthogonal), as implemented in yEd [46] and GraphViz [17]. HOLA [32] was considered, but its orthogonal design was deliberately based on human preferences (unlike the other algorithms), and so its inclusion would introduce a bias that could distort human judgements. We considered structure-specific algorithms (e.g., algorithms designed for planar graphs or trees), but for generality used generic algorithms that could handle all nine graphs, leaving specific algorithms for future work.

Table 2. The four graph layout algorithms used.

Algorithm ID	Algorithm type	Original name	Parameters
A_{FD}	Force-directed	Organic [46]	Default
A_{MDS}	Stress-based	MDS [17]	Default
A_C	Circular	Circular [46]	Default
A_O	Orthogonal	Orthogonal [46]	Classic, default

The Hand-Created Drawings. The process of creating hand-drawn graphs mimicked the context of a graph drawing researcher deciding whether to manually draw a small graph, or to use a well-established graph layout algorithm. Thus, the graphs were drawn in the knowledge they would compete against drawings created by algorithms, making the Turing test as hard as possible. This process was therefore a mini-experiment, with four of the authors (all with graph drawing expertise, called the 'drawers', D_1-D_4) as participants, the context of the study being clear to them. While the drawers might have recognised some of the graphs they were asked to draw, this scenario is comparable to a real-world situation where graph drawing researchers might know the nature of the graph to be drawn.

The first author asked the drawers to lay out the graphs using yEd [46], starting from a random layout (the yEd 'Random' tool). There were no other instructions: it was not specified, for example, that edges needed to be straight lines rather than splines or multiple segments, nor that nodes should not overlap, nor edges cross over nodes. To improve ecological validity, all drawers were told that they could use yEd tools to support their drawing process if they wished (as likely to happen in practice). However, somewhat surprisingly, they all drew the graphs without any yEd tool support (automatic layout or otherwise). The drawers suggested doing the exercise again on a 'manually-adjusted' basis; that is, using the output from a yEd layout algorithm of their choice as an initial starting point. However, once we paired the algorithmic drawings with their manually-adjusted versions, most of them were visually almost identical. We therefore only used the initial hand-drawn versions.

The mini-experiment output is a set of visual stimuli comprising 9 graphs $(G_1, ..., G_9)$, each with four layout algorithms applied $G_1 A_{FD}$, $G_1 A_{MDS}$, $\ldots$, $G_2 A_C$, $\ldots$, $G_9 A_O$) and each with four hand-drawn versions $(G_1 D_1, G_1 D_2, \ldots, G_2 D_1, \ldots, G_9 D_4)$, all represented in yEd. All 72 drawings were subject to the same automatic scaling process to ensure the same vertex size and edge thickness. After scaling, all drawings were automatically converted into jpeg images.

3.2   Experimental Design

Each experimental trial (Fig. 1) comprises two versions of the same graph, one hand-drawn, and one created by a layout algorithm. For each graph, we firstly paired the four algorithmic versions (on the left) with the four hand-drawn ver-

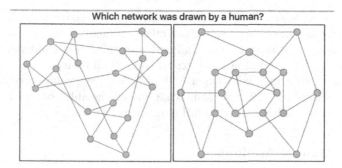

Fig. 1. Screen shot of the experimental system.

sions (right) (16 pairs). We then flipped the algorithmic versions along the y axis (reducing the possibility of participants remembering the algorithm drawings), and paired the flipped versions (right) with the four drawn versions (left) (32 pairs for each graph). Putting all graphs in one experiment means 288 trials, an unreasonably long experiment. The alternative of running a separate experiment for each graph means several very small experiments, greatly increasing the number of participants needed. As a compromise, we divided our 9 graphs into three sets, (loosely 'small', 'medium' and 'large' (Table 1)), a convenience decision so as to reduce the duration of each experiment while ensuring we would be able to recruit enough participants. We thus had three sub-experiments, one 'small' (128 trials), one 'medium' (96 trials) and one 'large' (64 trials).

Using a custom-built online experimental system, participants read instructions and information about graphs (referred to as 'networks') and indicated consent before proceeding. They were told it would always be the case that the two drawings presented were the same graph. Twelve practice trials used a different graph of similar size for familiarisation purposes. Experimental trials were presented in random order, with no distinction between graphs. Participants took a self-timed break every 20 trials, and demographic data was collected.

4   Results and Data Analysis

The experimental link was distributed to authors' colleagues, students, family and friends. Participants were considered outliers if their mean time over all trials was unreasonably low (less than 1 s, $n = 2$), or if they consistently responded one side for a large number of consecutive trials (e.g., always left, $n = 1$). No participants consistently alternated left and right. We removed the data from one participant who used a very small screen (198×332 pixels), unconvinced that the stimuli could be perceived sufficiently well. Although some participants did not complete the experiment, since the answer to each trial is a data point in its own right (i.e., it is independent and its value to the experiment does not depend on answers to any other trial), we retained all data for participants who completed at least 3/4 of the trials, inferring that those who did not do so ($n = 20$) might not have taken the experiment seriously.

Data from 46 participants was analysed; a total of 4364 independent decisions. We categorised participants as expert ($n = 21$) if their self-declared knowledge of network drawings was 'expert', 'highly knowledgeable', or 'knowledgeable', and novice ($n = 22$) for 'somewhat knowledgeable' or 'no knowledge'. Three participants did not provide full demographic details.

4.1 Data Analysis Methods

Our data was analysed in three parts: Part 1 investigates the extent to which 'human' was chosen over 'algorithm', comparing the proportion of responses with random selection. We look at overall responses, responses for each algorithm, for each graph size, for novice and expert participants, for planar and non-planar graphs, and consider the combination of graph size and algorithm. The Binomial distribution test compares observed proportion against the 'random' proportion of 0.5, where each trial is independent; its calculated p-value represents the probability that the mean of the population distribution (based on the observed samples) is equal to 0.5. A p-value < 0.05 indicates a significant result: that is, the observed choice proportion is so much greater than 0.5 that there is a very low probability that the hand-drawn and algorithmically drawn graphs cannot be distinguished; statistically, this means there is insufficient evidence to indicate Turing Test success. A p-value > 0.05 is a high probability that hand-drawn and algorithmically drawn graphs cannot be distinguished: thus, Turing Test success. We apply p-value Bonferroni corrections when dividing the data sets.

Part 2 considers response times with respect to different algorithms, sizes, expertise, and planarity, using non-parametric tests since our data is not normally distributed. Response time is considered as a proxy for the perception of difficulty of the task: participants will take longer if they find the task difficult.

Part 3 identifies trials with extreme responses (high or low response time, or extreme proportional choice).

A choice for a hand-drawn graph is scored as 1; a choice for an algorithmic drawings is 0. Thus, proportions > 0.5 indicate that the human drawing was selected more often on average. Proportions < 0.5 indicate that the algorithmic drawing was (incorrectly) selected with greater frequency.

4.2 Results

Choice of Drawing. Our hypotheses are:

- H_0: It is not possible to distinguish algorithmic drawings from hand-drawn ones; thus, the true proportion $= 0.5$; the algorithm passes the Turing test. This hypothesis is accepted if the Binominal p-value > 0.05.
- H_1: It is possible to distinguish algorithmic drawings from hand-drawn ones; thus, the true proportion $\neq 0.5$.

Table 3. Binomial test results for 'Which network was drawn by a human?' Accepting H_0 indicates Turing Test 'pass'. Although $0.049 < 0.05$, statistical correction means the MDS p-value threshold is $0.05/4 = 0.0125$. The corrected Novice p-value threshold is $0.05/2 = 0.025$, a significant result.

	Number of samples	Mean response time (s)	Observed proportion	Binomial p-value	Result
All trials	4364	3.14	0.56	$p < 0.001$	Reject H_0
Force-Directed (A_{FD})	1094	4.26	0.51	$p = 0.566$	Accept H_0
MDS (A_{MDS})	1090	3.32	0.53	$p = 0.049$	Reject H_0
Circular (A_C)	1090	2.85	0.56	$p < 0.001$	Reject H_0
Orthogonal (A_O)	1090	2.79	0.65	$p < 0.001$	Reject H_0
Small graphs (G_2, G_5, G_6, G_9)	1656	2.58	0.55	$p < 0.001$	Reject H_0
Medium graphs (G_4, G_7, G_8)	1817	3.08	0.55	$p < 0.001$	Reject H_0
Large graphs (G_1, G_3)	891	4.28	0.62	$p < 0.001$	Reject H_0
Expert participants	1915	3.99	0.63	$p < 0.001$	Reject H_0
Novice participants	2101	2.74	0.53	$p = 0.016$	Reject H_0
Planar graphs	2069	3.15	0.55	$p < 0.001$	Reject H_0
Non-planar graphs	2295	3.49	0.58	$p < 0.001$	Reject H_0

Binomial test results over all 4364 data points are shown in Table 3. Accepting H_0 means it is not possible to distinguish between hand-drawn and algorithmic drawings: the Turing Tests succeeds. Rejecting it means that there is insufficient support for the hypothesis; we infer that telling the difference is possible. There are no proportions < 0.5, so no cases where, on average, algorithmically-drawn graphs were incorrectly selected more often than hand-drawn ones.

Table 4. Binomial test results by graph size and algorithm; * indicates responses sufficiently close to random for Turing Test 'pass'.

	Force-Directed		MDS		Circular		Orthogonal	
	Proportion	P-value	Proportion	P-value	Proportion	P-value	Proportion	P-value
Small	0.52*	0.432	0.57*	0.006	0.51*	0.786	0.62	< 0.001
Medium	0.49*	0.851	0.52*	0.542	0.53*	0.205	0.64	< 0.001
Large	0.52*	0.640	0.49*	0.789	0.73	< 0.001	0.74	< 0.001

The results indicate that people can distinguish between algorithmic and hand-drawn graphs (over all graphs and algorithms), correctly choosing the hand-drawn graph 56% of the time ($p < 0.001$). This result applies equally well regardless of graph size, viewer expertise, or graph planarity: the tests all reveal significant difference between the observed proportion and 0.5. Thus, overall, the Turing test fails.

There is a difference, however, when the algorithm is taken into account: the observed proportion for Force-Directed algorithm trials was 0.51, sufficiently close to the random response proportion of 0.50 that we can accept H_0, and state that this algorithm passes the Turing Test. The proportion of 0.53 for MDS is very close (but not really close enough in statistical terms), and we clearly reject H_0 for circular and orthogonal algorithms.

The size/algorithm combination (threshold p-value = 0.05/12 = 0.0042) reveals additional results according to the size of the graph (Table 4). As expected, the Force-Directed algorithm gives proportions close to 0.5 for all graph sizes. The MDS results suggest Turing Test success for all three sizes when analysed separately (albeit a marginal result for the smallest graphs), even though the overall MDS result reported above (at $p = 0.049$) indicates rejection of the null hypothesis. The MDS result is therefore clearly on the boundary of success. There are Turing Test passes for small and medium graphs for the Circular algorithm.

Response Time. Non-parametric tests on response time for algorithm and graph size (Table 3) reveals that MDS decisions were slower than orthogonal ones (adjusted pairwise comparison after repeated measures Freidman, $p = 0.022$), decisions on large graphs were slower than on small graphs (adjusted pairwise comparison after independent measures Kruskal Wallis, $p = 0.039$), and experts made slower decisions than novices (independent measures Mann-Whitney, $p = 0.014$). There was no statistical difference between response times with respect to graph planarity.

Extreme Examples. Extreme trials (response time: Fig. 2; proportion: Fig. 3) are identified as $G_i A_j$ and $G_i D_k$: G_i (graph), A_j (algorithm), D_k (drawer). All experimental stimuli jpeg files can be found in the supplementary material (visit http://www.dcs.gla.ac.uk/~hcp/GD2020).

Three slow trials relate to a particular FD graph, suggesting that this form of drawing was seen by participants as possibly hand-drawn – it shows clusters and symmetry, while the drawers all attempted to remove crosses. The combinations of $G_4 A_{\mathrm{MDS}}/G_4 D_4$ and $G_7 A_C/G_7 D_4$ (top row of Fig. 2) are interesting because, for each, the overall shape of the human-drawn graph is similar to that produced by the algorithm: it is not hard to see why participants found this choice difficult. Three quick responses ($G_5 A_{\mathrm{FD}}/G_5 D_3$, $G_5 A_C/G_5 D_4$, $G_9 A_{\mathrm{MDS}}/G_9 D_1$, bottom row of Fig. 2) demonstrate effort on the part of the drawer to depict symmetry that is not highlighted by the algorithms; the other two relate to the orthogonal algorithm, which, as noted above, produced worst performance in making a human *vs* algorithm judgements.

Of the four combinations where participants gave mostly correct responses, it is not hard to see why for $G_1 A_C/G_1 D_2$ and $G_1 A_C/G_1 D_1$ (top row of Fig. 3), since the human-drawn graphs lack any clear structure or visual elegance in comparison with those created by the circular algorithm. The fact that $G_5 A_{\mathrm{MDS}}$ is geometrically precise in its node positioning (while $G_5 D_2$ has slight mispositionings) can explain the 0.92 accuracy for this combination, although we note that this decision still took above average time (32.4 s). More difficult to explain is the high proportion associated with $G_6 A_{\mathrm{FD}}/G_6 D_3$, since the human

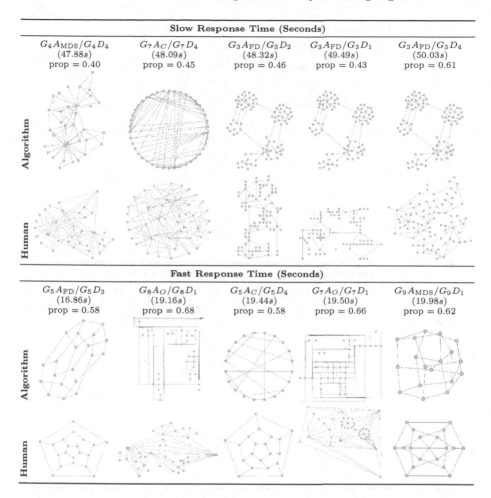

Fig. 2. Trials with slow response times (top) and quick response times (bottom). Time in seconds, and human-selection proportion shown.

drawing is highly structured and symmetrical. Of the combinations where the average accuracy is low, three algorithmic drawings depict some extent of symmetry (G_3A_{MDS}, G_9A_C, G_5A_{FD}, bottom row of Fig. 3), while the fourth is compared against a human drawing which used an approach that, if adopted by an algorithm, would have resulted in a more geometrically precise diagram. The examples in Fig. 3 (top and bottom rows) suggest that regular node and edge placements (that is, grid-like or evenly spaced on a circle), indicate an algorithmically-drawn graph.

Key factors affecting the human vs algorithm choice were thus depiction of symmetry (even if only approximate), and geometric precision (i.e. very precise node placement, with regular spacing or grid-like).

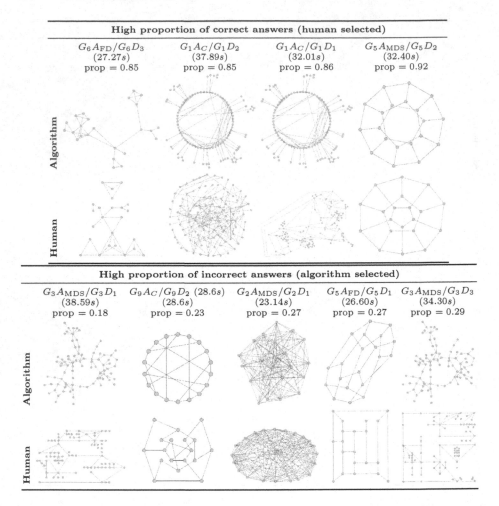

Fig. 3. Trials with a high proportion of correct (human drawing chosen, upper) and incorrect (algorithm drawing chosen, lower) answers.

5   Discussion

In general, over all graphs and algorithms, participants can correctly distinguish hand-drawn layouts from algorithmically created ones: graph drawing algorithms (in general) effectively fail the Turing Test. The only exception is the Force-Directed algorithm, where we did not find evidence that participants could reliably distinguish between the algorithmic and hand-drawn layouts. We speculate this might be because our drawers (consciously or unconsciously) created drawings with similar FD layout principles in mind: separating unconnected nodes, and clustering connected ones together. The MDS algorithm provided some evidence of passing the test (in particular for medium and large graphs); it produces similar shapes to FD.

Table 5. Results for the 'Which is better' question, by graph size and algorithm. * indicates statistically significant results ($p < 0.05/12 = 0.0042$)

	Force-Directed		MDS		Circular		Orthogonal	
	Proportion	P-value	Proportion	P-value	Proportion	P-value	Proportion	P-value
Small	0.83*	< 0.001	0.68*	< 0.001	0.55	0.040	0.62*	< 0.001
Medium	0.44	0.006	0.42*	0.001	0.62*	< 0.001	0.74*	< 0.001
Large	0.19*	< 0.001	0.42	0.009	0.41*	0.002	0.63*	< 0.001

We were not surprised that it was easy to distinguish circular (especially large circular) and orthogonal graph drawings from hand-drawn ones, since they make use of precise node placement: equal separation around the circle circumference, placement on equally-spaced horizontal lines or on an underlying unit grid. While the human drawers sometimes used such placements (G_2D_1 and G_5D_1 in Fig. 3), in many cases (G_8D_1 in Fig. 2, G_5D_2 in Fig. 3) they did not. We were also not surprised to find that larger graphs took more time than the smaller ones, but were surprised that experts took longer than novices – we had expected the converse; perhaps experts made more considered analytical decisions as opposed to novices' more spontaneous ones.

6   Subjective Quality of the Drawings

Our study shows that some graph drawing algorithms produce diagrams that are obviously perceived as different from those drawn by graph drawing experts. This raises the question: if algorithmic drawings are perceived as being different from hand-drawn ones, are they any better? And even if they are not perceived as different, is there a perceived difference in quality?

We followed our Turing experiment with a supplementary, almost identical study, using the same paired stimuli and experimental system. The only difference was the question asked: 'Which drawing is better?'. We deliberately did not give a definition for 'better', since (at least for this initial study), we wished to get an overall judgement, rather than, for example, one based on a particular task or defined aesthetic. 52 participants took part, producing a total of 4887 data points. As before, hand-drawn graphs are scored 1, and algorithmic drawings 0. Thus, proportions > 0.5 indicate the human drawing was, on average, considered better. Over all graphs and algorithms, the vote was for hand-drawn graphs (proportion $= 0.57$, $p < 0.001$). However, size and algorithm data show variations within this overall result (Table 5). Hand-drawn graphs were always preferred over orthogonal drawings; FD and MDS were preferred for medium and large graphs, and circular only for the large graphs.

Thus, even when hand-drawn and algorithmic drawings are indistinguishable (as shown for FD and MDS in the first experiment), subjective judgement (experiment 2) determines that the algorithmic versions are 'better', especially for the larger graphs. The orthogonal algorithm had no wins: it did not pass the Turing Test, and was always considered worse than the hand-drawn versions.

There were mixed results for the circular algorithm: easy to distinguish from hand-drawn layouts when small or medium, and only preferred when large.

7   Conclusions and Future Work

This is the first experiment that compares graphs drawn by graph drawing researchers to those produced by graph drawing algorithms as a Turing Test. Overall, we found that hand-drawn graphs could be reliably distinguished from those generated by algorithms – thus, on average, Turing Test failure. However, we did not find evidence that force-directed and (marginally) MDS algorithms could be reliably distinguished from hand-drawn layouts – they therefore effectively 'pass' the Turing Test. We speculate that this is the case because of the prevalence of these algorithms in the popular media (e.g., for depicting social networks); further studies could establish exactly why these two algorithms perform differently from the others.

The generalisability of our conclusions is, of course, limited by our experimental scope. While we used a good range of real-world and abstract graphs, differently sized graphs, planar and non-planar graphs, and good coverage of various graph metrics, our data set comprises nine experimental graphs. Using only 'small' graphs (15 to 108 nodes) was an obvious design decision when considering the feasibility of creating hand-drawn layouts. We chose four common layout algorithms representing different approaches, and four human drawers (experts in graph drawing research). Despite these experimental limitations, our results represent the first empirical attempt to compare perception of a range of hand-drawn versus algorithmic graph layouts as a 'Turing Test'.

Our motivation for these studies arose from a desire to determine whether algorithms depicting small graphs produce results that are similar to human efforts. Our results show that, in general, people notice when a graph has been hand-drawn. This result must, of course, be weighed against the length of time that it takes to draw a graph: we found that it takes much longer than we had anticipated to create drawings by hand. We also need to consider that, when considering the algorithmic approaches separately, some algorithmic versions were considered 'better' than the hand-drawn ones – the notable exception being the orthogonal algorithm.

Graph drawing algorithms are often inspired by assumptions about what a human would do in generating a drawing. Therefore, understanding what makes a drawing human-like will help inform future algorithm designers to make algorithms of higher quality. In future work, we would like to explore whether we get similar results if we explicitly match graph structure with graph algorithm (e.g., tree algorithms for trees, planar algorithms for planar graphs), use other less common algorithms (e.g., HOLA [32], Wang et al. [45]), and use graphs drawn by a wider range of people (including non-experts). In addition, gathering both quantitative and qualitative data in future studies will help determine those attributes of a graph drawing that suggest that it is human-like or machine-like.

Acknowledgement. We are grateful to all the experimental participants, to Drew Sheets who assisted with creating the graphs in yEd, and to John Hamer who implemented the online experimental system. Ethical approval was given by the University of Arizona Institutional Review Board (ref: 1712113015). This work is supported by NSF grants CCF-1740858, CCF-1712119, DMS-1839274, and FWF grant P 31119.

References

1. Bipartite 3-regular undirected graph (2019). http://reference.wolfram.com/language/ref/GraphData.html. Accessed 8 Aug 2019
2. Block graph. https://en.wikipedia.org/wiki/Block_graph (2019). Accessed 8 Aug 2019
3. Fullerene graphs. https://commons.wikimedia.org/wiki/Fullerene_graphs/ (2019). Accessed 8 Aug 2019
4. G_2. http://haikuanalytics.com/portfolio/ (2019). Accessed 8 Aug 2019
5. K-tree. http://citeseerx.ist.psu.edu/viewdoc/summary?doi=10.1.1.71.3875 (2019). Accessed 8 Aug 2019
6. Planar graph. https://arxiv.org/pdf/1707.08833.pdf (2019), page 18. Accessed 8 Aug 2019
7. Allender, S., et al.: A community based systems diagram of obesity causes. PLoS ONE **10**(7), e0129683 (2015). https://doi.org/10.1371/journal.pone.0129683
8. Bartel, G., Gutwenger, C., Klein, K., Mutzel, P.: An experimental evaluation of multilevel layout methods. In: Brandes, U., Cornelsen, S. (eds.) Proceedings of Graph Drawing (GD 2010). LNCS, vol. 6502, pp. 80–91 (2011). https://doi.org/10.1007/978-3-642-18469-7_8
9. Baur, M., Brandes, U.: Crossing reduction in circular layouts. In: Hromkovič, J., Nagl, M., Westfechtel, B. (eds.) Graph-Theoretic Concepts in Computer Science (WG 2004). LNCS, vol. 3353, pp. 332–343. Springer, Heidelberg (2004). https://doi.org/10.1007/978-3-540-30559-0_28
10. Biedl, T., Kant, G.: A better heuristic for orthogonal graph drawings. Comput. Geom. Theory Appl. **9**(3), 159–180 (1998). https://doi.org/10.1016/S0925-7721(97)00026-6
11. Biedl, T.C., Madden, B.P., Tollis, I.G.: The three-phase method: a unified approach to orthogonal graph drawing. Int. J. Comput. Geom. Appl. **10**(6), 553–580 (2000). https://doi.org/10.1142/S0218195900000310
12. Brandes, U., Pich, C.: Eigensolver methods for progressive multidimensional scaling of large data. In: Proceedings of Graph Drawing. LNCS, vol. 4372, pp. 42–53. Springer (2007). https://doi.org/10.1007/978-3-540-70904-6_6
13. Di Battista, G., Eades, P., Tamassia, R., Tollis, I.G.: Graph Drawing: Algorithms for the Visualization of Graphs. 1st edn. Prentice Hall, Upper Saddle River (1998)
14. Dwyer, T., et al.: A comparison of user-generated and automatic graph layouts. IEEE Trans. Vis. Comput. Graph. **15**(6), 961–968 (2009). https://doi.org/10.1109/TVCG.2009.109
15. Eades, P.: A heuristic for graph drawing. Congressus Numerantium **42**, 149–160 (1984)
16. Eiglsperger, M., Fekete, S.P., Klau, G.W.: Orthogonal graph drawing. In: Kaufmann, M., Wagner, D. (eds.) Drawing Graphs: Methods and Models, LNCS, vol. 2025, chap. 6, pp. 121–171. Springer-Verlag (2001). https://doi.org/10.1007/3-540-44969-8_6

17. Ellson, J., Gansner, E.R., Koutsofios, E., North, S.C., Woodhull, G.: Graphviz - open source graph drawing tools. In: Proceedings of Graph Drawing (GD 2001). LNCS, vol. 2265, pp. 483–484 (2002). https://doi.org/10.1007/3-540-45848-4_57

18. Frishman, Y., Tal, A.: Online dynamic graph drawing. IEEE Trans. Vis. Comput. Graph. **14**(4), 727–740 (2008). https://doi.org/10.1109/TVCG.2008.11

19. Fruchterman, T.M.J., Reingold, E.M.: Graph drawing by force directed placement. Softw. Pract. Experience **21**, 1129–1164 (1991). https://doi.org/10.1002/spe.4380211102

20. Fößmeier, U., Kaufmann, M.: Algorithms and area bounds for nonplanar orthogonal drawings. In: Di Battista, G. (ed.) Proceedings of Graph Drawing (GD1997). LNCS, vol. 1353, pp. 134–145. Springer (1997). https://doi.org/10.1007/3-540-63938-1_57

21. Gajer, P., Kobourov, S.: GRIP: graph dRawing with intelligent placement. J. Graph Algorithms Appl. **6**(3), 203–224 (2002). https://doi.org/10.7155/jgaa.00052

22. Gansner, E.R., Hu, Y., Krishnan, S.: COAST: a convex optimization approach to stress-based embedding. In: Wismath, S., Wolff, A. (eds.) GD 2013. LNCS, vol. 8242, pp. 268–279. Springer, Cham (2013). https://doi.org/10.1007/978-3-319-03841-4_24

23. Gansner, E.R., Hu, Y., North, S.: A maxent-stress model for graph layout. IEEE Trans. Vis. Comput. Graph. **19**(6), 927–940 (2013). https://doi.org/10.1109/TVCG.2012.299

24. Gansner, E.R., Koren, Y.: Improved circular layouts. In: Kaufmann, M., Wagner, D. (eds.) GD 2006. LNCS, vol. 4372, pp. 386–398. Springer, Heidelberg (2007). https://doi.org/10.1007/978-3-540-70904-6_37

25. Gibson, H., Faith, J., Vickers, P.: A survey of two-dimensional graph layout techniques for information visualisation. Inf. Vis. **12**(3–4), 324–357 (2013). https://doi.org/10.1177/1473871612455749

26. Hadany, R., Harel, D.: A multi-scale algorithm for drawing graphs nicely. Discrete Appl. Math. **113**(1), 3–21 (2001). https://doi.org/10.1016/S0166-218X(00)00389-9

27. van Ham, F., Rogowitz, B.E.: Perceptual organization in user-generated graph layouts. IEEE Trans. Vis. Comput. Graph. **14**(6), 1333–1339 (2008). https://doi.org/10.1109/TVCG.2008.155

28. Harel, D., Koren, Y.: A fast multi-scale method for drawing large graphs. J. Graph Algorithms Appl. **6**, 179–202 (2002). https://doi.org/10.7155/jgaa.00051

29. Hu, Y.: Efficient, high-quality force-directed graph drawing. Math. J. **10**(1), 37–71 (2006)

30. Isenberg, T., Neumann, P., Carpendale, S., Sousa, M.C., Jorge, J.A.: Non-photorealistic rendering in context: an observational study. In: Non-Photorealistic Animation and Rendering (NPAR 2006), pp. 115–126. ACM (2006). https://doi.org/10.1145/1124728.1124747

31. Kamada, T., Kawai, S.: An algorithm for drawing general undirected graphs. Inf. Process. Lett. **31**(1), 7–15 (1989). https://doi.org/10.1016/0020-0190(89)90102-6

32. Kieffer, S., Dwyer, T., Marriott, K., Wybrow, M.: HOLA: human-like orthogonal network layout. IEEE Trans. Vis. Comput. Graph. **22**(1), 349–358 (2016). https://doi.org/10.1109/TVCG.2015.2467451

33. Klawitter, J., Mchedlidze, T., Nöllenburg, M.: Experimental evaluation of book drawing algorithms. In: Frati, F., Ma, K.-L. (eds.) GD 2017. LNCS, vol. 10692, pp. 224–238. Springer, Cham (2018). https://doi.org/10.1007/978-3-319-73915-1_19

34. Koren, Y., Carmel, L., Harel, D.: ACE: A fast multiscale eigenvectors computation for drawing huge graphs. In: Proceedings of IEEE Symposium on Information Visualization, pp. 137–144. IEEE (2002). https://doi.org/10.1109/INFVIS.2002. 1173159

35. Kruskal, J.B., Seery, J.B.: Designing network diagrams. In: General Conference on Social Graphics, vol. 49, p. 22 (1980)

36. Kruskal, J.B., Wish, M.: Multidimensional Scaling. Sage Press, Thousand Oak (1978)

37. Masuda, S., Nakajima, K., Kashiwabara, T., Fujisawa, T.: Crossing minimization in linear embeddings of graphs. IEEE Trans. Comput. **39**(1), 124–127 (1990). https:// doi.org/10.1109/12.46286

38. Purchase, H.C., Cohen, R.F., James, M.: Validating graph drawing aesthetics. In: Brandenburg, F.J. (ed.) GD 1995. LNCS, vol. 1027, pp. 435–446. Springer, Heidelberg (1996). https://doi.org/10.1007/BFb0021827

39. Purchase, H.: Which aesthetic has the greatest effect on human understanding? In: DiBattista, G. (ed.) GD 1997. LNCS, vol. 1353, pp. 248–261. Springer, Heidelberg (1997). https://doi.org/10.1007/3-540-63938-1_67

40. Purchase, H.C., Pilcher, C., Plimmer, B.: Graph drawing aesthetics created by users, not algorithms. IEEE Trans. Vis. Comput. Graph. **18**(1), 81–92 (2012). https://doi.org/10.1109/TVCG.2010.269

41. Six, J.M., Tollis, I.G.: Circular drawing algorithms. In: Tamassia, R. (ed.) Handbook of Graph Drawing and Visualization, chap. 9, pp. 285–315. CRC Press, Boca Raton (2013)

42. Tamassia, R.: On embedding a graph in the grid with the minimum number of bends. SIAM J. Comput. **16**(3), 421–444 (1987). https://doi.org/10.1137/0216030

43. Torgerson, W.S.: Multidimensional scaling: I. theory and method. Psychometrika **17**(4), 401–419 (1952). https://doi.org/10.1007/BF02288916

44. Turing, A.: Computing machinery and intelligence. Mind **59**(236), 433–460 (1950). https://doi.org/10.1093/mind/LIX.236.433

45. Wang, Y., et al.: Revisiting stress majorization as a unified framework for interactive constrained graph visualization. IEEE Trans. Vis. Comput. Graph. **24**(1), 489–499 (2018). https://doi.org/10.1109/TVCG.2017.2745919

46. Wiese, R., Eiglsperger, M., Kaufmann, M.: yFiles: visualization and automatic layout of graphs. In: Proceedings of Graph Drawing (GD 2001), pp. 453–454. LNCS (2002). https://doi.org/10.1007/978-3-642-18638-7_8

47. Zachary, W.W.: An information flow model for conflict and fission in small groups. J. Anthropol. Res. **33**, 452–473 (1977). https://doi.org/10.1086/jar.33.4.3629752, https://www.cise.ufl.edu/research/sparse/matrices/Newman/karate.html

Plane Spanning Trees in Edge-Colored Simple Drawings of K_n

Oswin Aichholzer[1], Michael Hoffmann[2], Johannes Obenaus[3],
Rosna Paul[1], Daniel Perz[1], Nadja Seiferth[3], Birgit Vogtenhuber[1],
and Alexandra Weinberger[1][(✉)]

[1] Institute of Software Technology, Graz University of Technology, Graz, Austria
{oaich,ropaul,daperz,bvogt,aweinber}@ist.tugraz.at
[2] Department of Computer Science, ETH Zürich, Zürich, Switzerland
hoffmann@inf.ethz.ch
[3] Institut für Informatik, Freie Universität Berlin, Berlin, Germany
{johannes.obenaus,nadja.seiferth}@fu-berlin.de

Abstract. Károlyi, Pach, and Tóth proved that every 2-edge-colored straight-line drawing of the complete graph contains a monochromatic plane spanning tree. It is open if this statement generalizes to other classes of drawings, specifically, to *simple drawings* of the complete graph. These are drawings where edges are represented by Jordan arcs, any two of which intersect at most once. We present two partial results towards such a generalization. First, we show that the statement holds for cylindrical simple drawings. (In a *cylindrical* drawing, all vertices are placed on two concentric circles and no edge crosses either circle.) Second, we introduce a relaxation of the problem in which the graph is k-edge-colored, and the target structure must be *hypochromatic*, that is, avoid (at least) one color class. In this setting, we show that every $\lceil (n+5)/6 \rceil$-edge-colored monotone simple drawing of K_n contains a hypochromatic plane spanning tree. (In a *monotone* drawing, every edge is represented as an x-monotone curve.)

Keywords: Simple drawing · Cylindrical drawing · Monotone drawing · Plane subdrawing

1 Introduction

A *simple drawing* of a graph represents vertices by pairwise distinct points (in the Euclidean plane) and edges by Jordan arcs connecting their endpoints such

We are particularly grateful to Irene Parada for bringing this problem to our attention. We also thank the organizers of the 4th DACH Workshop on Arrangements, that took place in February 2020 in Malchow and was funded by Deutsche Forschungsgemeinschaft (DFG), the Austrian Science Fund (FWF) and the Swiss National Science Foundation (SNSF). M. H. is supported by SNSF Project 200021E-171681. R. P. and A. W. are supported by FWF grant W1230. J. O. is supported by ERC StG 757609. N. S. is supported by DFG Project MU3501/3-1. D. P. and B. V. are supported by FWF Project I 3340-N35.

D. Auber and P. Valtr (Eds.): GD 2020, LNCS 12590, pp. 482–489, 2020.
https://doi.org/10.1007/978-3-030-68766-3_37

that (1) no (relative interior of an) edge passes through a vertex and (2) every pair of edges intersect at most once, either in a common endpoint or in their relative interior, forming a proper crossing. Simple drawings (also called *good drawings* [8] or *simple topological graphs* [12]) have been well studied, amongst others, in the context of crossing minimization (see e.g. [15]), as it is known that every crossing-minimal drawing of a graph is simple. Also every straight-line drawing is simple. Further well-known classes of simple drawings relevant for this work are *pseudolinear drawings*, where every edge can be extended to a bi-infinite Jordan arc such that every pair of them intersects exactly once; *cylindrical simple drawings*, where all vertices are placed on two concentric circles, no edge crosses either circle, and edges between two vertices on the outer (inner) circle lie completely outside (inside) that circle; *2-page book drawings*, where all vertices lie on a line and no edge crosses that line; and *monotone simple drawings*, where all edges are x-monotone curves. Unless explicitly mentioned otherwise, all considered drawings are simple, and the term simple is mostly omitted.

In this paper we are concerned with finding plane substructures in simple drawings. Specifically, we study the existence of plane spanning trees in edge-colored simple drawings of the complete graph K_n. A *k-edge-coloring* of a graph is a map from its edge set to a set of k colors.[1] A subgraph H of a k-edge-colored graph G is *hypochromatic* if the edges of H use at most $k-1$ colors, that is, H avoids at least one of the k color classes. If all edges of H have the same color, then H is *monochromatic*. We are inspired by the following conjecture.

Conjecture 1. Every 2-edge-colored simple drawing of K_n contains a monochromatic plane spanning tree.

Károlyi, Pach, and Tóth [10] proved the statement for straight-line drawings, where the 2-edge-coloring can also be interpreted as a Ramsey-type setting, where one color corresponds to the edges of the graph and the other color to the edges of its complement. Such an interpretation is less natural in the topological setting, where the edges are not implicitly defined by placing the vertices.

Unfortunately, a proof of Conjecture 1 seems elusive. However, we show that it holds for specific classes of simple drawings, such as 2-page book drawings, pseudolinear drawings, and cylindrical drawings. The result for 2-page book drawings can be shown straightforwardly. The statement for pseudolinear drawings follows from generalizing the proof for straight-line drawings by Károlyi, Pach, and Tóth [10] to this setting.

Proposition 1. *Every 2-edge-colored 2-page book drawing of K_n contains a plane monochromatic spanning tree.*

Proposition 2. *Every 2-edge-colored pseudolinear drawing of K_n contains a plane monochromatic spanning tree.*

See Appendix A in the full version of this paper [3] for proofs of those statements.

---

[1] Note that the coloring need *not* be proper nor have any other special properties.

The result for cylindrical drawings is more involved; it forms our first main contribution.

Theorem 1. *Every 2-edge-colored cylindrical simple drawing of K_n contains a monochromatic plane spanning tree.*

In light of the apparent challenge in attacking Conjecture 1, we also consider the following generalized formulation, which uses more colors.

Conjecture 2. For $k \geq 2$, every k-edge-colored simple drawing of K_n contains a hypochromatic plane spanning tree.

Note that both conjectures are in fact equivalent: On the one hand, Conjecture 2 implies Conjecture 1 by setting $k = 2$. On the other hand, assuming Conjecture 2 holds for some k, it also holds for every larger k' because we can simply merge color classes until we are down to k colors. Avoiding any one of the resulting color classes also avoids at least one of the original color classes.

Our second result is the following statement about *monotone* drawings.

Theorem 2. *Every $\lceil (n + 5)/6 \rceil$-edge-colored monotone simple drawing of K_n contains a hypochromatic plane spanning tree.*

Finally, note that *some* assumptions concerning the drawing are necessary to obtain any result on the existence of plane substructures. Without any restriction, every pair of edges may cross. The class of simple drawings is formed by two restrictions: forbid adjacent edges to cross and forbid independent edges to cross more than once. Both restrictions are necessary in the statement of Conjecture 1. If adjacent edges may cross, then one can construct drawings where every pair of adjacent edges crosses (e.g., in the neighborhood of the common vertex), implying that no plane substructure can have a vertex of degree more than one. And for *star-simple* drawings, where adjacent edges do not cross but independent edges may cross more than once, already K_5 admits 2-edge-colored star-simple drawings without any monochromatic plane spanning tree; see Fig. 1.

Fig. 1. Star-simple drawings of K_5 without monochromatic plane spanning tree.

Related Work. The problem of finding plane subdrawings in a given drawing has gained some attention over the past decades. We mention only a few results from the vast literature on plane substructures. In 1988, Rafla [13] conjectured that every simple drawing of K_n contains a plane Hamiltonian cycle. By now the conjecture is known to be true for $n \leq 9$ [1] and several classes of simple drawings (e.g., 2-page book drawings, monotone drawings, cylindrical drawings), but remains open in general. See also [4–6,11,14] for some results about plane spanning trees in straight-line drawings of complete graphs. In an edge-colored setting, many other coloring schemes were studied in this context, see e.g. [7,9].

Observe that if one color class of a drawing is not spanning, the drawing of the remaining colors contains a complete bipartite graph as a subdrawing. Recently, it has been shown that every simple drawing of the complete bipartite graph contains a plane spanning tree [2]. Consequently, this implies the following lemma, which turns out to be useful later on (see Appendix A of [3] for the proof).

Lemma 1. *Let D be a k-edge-colored simple drawing of K_n, for $k \geq 2$. If one of the color classes is not spanning, then D contains a hypochromatic plane spanning tree.*

2   Cylindrical Drawings

This section is devoted to Theorem 1, which states that every 2-edge-colored cylindrical drawing of K_n contains a monochromatic plane spanning tree. We give a detailed outline of the proof. The full proof can be found in Appendix B of [3].

For easier readability, we introduce some names for the different elements of a cylindrical drawing (cf. Fig. 2). We call the vertices on the inner (outer) circle *inner (outer) vertices*. Similarly, we call edges connecting two inner (outer) vertices *inner (outer) edges*; the remaining edges are called *side edges*. The edges between consecutive vertices on the inner (outer) circle are called *cycle edges* and the union of all inner (outer) cycle edges are called inner (outer) *cycle*. The definition of cylindrical drawings implies that all cycle edges are uncrossed. The *rotation* of a vertex v is the circular ordering of all edges incident to v. In this ordering, the cycle edges separate the inner (outer) edges from the side edges. Hence, the rotation of v induces a linear order on the side edges incident to v.

Proof (sketch). Our proof consists of two steps. In Step 1, we restrict considerations to drawings fulfilling two properties, for which we compute a monochromatic plane spanning subgraph using a multi-stage sweep algorithm. In Step 2, we show how to handle drawings that do not fulfill all properties from Step 1.

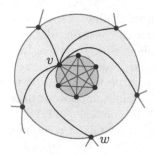

Fig. 2. Sketch of a cylindrical drawing. Inner edges are drawn blue, outer edges red, and side edges black. vw is the first side edge in the clockwise rotation of v. (Color figure online)

Step 1. Let D be a 2-edge-colored cylindrical drawing that fulfills the following properties:

(P1) D has inner and outer vertices, and
(P2) D's inner and outer cycle are both monochromatic, but of different color.

Assume without loss of generality that the inner cycle of D is blue and hence the outer cycle is red. We will refer to them as the blue and red cycle and to the vertices on them as blue and red vertices, respectively.

We use the following algorithm to compute a (bichromatic) subdrawing H of D consisting of some side edges of D and their endpoints (cf. Fig. 3).

Phase 0. Initially, let H be empty. Choose an arbitrary inner vertex as initial *rotation vertex* v_{cur}, set the *rotation direction* to clockwise, and set the first side edge of v_{cur} in the rotation direction as initial *current edge* e_{cur}.

Phase 1. We repeat the following process while e_{cur} is a side edge and while H is still missing vertices from the cycle of D not containing v_{cur}: Add e_{cur} to H; If e_{cur} does not have the same color as v_{cur}, set v_{cur} to be the other endpoint of e_{cur} and reverse the rotation direction (clockwise $\leftrightarrow$ counterclockwise); In any case, set e_{cur} to be the next edge incident to v_{cur} after e_{cur} in the (possibly changed) rotation direction.

Phase 2. If H contains all vertices of D from the cycle not containing v_{cur}: Return H.

Phase 3. Otherwise: Set $H_{\text{prev}} = H$, reset H to be empty, reverse the rotation direction, set e_{cur} to be the first side edge of v_{cur} in the new rotation direction, and restart with Phase 1.

The following invariants hold for the algorithm (see Appendix B of [3] for a proof):

(J1)   At any time, the union of H and the two cycles of D forms a plane drawing.
(J2)   Any blue (red) vertex in H is incident to a red (blue) edge in H, except for the current rotation vertex.

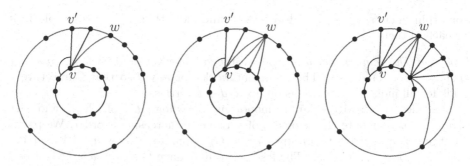

Fig. 3. The first steps of the algorithm. The black arc at vertex v indicates that vv' is the first side edge of v in clockwise order (the initial rotation direction). (Color figure online)

(J3) Assume that Phase 1 is performed more than once and let $V(H)$ be the set of vertices of H. Then for any $i \geq 2$, after round i of Phase 1, either $V(H)$ is a strict superset of $V(H_{\text{prev}})$ or H contains all vertices from the cycle not containing v_{cur}, the current rotation vertex (or both conditions hold).

Using those invariants, we can now complete Step 1: By (J3), the algorithm terminates. And by (J1) and (J2), at least one of the color classes of the union of H and the two cycles of D is a monochromatic plane spanning graph for D.

Step 2. Now assume that D violates at least one of the properties (P1) and (P2)

If it violates (P1), then D is isomorphic to a 2-page book drawing and hence contains a monochromatic plane spanning tree (see Proposition 1).

If D does not fulfill (P2), then we remove vertices whose cycle edges are of different color until we reach a subdrawing D' where both cycles are monochromatic, find a plane monochromatic spanning tree on D' by either Step 1 or Lemma 1, and then extend it to a monochromatic spanning tree on D.   □

3   Monotone Drawings

In this section, we prove the existence of hypochromatic plane spanning trees in k-edge-colored monotone drawings of K_n, for k linear in n.

Lemma 2. *Conjecture 1 holds for any simple drawing of K_n with $n \leq 7$ vertices.*

For $n \leq 4$ this can easily be observed by hand. For $n = 5, \ldots, 7$ we considered all weak isomorphism classes[2] of simple drawings of K_n [1] and checked for all possible 2-edge colorings that there exists a monochromatic plane spanning tree. Computations for $n = 8$ are currently out of reach, as there are 5,370,725 weak

---

[2] Two simple drawings of K_n are weakly isomorphic iff they have the same crossing edge pairs.

isomorphism classes of simple drawings [1] and more than 10^8 possible colorings for each of them.

Proof (of Theorem 2). Let $d \geq 2$ be an integer constant, and let $k = \lceil (n + d - 1)/d \rceil = \lceil (n-1)/d \rceil + 1$. The argument works for any d so that Conjecture 1 holds for all monotone drawings on up to $d + 1$ vertices.

Consider a k-edge-colored monotone drawing D of K_n, and let $v_0, v_1, \ldots, v_{n-1}$ denote the sequence of vertices in increasing x-order. We partition the vertices into $k - 1$ groups $G_0, \ldots, G_{k-2}$ of size at most $d + 1$ by setting $G_i = (v_{di}, v_{di+1}, \ldots, v_{di+d})$. (The last group may have less than $d + 1$ vertices.) Observe that $G_i \cap G_{i+1} = \{v_{d(i+1)}\}$.

We proceed in two phases. In both phases we consider each group separately. At the end of the first phase, we choose which color to remove. At the end of the second phase, we have an induced plane spanning tree T_i for G_i that avoids the chosen color, for each $i \in \{0, \ldots k - 2\}$. As D is monotone, the union $\bigcup_{i=0}^{k-2} T_i$ forms a hypochromatic plane spanning tree in D.

In the first phase, we consider each group G_i, and check whether it has a monochromatic plane spanning tree in some color c. If so, we put c in a set S of colors to keep. If not, then by Conjecture 1 (which we assume to hold for G_i, as G_i has at most $d + 1$ vertices) we can remove *any* single color and still find a monochromatic plane spanning tree in G_i. (If c is the color to be removed, then consider the bicoloring where all colors other than c are merged into a single second color.) As $|S| \leq k - 1$, we can choose a color not in S to be removed at the end of the first phase.

In the second phase, for each group G_i we either select a monochromatic plane spanning tree (if it exists), or find a plane spanning tree that avoids the chosen color.

To obtain the statement of Theorem 2, we use the result of Lemma 2.   □

4   Open Problems

Besides resolving the conjectures in full generality, it would be interesting to prove them for other specific classes of drawings (e.g., monotone). A useful step in this direction would be to expand the range of k for which Conjecture 2 holds.

References

1. Ábrego, B., et al.: All good drawings of small complete graphs. In: Abstracts 31st European Workshop on Computational Geometry (EuroCG'15), pp. 57–60 (2015)
2. Aichholzer, O., García, A., Parada, I., Vogtenhuber, B., Weinberger, A.: Simple drawings of $K_{m,n}$ contain shooting stars. In: Abstracts 35th European Workshop on Computational Geometry (EuroCG'20), pp. 36:1–36:7 (2020)
3. Aichholzer, O., et al.: Plane spanning trees in edge-colored simple drawings of K_n. ArXiv e-Prints (2020). http://arxiv.org/abs/2008.08827v1
4. Bernhart, F., Kainen, P.C.: The book thickness of a graph. J. Comb. Theory Ser. B **27**(3), 320–331 (1979). https://doi.org/10.1016/0095-8956(79)90021-2

5. Biniaz, A., García, A.: Partitions of complete geometric graphs into plane trees. Comput. Geom. **90**, 101653 (2020). https://doi.org/10.1016/j.comgeo.2020.101653
6. Bose, P., Hurtado, F., Rivera-Campo, E., Wood, D.R.: Partitions of complete geometric graphs into plane trees. Comput. Geom. **34**(2), 116–125 (2006). https://doi.org/10.1016/j.comgeo.2005.08.006
7. Brualdi, R.A., Hollingsworth, S.: Multicolored trees in complete graphs. J. Comb. Theory Ser. B **68**(2), 310–313 (1996). https://doi.org/10.1006/jctb.1996.0071
8. Erdős, P., Guy, R.: Crossing number problems. Am. Math. Monthly **88**, 52–58 (1973)
9. Erdős, P., Nešetril, J., Rödl, V.: Some problems related to partitions of edges of a graph. Graphs and other combinatorial topics, Teubner, Leipzig 5463 (1983)
10. Károlyi, G., Pach, J., Tóth, G.: Ramsey-type results for geometric graphs, I. Discrete Comput. Geom. **18**(3), 247–255 (1997). https://doi.org/10.1007/PL00009317
11. Keller, C., Perles, M.A., Rivera-Campo, E., Urrutia-Galicia, V.: Blockers for noncrossing spanning trees in complete geometric graphs. In: Pach, J. (ed.) Thirty Essays on Geometric Graph Theory, pp. 383–397. Springer, New York (2013). https://doi.org/10.1007/978-1-4614-0110-0_20
12. Kynčl, J.: Enumeration of simple complete topological graphs. Eur. J. Comb. **30**, 1676–1685 (2009). https://doi.org/10.1016/j.ejc.2009.03.005
13. Rafla, N.H.: The good drawings D_n of the complete graph K_n. Ph.D. thesis, McGill University, Montreal (1988)
14. Rivera-Campo, E., Urrutia-Galicia, V.: A sufficient condition for the existence of plane spanning trees on geometric graphs. Comput. Geom. **46**(1), 1–6 (2013). https://doi.org/10.1016/j.comgeo.2012.02.006
15. Schaefer, M.: The graph crossing number and its variants: a survey. Electron. J. Comb. Dyn. Surv. **21**(4) (2020). https://doi.org/10.37236/2713

Augmenting Geometric Graphs with Matchings

Alexander Pilz[1], Jonathan Rollin[2]([✉])[ID], Lena Schlipf[3][ID], and André Schulz[2][ID]

[1] Graz University of Technology, Graz, Austria
apilz@ist.tugraz.at
[2] FernUniversität in Hagen, Hagen, Germany
{jonathan.rollin,andre.schulz}@fernuni-hagen.de
[3] Universität Tübingen, Tübingen, Germany
lena.schlipf@uni-tuebingen.de

Abstract. We study noncrossing geometric graphs and their disjoint compatible geometric matchings. Given a cycle (a polygon) P we want to draw a set of pairwise disjoint straight-line edges with endpoints on the vertices of P so that these new edges neither cross nor contain any edge of the polygon. We prove NP-completeness of deciding whether there is such a perfect matching. For any n-vertex polygon, with $n \geq 4$, we show that such a matching with $< n/7$ edges is not maximal, that is, it can be extended by another compatible matching edge. We also construct polygons with maximal compatible matchings with $n/7$ edges, demonstrating the tightness of this bound. Tight bounds on the size of a minimal maximal compatible matching are also obtained for the families of d-regular geometric graphs for each $d \in \{0, 1, 2\}$. Finally we consider a related problem. We prove that it is NP-complete to decide whether a noncrossing geometric graph G admits a set of compatible noncrossing edges such that G together with these edges has minimum degree five.

Keywords: Geometric graph · Compatible matching · Graph augmentation

1 Introduction

A geometric graph is a graph drawn in the plane with straight-line edges. Throughout this paper we additionally assume that all geometric graphs are noncrossing. Let G be a given (noncrossing) geometric graph G. We want to augment G with a geometric matching on the vertices of G such that no edges cross in the augmentation. We call such a (geometric) matching *compatible* with G. Note that our definition of a compatible matching implies that the matching is noncrossing and avoids the edges of G. Questions regarding compatible matchings were first studied by Rappaport et al. [14,15]. Rappaport [14] proved that it

L. Schlipf—This research is supported by the Ministry of Science, Research and the Arts Baden-Württemberg (Germany).

D. Auber and P. Valtr (Eds.): GD 2020, LNCS 12590, pp. 490–504, 2020.
https://doi.org/10.1007/978-3-030-68766-3_38

is NP-hard to decide whether for a given geometric graph G there is a compatible matching M such that $G + M$ is a (spanning) cycle. Recently Akitaya et al. [3] confirmed a conjecture of Rappaport and proved that this holds even if G is a perfect matching. Note that in this case also M is necessarily a perfect matching. However, for some compatible perfect matchings M the union $G + M$ might be a collection of several disjoint cycles. There are graphs G that do not admit any compatible perfect matching, even when G is a matching. Such matchings were studied by Aichholzer et al. [1] who proved that each m-edge perfect matching G admits a compatible matching of size at least $\frac{4}{5}m$. Ishaque et al. [10] confirmed a conjecture of Aichholzer et al. [1] which says that any perfect matching G with an even number of edges admits a compatible perfect matching. For a geometric graph G let $d(G)$ denote the size of a largest compatible matching of G and for a family $\mathcal{F}$ of geometric graphs let $d(\mathcal{F}) = \min\{d(G) \mid G \in \mathcal{F}\}$. Aichholzer et al. [2] proved that for the family T_n of all n-vertex geometric trees $\frac{1}{10}n \leq d(T_n) \leq \frac{1}{4}n$ holds and for the family P_n of all n-vertex simple polygons $\frac{n-3}{4} \leq d(P_n) \leq \frac{1}{3}n$ holds.

We continue this line of research and consider the following problems. Given a polygon, we first show that it is NP-complete to decide whether the polygon admits a compatible perfect matching. Then we ask for the "worst" compatible matchings for a given polygon. That is, we search for small maximal compatible matchings, where a compatible matching M is maximal if there is no compatible matching M' that contains M. We study such matchings also for larger families of d-regular geometric graphs.

The first studied problem can also be phrased as follows: Given a geometric cycle, can we add edges to obtain a cubic geometric graph? In the last section, we consider a related augmentation problem. Given a geometric graph, we show that it is NP-complete to decide whether the graph can be augmented to a graph of minimum degree five. The corresponding problem for the maximum vertex degree asks to add a *maximal* set of edges to the graph such that the maximum vertex degree is bounded from above by a constant. This problem is also known to be NP-complete for maximum degree at most seven [11].

A survey of Hurtado and Tóth [9] discusses several other augmentation problems for geometric graphs. Specifically it is NP-hard to decide whether a geometric graph can be augmented to a cubic geometric graph [13] and also whether an abstract planar graph can be augmented to a cubic planar graph (not preserving any fixed embedding) [8]. Besides the problems mentioned in that survey, decreasing the diameter [6] and the continuous setting (where every point along the edges of an embedded graph is considered as a vertex) received considerable attention [4, 7].

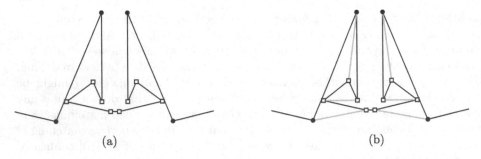

Fig. 1. (a) This gadget al.lows for simulating a "bend" in the polygon without a vertex that needs to be matched. The construction is scaled such that the eight points marked with squares do not see any other point outside of the gadget (in particular, narrowing it horizontally). (b) A possible matching is shown in red. (Color figure online)

2   Compatible Perfect Matchings in Polygons

Theorem 1. *Given a simple polygon, it is* NP-*complete to decide whether it admits a compatible perfect matching.*

Proof. The problem is obviously in NP, as a certificate one can merely provide the added edges. NP-hardness is shown by a reduction from POSITIVE PLANAR 1-IN-3-SAT. In this problem, shown to be NP-hard by Mulzer and Rote [12], we are given an instance of 3-SAT with a planar variable–clause incidence graph (i.e., the graph whose vertices are the variables and clauses, which are connected by an edge if and only if the variable occurs in the clause) and no negative literals; the instance is considered satisfiable if and only if there is exactly one true variable per clause.

For a given 1-in-3-SAT formula, we take an embedding of its incidence graph and replace its elements by gadgets. We first show that finding compatible matchings for a set of disjoint simple polygons is hard and we then show how to connect the individual polygons to obtain a single polygon.

Our construction relies on a gadget that restricts the possible matching edges of vertices. In particular, we introduce a polygonal chain, whose vertices need to be matched to each other in any perfect matching. This is achieved by the *twin-peaks gadget* as shown in Fig. 1. The gadget is scaled such that the eight vertices in its interior (which are marked with squares in Fig. 1) do not see any edges outside of the gadget. (We say that a vertex *sees* another vertex if the relative interior of the segment between them does not intersect the polygon.) The two topmost vertices must have an edge to the vertices directly below as the vertices below do not see any other (nonadjacent) vertices. The remaining six "square" vertices do not have a geometric perfect matching on their own, so any geometric perfect matching containing them must connect them to the two bottommost vertices. Clearly, there is such a matching.

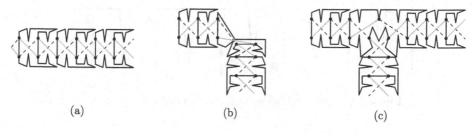

(a)   (b)   (c)

Fig. 2. (a) A wire gadget and its two truth states (one in dashed, the other in solid red). (b) A bend in a wire gadget. (c) A split gadget that transports the truth setting of one wire to two other ones. This is used for representing the variables.

We now present the remaining gadgets (*wire*, *split*, and *clause*) for our reduction. The ideas are inspired by the reduction of Pilz [13] who showed that augmenting an arbitrary geometric graph to a crossing-free cubic graph is NP-complete. In the following illustrations, vertices of degree two are drawn as a dot. Other vertices in the figures represent a sufficiently small twin-peaks gadget.

The *wires* propagate the truth assignment of a variable. A wire consists of a sequence of polygons, each containing four vertices of degree two (ignoring twin-peak vertices). There are only two possible global matchings for these vertices; see Fig. 2(a). A *bend* in a wire can be drawn as shown in Fig. 2(b). The truth assignment of a wire can be duplicated by a *split gadget*; see Fig. 2(c). A variable is represented by a cyclic wire with split gadgets. Recall that in our reduction, we do not need negated variables. The *clause gadget* is illustrated in Fig. 3, where

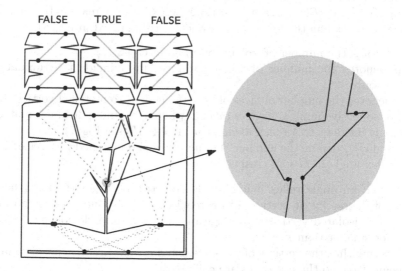

Fig. 3. The clause gadget. The visibility among the vertices of degree two is indicated by the lighter lines. Exactly one vertex of degree two of the part in the circle must be connected to a wire above that carries the true state.

Fig. 4. Merging neighboring polygons to a single polygon.

the wires enter from the top. The vertices there can be matched if and only if one of the vertices is connected to a wire that is in the true state. The vertices at the bottom of the gadget make sure that if there are exactly two wires in the false state, then we can add an edge to them. Hence, this set of polygons has a compatible perfect matching if and only if the initial formula was satisfiable.

It remains to "merge" the polygons of the construction to one simple polygon. Observe that two neighboring polygons can be merged by a small tunnel using four new bends with twin-peaks gadgets line in Fig. 4, without affecting the possible compatible perfect matchings of the other vertices. We can consider the incidence graph to be connected (otherwise the reduction splits into disjoint problems). Hence, we can always merge two distinct neighboring polygons, until there is only a single polygon left. □

3   Compatible Maximal Matchings in Geometric Graphs

For a geometric graph G let $\mathrm{mm}(G)$ denote the size of a minimal maximal compatible matching of G and for a family $\mathcal{F}$ of geometric graphs let $\mathrm{mm}(\mathcal{F}) = \min\{\mathrm{mm}(G) \mid G \in \mathcal{F}\}$. For a geometric graph G and a maximal compatible matching M we define the following parameters (illustrated in Fig. 5):

- i_{GM} denotes the number of isolated vertices in $G + M$,
- Δ_{GM} denotes the number of triangular faces in G incident to unmatched vertices only,
- σ_{GM} denotes the number of faces of $G + M$ incident to matched vertices only,
- ν_{GM} denotes the number of edges uv in G where u is unmatched, v is matched, and uv is incident to a reflex angle at u in $G + M$ (see Fig. 7),
- r^u_{GM} and r^m_{GM} denote the number of unmatched and matched vertices incident to a reflex angle in $G + M$, respectively.

Here, we call an angle reflex if it is of degree strictly larger than π (there is an angle of degree 2π at vertices of degree 1 in $G + M$ and there is no angle considered at isolated vertices). Analogically, we call an angle convex if it is of degree π or smaller than π.

We assume that the vertices of the considered graphs are in general position. That means that no three vertices are collinear.

The following lemma gives a general lower bound on the size of any maximal matching in terms of the parameters introduced above. We use this bound later to derive specific lower bounds for various classes of geometric graphs below.

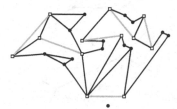

Fig. 5. A geometric graph G (black) and a maximal compatible matching M (red). Here $i_{GM} = \Delta_{GM} = 1$, $\sigma_{GM} = 2$, $\nu_{GM} = 10$, $r_{GM}^u = 11$, and $r_{GM}^m = 10$. (Color figure online)

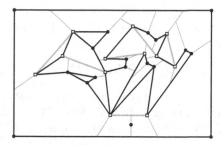

Fig. 6. The geometric graph (black) with maximal matching (red) from Fig. 5 where each reflex angle is cut by a gray edge. (Color figure online)

Lemma 1. *For each geometric graph G and each maximal compatible matching M of G we have*

$$2\,|V(G)| + \nu_{GM} + 2\,\sigma_{GM} - r_{GM}^u - 2\,r_{GM}^m - \sum_{u \in V(M)} d_G(u) - \Delta_{GM} - 2 \le 2\,|E(M)|\,.$$

Proof. We subdivide the plane into cells as follows. First draw a rectangle enclosing G in the outer face (with four vertices and four edges). For each isolated vertex in $G + M$ (one after the other) draw two collinear straight-line edges, both starting at that vertex and until they hit some already drawn edges e and e'. The direction of these new edges is arbitrary as long as they do not hit any vertex. Their endpoints become new vertices (subdividing e and e'). Similarly, for each vertex $u \in V(G)$ incident to some reflex angle in the resulting drawing we draw (one after the other) a straight-line edge starting at u. The direction of this new edge is chosen such that it cuts the reflex angle at u into two convex angles and such that it stops on some already drawn edge (but not a vertex) which is then subdivided by a new vertex. Avoiding to hit vertices is possible as the points are in general position. See Fig. 6. Let D denote the final plane graph. Then each bounded face in D is convex and D is connected. Further, D has exactly $|V(G)| + r_{GM}^u + r_{GM}^m + 2\,i_{GM} + 4$ vertices and $|E(G)| + |E(M)| + 2(r_{GM}^m + r_{GM}^u + 2\,i_{GM}) + 4$ edges (each edge starting at an isolated vertex and each edge cutting a reflex angle creates a new vertex and

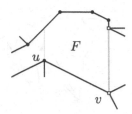

Fig. 7. An edge $uv \in E(G)$ where $u \in V(G) \setminus V(M)$ and $v \in V(M)$ with a reflex angle at u (in $G + M$). Then u is the only vertex from $V(G) \setminus V(M)$ incident to the face F (obtained by cutting the reflex angle at u) since M is maximal.

subdivides an existing edge into two parts). By Euler's formula the number F_D of faces in D is exactly

$$F_D = |E(D)| - |V(D)| + 2 = |E(G)| - |V(G)| + |E(M)| + r_{GM}^m + r_{GM}^u + 2\,i_{GM} + 2.$$

Let $U = V(G) \setminus V(M)$ denote the set of unmatched vertices of G and let F_i denote the number of faces in D with exactly i vertices of U in their boundary. Each isolated vertex in $G + M$ is incident to exactly two faces of D, each vertex $u \in U$ not incident to a reflex angle in $G + M$ is incident to exactly $d_G(u)$ faces of D, and each remaining vertex $u \in U$ is incident to exactly $d_G(u) + 1$ faces of D. Therefore

$$2\,i_{GM} + r_{GM}^u + \sum_{u \in U} d_G(u) = \sum_{i \geq 1} i\, F_i. \qquad (1)$$

Consider two vertices in U incident to a common face F in D. The line segment connecting these two vertices is an edge of G, otherwise M is not maximal. So either F has at most two vertices from U or F is a triangular face of G incident to vertices from U only. This shows that $F_3 = \Delta_{GM}$ and $F_i = 0$ for each $i \geq 4$. Further, each face incident to a vertex that is isolated in $G + M$ is not incident to any other unmatched vertex. Similarly, for each edge counted by ν_{GM} there is a face in D with only one unmatched vertex in its boundary, see Fig. 7. Hence $F_1 \geq 2\,i_{GM} + \nu_{GM}$. The outer face does not contain any vertices of U and hence $F_0 \geq 1 + \sigma_{GM}$. Combining these observations with (1) and $F_2 = F_D - F_0 - F_1 - F_3$ yields

$$
\begin{aligned}
& 2\,i_{GM} + r_{GM}^u + \sum_{u \in U} d_G(u) \\
= \ & F_1 + 2\,F_2 + 3\,\Delta_{GM} \\
= \ & 2\,F_D - 2\,F_0 - F_1 + \Delta_{GM} \\
\leq \ & 2\,|E(G)| - 2\,|V(G)| + 2\,|E(M)| \\
& + 2\,i_{GM} + 2\,r_{GM}^m + 2\,r_{GM}^u + \Delta_{GM} - \nu_{GM} - 2\,\sigma_{GM} + 2.
\end{aligned}
$$

Now the desired result follows using $\sum_{u \in U} d_G(u) = 2\,|E(G)| - \sum_{u \in V(M)} d_G(u)$. $\square$

The bound of Lemma 1 is particularly applicable for regular graphs.

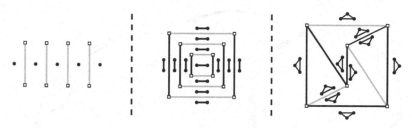

Fig. 8. Geometric graphs (black) with minimal maximal compatible matchings (red). (Color figure online)

Theorem 2. *Consider an n-vertex geometric graph G.*

a) If G is 0-regular (a point set) we have $\mathrm{mm}(G) \geq \frac{n-1}{3}$.
b) If G is 1-regular (a perfect matching) we have $\mathrm{mm}(G) \geq \frac{n-2}{6}$.
c) If G is 2-regular (disjoint polygons) we have $\mathrm{mm}(G) \geq \frac{n-3}{11}$.

All these bounds are tight for infinitely many values of n.

Proof. First consider a 0-regular n-vertex graph G (a point set). Then $r_{GM}^u = 0$, $r_{GM}^m = 2|E(M)|$, $\nu_{GM} = \Delta_{GM} = 0$, and $\sigma_{GM} \geq 0$ for any maximal compatible matching M of G. By Lemma 1 we have $2n - 4|E(M)| - 2 \leq 2|E(M)|$. This shows $\mathrm{mm}(G) \geq (n-1)/3$. This is tight due to the graphs G and the maximal matchings given in Fig. 8 (left).

Next consider a 1-regular n-vertex graph G. Each vertex in G is reflex in $G + M$. Then $\Delta_{GM} = 0$, $\nu_{GM} \geq 0$, $r_{GM}^u = n - 2|E(M)|$, $r_{GM}^m = 2|E(M)|$, and $\sigma_{GM} \geq 0$ for any maximal compatible matching M of G. By Lemma 1 we have $n - 4|E(M)| - 2 \leq 2|E(M)|$. This shows $\mathrm{mm}(G) \geq (n-2)/6$. This is tight due to the graphs G and the maximal matchings given in Fig. 8 (middle).

Finally consider a 2-regular n-vertex geometric graph G. Each vertex in $V(G) \setminus V(M)$ is reflex in $G + M$. Then $\nu_{GM} \geq 0$, $r_{GM}^u = n - 2|E(M)|$, $r_{GM}^m \leq 2|E(M)|$, $\sigma_{GM} \geq 0$, and $\Delta_{GM} \leq (n - 2|E(M)|)/3$ for any maximal compatible matching M of G. By Lemma 1 we have $n - 6|E(M)| - (n - 2|E(M)|)/3 - 2 \leq 2|E(M)|$. This shows $\mathrm{mm}(G) \geq (n-3)/11$. This is tight due to the graph G and the maximal matching M given in Fig. 8 (right), as an infinite family is obtained by repeatedly replacing an arbitrary triangle with a (scaled) copy of $G + M$. $\square$

Theorem 3. *Let $n \geq 4$ and let P_n denote the family of all n-vertex polygons. Then $\mathrm{mm}(P_n) \geq \frac{1}{7}n$ for all n and this bound is tight for infinitely many values of n.*

Proof. The construction in Fig. 9 shows that for infinitely many values of n there is an n-vertex polygon with a compatible maximal matching of size $\frac{n}{7}$. This shows $\mathrm{mm}(P_n) \leq \frac{n}{7}$ for infinitely many values of n.

It remains to prove the lower bound. Let P be an n-vertex polygon with a maximal compatible matching M. Since $n \geq 4$, we have $|E(M)| \geq 1$, $\Delta_{PM} = 0$, $r_{PM}^m \leq 2|E(M)|$, and $\sigma_{PM} \geq 0$. Let $U = V(P) \setminus V(M)$ denote the unmatched

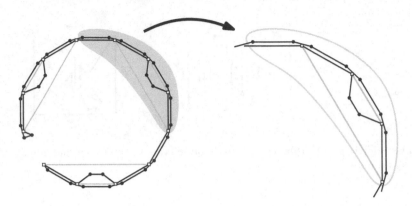

Fig. 9. A polygon (black) with a maximal matching (red) with $\frac{n}{7}$ edges (here $n = 42$). Note that there are exactly two matching edges between the 14 vertices in the gray area which can be repeated along a cycle arbitrarily often. (Color figure online)

vertices of P and let E_{UM} denote the set of edges uv in P where $u \in U$ and $v \in V(M)$. Each vertex in U has a reflex angle. Hence $r_{PM}^u = n - 2\,|E(M)|$ and $\nu_{PM} = |E_{UM}|$. There are $2 + |E(M)|$ faces in $P + M$. Each of them either has no vertex from U in its boundary or at least two edges from E_{UM}. So $P + M$ has $2 + |E(M)| - \sigma_{PM}$ faces incident to at least two edges from E_{UM} each. Each edge in E_{UM} is on the boundary of two faces of $P + M$. Together we have $2\,|E_{UM}| \geq 2(2+|E(M)|-\sigma_{PM})$ and hence $\nu_{PM}+\sigma_{PM} \geq 2+|E(M)|$. Combining these observations with Lemma 1 yields $|E(M)| \geq n/7$, because

$$2n + 2 + |E(M)| - n - 2\,|E(M)| - 4\,|E(M)| - 4\,|E(M)| \leq$$
$$2n + \nu_{PM} + 2\,\sigma_{PM} - r_{PM}^u - 2\,r_{PM}^m - \sum_{u \in V(M)} d_P(u) - \Delta_{PM} - 2 \leq 2\,|E(M)|$$

$\square$

For nonregular (abstract) graphs $\hat{G}$ determining a geometric drawing G minimizing $\mathrm{mm}(G)$ seems harder. For an integer n and a real number d with $0 \leq d \leq 3$, let $\mathcal{F}_d^n$ denote the family of all (noncrossing) geometric graphs with n vertices and at most dn edges. Further let $\mathrm{mm}(d) = \liminf_{n \to \infty} \min\{\mathrm{mm}(G)/n \mid G \in \mathcal{F}_d^n\}$. For each n and each $d \geq 2$ the set $\mathcal{F}_d^n$ contains a triangulation of a convex polygon (on $2n - 3$ edges). This shows $\mathrm{mm}(d) = 0$ for $d \geq 2$. Theorem 2 shows $\mathrm{mm}(0) = 1/3$ and $\mathrm{mm}(1/2) \leq 1/6$. The construction in the following lemma shows $\mathrm{mm}(d) \leq (2 - d)/13$ for $7/10 < d < 2$.

Lemma 2. *For any integers m, n with $n \geq 5$, $\frac{7n+95}{10} \leq m \leq 2n + 2$ there is a geometric graph on n vertices and m edges with a maximal compatible matching of size $\lceil \frac{2n-m+3}{13} \rceil$.*

Proof. Let $k = \lceil \frac{2n-m+3}{13} \rceil$. Then $k \geq 1$ since $m \leq 2n + 2$. First suppose that $2n - m + 3$ is divisible by 13, that is, $m = 2n + 3 - 13k$. We shall construct a

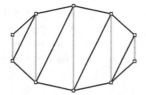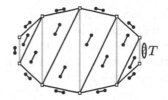

Fig. 10. A geometric graph (black) with a maximal compatible matching (red). (Color figure online)

geometric graph on n vertices and m edges with a maximal compatible matching of size k.

Choose a (noncrossing, geometric) perfect matching M of $2k$ points in convex position and an (inner) triangulation of that geometric graph. See Fig. 10 (left). There are $2k - 2$ triangular faces and $2k$ edges in the boundary of the outer face. Place an isolated edge in the interior of each triangular face. Further for all but one of the outer edges e place another (tiny) isolated edge close to e in the outer face (so that there are no visibilities between these). So far there are $2k + (4k - 4) + (4k - 2) = 10k - 6$ vertices and $(3k - 3) + (2k - 2) + (2k - 1) = 7k - 6$ edges not in M. Close to the remaining outer edge we place a triangulation T of a convex polygon on $n - 10k + 6$ vertices (so that there are no visibilities between these vertices and the isolated edges not in M). See Fig. 10 (right). Note that $n - 10k + 6 \geq 2$ since $m \geq \frac{7n + 95}{10}$. So the graph T contains $2n - 20k + 9 = m - 7k + 6$ edges. The final graph has in total n vertices and m edges not in M. Further M is a maximal matching by construction.

It remains to consider the case that $2n - m + 3$ is not divisible by 13. In this case we apply the construction above with $m' = 2n + 3 - 13k$ edges. To add the remaining $m - m' \leq 12$ edges we replace the triangulation T by an appropriate triangulation of another point set that has some interior points (and hence has more edges). □

4   Augmenting to Minimum Degree Five

In this section, we show that augmenting to a geometric graph with minimum degree five is NP-complete.

Theorem 4. *Given a geometric crossing-free graph G, it is NP-complete to decide whether there is a set of compatible edges E such that $G + E$ has minimum degree five.*

Proof. The problem is obviously in NP, a certificate provides the added edges. NP-hardness is shown by a reduction from MONOTONE PLANAR RECTILINEAR 3-SAT.

In this problem, shown to be NP-hard by de Berg and Khosravi [5], we are given an instance of monotone (meaning that each clause has only negative or only positive variables) 3-SAT with a planar variable-clause incidence graph. In this graph, the variable and clause gadgets are represented by rectangles. All variable rectangles lie on a horizontal line. The clauses with positive variables lie above the variables and the clauses with negative variables below. The edges connecting the clause gadgets to the variable gadgets are vertical line segments and no edges cross. See Fig. 11.

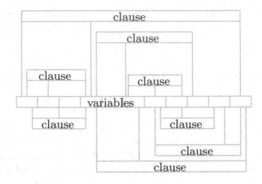

Fig. 11. A monotone planar 3-SAT instance with a corresponding embedding.

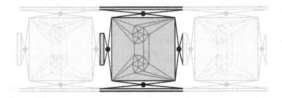

Fig. 12. A (geometric) subgraph whose copies will form a wire gadget.

For a given monotone planar 3-SAT formula, we take an embedding of its incidence graph (as discussed) and replace its elements by gadgets. Note that the corresponding rectilinear layout can be computed in polynomial time and has coordinates whose size is bounded by a polynomial [16]. We use a *wire gadget* that propagates the truth assignments; see Fig. 12. It consists of a linear sequence of similar subgraphs, each containing exactly four vertices of degree four (the other vertices have at least degree five). The gray areas contain subgraphs where all vertices have at least degree five. The main idea is that we need to add an edge to each of the vertices of degree four surrounding the big gray squares. But due to blocked visibilities this can only be achieved by a "windmill" pattern, which has to synchronize with the neighboring parts; see Fig. 13. Thus, we have exactly two ways to add edges in order to augment the wire to a graph with minimum degree five.

A bend in a wire is shown in Fig. 14. The truth assignment of a wire can be duplicated by the *split gadget* as shown in Fig. 14.

(a)   (b)

Fig. 13. The wire gadget with its only true possible augmentations, associated with the assignment true (a) and false (b).

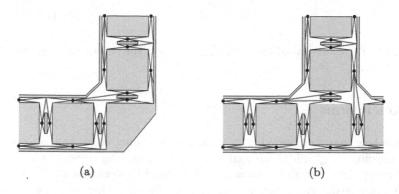

(a)   (b)

Fig. 14. A bended wire (a) and the split gadget (b).

A variable is represented by a long wire with split gadgets. Recall that in our reduction, all variables lie on a horizontal line. The clauses with positive variables lie above and the ones with negated variables lie below this line. We can control whether a variable or a negated variable is transmitted to the clause gadget by choosing appropriate positions for the corresponding split gadgets. In particular, if we translate the split gadget at the wire by one position to the left or right and keep the truth assignment for the wire, the orientation of the augmentation at the position of the new split gadget is flipped.

The *clause gadget* is illustrated in Fig. 15. The wires enter from left, right and below (respectively above). The 7-gon in the middle of the clause gadget can be augmented to a subgraph with minimum degree five if and only if it is connected to at least one wire in the true state. See also Fig. 16.   □

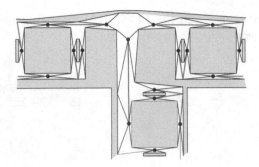

Fig. 15. A clause gadget, the three bold segments represent that the corresponding literals are set to true. The central 7-gon (blue) can be augmented to a subgraph of degree at least five if and only if at least one literal is true. (Color figure online)

Fig. 16. The three valid possibilities to augment the 7-gon in the clause gadget if one literal is true.

5   Conclusions

We study how many noncrossing straight-line matching edges can be drawn on top of a geometric graph G without crossing or using the edges of G. From an algorithmic point of view we show that it is hard to decide whether a perfect matching can be drawn on top of a polygon in this way. Our results on minimal maximal matchings show that a greedy algorithm will always draw at least $\frac{n}{7}$ edges on top of any n-vertex polygon. However, there are instances where it may draw not more than this amount of edges, although larger compatible matchings exist.

We are interested in how the function $\mathrm{mm}(G)$ (the size of a minimal maximal compatible matching of G) behaves among all geometric graphs G on n vertices and at most dn edges for any value $d \in [0, 3]$. Our results show that degree constraints (like d-regularity) help to determine $\mathrm{mm}(G)$ and also increase the value of $\mathrm{mm}(G)$ (compared to graphs on the same average degree). Indeed, we show that any 2-regular graph has at least $(n - 3)/11$ edges in any maximal compatible matching while the construction in Lemma 2 shows that there is a geometric graph G on n vertices with n edges and $\mathrm{mm}(G) = (n + 3)/13$. We do not know whether there is a family of such geometric graphs with values of $\mathrm{mm}(G)$ (asymptotically) even smaller than $n/13$. It is also not clear for which graphs $\mathrm{mm}(G)$ is maximized. For some drawings of empty graphs G we have $\mathrm{mm}(G) = \lceil \frac{n}{3} \rceil$. Is this the (asymptotically) largest possible value?

Acknowledgments. This work was initiated during the 15th European Research Week on Geometric Graphs (GG Week 2018) in Pritzhagen, Germany. We thank Kevin Buchin, Michael Hoffmann, Wolfgang Mulzer and Nadja Seiferth for helpful discussions.

References

1. Aichholzer, O., et al.: Compatible geometric matchings. Comput. Geom. **42**(6–7), 617–626 (2009). https://doi.org/10.1016/j.endm.2008.06.040
2. Aichholzer, O., García, A., Hurtado, F., Tejel, J.: Compatible matchings in geometric graphs. In: Proceeding of XIV Encuentros de Geometría Computacional, pp. 145–148. Alcalá, Spain (2011)
3. Akitaya, H.A., Korman, M., Rudoy, M., Souvaine, D.L., Tóth, C.D.: Circumscribing polygons and polygonizations for disjoint line segments. In: Proceedings of the 35th International Symposium on Computational Geometry (SoCG 2019), LIPIcs. Leibniz International Proceedings in Informatics, vol. 129, Art. No. 9, 17. Schloss Dagstuhl (2019). https://doi.org/10.4230/LIPIcs.SoCG.2019.9
4. Bae, S.W., de Berg, M., Cheong, O., Gudmundsson, J., Levcopoulos, C.: Shortcuts for the circle. Comput. Geom. **79**, 37–54 (2019). https://doi.org/10.1016/j.comgeo.2019.01.006
5. de Berg, M., Khosravi, A.: Optimal binary space partitions in the plane. In: Thai, M.T., Sahni, S. (eds.) COCOON 2010. LNCS, vol. 6196, pp. 216–225. Springer, Heidelberg (2010). https://doi.org/10.1007/978-3-642-14031-0_25
6. Cohen, N., et al.: A polynomial-time algorithm for outerplanar diameter improvement. J. Comput. System Sci. **89**, 315–327 (2017). https://doi.org/10.1016/j.jcss.2017.05.016
7. De Carufel, J.-L., Grimm, C., Schirra, S., Smid, M.: Minimizing the continuous diameter when augmenting a tree with a shortcut. WADS 2017. LNCS, vol. 10389, pp. 301–312. Springer, Cham (2017). https://doi.org/10.1007/978-3-319-62127-2_26
8. Hartmann, T., Rollin, J., Rutter, I.: Regular augmentation of planar graphs. Algorithmica **73**(2), 306–370 (2014). https://doi.org/10.1007/s00453-014-9922-4
9. Hurtado, F., Tóth, C.D.: Plane geometric graph augmentation: a generic perspective. In: Pach, J. (eds.) Thirty Essays on Geometric Graph Theory, pp. 327–354. Springer, New York (2013). https://doi.org/10.1007/978-1-4614-0110-0_17
10. Ishaque, M., Souvaine, D.L., Tóth, C.D.: Disjoint compatible geometric matchings. Discrete Comput. Geom. **49**(1), 89–131 (2012). https://doi.org/10.1007/s00454-012-9466-9
11. Jansen, K.: One strike against the min-max degree triangulation problem. Comput. Geom. **3**, 107–120 (1993). https://doi.org/10.1016/0925-7721(93)90003-O
12. Mulzer, W., Rote, G.: Minimum-weight triangulation is NP-hard. J. ACM **55**(2), 11:1–11:29 (2008). https://doi.org/10.1145/1346330.1346336
13. Pilz, A.: Augmentability to cubic graphs. In: Proceedings of 28th European Workshop on Computational Geometry (EuroCG 2012), pp. 29–32. Assisi, Italy, March 2012
14. Rappaport, D.: Computing simple circuits from a set of line segments is NP-complete. SIAM J. Comput. **18**(6), 1128–1139 (1989). https://doi.org/10.1137/0218075

15. Rappaport, D., Imai, H., Toussaint, G.T.: On computing simple circuits on a set of line segments. In: Proceedings of the 2nd Annual Symposium on Computational Geometry (SoCG 1986), pp. 52–60. ACM (1986). https://doi.org/10.1145/10515.10521
16. Tamassia, R., Tollis, I.G.: Planar grid embedding in linear time. IEEE Trans. Circ. Syst. **36**(9), 1230–1234 (1989)

Graph Drawing Contest

Graph Drawing Contest Report

Philipp Kindermann[1]([✉]), Tamara Mchedlidze[2], Wouter Meulemans[3],
and Ignaz Rutter[4]

[1] Universität Würzburg, Würzburg, Germany
`philipp.kindermann@uni-wuerzburg.de`
[2] Universiteit Utrecht, Utrecht, The Netherlands
`t.mtsentlintze@uu.nl`
[3] Eindhoven University of Technology, Eindhoven, The Netherlands
`w.meulemans@tue.nl`
[4] Universität Passau, Passau, Germany
`rutter@fim.uni-passau.de`

Abstract. This report describes the 27th Annual Graph Drawing Contest, held in conjunction with the 28th International Symposium on Graph Drawing and Network Visualization (GD'20) in Vancouver, BC, Canada. Due to the global COVID-19 pandemic, the contest was held completely online. The mission of the Graph Drawing Contest is to monitor and challenge the current state of the art in graph-drawing technology.

1 Introduction

Following the tradition of the past years, the Graph Drawing Contest was divided into two parts: the *creative topics* and the *live challenge*.

Creative topics were comprised by two data sets. The first data set described the genealogy and interactions in the Icelandic Saga *Hrafnkels Saga*. The second data set modeled various types of relationships between Korean artists, bands, and management and recording companies. Both data sets were provided by Timothy R. Tangherlini, whose research interests concern among others folk narrative and popular culture. The data sets were published a year in advance, and contestants submitted their visualizations before the conference started.

The live challenge took place during the conference in a format similar to a typical programming contest. Teams were presented with a collection of *challenge graphs* and had one hour to submit their highest scoring drawings. This year's topic was to minimize the number of crossings an upward straight-line drawing of a graph with vertex locations restricted to a grid.

Overall, we received 29 submissions: 9 submissions for the creative topics and 20 submissions for the live challenge.

2 Creative Topics

The general goal of the creative topics was to model each data set as a graph and visualize it with complete artistic freedom, and with the aim of communicating

© Springer Nature Switzerland AG 2020
D. Auber and P. Valtr (Eds.): GD 2020, LNCS 12590, pp. 507–519, 2020.
https://doi.org/10.1007/978-3-030-68766-3_39

as much information as possible from the provided data in the most readable and clear way.

We received 5 submissions for the first topic, and 4 for the second. Submissions were evaluated according to their aesthetic quality, domain-specific requirements, readability and clarity of the visualization, faithfulness of the data representation, and novelty of the visualization concept. We noticed overall that it is a complex combination of several aspects that make a standing out submission. These aspects include but are not limited to the understanding of the structure of the data, investigation of the additional data sources, applying intuitive and powerful for the case of the data visual metaphors, careful design choices, hand post-processing of the automatically created visualizations, as well as finding the thin balance between the amount of data to be represented and the clarity of the visualization. We made all the submissions available on the contest website in the form of a virtual poster exhibition. During the conference, we presented these submissions and announced the winners. We will now review the top three submissions for each topic (for a complete list of submissions, refer to http://www.graphdrawing.org/gdcontest/contest2020/results.html).

2.1   Hrafnkels Saga

Hrafnkels saga is one of the Icelanders' sagas, which tells of struggles between chieftains and farmers in the east of Iceland in the 10th century. The provided Hrafnkel Saga Network models relationships between the actants of the saga. The network was available as an Excel file with a spreadsheet for nodes, a spreadsheet for edges, and two spreadsheets for code references. The graph consisted of 43 nodes and 110 edges.

The Hrafnkel Saga network contained two disconnected components, one of them having only 6 vertices. We decided to leave this small component in the graph as it is typical for sagas, representing a "time stamp".

Third Place: Henry Fürster, Axel Kuckuk, and Lena Schlipf (University of Tübingen). The authors have chosen to present Hrafnkels Saga in a story-line visualization. They have carefully studies the text of the Saga and augmented their visualization with pieces of text describing the corresponding events. The authors designed the visualization in such a way that the drawing can be printed, cut out, folded, and glued together to form a rune stone where every face corresponds to a chapter of the saga. The committee was impressed by the understanding of the data that the authors had developed and the creativity of the design choices.

Second Place: Fabian Jogl, Melanie Paschinger, and Anna Chmurovic (TU Wien). The authors have chosen to represent the actants by concentric circular arcs, and the interactions among the actants by directed segments between the arcs. The types of the interactions were made clear with a variety of easy to understand glyphs, while the family trees were presented separately in a classical layout. The committee especially valued the readability of the drawing and the careful design choices leading to an intuitive visualization.

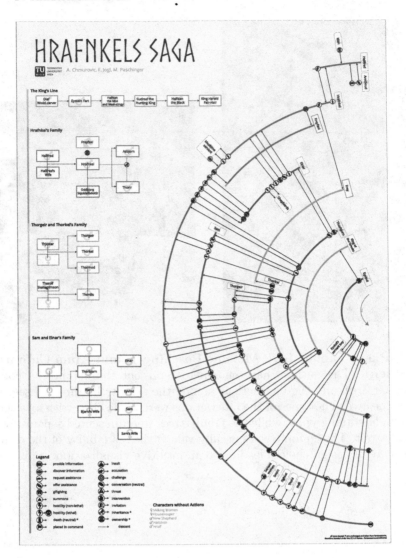

Winner: Tamara Drucks, Moritz Leidinger, and Giulio Pace (TU Wien). The authors have chosen a metro-map metaphor for their visualization. The characters were represented by metro-lines and the interactions by metro-stations, the type of which was additionally marked by glyphs. The committee was impressed by the aesthetics and the clarity of the visualization. We especially valued the design choices such as the spiral layout of the story-line visualization – giving the poster a creative look, the color palette, the carefully chosen glyphs, and the clever usage of the background image of Iceland in combination with the careful placement of text on it.

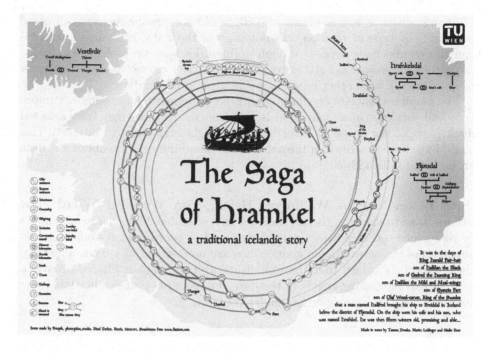

To represent the cyclic nature of the saga, we shaped the graph as a spiral where the development of the story is shown by character lines that pass through action nodes. The colors of the character lines stand for the different families, whereas the action types are represented by a set of different icons. The spiral is superimposed onto a map of Iceland to give, where possible, an approximate localization to the events.

Giulio Pace

2.2   K-Pop

K-pop, short for Korean pop, is an umbrella term for popular music originating from South Korea. The provided K-pop network models relations between Korean artists, bands (groups), and management & recording companies (labels). The network consisted of 4674 nodes and 5094 edges.

Though given as source-target, the edges did not always represent directed relations. The type of relation was not part of the data, but can be inferred from the related node types – though ambiguities may be present. Primarily, relations are *management* (label to group or artist) or *membership* (artist or group to group), with other relations being general *association*.

The graph consisted of one large connected component, a few smaller components and various triples, pairs and isolated nodes. Authors may have selected which parts of the graph to visualize to highlight structures of interest. This should have been described in the submission. More information about the data can be found in Broadwell et al.[1]

Third Place: Markus Wallinger, Hsiang-Yun Wu (TU Wien). The authors chose to visualize the big data set as a so-called *Hive Plot*, where nodes are represented on a radially oriented axis with a coordinate system based on properties of the network, and connections are represented as curves. The committee especially valued that the visualization, if the data is filtered appropriately, reveals interesting trends in the data, for example that smaller labels are more inclined to take male artists under contract.

---

[1] Peter M. Broadwell, Timothy R. Tangherlini, and Hyun Kyong Hannah Chang. *Online Knowledge Bases and Cultural Technology: Analyzing Production Networks in Korean Popular Music.* Series on Digital Humanities 7 (2016): 369–394.

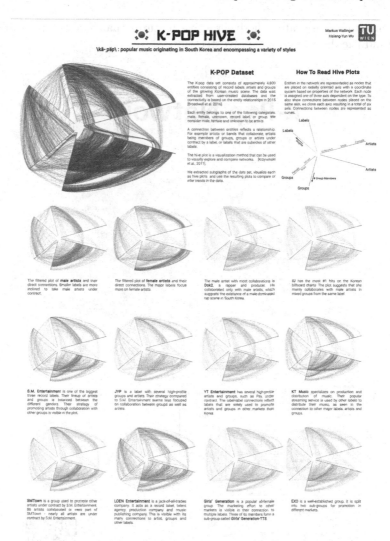

Second Place: Philipp Christoph Hekeler, Lucas Joos, Merle Kammer, Marco Piechotta, Markus Schramm, Kathrin Seßler, Friederike Maite Siemers, Andreas Tanner, Felix Weckesser, Henry Förster, and Axel Kuckuk (University of Tübingen). The authors have designed and implemented an interactive system which allowed to investigate the data set in details: in *overview mode*, the user sees the whole data and can see links between groups by hovering over nodes. By selecting a subset of the data with a lasso, the user can enter *Exploration Mode* to see the selected data in more detail. If one node is selected, some additional information about the node is shown in an infobox. The interactive tool is available online: http://algo.inf.uni-tuebingen.de/kpop/.

The committee appreciated the interactive features that give the possibility to understand the data after some investigations.

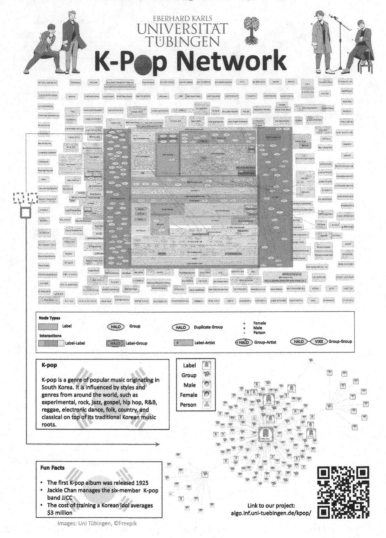

Winner: Rupert Ettrich, Julian Haumer, and Samantha Fuchs (TU Wien). The authors used a hierarchical layout in the style of a vintage star map, depicting labels as circles, and groups and artists that are signed by the label as smaller circles and icons inside. The authors decided to restrict the visualization to the largest component of the graph, consisting of 2422 nodes and 4190 edges. The committee appreciated the careful choice of the subset of data to be visualized which was sufficient to give an overview over the biggest labels in Korean pop music and the careful design choices which leaded to an aesthetically pleasing, readable and intuitive visualization.

To deal with the large dataset, we decided early that we want to focus only on the largest connected component. This allowed us to manually research all remaining label to label relations in order to further simplify the graph. For example, we noticed that some relations simply represent a renaming of a label. Thus, we were able to remove these kind of edges and display this information in another way. While initially being a force based layout, we manually adjusted label vertices to remove edge crossings and have only downward pointing edges to suggest hierarchy in-between labels. Within label vertices we display subgraphs of corresponding artists and groups with a force based layout with additional manual adjustments to improve legibility. The size of a group is reflected by the radius of its node. To achieve our Map of K-Pop Stars, we used D3 and built a custom web-based editor.   ""

Julian Haumer

3   Live Challenge

The live challenge took place during the conference and lasted exactly one hour. During this hour, local participants of the conference could take part in the manual category (in which they could attempt to draw the graphs using a supplied tool[2]), or in the automatic category (in which they could use their own software to draw the graphs). Because of the global COVID-19 pandemic, we allowed everybody in both categories to participate remotely. To coordinate the contest, give a brief introduction, answering questions, and giving participants the possibility to form teams, we setup a gather.town space[3].

The challenge focused on minimizing the number of crossings in an upward straight-line embedding of a given directed graph, with vertex locations restricted to a grid. The results were judged solely with respect to the number of crossings; other aesthetic criteria were not taken into account. This allows an objective way to evaluate each drawing.

3.1   The Graphs

In the manual category, participants were presented with six graphs. These were arranged from small to large and chosen to contain different types of graph structures. In the automatic category, participants had to draw the same six graphs as in the manual category, and in addition another seven larger graphs. Again, the graphs were constructed to have different structure.

For illustration, we include the sixth graph, which depicts a map of Western Europe, in the form we created the graph, in its initial state with vertices moved around randomly, the best manual solution we received (by team upwürz), and the best automatic solution we received (by team wildcat).

---

[2] http://graphdrawing.org/gdcontest/tool/.
[3] https://gather.town/eIfIsr1xGfJm1HCW/gdcontest2020.

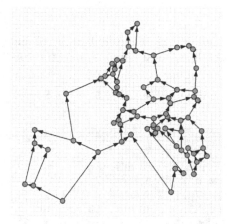

The sixth graph

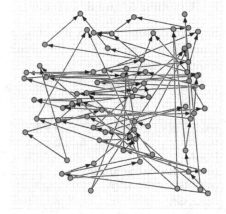

The provided drawing

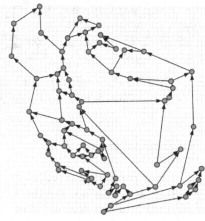

The best manual solution (upwürz)

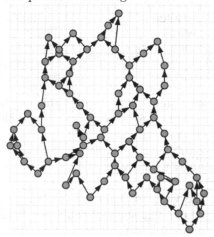

The best automatic solution (wildcat)

For the complete set of graphs and submissions, refer to the contest website at http://www.graphdrawing.org/gdcontest/contest2020/results.html. The graphs are still available for exploration and solving Graph Drawing Contest Submission System[4].

Similarly to the past years, the committee observed that manual (human) drawings of graphs often display a deeper understanding of the underlying graph structure than automatic and therefore gain in readability. The committee was also impressed by the fact that for four of the six small graphs the manual drawings were as good than the automatic drawings, while for one graph the manual drawings were better, and for one graph the automating drawings performed better.

---

[4] https://graphdrawingcontest.appspot.com.

3.2 Results: Manual Category

Below we present the full list of scores for all teams. The numbers listed are the numbers of crossings in the drawings; the horizontal bars visualize the corresponding scores.

graph	1	2	3	4	5	6	
Good luck	3	9	9	8	4	0	**3.**
elyjan	3	9	12	9	4	0	
ElDeMi	4	9	9	8	2	12	
New keyboard, who dis?	3	13	8	8	7	0	
#OnlyGoodCrossings	3	9	9	7	4	6	
Barat OK	4	10	15	10	10	18	
This time we brought 3 chargers!	3	9	9	8	4	0	**3.**
MinCrossing	4	20	15	264	33		
NowDutch	4	11	9	8	4	1	
upwürz	3	10	8	0	4	0	**1.**
Lakeside Inferno	4	10	10	8	4	6	
↑	4	10	10	0	4	0	**2.**
justforfun	4	31	17	126	5	13	

Shared third place: **This time we brought 3 chargers!**, consisting of Fouli Argyriou, Martin Gronemann, and Henry Förster; and **Good luck**, consisting of Oksana Firman.

Second place: ↑, consisting of Maarten Löffler.

Winner: **upwürz**, consisting of Jonathan Klawitter and Johannes Zink.

> " We want to thank all the fallen crossings for their support and without whom this success would not have been possible. If it was not for their honorable self-sacrifice, we would not have been able to achieve our winning scores with our strategy of merely searching for more comfy places for vertices. "
>
> *Jonathan Klawitter*

3.3 Results: Automatic Category

In the following we present the full list of scores for all teams that participated in the automatic category. The numbers listed are the numbers of crossings in the drawings; the horizontal bars visualize the corresponding scores.

graph	1	2	3	4	5	6	7	8	9	10	11	12	13	
Calvin	4	11	14	17	21	32	0							
wildcat	3	18	15	11	22	0	0	-	-					
NEKO#φωφ	5	9	11	14	13	21	0	1.2k	9.9k	8.6k	21k	154k	691k	
Simple is best	3	9	8	7	4	0	0	772	5.6k	31	90	486	26k	**2.**
JustForFun	6	9	14	7	4	13	0	-	41k	14k	74k	316k	1.4M	
Team ex-UBC	3	9	8	11	4	7	0		13k	1.9k	3.8k	84k	278k	
graphX	5	9	10	8	4	1	0	1.7k	13k	60	224	3.1k	1M	**3.**
NothingChangedFromGD19	3	9	7	2	2	0	0	772	5.2k	7	89	504	34k	**1.**

Third place: **graphX**, consisting of Luca Castelli Aleardi, Victor Bourdeaudhui, Lucas Guirardel, Auguste Poiroux, Geoffrey Saunois, Thomas Sepulchre, and Antoine Stark.

Second place: **Simple is best**, consisting of Sebastian Benner and Dominik Dürrschnabel.

Winner: **NothingChangedFromGD19**, consisting of Solveig Klepper, Axel Kuckuk, Paul Palomero Bernardo, Maximilian Pfister, Patrizio Angelini, Michalis Bekos, Henry Förster, and Michael Kaufmann.

> The hill-climbing probabilistic approach that we adopted last year was efficient enough to give us the first place in the contest even though the competition was stronger this year. We believe that we managed to win over the second placed team because we had a close look at the ranking formula and adjusted our optimization tactic over the input graphs accordingly.
>
> *Maximilian Pfister*

Acknowledgments. The contest committee would like to thank the organizing and program committee of the conference; the generous sponsors of the symposium; and all the contestants for their participation. We especially thank Timothy R. Tangherlini for providing us the data for this year's creative topics. Further details including all submitted drawings and challenge graphs can be found at the contest website:

http://www.graphdrawing.org/gdcontest/contest2020/results.html

Short Poster Papers

Algorithms and Experiments Using the Path Based Hierarchical Drawing Framework

Panagiotis Lionakis$^{(\boxtimes)}$, Giorgos Kritikakis, and Ioannis G. Tollis

Computer Science Department, University of Crete, Heraklion, Greece
{lionakis,gkrit,tollis}@csd.uoc.gr

Hierarchical graphs are important for many applications [4, 5] in several areas of research and business. They are directed (often acyclic) graphs and their visualization has received significant attention recently [3, 6, 7]. We present algorithms that are based on the concepts of path and channel decomposition of such graphs as proposed in the framework presented in [8, 9]. We present an extension of the framework of [9] by (a) compacting the drawing in the vertical direction, and (b) drawing the path transitive edges that were not drawn in [9]. This approach naturally splits the edges of G into: (a) *path edges* that connect consecutive vertices in the same path, (b) *cross edges* that connect vertices that belong to different paths, and (c) *path transitive edges* that connect non-consecutive vertices in the same path. A user may use a different color for each category. Our algorithms run in $O(km)$ time, where k is the number of paths, and provide better upper bounds than the ones given in [9] e.g.., the height of the resulting drawings is equal to the length of the longest path of G. Additionally, we bundle and draw all edges of the DAG in $O(m + n \log n)$ time, using minimum width per path.

The Path Based Hierarchical Drawing Framework (PBF) is based on the idea of partitioning the vertices of a graph G into (a minimum number of) *channels/paths*, that we call *channel/path decomposition* of G, which can be computed in polynomial time. PBF is orthogonal to Sugiyama framework in the sense that it is a vertical (instead of horizontal) decomposition of G into (vertical) paths/channels. Figure 1 shows two different hierarchical drawings of G with 31 nodes and 69 edges: Part (a) shows a drawing Γ_1 of G which follows our framework; part (b) shows a drawing Γ_2 of G as computed by OGDF [2] that follows the Sugiyama framework [10]. Drawing Γ_1 has 74 crossings, 33 bends, width 14, and height 16 (area 224). On the other hand, drawing Γ_2 has 72 crossings, 64 bends, width 42 and height 16 (area 672). The width and height reported by OGDF are 961 and 2273, respectively. We normalized all such figures (also in Fig. 2) in order to provide a reasonable comparison.

Figure 2 results regarding the number of crossings, bends, width, height and area of the drawings. The experiments show that the drawings produced by our algorithms have a significantly lower number of bends and are much smaller in area than the ones produced by OGDF. On the other hand, the drawings of OGDF have a lower number of crossings when the input graphs are relatively sparse. As expected, OGDF is better than our algorithms in the number of crossings since OGDF places a significant weight in minimizing crossings, whereas we

© Springer Nature Switzerland AG 2020
D. Auber and P. Valtr (Eds.): GD 2020, LNCS 12590, pp. 523–525, 2020.
https://doi.org/10.1007/978-3-030-68766-3

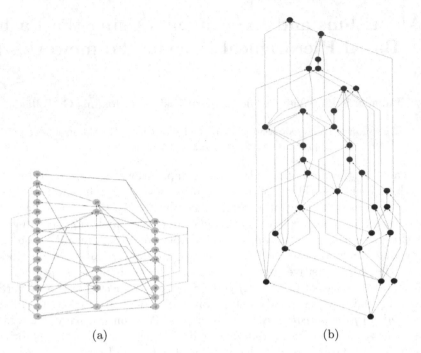

(a)   (b)

Fig. 1: (a) Drawing Γ_1 of G computed by our algorithms and visualized by Tom Sawyer Perspectives [1]. (b) Drawing Γ_2 of G as computed by OGDF.

n=50	m=62		m=87		m=150		m=250		m=500	
	PBF	OGDF	PBF	OGDF	PBF	OGDF	PBF	OGDF	PBF	OGDF
Crossings	17	6	126	92	839	703	2469	2585	8061	14479
Bends	15	25	22	69	54	188	91	380	176	863
Width	12	36	13	59	18	116	24	206	33	442
Height	13	16	17	21	20	23	21	28	24	33
Area	156	576	221	1239	360	2668	504	5768	792	14586

n=100	m=125		m=175		m=300		m=500		m=1000	
	PBF	OGDF	PBF	OGDF	PBF	OGDF	PBF	OGDF	PBF	OGDF
Crossings	105	29	705	430	3749	3366	13068	12890	42934	62695
Bends	29	50	59	143	108	388	194	757	324	1737
Width	18	60	20	103	26	230	36	414	51	912
Height	22	27	22	32	26	30	27	28	38	45
Area	396	1620	440	3296	676	6900	972	11592	1938	41040

n=200	m=250		m=350		m=600		m=1000		m=2000	
	PBF	OGDF	PBF	OGDF	PBF	OGDF	PBF	OGDF	PBF	OGDF
Crossings	594	278	3094	1929	16357	12490	52095	49278	209446	266260
Bends	48	100	128	288	226	763	350	1519	597	3498
Width	23	107	32	216	37	450	52	830	83	1813
Height	27	46	33	48	39	40	38	42	49	60
Area	621	4922	1056	10368	1443	18000	1976	34860	4067	108780

n=500	m=625		m=875		m=1500		m=2500		m=5000	
	PBF	OGDF	PBF	OGDF	PBF	OGDF	PBF	OGDF	PBF	OGDF
Crossings	2746	1501	15474	11221	102195	81537	389241	327017	1486777	1636057
Bends	123	246	280	730	544	1916	911	3909	1482	8802
Width	41	260	51	531	71	1142	96	2103	138	4565
Height	42	78	39	71	50	57	64	74	79	90
Area	1722	20280	1989	37701	3550	65094	6144	155622	10902	410850

Fig. 2: Results on *number of crossings, bends, width, height* and *area* for *PBF* and *OGDF* using different DAGs.

do not explicitly minimize crossings. Hence, this approach offers an interesting alternative to visualize hierarchical graphs.

References

1. Sawyer, T.: Software. www.tomsawyer.com
2. Chimani, M., Gutwenger, C., Jünger, M., Klau, G.W., Klein, K., Mutzel, P.: The open graph drawing framework (OGDF). In: Tamassia, R. (ed.) Handbook on Graph Drawing and Visualization, pp. 543–569. CRC Press, Boca Raton (2013)
3. Di Battista, G., Eades, P., Tamassia, R., Tollis, I.G.: Graph Drawing: Algorithms for the Visualization of Graphs. Prentice-Hall, Upper Saddle River (1999)
4. Di Battista, G., Pietrosanti, E., Tamassia, R., Tollis, I.G.: Automatic layout of pert diagrams with x-pert. In: Proceedings of 1989 IEEE Workshop on Visual Languages, pp. 171–176. IEEE (1989)
5. Fisher, D.L., Goldstein, W.M.: Stochastic PERT networks as models of cognition: derivation of the mean, variance, and distribution of reaction time using order-of-processing (op) diagrams. J. Math. Psychol. **27**(2), 121–151 (1983)
6. Kaufmann, M., Wagner, D. (eds.): Drawing Graphs. LNCS, vol. 2025. Springer, Heidelberg (2001). https://doi.org/10.1007/3-540-44969-8
7. Nikolov, N.S., Healy, P.: In: Tamassia, R. (ed.) Handbook of Graph Drawing and Visualization, pp. 409–453. CRC Press, Boca Raton (2014)
8. Ortali, G., Tollis, I.G.: Algorithms and bounds for drawing directed graphs. In: Biedl, T., Kerren, A. (eds.) GD 2018. LNCS, vol. 11282, pp. 579–592. Springer, Cham (2018). https://doi.org/10.1007/978-3-030-04414-5_41
9. Ortali, G., Tollis, I.G.: A new framework for hierarchical drawings. J. Graph Algorithms Appl. **23**(3), 553–578 (2019). https://doi.org/10.7155/jgaa.00502
10. Sugiyama, K., Tagawa, S., Toda, M.: Methods for visual understanding of hierarchical system structures. IEEE Trans. Syst. Man Cybern. **11**(2), 109–125 (1981)

Drawing Outer-1-planar Graphs Revisited

Therese Biedl$^{(\boxtimes)}$ 🔟

David R. Cheriton School of Computer Science, University of Waterloo,
Waterloo, ON N2L 1A2, Canada
biedl@uwaterloo.ca

1 Introduction

A *1-planar* graph is a graph that can be drawn in the plane such that every edge
has at most one crossing. Many graph-theoretic and graph-drawing results are
known for 1-planar graphs, see for example [5]. One subclass is the class of *outer-
1-planar (o1p) graphs*, which have a 1-planar drawing such that additionally
every vertex is on the outer-face (the unbounded region of the drawing).

Outer-1-planar graphs were introduced by Eggleton [4] and later studied
extensively by Auer, Bachmeier, Brandenburg, Gleißner, Hanauer, Neuwirth and
Reislhuber [1]. Among others, they characterize the forbidden minors of outer-
1-planar graphs, give a recognition algorithm, and give bounds on various graph
parameters such as number of edges, treewidth, stack number and queue number.
Finally they turn to drawing algorithms for outer-1-planar graphs, and here claim
the following result: "Every o1p graph has a planar visibility representation in
$O(n \log n)$ area." (Theorem 8).

2 Lower Bound

We show that the claim by Auer et al. is **incorrect**, and construct an outer-
1-planar graph that requires $\Omega(n^2)$ area in any planar poly-line drawing (this
implies the lower bound for visibility representations as well [3]). Our lower-
bound graph G_L (for $L \geq 2$) consists of a $2 \times L$-grid with every inner face filled
with a crossing. Clearly this is an outer-1-planar graph, see Fig. 1.

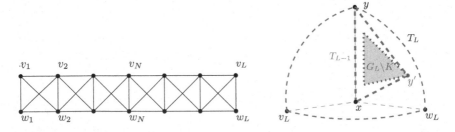

Fig. 1. The outer-1-planar graph G_7, and three 1-fused stacked triangles.

Supported by NSERC.

D. Auber and P. Valtr (Eds.): GD 2020, LNCS 12590, pp. 526–528, 2020.
https://doi.org/10.1007/978-3-030-68766-3

Call a set of triangles $T_1, \ldots, T_\ell$ (in a planar embedding) *1-fused stacked triangles* [2] if for $i = 2, \ldots, \ell$ the region bounded by T_i includes every vertex of T_{i-1}, and T_i and T_{i-1} have at most one vertex in common.

Enumerate the vertices of G_L as in Fig. 1.

Lemma 1. *Fix $L \geq 2$. Any planar embedding Γ of G_L with (v_L, w_L) on the outer-face contains $L - 1$ 1-fused stacked triangles, of which one is the outer-face and has vertices in $K := \{v_L, w_L, v_{L-1}, w_{L-1}\}$.*

Proof. The vertices of K form a K_4; its induced embedding Γ_K is hence unique up to renaming. By assumption the outer-face T_L of Γ_K contains v_L, w_L and one vertex $y \in \{v_{L-1}, w_{L-1}\}$; set $x = \{v_{L-1}, w_{L-1}\} \setminus y$.

If $L = 2$, then we are done (use triangle T_L). If $L > 2$, then graph $G_L \setminus K$ is connected, so must reside entirely within one face of Γ_K. It contains neighbours of x and y, so it must reside within one of the interior faces of Γ_K adjacent to (x, y). Thus T_L is the outer-face not only of Γ_K but also of Γ.

The graph $G' := G_L \setminus \{v_L, w_L\}$ is a copy of G_{L-1}. Since $G_L \setminus K$ resides within one interior face of Γ_K, edge $(v_{L-1}, w_{L-1}) = (x, y)$ is on the outer-face of the induced drawing Γ' of G'. Applying induction to Γ', we hence obtain $L - 2$ 1-fused stacked triangles $T_2, \ldots, T_{L-1}$, where $T_{L-1} = \{v_{L-1}, w_{L-1}, y'\}$ for some $y' \in \{v_{L-2}, w_{L-2}\}$. Adding T_L to this gives the desired set of 1-fused stacked triangles for G, since T_L and T_{L-1} have only vertex y in common.

Fix an integer N. The graph G_L (for $L = 2N+1$) contains two copies of G_N, and in any planar drawing one copy has (v_N, w_N) on the outer-face. In this copy we can find $N - 1$ 1-fused stacked triangles, which require width and height at least N in any planar drawing [2]. Since G_L has $4N - 2$ vertices, we have:

Theorem 1. *There exists an n-vertex outer-1-planar graph that requires width and height at least $(n + 2)/4$ in any planar poly-line grid-drawing.*

3   Outlook

Where is the error in [1]? They used a visibility representation of area $O(n \log n)$ of an outer-planar subgraph G' [2], and added the edges of $G \setminus G'$. The drawing of [2] is created by splitting the graph, drawing parts recursively, and putting them together. Auer et al. assume that the edges of $G \setminus G'$ occur in one particular way relative to this graph-split. But the graph-split is determined by the size of the sub-graphs, and so they have missed some cases where edges of $G \setminus G'$ could be. (As our results show, it would be impossible to do the other cases without adding crossings or increasing the area.)

We *can* achieve $O(n \log n)$ area if we allow crossings and bends, and even exactly reflect the outer-1-planar drawing. To do so, use again the visibility representation of [2] of an outer-planar subgraph G'. Adding edges of $G \setminus G'$ can easily be done if we allow crossings and up to 4 bends per edge; the area at most doubles. As we will show in a forthcoming paper, up to 2 bends per edge is sufficient if we modify the algorithm of [2] a bit. Achieving $O(n \log n)$ area and 0 bends remains an open problem.

References

1. Auer, C., Bachmaier, C., Brandenburg, F.J., Gleißner, A., Hanauer, K., Neuwirth, D., Reislhuber, J.: Outer 1-planar graphs. Algorithmica **74**(4), 1293–1320 (2016)
2. Biedl, T.: Small drawings of outerplanar graphs, series-parallel graphs, and other planar graphs. Discrete Comput. Geom. **45**(1), 141–160 (2011)
3. Biedl, T.: Height-preserving transformations of planar graph drawings. In: Duncan, C., Symvonis, A. (eds.) GD 2014. LNCS, vol. 8871, pp. 380–391. Springer, Heidelberg (2014). https://doi.org/10.1007/978-3-662-45803-7_32
4. Eggleton, R.: Rectilinear drawings of graphs. Utilitas Mathematica **29**, 149–172 (1986)
5. Kobourov, S., Liotta, G., Montecchiani, F.: An annotated bibliography on 1-planarity. Comput. Sci. Rev. **25**, 49–67 (2017)

Contacts of Circular Arcs Representations of Tight Surface Graphs

James Curickshank[1] and Qays Shakir[2](✉) (iD)

[1] National University of Ireland, Galway, Ireland
james.cruickshank@nuigalway.ie
[2] Middle Technical University, Baghdad, Iraq
qays.shakir@gmail.com

Abstract. We present contacts of circular arcs representations (CCA-representation) of tight surface graphs where the vertices are interior disjoint circular arcs in the flat surface and each edge is realised by an endpoint of one arc touching the interior of another.

Keywords: Surface graphs · Tight graph · Geometric representation

1 Tight Surface Graphs

Let G be a surface graph [1] and let V and E be the set of the vertices and edges, respectively, of G. G is called $(2, 2)$-sparse if for every nonempty surface subgraph H of G, $|E_H| \leq 2|V_H| - 2$ where V_H respectively E_H is the set of the vertices respectively edges of H. G is called $(2, 2)$-tight if it is $(2, 2)$-sparse and $|E| = 2|V| - 2$. We mention that there are many other kinds of tight and sparse graphs, see for example [2]. However, we focus on $(2, 2)$-tight surface graphs. In [3], we considered studying constructions of higher genus surface graphs which are tight. Such a study was partly motivated by the connection to gain graphs [4] and related sparsity counts. We were interested in investigating torus graphs that correspond via the universal covering construction [5] to doubly periodic plane graphs. Thus, our geometric application can be viewed as a partial characterisation of a certain type of crystallographic structure in the flat plane, [6]. In the following we exhibit inductive constructions for the classes of $(2, 2)$-tight cylinder and torus graphs. In general, an inductive construction consists of two main tools; a set of inductive operations, Fig. 1, and a set of small graphs in which no contraction is possible, we call such graphs irreducible.

Theorem 1. *[3] Every $(2, 2)$-tight cylinder and torus graph can be constructed from one of 2 and 116, respectively, irreducible $(2, 2)$-tight cylinder and torus graphs by a sequence of digon, triangle or quadrilateral splitting operations.*

2 Contacts of Circular Arcs Representations

One of the fundamental topics in geometric graph theory is the geometric representations of graphs. Such a topic investigates whether a given graph admits a

D. Auber and P. Valtr (Eds.): GD 2020, LNCS 12590, pp. 529–531, 2020.
https://doi.org/10.1007/978-3-030-68766-3

certain kind of geometric representation. Many kinds of geometric contexts have been used in geometric representations of graphs such as triangles [7] and polygons [8]. For tight graphs, representations of such graphs have been considered in various studies, see [9, 10] and [11]. We are interested in geometric contexts that are circular arcs such that contacts are allowed while crossing are not and there are not two circular arcs have interior points in common. A CCA-representation of a surface graph G is a configuration of circular arcs embedded in the surface so that the graph induced by the contacts between the arcs is isomorphic to G. In [12], it is shown that $(2,2)$-tight plane graphs admit CCA-representations in the flat plane. Our main result, Theorem 2, can be proved by showing that each irreducible $(2,2)$-tight cylinder and torus graph admits CCA-representation and the inductive operations in Fig. 1 are CCA-representable.

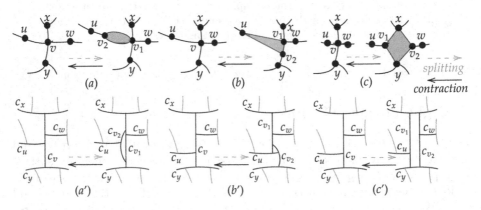

Fig. 1. (a), (b) and (c), respectively, illustrates a digon, triangle and quadrilateral splitting, respectively, and its CCA-rep. is given in (a'), (b') and (c'), respectively.

Theorem 2. *Every $(2,2)$-tight cylinder and torus graph admits a CCA-representation in the flat cylinder and torus, respectively.*

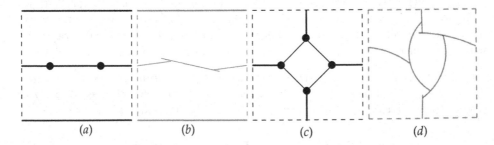

Fig. 2. (a) $(2,2)$-tight cylinder graph and (b) its CCA-representation in the flat cylinder. (c) $(2,2)$-tight torus graph and (d) its CCA-representation in the flat torus.

References

1. Mohar, B., Thomassen, C.: Graphs on surfaces (Johns Hopkins Studies in the Mathematical Sciences). Johns Hopkins University Press, Baltimore (2001)
2. Lee, A., Streinu, I.: Pebble game algorithms and sparse graphs. Discrete Math. **308**(8), 1425–1437 (2008)
3. Cruickshank, J., Kitson, D., Power, S.C., Shakir, Q.: Topological constructions for tight surface graphs. http://arxiv.org/abs/1909.06545 (2019)
4. Zaslavsky, T.: A mathematical bibliography of signed and gain graphs and allied areas. Electron. J. Combin. **5**, 124 (1998). Dynamic Surveys 8, (electronic), Manuscript prepared with Marge Pratt
5. Dillen, F., Verstraelen, L.: Handbook of Differential Geometry. Elsevier, North-Holland, Amsterdam (2000)
6. Veech, W.A.: Flat surfaces. Amer. J. Math. **115**, 589–689 (1993)
7. de Fraysseix, H., Ossona, P., Rosenstiehl, P.: On triangle contact graphs. Comb. Probab. Comput. **3**, 233–246 (1994)
8. Christian, A., Emden, R., Hu, Y., Kaufmann, M., Kobourov, G.: Optimal polygonal representation of planar graphs. Algorithmica **63**(3), 672–691 (2012)
9. Chaplick, S., Ueckerdt, T.: Planar graphs as VPG-graphs. J. Graph Algorithms Appl. **17**(4), 475–494 (2013)
10. Goncalves, D., Isenmann, L., Pennarun, C.: Planar graphs as L-intersection or L-contact graphs. In: SODA, pp. 172–184 (2018)
11. Kobourov, S.G., Ueckerdt, T., Verbeek, K.: Combinatorial and geometric properties of planar Laman graphs. In: SODA, pp. 1668–1678. SIAM (2013)
12. Alam, M.J., et al.: Contact graphs of circular arcs. In: Dehne, F., Sack, J.-R., Stege, U. (eds.) WADS 2015. LNCS, vol. 9214, pp. 1–13. Springer, Cham (2015). https://doi.org/10.1007/978-3-319-21840-3_1

Visualizing Communities and Structure in Dynamic Networks

Evan Ezell[1,2]($\boxtimes$) (ID), Seung-Hwan Lim[1,2] (ID), David Anderson[3],
and Robert Stewart[1,2] (ID)

[1] University of Tennessee, Knoxville, TN 37996, USA
{ezellec,lims1,stewartrn}@ornl.gov
[2] Oak Ridge National Laboratory, Oak Ridge, TN 37830, USA
[3] Cadre5, Knoxville, TN 37932, USA
david.anderson@cadre5.com

1 Introduction

Communities in dynamic networks evolve and exhibit different behavioral actions through time. These community actions: birth, death, growth, contraction, merging, splitting, continuity, and resurgence are fundamentally driven by changes in the underlying network topology. While it is interesting to visualize community evolution and network topology on their own, there is often a need to visualize each in coordination together to further understand network characteristics. We present a novel visualization technique for viewing the change in communities while simultaneously visualizing its relationship with the underlying network structure of a dynamic graph. We showcase the efficacy of our technique in supporting dynamic network analysis by presenting an example from economic trade data [1].

2 Motivation

Existing visualization methods attempt to preserve network structure by adjoining node-link diagrams next to community alluvial diagrams [3]. This approach requires the use of limited force-directed layouts by restricting the horizontal axis on which a node can appear. There are a few issues with this approach. Node-link diagrams do not scale well to large networks and will likely cause the observer to be overwhelmed, especially in cases when individual communities are large. Also, individual node-link diagrams for each community make it difficult to show inter-community edges. Because each node-link diagram's layout is oblivious to other communities, connecting all of the nodes between communities is not feasible without cluttering the visualization. Traditional approaches have tried to remedy this issue by displaying inter-community edges via overlaid arcs or curved links. While these curved links allow the user to see inter-community relationships, this approach creates dissonance between the visualization of inter-community edges and intra-community edges.

D. Auber and P. Valtr (Eds.): GD 2020, LNCS 12590, pp. 532–534, 2020.
https://doi.org/10.1007/978-3-030-68766-3

3 Our Approach

We employ a novel strategy for dealing with these issues. Our visualization technique superimposes a static Biofabric-like diagram [2] atop each timestep of a traditional alluvial community flow diagram. Our technique improves upon existing work by allowing the visualization of communities in larger networks all while preserving the awareness of the underlying structure of the graph topology. Visualizing the structure of networks in addition to community evolution is important in understanding overall network dynamics and the concrete underpinnings for specific changes in individual communities. Our approach consolidates visualizing inter-community edges and intra-community edges by displaying all types of edges in a congruent manner, eliminating the need to employ two different strategies to visualize each edge type, creating a more cohesive display of network characteristics. This new visualization approach allows researchers to more adequately verify different community detection algorithms, locate noisy nodes and anomalous data, and find useful trends and patterns in the dynamics of an evolving network. Figure 1 shows an example of our approach on vehicle trade data [1].

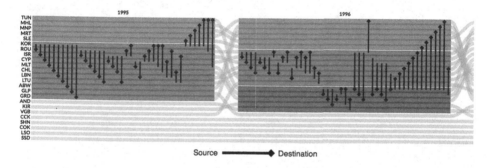

Fig. 1. An example of our visualization technique on vehicle trade data [1] for the years 1995 and 1996.

Acknowledgements. This manuscript was authored by UT-Battelle LLC under Contract No. DE-AC05-00OR22725 with DOE. The US government retains—and the publisher, by accepting the article for publication, acknowledges that the US government retains—a nonexclusive, paid-up, irrevocable, world-wide license to publish or reproduce the published form of the manuscript or allow others to do so for US government purposes. DOE will provide public access to these results of federally sponsored research in accordance with the DOE Public Access Plan http://energy.gov/downloads/doe-public-access-plan.

References

1. at Harvard University, T.G.L.: International Trade Data (SITC, Rev. 2) (2019). https://doi.org/10.7910/DVN/H8SFD2
2. Longabaugh, W.: Combing the hairball with biofabric: a new approach for visualization of large networks. BMC Bioinform. **13**, 275 (2012). https://doi.org/10.1186/1471-2105-13-275
3. Vehlow, C., Beck, F., Auwärter, P., Weiskopf, D.: Visualizing the evolution of communities in dynamic graphs. Comput. Graphics Forum **34**(1), 277–288 (2015). https://doi.org/10.1111/cgf.12512. https://onlinelibrary.wiley.com

The Complexity of Finding Tangles

Oksana Firman[1]($\boxtimes$), Stefan Felsner[2], Philipp Kindermann[1],
Alexander Ravsky[3], Alexander Wolff[1], and Johannes Zink[1]

[1] Institut für Informatik, Universität Würzburg, Würzburg, Germany
coksana.firman@uni-wuerzburg.de
[2] TU Berlin, Berlin, Germany
felsner@math.tu-berlin.de
[3] Pidstryhach Institute for Applied Problems of Mechanics and Mathematics,
National Academy of Sciences of Ukraine, Lviv, Ukraine
alexander.ravsky@uni-wuerzburg.de

This work is devoted to the visualization of so-called *chaotic attractors*, describing chaotic dynamic systems. Such systems arise in physics, celestial mechanics, electronics, fractals theory, chemistry, biology, genetics, and population dynamics. Birman and Williams [1], investigating how the orbits of attractors are knotted, described their topological structure via tangles (described below). Mindlin et al. [4] characterized attractors via integer matrices containing numbers of swaps between the orbits.

Olszewski et al. [5] studied computational aspects of visualizing chaotic attractors. In the framework of their paper, one is given a set of y-monotone curves called *wires* that hang off a horizontal line in a fixed order, and a multiset of swaps between the wires (called *list*). A *tangle* is a visualization of these swaps, i.e., a sequence of permutations of the wires such that consecutive permutations differ only in swaps of neighboring wires (but disjoint swaps can be done simultaneously). We call a list *feasible* if there is a tangle realizing it. Olszewski et al. gave an exponential-time algorithm for minimizing the *height* of a tangle (that is, the number of permutations) and tested it on a benchmark set.

Later, we [3] showed that tangle-height minimization is NP-hard (by reduction from 3-PARTITION). We also presented another (exponential-time) algorithm for the problem. Using an extended benchmark set, we showed that in most cases our algorithm is faster than the algorithm of Olszewski et al.

In this work, we strengthen our previous complexity result and show that it is even NP-hard to decide, given a multiset of swaps and a start permutation of the wires, whether there is *any* tangle that realizes the given swaps. We call this problem LIST-FEASIBILITY.

Theorem 1. LIST-FEASIBILITY *is NP-hard (even if every pair of wires has at most eight swaps).*

In the proof we use a reduction from POSITIVE NAE 3-SAT DIFF, a variant of the NP-hard problem NOT-ALL-EQUAL 3-SAT, where no negative literals are

---

This abstract is based on the paper [2].

© Springer Nature Switzerland AG 2020
D. Auber and P. Valtr (Eds.): GD 2020, LNCS 12590, pp. 535–537, 2020.
https://doi.org/10.1007/978-3-030-68766-3

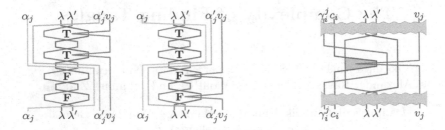

Fig. 1. Left: a variable gadget (in true/false state); right: a clause gadget

admitted and each clause contains three different variables. In the full version of this paper [2], we show that POSITIVE NAE 3-SAT DIFF is also NP-hard.

We sketch the idea behind our reduction. Given an instance F of POSITIVE NAE 3-SAT DIFF, we construct in polynomial time a list L of swaps such that there is a tangle T realizing L if and only if F has a satisfying truth assignment.

In L we have two "central" wires λ and λ' that swap eight times; see Fig. 1. This yields two types of loops: four λ'–λ loops, where λ' is on the left and λ is on the right side, and three λ–λ' loops with λ on the left and λ' on the right side. Each variable gadget contains a specific wire, the *variable wire*, that represents the variable, and each clause gadget contains a specific wire, the *clause wire*, that represents the clause. For each occurrence of a variable in a clause, the corresponding variable and clause wires swap twice in one of the four λ'–λ loops. We call the first two λ'–λ loops *true-loops*, and the last two λ'–λ loops *false-loops*. If the variable is true, then the corresponding variable wire swaps twice with the corresponding clause wires in a true-loop, otherwise in a false-loop.

We first describe the variable gadgets. In order to prevent a variable wire v_j from intersecting both a true- and a false-loop, we introduce two wires α_j and α'_j. These wires neither swap with v_j nor with each other, but they have two swaps with both λ and λ'. Using additional wires, we force α_j and α'_j to have the two true-loops on their right and the two false-loops on their left, or vice versa. This ensures that v_j cannot reach both a true- and a false-loop.

We now turn to the clause gadgets. We force each clause wire c_i to appear in all λ'–λ loops by using additional wires. Since every clause has exactly three different positive variables, we want to force variable wires that belong to the same clause to swap with the corresponding clause wire in different λ'–λ loops. This way, every clause contains at least one true and at least one false variable if F is satisfiable. We call a part of a clause wire c_i that is inside a λ'–λ loop— i.e., a λ'–c_i loop—an *arm of the clause* c_i. In order to "*protect*" the arm that is intersected by a variable wire from other variable wires, we use the wire γ_i^j. Three additional wires help us to force γ_i^j to protect the correct arm.

Finally, we argue why our reduction is correct. If F is satisfiable, we obtain a tangle from F as described above. The tangle realizes L, so L is feasible.

On the other hand, if there is a tangle that realizes the list L that we obtain from the reduction, then F is satisfiable. This follows from the rigid structure of a tangle that realizes L. The only flexibility is in which type of loop (true or false) a variable wire swaps with the corresponding clause wire.

References

1. Birman, J.S., Williams, R.F.: Knotted periodic orbits in dynamical systems–I: Lorenz's equation. Topology **22**(1), 47–82 (1983). https://doi.org/10.1016/0040-9383(83)90045-9
2. Firman, O., Felsner, S., Kindermann, P., Ravsky, A., Wolff, A., Zink, J.: The complexity of finding tangles. Arxiv report (2020). http://arxiv.org/abs/2002.12251
3. Firman, O., Kindermann, P., Ravsky, A., Wolff, A., Zink, J.: Computing height-optimal tangles faster. In: Archambault, D., Tóth, C.D. (eds.) GD 2019. LNCS, vol. 11904, pp. 203–215. Springer, Cham (2019). https://doi.org/10.1007/978-3-030-35802-0_16
4. Mindlin, G., Hou, X.J., Gilmore, R., Solari, H., Tufillaro, N.B.: Classification of strange attractors by integers. Phys. Rev. Lett. **64**, 2350–2353 (1990). https://doi.org/10.1103/PhysRevLett.64.2350
5. Olszewski, M., et al.: Visualizing the template of a chaotic attractor. In: Biedl, T., Kerren, A. (eds.) GD 2018. LNCS, vol. 11282, pp. 106–119. Springer, Cham (2018). https://doi.org/10.1007/978-3-030-04414-5_8

MetroSets: Visualizing Hypergraphs as Metro Maps

Ben Jacobsen[1] , Markus Wallinger[2(✉)] , Stephen Kobourov[1] ,
and Martin Nöllenburg[2]

[1] University of Arizona, Tucson, AZ, USA
bjacobsen@email.arizona.edu,kobourov@cs.arizona.edu
[2] TU Wien, Vienna, Austria
{mwallinger,noellenburg}@ac.tuwien.ac.at

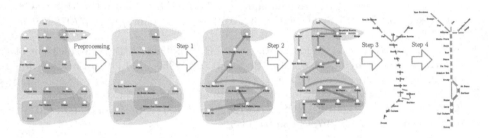

Fig. 1. The MetroSets pipeline. The input set system (here: characters of *The Simpsons*) is first compressed into a combinatorially equivalent smaller instance; Step 1 creates an optimized path support graph; Step 2 reinserts temporarily discarded elements; Step 3 creates an initial layout of the support graph; finally Step 4 schematizes the layout as a metro map and places the labels.

A path-based support of a hypergraph $\mathcal{H} = (V, \mathcal{S})$ is a simple graph $G = (V, E)$ with the same vertex set V, with the property that the subgraph of G induced by each hyperedge in $\mathcal{S}$ contains a Hamiltonian path. A metro map drawing of a hypergraph consists of a path-based support, which has been embedded in the plane and adheres to the metro map design goals. Each path corresponding to a hyperedge is drawn as a colored curve traversing its vertices, analogous to the way that subway lines traverse subway stations. This style of visualization has been applied in scientific works [11, 14] and information graphics [2, 5, 15].

We have created MetroSets [9], a flexible, online system for visualizing hypergraphs as metro-map drawings. MetroSets decomposes this task into a four-step pipeline, see Fig. 1, with additional pre- and post-processings steps. Because many of the underlying problems involved are NP-hard, we implement multiple heuristic algorithms for each step, which can be freely mixed and matched to produce different visualizations of the same data.

We begin by creating a combinatorially equivalent smaller instance of the input hypergraph $\mathcal{H}$, called $\mathcal{H}'$. For this we set aside all vertices that are only

D. Auber and P. Valtr (Eds.): GD 2020, LNCS 12590, pp. 538–540, 2020.
https://doi.org/10.1007/978-3-030-68766-3

incident to one hyperedge and afterwards merge all sets of vertices that are incident to the exact same set of hyperedges. Then, we create a path-based support of $\mathcal{H}'$, which is equivalent to determining a total order for the vertices along each path. We implement two methods for this task. The first method tries to minimize the total number of edges needed to draw the support graph, such that whenever the intersection of two hyperedges contains two or more vertices, the hyperedges should be drawn in parallel, rather than branching of and rejoining. The second approach assigns a cost to each edge in the support graph, which is lower for pairs of nodes that share many hyperedges, and attempts to find an ordering which minimizes this cost for each path.

The next step reintroduces the previously discarded vertices. The merged vertices can be expanded trivially, while vertices belonging to single hyperedges present a more interesting choice, as they could in principle be introduced at any position in the given path. The first-viable approach introduces vertices at the beginning or end of their respective support path, leading to long, trailing lines of single-hyperedge vertices. The split-insert approach instead places half of the vertices at the beginning of the line, and distributes the others evenly throughout all of the candidate positions along the path.

After the support graph is constructed, we determine an initial layout for it. The simpler approach is to use a force-based approach [8], which embeds the graph purely based on the structure determined in the previous steps. The second approach instead iterates between first laying out the graph, using either Kamada-Kawai [10] or Neato [7], and then secondly revising the ordering of vertices each path, incorporating the Euclidean distance between vertices into the cost function.

This embedding is then used as input for the schematization algorithms. We implemented the force-based schematization approach of Chivers and Rodgers [4], as well as an adapted version of Lutz et al. [6] least squares approximation. Finally, in a post-processing step we optimize line orders [3, 13] and apply the greedy labeling algorithm of Niedermann and Haunert [12].

Having implemented multiple algorithms for each step of the pipeline, we naturally wanted to determine which choices of steps were optimal for which purposes. To this end, we defined different quality criteria (e.g., octolinearity, edge length uniformity) and evaluated every possible pipeline configuration with a dataset of 4096 hypergraphs, which are subgraphs of a large (6,714 vertices, 39,774 hyperedges) hypergraph of recipes [1]. Based on our analysis of this data, we found three pipeline presets which worked particularly well for broad optimization goals (e.g., the speed with which the map is generated, or the simplicity of the drawing produced), and included these prominently in our system's interface, with the goal of increasing its usability.

The final map is presented interactively in the browser. MetroSets is fully implemented, and can be accessed online at https://metrosets.ac.tuwien.ac.at. Thanks to the modular design of the MetroSets pipeline, it is possible in the future to develop and incorporate more sophisticated algorithms in place of some of those sketched above.

References

1. Amburg, I., Veldt, N., Benson, A.R.: Clustering in graphs and hypergraphs with categorical edge labels. In: The Web Conference (WWW 2020), pp. 706–717. ACM (2020). https://doi.org/10.1145/3366423.3380152
2. Alberto, A.: Metro map of Rock'n'Roll (2010). https://www.flickr.com/photos/smoy/4413987999/
3. Asquith, M., Gudmundsson, J., Merrick, D.: An ILP for the metro-line crossing problem. In: Computing: Australasian Theory Symposium (CATS 2008) CRPIT, vol. 77, pp. 49–56 (2008)
4. Chivers, D., Rodgers, P.: Octilinear force-directed layout with mental map preservation for schematic diagrams. In: Dwyer, T., Purchase, H., Delaney, A. (eds.) Diagrams 2014. LNCS (LNAI), vol. 8578, pp. 1–8. Springer, Heidelberg (2014). https://doi.org/10.1007/978-3-662-44043-8_1
5. De Groot, S., Roberts, M.: Brexit mapping (2019). http://www.brexitmapping.com
6. van Dijk, T.C., Lutz, D.: Realtime linear cartograms and metro maps. In: Advances in Geographic Information Systems (SIGSPATIAL 2018), pp. 488–491. ACM (2018). https://doi.org/10.1145/3274895.3274959
7. Ellson, J., Gansner, E., Koutsofios, L., North, S.C., Woodhull, G.: Graphviz—open source graph drawing tools. In: Mutzel, P., Jünger, M., Leipert, S. (eds.) GD 2001. LNCS, vol. 2265, pp. 483–484. Springer, Heidelberg (2002). https://doi.org/10.1007/3-540-45848-4_57
8. Fruchterman, T.M., Reingold, E.M.: Graph drawing by force-directed placement. Softw. Pract. Exp. **21**(11), 1129–1164 (1991). https://doi.org/10.1002/spe.4380211102
9. Jacobsen, B., Wallinger, M., Kobourov, S., Nöllenburg, M.: Metrosets: visualizing sets as metro maps. In: IEEE Symposium on Information Visualization (Oct 2020). https://arxiv.org/abs/2008.09367
10. Kamada, T., Kawai, S.: An algorithm for drawing general undirected graphs. Inf. Proc. Lett. **31**(1), 7–15 (1989). https://doi.org/10.1016/0020-0190(89)90102-6
11. Nesbitt, K.V.: Getting to more abstract places using the metro map metaphor. In: Information Visualisation (IV 2004), pp. 488–493. IEEE (2004). https://doi.org/10.5555/1018435.1021663
12. Niedermann, B., Haunert, J.-H.: An algorithmic framework for labeling network maps. Algorithmica **80**(5), 1493–1533 (2018). https://doi.org/10.1007/s00453-017-0350-0
13. Nöllenburg, M.: An improved algorithm for the metro-line crossing minimization problem. In: Eppstein, D., Gansner, E.R. (eds.) GD 2009. LNCS, vol. 5849, pp. 381–392. Springer, Heidelberg (2010). https://doi.org/10.1007/978-3-642-11805-0_36
14. Shahaf, D., Guestrin, C., Horvitz, E., Leskovec, J.: Information cartography. Commun. ACM **58**(11), 62–73 (2015). https://doi.org/10.1145/2735624
15. Jonathan, S.: An illustrated subway map of human anatomy (2018). www.visualcapitalist.com/subway-map-human-anatomy/

Visualizing Massive Networks by GPU Accelerated Streaming Algorithms

Ehsan Moradi$^{(\boxtimes)}$ ⑩ and Debajyoti Mondal ⑩

Visualization, Geometry and Algorithms Lab
Department of Computer Science, University of Saskatchewan, Saskatoon, Canada
{e.moradi,d.mondal}@usask.ca

Introduction. The data-driven business analytics and technologies are giving rise to petabyte-scale data repositories. Graphs with millions (even billions) of nodes and edges are becoming increasingly common in social media databases [14]. Therefore, the need for improving the scalability of the network layout algorithms is being felt more today than ever before [15]. A natural question in this context is how much can we push the boundary of scalability for network layout algorithms by leveraging cutting edge hardware technologies and ideas of network summaries maintaining layout accuracy or trustworthiness.

Traditional layout algorithms [7] may take hours in single-threaded execution. Graph summarizing and then creating the layout [6, 13] is a common approach to achieve scalability. However, dealing with massive networks would require implementing these techniques in parallel [12] or leverage distributed computing [1]. Graph parallel processing is often challenging to implement efficiently, and hence most graph algorithms partition the graphs to be processed by active threads. However, graph partitioning algorithms are also costly for massive graphs. Increasing the number of processing units may speed-up the computation, but it may also require more time (while synchronizing the results) as the number of partitions increase.

We attempted to leverage GPU technologies and streaming community detection algorithms to create approximate visualizations of networks with millions of nodes and edges. GPUs can create enormous parallel processing threads with the least possible overhead, but it also makes the implementation challenging as some computations (e.g., frequency counting) are difficult to handle using GPU while fully utilizing the power of parallelism. Since we use streaming algorithms, we trade layout quality with time, however, for large networks, this trade-off is beneficial for the following reasons: (a) Big communities take most display space in a traditional ForceAtlas2 layout [8] and those communities are also revealed by our approach. (b) Since our method takes a few seconds, it helps interactive exploration. In addition, users can quickly create many layouts by changing layout parameters and choose the one that best suits their need.

Our Approach and Results. ForceAtlas2 [8] is a well-known graph layout algorithm designed for social network visualization in Gephi [2]. A recent GPU

This Work Is Supported by the Natural Sciences and Engineering Research Council of Canada (NSERC), and by Two CFREF Grants Coordinated by GIFS and GIWS

© Springer Nature Switzerland AG 2020
D. Auber and P. Valtr (Eds.): GD 2020, LNCS 12590, pp. 541–543, 2020.
https://doi.org/10.1007/978-3-030-68766-3

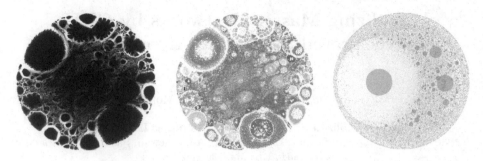

Fig. 1. (left) web-BerkStan [10] network with 685,230 nodes and 7,600,595 edges produced in 138 s by [3]. (middle) Communities are colored by a streaming community detection algorithm. (right) A layout of the compressed network by our method in 6 s.

accelerated implementation of ForceAtlas2 [3] showed the GPU assisted layout can be over 40 times faster compared to existing CPU implementations. However, for large networks it still takes a few minutes. Furthermore, the output does not clearly distinguish the communities (Fig. 1(left)).

Instead of adapting complex network summarization algorithms [11] or graph sampling methods [6], we leveraged streaming algorithms for community detection [5]. This is a one-pass algorithm which reads the edges exactly once to label the vertices with their community labels. The crux of our GPU implementation is to compress the network using an approximate counting technique (count–min sketch [4]) that produces a compressed network in another one-pass reading of the edges. The nodes of the compressed network represents communities, and they are weighted based on the number of edges in the corresponding communities. The edges are weighted by the number of edges between the corresponding pair of communities. We draw the compressed graph using the GPU accelerated ForceAtlas2 algorithm [3] (Fig. 1(right)).

The communities revealed by the streaming community detection algorithm were highly consistent with the ones found by ForceAtlas2 (Fig. 1(middle)). As shown in the following table, with a compression rate of above 80%, we have been able to achieve a speedup above 83% for real life graphs with millions of edges (when compared to the existing GPU accelerated implementation of ForceAtlas2). The visual inspection shows that the compressed graph layout retains most big communities detected by ForceAtlas2, and the relative distances between communities provide a better understanding of how they interact.

Network	Nodes	Edges	Time (ms)	Avg. Deg	Speedup (%)	Compr. (%)
web-Google [10]	916,427	5,105,039	10905	11.14	91.72	91.74
web-BerkStan [10]	685,230	7,600,595	6137	22.18	95.56	93.99
Com-youtube [16]	1,157,827	2,987,624	39050	5.16	83.30	81.91
WikiTalk [9]	2,394,384	5,021,410	15754	4.19	96.07	95.24

References

1. Arleo, A., Didimo, W., Liotta, G., Montecchiani, F.: A distributed multilevel force-directed algorithm. IEEE Trans. Parallel Distrib. Syst. **30**(4), 754–765 (2019)
2. Bastian, M., Heymann, S., Jacomy, M.: Gephi: An open source software for exploring and manipulating networks. In: Proceedings of the Third International Conference on Weblogs and Social Media (ICWSM) (2009)
3. Brinkmann, G.G., Rietveld, K.F.D., Takes, F.W.: Exploiting GPUs for fast force-directed visualization of large-scale networks. In: Proceedings of the 46th International Conference on Parallel Processing (ICPP), pp. 382–391 (2017)
4. Cormode, G., Muthukrishnan, S.: An improved data stream summary: the count-min sketch and its applications. J. Algorithms **55**(1), 58–75 (2005)
5. Hollocou, A., Maudet, J., Bonald, T., Lelarge, M.: A streaming algorithm for graph clustering. In: NIPS 2017 - Workshop on Advances in Modeling and Learning Interactions from Complex Data, pp. 1–12 (2017)
6. Hong, S., Lu, S.: Graph sampling methods for big complex networks integrating centrality, k-core, and spectral sparsification. In: Proceedings of the 35th ACM/SIGAPP Symposium on Applied Computing (SAC), pp. 1843–1851 (2020)
7. Hu, Y., Shi, L.: Visualizing large graphs. WIREs. Comput. Stat. **7**(2), 115–136 (2015)
8. Jacomy, M., Venturini, T., Heymann, S., Bastian, M.: ForceAtlas2, a continuous graph layout algorithm for handy network visualization designed for the Gephi software. PLoS ONE **9**(6), 1–12 (2014)
9. Leskovec, J., Huttenlocher, D., Kleinberg, J.: Predicting positive and negative links in online social networks. In: Proceedings of the 19th international conference on World wide web, pp. 641–650 (2010)
10. Leskovec, J., Lang, K.J., Dasgupta, A., Mahoney, M.W.: Community structure in large networks: natural cluster sizes and the absence of large well-defined clusters. Internet Math. **6**(1), 29–123 (2009)
11. Liu, Y., Safavi, T., Dighe, A., Koutra, D.: Graph summarization methods and applications: a survey. ACM Comput. Surv. **51**(3), 61:1–62:34 (2018)
12. Meyerhenke, H., Nöllenburg, M., Schulz, C.: Drawing large graphs by multilevel maxent-stress optimization. IEEE Trans. Vis. Comput. Graph. **24**(5), 1814–1827 (2018)
13. Mondal, D., Nachmanson, L.: A new approach to graphmaps, a system browsing large graphs as interactive maps. In: Proceedings of the 13th International Joint Conference on Computer Vision, Imaging and Computer Graphics Theory and Applications (VISIGRAPP-IVAPP), pp. 108–119. SciTePress (2018)
14. Moradi, E., Fazlali, M., Malazi, H.T.: Fast parallel community detection algorithm based on modularity. In: Proceedings of the 18th CSI International Symposium on Computer Architecture and Digital Systems (CADS), pp. 1–4. IEEE (2015)
15. Perrot, A., Auber, D.: Cornac: tackling huge graph visualization with big data infrastructure. IEEE Trans. Big Data **6**(1), 80–92 (2020)
16. Yang, J., Leskovec, J.: Defining and evaluating network communities based on ground-truth. Knowl. Inf. Syst. **42**(1), 181–213 (2013). https://doi.org/10.1007/s10115-013-0693-z

Author Index

Printed in the United States
By Bookmasters